Texts and Monographs in Physics

Springer-Verlag Berlin Heidelberg GmbH

O. M. Braun   Y. S. Kivshar

The Frenkel–Kontorova Model

Concepts, Methods, and Applications

With 161 Figures

Springer

Professor Oleg M. Braun
Ukrainian Academy of Sciences
Institute of Physics
03650 Kiev, Ukraine

Professor Yuri S. Kivshar
Australian National University
Research School of Physical Sciences and Engineering
0200 Canberra ACT, Australia

Library of Congress Cataloging-in-Publication Data: Braun, O.M. (Oleg M.), 1949– The Frenkel-Kontorova model: concepts, methods, and applications / O.M. Braun, Y.S. Kivshar. p. cm. – (Texts and monographs in physics, ISSN 0172-5998) Includes bibliographical references and index. ISBN 3-540-40771-5 (alk. paper) 1. Solid state physics. 2. Nonlinear theories. I. Kivshar, Y. S. (Yuri S.), 1959– II. Title, III. Series. QC176.B68 2003 530.4'1–dc22 2003059072

ISSN 0172-5998
ISBN 978-3-642-07397-7 ISBN 978-3-662-10331-9 (eBook)
DOI 10.1007/978-3-662-10331-9

springeronline.com

Originally published by Springer-Verlag Berlin Heidelberg New York in 2004
Softcover reprint of the hardcover 1st edition 2004

Typesetting: EDV-Beratung, Frank Herweg, Leutershausen
Cover design: *design & production* GmbH, Heidelberg

Printed on acid-free paper SPIN 10826296 55/3141/ba 5 4 3 2 1 0

For Our Parents and Families

Preface

Theoretical physics deals with physical models. The main requirements for a good physical model are *simplicity* and *universality*. Universal models which can be applied to describe a variety of different phenomena are very rare in physics and, therefore, they are of key importance. Such models attract the special attention of researchers as they can be used to describe underlying physical concepts in a simple way. Such models appear again and again over the years and in various forms, thus extending their applicability and educational value. The simplest example of this kind is the model of a pendulum; this universal model serves as a paradigm which encompasses basic features of various physical systems, and appears in many problems of very different physical context.

Solids are usually described by complex models with many degrees of freedom and, therefore, the corresponding microscopic equations are rather complicated. However, over the years a relatively simple model, known these days as *the Frenkel-Kontorova model*, has become one of the fundamental and universal tools of low-dimensional nonlinear physics; this model describes a chain of classical particles coupled to their neighbors and subjected to a periodic on-site potential. Although links with the classical formulation are not often stated explicitly in different applications, many kinds of nonlinear models describing the dynamics of discrete nonlinear lattices are based, directly or indirectly, on a 1938 classical result of Frenkel and Kontorova, who applied a simple one-dimensional model for describing the structure and dynamics of a crystal lattice in the vicinity of a dislocation core. This is one of the first examples in solid-state physics when the dynamics of an extended defect in a bulk was modelled by a simple one-dimensional model. Over the years, similar ideas have been employed in many different physical problems, also providing a link with the mathematical theory of solitons developed later for the continuum analog of the Frenkel-Kontorova (FK) model.

In the continuum approximation, the FK model is known to reduce to *the exactly integrable sine-Gordon (SG) equation*, and this explains why the FK model has attracted much attention in nonlinear physics. The SG equation gives an example of a fundamental nonlinear model for which we know everything about the dynamics of nonlinear excitations, namely *phonons*, *kinks* (topological solitons), and *breathers* (dynamical solitons); and their multi-particle dynamics determines the global behavior of a nonlinear system as a whole. Although the FK model is inherently *discrete* and is not integrable, one may get a deep physical insight and simplify one's understanding of the nonlinear dynamics using the language of the SG nonlinear modes as weakly interacting effective quasi-particles. The discreteness of the FK model manifests itself in such phenomena as the existence of an effective periodic energy known as the Peierls-Nabarro potential.

The simplicity of the FK model, due to the assumptions of linear interatomic forces and a sinusoidal external potential, as well as its surprising richness and capability to describe a range of important nonlinear phenomena, has attracted a great deal of attention from physicists working in solid-state physics and nonlinear science. Many important physical phenomena, ranging from solitons to chaos as well as from the commensurate-incommensurate phases to glass-like behavior, present complicated sub-fields of physics each requiring a special book. However, the FK model provides a unique opportunity to combine many such concepts and analyze them together in a unified and consistent way.

The present book aims to describe, from a rather general point of view, *the basic concepts and methods of low-dimensional nonlinear physics* on the basis of the FK model and its generalizations. We are not restricted by the details of specific applications but, instead, try to present *a panoramic view* on the general properties and dynamics of solid-state models and summarize the results that involve fundamental physical concepts.

Chapter 1 makes an introduction into the classical FK model, while Chap. 2 discusses in more detail the applicability of the FK model to different types of physical systems. In Chap. 3 we introduce one of the most important concepts, the concept of kinks, and describe the characteristics of the kink motion in discrete chains, where kinks are affected by the Peierls-Nabarro periodic potential. In Chap. 4 we analyze another type of nonlinear mode, the spatially localized oscillating states often called *intrinsic localized modes* or *breathers*. We show that these nonlinear modes may be understood as a generalization of the SG breathers but exist in the case of strong discreteness. Chapters 3 and 4 also provide an overview of the dynamical properties of the generalized FK chains which take into account more general types of on-site potential as well as anharmonic interactions between particles in the chain. The effect of impurities on the dynamics of kinks as well as the dynamics and structure of nonlinear impurity modes are also discussed there. Chapter 5 gives a simple introduction to the physics of commensurate and incommensurate systems, and it discusses the structure of the ground state of the discrete FK chain. We show that the FK model provides probably the simplest approach for describing systems with two or more competing spatial periods. While the interaction between the atoms favors their equidistant separation with a period corresponding to the minimum of the interatomic potential, the interaction of atoms with the substrate potential (having its own period) tends to force the atoms into a configuration where they are regularly spaced. In Chap. 5 we employ two methods for describing the properties of the FK model: first, in the continuum approximation we describe the discrete model by the exactly integrable SG equation, and second, we study the equations for stationary configurations of the discrete FK model reducing it to the so-called *standard map*, one of the classical models of stochastic theory. The statistical mechanics of the FK model is discussed in Chap. 6, which also includes the basic results of the transfer-integral method. Here, the FK model again appears to be unique because, on the one hand, it allows the derivation of exact results in the one-dimensional case and, on the other hand, it allows for the introduction of weakly interacting quasi-particles (kinks and phonons) for describing the statistical mechanics of systems of strongly interacting particles. Chapter 7 gives an overview of the dynamical properties of the FK model at nonzero temperatures, including kink diffusion and mass transport in nonlinear discrete systems. Chapter 8 discusses the dynamics of nonlinear chains under the action of dc and ac forces when the system is far from its equilibrium state. Chapter 9 discuses ratchet dynamics in driven systems with broken spatial or temporal symmetry when a directed motion is induced. The properties of finite-length chains are discussed in Chap. 10, whereas two-dimensional generalizations of the FK model are introduced and described in Chap. 11, for both scalar and vector models. In the concluding Chap. 12 we present more examples where the basic concepts and physical effects, demonstrated above for simple versions of the FK chain, may find applications in a broader context. At last, the final chapter includes some interesting historical remarks written by *Prof. Alfred Seeger*, one of the pioneers in the study of the FK model and its applications.

We thank our many colleagues and friends around the globe who have collaborated with us on different problems related to this book, or contributed to our understanding of the field. It is impossible to list all of them, but we are particularly indebted to A.R. Bishop, L.A. Bolshov, D.K. Campbell, T. Dauxois, S.V. Dmitriev, S. Flach, L.M. Floria, R.B. Griffiths, Bambi Hu, B.A. Ivanov, A.M. Kosevich, A.S. Kovalev, I.F. Lyuksyutov, B.A. Malomed, S.V. Mingaleev, A.G. Naumovets, M.V. Paliy, M. Peyrard, M. Remoissenet, J. Röder, A. Seeger, S. Takeno, L.-H. Tang, A.V. Ustinov, I.I. Zelenskaya, and A.V. Zolotaryuk.

Canberra, Australia
May 2003

Oleg Braun
Yuri Kivshar

Table of Contents

List of Abbreviations

CP	central peak
DW	double well (substrate potential); domain wall
DB	double barrier (substrate potential)
DSG	double sine-Gordon (equation)
FPK	Fokker-Planck-Kramers (equation)
FK	Frenkel-Kontorova (model)
FvdM	Frank – van der Merwe (limit)
GS	ground state
IC	incommensurate (phase)
LJ	Lennard-Jones (potential)
NS	nonsinusoidal (substrate potential)
SG	sine-Gordon (equation)
TI	transfer integral (method)
B	mobility
D	diffusion coefficient (D_k, D_η, D_a, D_s, D_μ, D_c)
E	(total) system energy
F	free energy; force
G	Gibbs free energy; Green's function
H	Hamiltonian
J	atomic flux
K	kinetic energy; transfer matrix; Chirikov's constant
L	length of a chain
M	number of wells of the substrate potential; memory function
N	number of atoms, kinks, breathers
P	kink momentum; misfit parameter
Q	correlation function; number of atoms in the cnoidal wave per period
R	distance between kinks
S	entropy
T	temperature; Chirikov map
U	total potential energy (U_{sub}, U_{int})
V	potential energy (V_{sub}, V_{int}, V_{PN})
W	enthalpy
X	kink coordinate
Y	statistical sum $Y(T, \Pi, N)$; center of mass coordinate; point in the tour
Z	statistical sum $Z(T, L, N)$

a	lattice constant (a_s, a_A, $a_{\min}$, a_{FM})
c	sound speed
d	kink width
f	force; distribution function
g	elastic constant (g_{Aubry}, g_a, g_k)
h	discreteness parameter; hull function; scaling function
j	flux density
k	momentum; modulus (of elliptic function); wavevector
l	atomic index
m	kink mass
n	concentration of kinks (n_{tot}, n_k, n_w, n_{pair})
q	period of C-phase
r	used in window number, $w = r/s$
s	number of atoms in the unit cell, $\theta = s/q$; entropy per particle; spin
t	time
u	atomic displacement
v	kink velocity; interaction between kinks, v_{int}
w	displacement; window number, $w = r/s$
x	atomic coordinate
α	anharmonicity parameter
β	Boltzmann factor, $\beta \equiv (k_B T)^{-1}$
β	parameter for exponential interaction
γ	Lorentz factor; Euler constant
δ	phase shift (in collisions of kinks)
ε	energy (ε_s, ε_k, $\varepsilon_{\text{pair}}$, ε_{PN})
ψ	order parameter
κ	phonon momentum; parameter for Morse potential
χ	susceptibility; small displacement
λ	eigenvalues
ξ	correlation length; canonical variable
μ	chemical potential
η	friction coefficient; Bloch function
ρ	density (of atoms, phonons, kinks)
τ	dimensionless temperature, $\tau = k_B T/\varepsilon_k$
θ	dimensionless concentration (coverage), $\theta = N/M = s/q$
ω	frequency
$\omega_{\min}$	minimal phonon frequency of the pinned FK chain
σ	kink topological charge
ϵ	eigenvalues
$\varGamma$	phase volume
$\varDelta$	gap in spectrum
$\varPi$	pressure
$\varXi$	great statistical sum $\varXi(T, L, \mu)$
$\varOmega$	number of states; phase-space sharing; Mori function
$\varTheta$	step function
$\mathcal{N}$	response function
$\mathcal{L}$	Liouville operator
$\mathcal{P}$	projection operator

1 Introduction

This introductory chapter is intended to provide a general overview of the classical formulation of the Frenkel-Kontorova model and its continuum version, the sine-Gordon equation. The chapter introduces also the fundamental modes of the model, phonons, kinks, and breathers, and describes some of their general properties. It also provides the background for the subsequent discussion of the basic physical systems where the nonlinear dynamics is described by the Frenkel-Kontorova model and its generalizations.

1.1 The Frenkel-Kontorova Model

A simple model that describes the dynamics of a chain of particles interacting with the nearest neighbors in the presence of an external periodic potential was firstly mentioned by Prandtl [1] and Dehlinger [2], see the historical notes of Prof. Alfred Seeger at the end of the book (see Chap. 13). This model was then independently introduced by Frenkel and Kontorova [3]–[6]. Such a chain of particles is presented schematically in Fig. 1.1. The corresponding mechanical model can be derived from the standard Hamiltonian,

$$\mathcal{H} = K + U, \tag{1.1}$$

where K and U are the kinetic and potential energies, respectively. The kinetic energy K is defined in a standard way,

$$K = \frac{m_a}{2} \sum_n \left(\frac{dx_n}{dt} \right)^2, \tag{1.2}$$

where m_a is the particle mass and x_n is the coordinate of the n-th particle in the chain. The potential energy U of the chain shown in Fig. 1.1 consists of two parts,

$$U = U_{\text{sub}} + U_{\text{int}}. \tag{1.3}$$

The first term U_{sub} characterizes the interaction of the chain with an external periodic on-site potential, taken in the simplest form,

$$U_{\text{sub}} = \frac{\varepsilon_s}{2} \sum_n \left[1 - \cos\left(\frac{2\pi x_n}{a_s} \right) \right], \tag{1.4}$$

where ε_s is the potential amplitude and a_s is its period. The second term U_{int} in Eq. (1.3) takes into account a linear coupling between the nearest neighbors of the chain,

$$U_{\text{int}} = \frac{g}{2} \sum_n (x_{n+1} - x_n - a_0)^2, \tag{1.5}$$

and it is characterized by the elastic constant g and the equilibrium distance of the inter-particle potential, a_0, in the absence of the on-site potential (see Fig. 1.1). The model introduced by the Hamiltonian (1.1)–(1.5) can be justified under the following simplifying assumptions on the corresponding physical system:

(i) The particles of the chain can move along one direction only;

(ii) In the general expression for the substrate potential energy written as

$$U_{\text{sub}} = \sum_n V_{\text{sub}}(x_n), \tag{1.6}$$

the function $V_{\text{sub}}(x)$ is expanded into the Fourier series, and only the first term is taken into account;

(iii) The energy of the interparticle interaction includes only the coupling between the nearest neighbors,

$$U_{\text{int}} = \sum_n V_{\text{int}}(x_{n+1} - x_n), \tag{1.7}$$

and, expanding $V_{\text{int}}(x)$ in a Taylor series, only the harmonic interaction is taken into account, so that $g = V''_{\text{int}}(a_0)$.

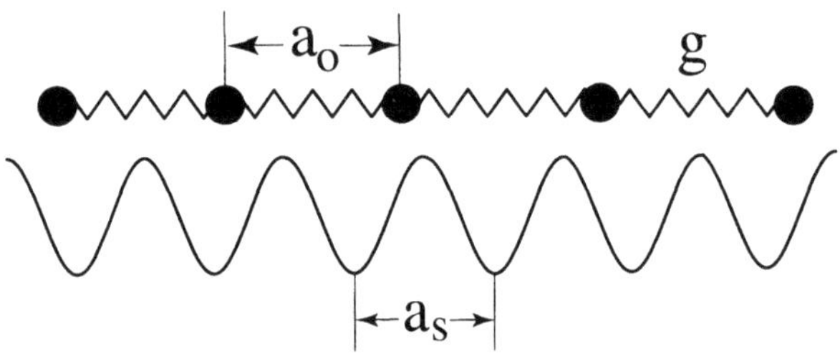

Fig. 1.1. Schematic presentation of the Frenkel-Kontorova model: A chain of particles interacting via harmonic springs with elastic coupling g is subjected to the action of an external periodic potential with period a_s.

Introducing the dimensionless variables, we re-write the Hamiltonian (1.1)-(1.5) in the conventional form ($H = 2\mathcal{H}/\varepsilon_s$)

$$H = \sum_n \left\{ \frac{1}{2} \left(\frac{dx_n}{dt} \right)^2 + (1 - \cos x_n) + \frac{g}{2} (x_{n+1} - x_n - a_0)^2 \right\}, \tag{1.8}$$

where $a_0 \to (2\pi/a_s)a_0$, $x_n \to (2\pi/a_s)x_n$, $t \to (2\pi/a_s)(\varepsilon_s/2m_a)^{1/2}t$, and the dimensionless coupling constant is changed according to the following, $g \to (a_s/2\pi)^2 g(\varepsilon_s/2)^{-1}$. In such a renormalized form, the Hamiltonian (1.8) describes a harmonic chain of particles of equal unit mass moving in a sinusoidal external potential with period $a_s = 2\pi$ and amplitude $\varepsilon_s = 2$. In order to obtain all the physical values in the corresponding dimensional form, one should multiply the spatial variables by $(a_s/2\pi)$, the frequencies by $(2\pi/a_s)\sqrt{\varepsilon_s/2m_a}$, the masses by m_a, and the energies by $(\varepsilon_s/2)$.

From the Hamiltonian (1.8), we obtain the corresponding equation of motion of a discrete chain,

$$\frac{d^2x_n}{dt^2} + \sin x_n - g\,(x_{n+1} + x_{n-1} - 2x_n) = 0. \tag{1.9}$$

It is important to notice here that Eq. (1.9) does not include explicitly the equilibrium lattice spacing in the absence of the on-site potential, a_0.

In this chapter, as well as in two subsequent chapters 3 and 4 devoted to the dynamics of kinks and breathers, we consider an infinite chain with $a_0 = a_s$ when the ground state of the chain (i.e., its stationary state that corresponds to the absolute minimum of the potential energy) describes a *commensurate array of particles*. This means that each minimum of the substrate potential is occupied by one particle, so that the "coverage" parameter θ (defined for a finite chain as the ratio of the number of particles to the number of minima of the substrate potential) is $\theta = 1$. In this case it is convenient to introduce new variables for the particle displacements, u_n, defined by the relation $x_n = na_s + u_n$. When the particle displacements u_n are small, $|u_n| \ll a_s$, we can linearize the motion equation and obtain

$$\frac{d^2u_n}{dt^2} + u_n - g\,(u_{n+1} + u_{n-1} - 2u_n) = 0. \tag{1.10}$$

Equation (1.10) describes linear modes of the chain, also called *phonons*, $u_n(t) \propto \exp[i\omega_{\rm ph}(\kappa)t - i\kappa n]$, which are characterized by the *dispersion relation*

$$\omega_{\rm ph}^2(\kappa) = \omega_{\rm min}^2 + 2g\,(1 - \cos\kappa), \tag{1.11}$$

where κ is the dimensionless wavenumber ($|\kappa| \le \pi$). According to the dispersion relation (1.11), the frequency spectrum of the FK chain is characterized by a finite band separated from zero by a gap $\omega_{\rm min} \equiv \omega_{\rm ph}(0) = 1$, and has the cut-off frequency $\omega_{\rm max} \equiv \omega_{\rm ph}(\pi) = \sqrt{\omega_{\rm min}^2 + 4g}$.

When particles' displacements are not small, the linear approximation is no longer valid, and the primary nonlinear motion equation (1.9) may support both extended modes (which generalize the linear modes – phonons) and localized modes, a new type of excitations of the chain. The simplest way to introduce and describe these nonlinear modes is to consider the continuum approximation for the dynamics of a discrete chain. A correct procedure for deriving the motion equations in the continuum limit from a lattice model was suggested by Rosenau [7], and his approach can be applied to the dynamics of chains with arbitrary inter-particle and substrate potentials.

In order to obtain the equations of the continuum model, we re-write the motion equations that follow from the Hamiltonian (1.1)–(1.5) in a general form

$$\frac{d^2 u_n}{dt^2} = F_{\text{int}}(a_s + u_n - u_{n-1}) - F_{\text{int}}(a_s + u_{n+1} - u_n) + F_{\text{sub}}(u_n), \tag{1.12}$$

where

$$F_{\text{int}}(u) = -\frac{dV_{\text{int}}(u)}{du}, \quad \text{and } F_{\text{sub}}(u) = -\frac{dV_{\text{sub}}(u)}{du}.$$

Equation (1.12) can be modified further by introducing the so-called n-th bond lengths defined as $v_n = (u_n - u_{n-1})/a_s$. As a result, we obtain

$$\begin{aligned} a_s \frac{d^2 v_n}{dt^2} &= -[F_{\text{int}}(a_s + a_s v_{n+1}) + F_{\text{int}}(a_s + a_s v_{n-1}) \\ &\quad -2F_{\text{int}}(a_s + a_s v_n)] + [F_{\text{sub}}(u_n) - F_{\text{int}}(u_{n-1})]. \end{aligned} \tag{1.13}$$

Now, we treat the discrete values v_n as functions of the continuous variable na_s, and expand the function v_{n+1} in the vicinity of v_n. Formally, this can be done by introducing a new variable x and continuous function $v(x)$ by changing $n \to x = na_s$ and $v_n \to \partial u/\partial x|_{x=na_s}$. Keeping only the lowest-order terms in the expansions, we obtain [7]

$$\frac{\partial^2 u}{\partial t^2} + a_s \hat{L}_A \frac{\partial}{\partial x} \left\{ F_{\text{int}} \left[a_s \left(1 + \frac{\partial u}{\partial x} \right) \right] \right\} - F_{\text{sub}}(u) = 0, \tag{1.14}$$

where

$$\hat{L}_A = 1 + \frac{a_s^2}{12} \frac{\partial^2}{\partial x^2} + \dots .$$

Then, letting $\hat{L}_A^{-1}$ act on Eq. (1.14), we obtain the continuous equation of the order of $\mathcal{O}(a_s^4)$,

$$\frac{\partial^2 u}{\partial t^2} - \frac{a_s^2}{12} \frac{\partial^4 u}{\partial^2 x \partial^2 t} - F_{\text{sub}}(u) + a_s \frac{\partial}{\partial x} \left\{ F_{\text{int}} \left[a_s \left(1 + \frac{\partial u}{\partial x} \right) \right] - \frac{a_s}{12} \frac{\partial}{\partial x} F_{\text{sub}}(u) \right\} = 0.$$

The method described above takes into account all terms of the lowest order which appear due to the effect of the lattice discreteness. In particular, applied to the FK model discussed above, this method yields a perturbed SG equation,

$$\frac{\partial^2 u}{\partial t^2} - d^2 \frac{\partial^2 u}{\partial x^2} + \sin u = \epsilon f(u), \tag{1.15}$$

where $d = a_s \sqrt{g}$, and the function

$$\epsilon f(u) = \frac{a_s^2}{12}\left[\frac{\partial^4 u}{\partial^2 x \partial^2 t} + \left(\frac{\partial u}{\partial x}\right)^2 \sin u - \frac{\partial^2 u}{\partial x^2}\cos u\right] \tag{1.16}$$

takes into account, in the first-order approximation, the effects produced by the chain discreteness.

1.2 The Sine-Gordon Equation

Neglecting the discreteness effects in the standard FK model, we obtain the well-known sine-Gordon (SG) equation,

$$\frac{\partial^2 u}{\partial t^2} - d^2\frac{\partial^2 u}{\partial x^2} + \sin\, u = 0, \tag{1.17}$$

where $d = a_s\sqrt{g}$ and $g = V''_{\text{int}}(a_s)$. Changing the spatial variable as follows, $x \to x/d$, we transform Eq. (1.17) to its canonical form,

$$\frac{\partial^2 u}{\partial t^2} - \frac{\partial^2 u}{\partial x^2} + \sin u = 0. \tag{1.18}$$

The most remarkable property of the nonlinear partial differential equation (1.18) is its complete integrability. Indeed, the SG equation was one of the first equations whose *multi-soliton properties* were recognized. In its transformed form, Eq. (1.18) was originally considered by Enneper [8] in the differential geometry of surfaces of a constant negative Gaussian curvature. The study of Eq. (1.18) in the context of differential geometry revealed very interesting properties, including the possibility to generate from one known solution of Eq. (1.18) a new (unknown) solution by means of the so-called Bäcklund transformation [9].

In physics, Eq. (1.18) found its first applications in the simplified dislocation models (see Chap. 2 below), and kinks and breathers of the SG equation were first introduced by Seeger and co-workers more than forty years ago [10]–[12] (see also Refs. [13, 14]). The original German names for the kinks and breathers were *translatorische and oszillatorische Eigenbewegungen*, and from a historical point of view it is interesting to note that this preceded the discovery of the soliton properties of the most familiar Korteweg–de Vries equation [15, 16] by more than a decade. Independently, Perring and Skyrme [17] introduced the SG equation as a simple one-dimensional model of the scalar field theory modelling a classical particle. Almost simultaneously, the SG equation appeared in the theory of weak superconductivity to be the main nonlinear equation describing the so-called long Josephson junctions [18], where the kink solution describes a quantum of the magnetic field, *a fluxon.* The two next important steps of the history of the SG equation were the emphasis on its pedagogical power by use of the very simple chain of coupled pendulums (the mechanical analog of the FK chain) made by Scott [19], and the solution of the related inverse scattering transform problem obtained by Ablowitz and co-authors [20].

Later, the SG equation (1.18) was proved to be completely integrable with the canonical variables introduced through the auxiliary scattering data, and its properties have been described in many survey papers and books (see, e.g., Ref. [21]). Here, we only mention the main properties of Eq. (1.18) and its solutions in the form of *nonlinear modes*, which are important for us in the study of a discrete FK model.

Elementary excitations of the SG system are known as: *phonons*, *kinks*, and *breathers*.

Phonons are extended periodic solutions of the SG model that describe, in the linear limit, the familiar continuous monochromatic waves, $u(x,t) \propto \exp(i\omega t - ikx)$. The phonons are characterized by the dispersion relation, the dependence of the wave frequency ω on its wave number k, $\omega_{\rm ph}^2(k) = 1 + k^2$, which is a long-wave expansion of the expression (1.11). A nonlinear model possess a set of periodic solutions that generalize the linear waves, but have their frequency dependent on the amplitude of the wave.

Kinks, also called *topological solitons*, appear due to an inherent degeneracy of the system ground state. A single kink can be understood as a solution connecting two nearest identical minima of the periodic on-site potential,

$$u_k(x,t) = 4\tan^{-1}\exp[-\sigma\gamma(v)(x - vt)], \tag{1.19}$$

where $\sigma = \pm 1$ stands for the kink's so-called *topological charge*. We call the solution (1.19) a *kink* for the case $\sigma = +1$, and an *antikink* for the case $\sigma = -1$. The parameter v is the kink's velocity that cannot exceed its maximum value c, the sound velocity. The kink's velocity determines the kink's width, $d_{\rm eff} = d/\gamma(v)$, where $d = c = a_s\sqrt{g}$, and the factor $\gamma(v) = 1/\sqrt{1-(v/c)^2}$ can be treated as a Lorentz contraction of the kink's width which follows from the relativistic invariance of the SG model.

Breathers, also called *dynamical solitons*, are *spatially localized* oscillating nonlinear modes. A SG breather has the form

$$u_{\rm br}(x,t) = 4\tan^{-1}\left\{\left(\frac{\sqrt{1-\Omega^2}}{\Omega}\right)\frac{\sin(\Omega t)}{\cosh\left(x\sqrt{1-\Omega^2}\right)}\right\}, \tag{1.20}$$

and it describes a nonlinear state with internal frequency Ω, $0 < \Omega < \omega_{\rm min}$ and amplitude $u_{\rm max} = 4\tan^{-1}\left(\sqrt{1-\Omega^2}/\Omega\right)$ localized on the spatial scale, $b = d/\sqrt{1-\Omega^2}$.

The kink's energy, expressed in dimensionless units, can be found in the form $E_k(v) = mc^2\gamma(v) \approx mc^2 + \frac{1}{2}mv^2$. Such an approximation allows introducing the kink's rest mass, $m = 2/(\pi^2\sqrt{g})$, and the kink's energy, $\varepsilon_k = mc^2 = 8\sqrt{g}$. The breather energy can be found as $\varepsilon_{\rm br} = 2\varepsilon_k\sqrt{1-\Omega^2}$, so that $0 < \varepsilon_{\rm br} < 2\varepsilon_k$. In the limit of low frequencies, $\Omega \ll 1$, the breather can be qualitatively treated as a weakly coupled pair of a kink and an antikink.

When two kinks are placed close to each other, they attract or repel one another depending on their relative topological charge $\sigma_1\sigma_2$. The energy of the repulsion between two static kinks separated by a distance R is known to be equal to (see, e.g., Ref. [22]) $v_{\rm int}(R) \approx \varepsilon_k \sinh^{-2}(R/2d)$, while the kink and antikink attract each other according to the interaction potential $v_{\rm int}(R) \approx -\varepsilon_k \cosh^{-2}(R/2d)$. Thus, the energy of the interaction between two static kinks with topological charges σ_1 and σ_2 separated by the distance R ($R \gg d$) can be approximately represented as $v_{\rm int}(R) \approx 32\,\sigma_1\sigma_2\sqrt{g}\exp(-R/d)$.

In the framework of the SG equation, kinks and breathers move freely along the chain without loss of their energy for dissipation. Indeed, the corresponding solution for a moving breather can be easily obtained from Eq. (1.20) by applying the Lorentz transformation. The SG equation is exactly integrable, i.e. it allows elastic interactions between all excitations, and the only effect of such collisions is a phase shift (see, e.g., Ref. [21], for more details). That is why kinks and breathers can be treated as *nonlinear eigenmodes*, or quasi-particles of the SG model. Such an approach remains approximately valid for the nearly integrable modifications of the SG equation, when the model includes small perturbations such as those which appear when deriving the SG equation from the primary FK model in the quasi-continuum approximation, assuming the effects of the model discreteness to be small. In fact, being perturbed by small (conservative or nonconservative) perturbations, kinks behave like *deformable* quasi-particles, i.e. their shapes may be perturbed as well. Besides that, some new features may appear even in the presence of small perturbations, e.g. a kink and an anti-kink may collide inelastically producing a long-lived breather mode.

For the integrable SG model, any localized excitation can be presented as an *asymptotic superposition* of elementary excitations of *three kinds*, i.e. kinks, breathers, and phonons. Many of the nonlinear periodic solutions of the SG equation can be also found in an analytical form. For example, a chain of kinks is described by the following solution of the SG equation known as a *cnoidal wave*,

$$u(x,t) = -\sin^{-1}[\mathrm{cn}(x/k;k)], \tag{1.21}$$

where cn is the Jacobi elliptic function. Equation (1.21) describes a periodic sequence of identical kinks with width kd (where $k < 1$), separated by the equal distance, $L = 4dk\ \mathrm{K}(k)$, $\mathrm{K}(k)$ being the complete elliptic integral of the first kind. The interaction of a kink and large-amplitude (anharmonic) phonons is also elastic in the framework of the SG model, and it is described by an exact solution of the SG equation [23, 24]. In the small-amplitude limit, this general kink-phonon interaction describes a phase shift in the elastic scattering of linear phonons by a SG kink [25] (see also Ref. [26]).

As has been mentioned above, in the framework of the exactly integrable SG model collisions of solitons are elastic, i.e. their shapes, velocities and energies remain unchanged after collisions and the only effect produced by the interactions is the phase shifts of the colliding solitons. For example, if we take the initial kink at rest and another kink coming from an infinity with the initial velocity v_{in}, then after their interaction the first kink will remain at rest but the kinks' coordinates will be shifted by a constant value Δx_1, where (see, e.g., Ref. [21])

$$\Delta x_1 = \frac{d}{\gamma} \ln \left(\frac{\gamma + 1}{\gamma - 1} \right), \quad \gamma = \sqrt{1 - v_{\text{in}}^2/c^2}.$$

Analogously, collisions of other nonlinear excitations (phonons, kinks, and breathers) in the framework of the SG model are accompanied by phase shifts only. One of the main features of soliton collisions in integrable models is their *two-particle nature*: When several solitons collide, a shift of any soliton involved in the interaction is equal to the sum of the shifts caused by its independent pair-wise interaction with other solitons. Such a two-particle nature of the soliton interactions is a specific property of integrable systems. When the primary SG system is modified by external (even conservative) perturbations, many (in particular, three) particle effects can emerge [27].

2 Physical Models

In this chapter we discuss several examples of the physical systems where the FK model plays an important role. We demonstrate how to introduce a simplified model accounting for the basic features of the system's dynamics, and also mention why in many cases the standard FK model should be generalized to take into account other important features, such as nonsinusoidal external potential, anharmonic coupling between the particles, thermal effects, higher dimension effects, etc. However, the main purpose of this chapter is not only to list examples of different physical systems that can be analyzed effectively with the help of the FK model but also provide *an important guideline* for deriving low-dimensional models in other applications of nonlinear physics.

2.1 General Approach

The basic approach for deriving the FK model is rather simple. First of all, from an original (usually rather complicated) discrete nonlinear system one should extract a low-dimensional sub-system and describe the remaining part as *a substrate* by introducing an effective potential to account for its action. The elements of the effective one-dimensional discrete array play the role of *effective atoms* in the FK model. In many cases, such elements correspond to real atoms, although they may model clusters of atoms, as in the case of the DNA-like chains, may correspond to spins in magnetic chains, or may even describe some complex objects such as point-like Josephson junctions in an array. In the model, the effective atoms interact with their neighbors, and the simplest interaction is a linear coupling between the nearest neighbors. In the systems described by the FK-type models, the effective substrate should have a crystalline structure. In the simplest case, when the structure corresponds to the Bravais elementary cell, i.e. it has one atom per an unit cell, one may keep only the first term in the Fourier expansion of the periodic potential and obtain a sinusoidal substrate potential.

In the framework of this approach, the standard FK model can describe, more or less rigorously, a realistic physical system only when the one-dimensional sub-system and substrate are of a different origin, e.g. the FK atoms are light particles, while the substrate potential is composed by heavy particles which can be treated as being "frozen" so that their motion can be neglected. As a matter of fact, this corresponds to many physical systems. For example, this is the case of surface physics where the model atoms are adsorbed atoms and the substrate corresponds to a crystal surface, or hydrogen-bonded chains where atoms are light hydrogen atoms while the substrate potential is created by heavy oxygen atoms. However, even in other cases, such as in the dislocation theory where the atoms and substrate are of the same origin, the FK model remains a very useful approximation, and it allows to describe many nontrivial phenomena of the system dynamics, often even at a simple qualitative level.

One more class of important problems described by the FK model is associated with *the physics of incommensurate systems*. Namely, if we consider a free-end chain and assume that the equilibrium distance introduced by the inter-particle interaction does not coincide with the period of the substrate potential, then we naturally arrive at one of the simplest physical models with two (or more) competing length scales. In this case, the richness of the FK model increases dramatically due to a possibility of two distinct types of the system ground states, *commensurate* and *incommensurate* ones. Thus, the FK model appears also in many applications dealing with the physics of incommensurable systems.

Below, we present several examples of the physical models for which the FK model and/or its generalizations play an important role. These models are of three distinct levels. *The first level* of the models, such as a simple mechanical chain of pendulums, is reduced directly to the FK model. *The second level* of models, such as that for describing a lattice crowdion, includes the FK model with a more complicated type of interaction and more complex type of the substrate potential. At last, *the third level* of models, such as the models for describing the DNA dynamics, are usually more complicated and they are reduced to the FK-like models under some additional assumptions.

2.2 A Mechanical Model

One of the simplest *macroscopic models* describing the dynamics of the FK model was first introduced by Scott [19]. This is an experimental mechanical transmission line which is an efficient and pedagogical system for observing kinks and studying their remarkable properties [28]. Following Scott [19], such a mechanical system can be easily constructed as a line of screws of length L, which are fixed to brass cylinders connected by a steel spring and supported horizontally by a piano wire. The stiffness of the springs can be selected in such a way that the chain will be described by either quasi-continuous or strongly discrete FK model. A useful pocket version of the mechanical transmission line can be constructed in a simpler way, by using a rubber band and dressmakers' pins [29].

The system consists of a chain of pendulums, each pendulum being elastically connected to its neighbors by elastic springs, as sketched in Fig. 2.1. The main variable characterizing the state of a pendulum with the number n is its angle of rotation θ_n. The equation of motion for the rotational degrees of freedom can be written in the form

$$I\frac{d^2\theta_n}{dt^2} = \beta\left(\theta_{n+1} + \theta_{n-1} - 2\theta_n\right) - mgL\,\sin\theta_n, \tag{2.1}$$

where $I = mL^2$ is the moment of inertia of an isolated pendulum of mass m and length L, β is the torque constant of a connecting spring between two neighboring pendulums, and g is the gravitation constant. Here, the first term corresponds to the inertial effects of the pendulums, the second term represents the restoring torque owing to the coupling between the pendulums and the last term represent the gravitational torque.

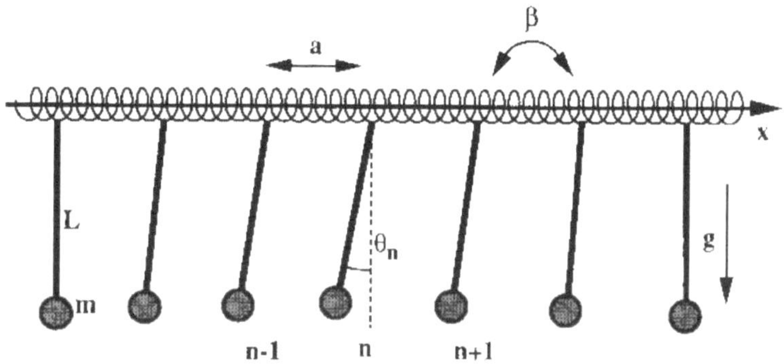

Fig. 2.1. Sketch of the mechanical line which consists of elastically coupled pendulums. The line is subjected to the gravity field that creates an external potential periodic in the rotation angle θ_n [29].

Introducing the pendulum frequency, $\omega_0^2 = mgL/I$, and the speed of sound in the chain, $c_0^2 = \beta a^2/I$, we can normalize the motion equation to one of the forms discussed in Chap. 1. In the limit $\beta \gg mgl$ or $c_0^2/a^2 \gg \omega_0^2$, the motion equations (2.1) reduce to the SG equation. The latter condition can be rewritten in the form $d = c_0/\omega_0 \gg a$, so that the parameter d characterizes the lattice discreteness , and it defines also the width of the kink.

If the right end of the mechanical line makes only small excursions around the down position at $\theta = 0$, then only small-amplitude waves are excited in the chain. If the initial disturbance is a complete rotation of the first pendulum, this rotation propagates collectively along the chain in the form of a kink. In the experimental line shown in Fig. 2.1, one can create a kink moving from the right to the left, then being reflected at the opposite free end and travelling from the left to the right in the form of an antikink.

Since the kink's width is defined by the discreteness parameter d, the case $d \gg a$ corresponds to the continuous model described by the SG equation. However, if $d \sim a$, the discreteness effects become important, and they may affect the kink's motion. Correspondingly, in the case $d \ll a$ the kink's width become of the order of the lattice spacing and, in particular, the kink is strongly pinning by the lattice potential, as shown in Fig. 2.2. In such a strongly discrete chain, one can observe a sequence of kinks and antikinks since they are pinned on the lattice.

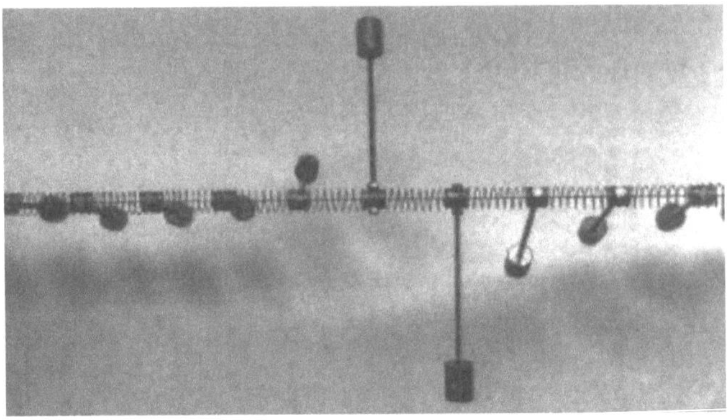

Fig. 2.2. A narrow kink in a strongly discrete chain of pendulums observed with very weak coupling. Photo by R. Chaux [29].

The mechanical transmission line can be further generalized to the case when the substrate potential has two equilibrium states only. Such a mechanical analog was constructed in 1998 by Dusuel *et al.* [30] who modelled the dynamics of the compacton kinks, the kink-like modes characterized by the absence of long tails being localized on a compact support.

2.3 Dislocation Dynamics

One of the first applications of the FK model was suggested in the theory of dislocations in metals [3]–[5], [11, 12], [31]–[33]. The importance of this problem cannot be overestimated since namely dislocations are responsible for most of the mechanical properties of solids. The FK model was the first model that explained the dynamics of a dislocation core on an atomistic level, and it resulted in simple formulas valid even quantitatively. We should notice that large-scale first-principle simulation of dislocation dynamics is not a trivial problem even for the modern parallel-computer technique.

In the dislocation theory, the FK model has a simple physical meaning. Indeed, let us consider an additional semi-infinite plane of atoms inserted into a perfect crystal lattice, as shown in Fig. 2.3. After relaxation to an equilibrium state, the lattice has one *edge dislocation*. The layer of atoms perpendicular to the inserted plane divides the crystal into two different parts. The atoms belonging to the interface layer are subjected to an external periodic potential produced by the surrounding atoms of the lattice; this gives a birth to the FK model. Note that similar models arise in the description of dynamics of plane defects such as *twin boundaries* and *domain wall* in *felloelectrics*, and *ferro-* or *antiferromagnetics* which will be discussed below.

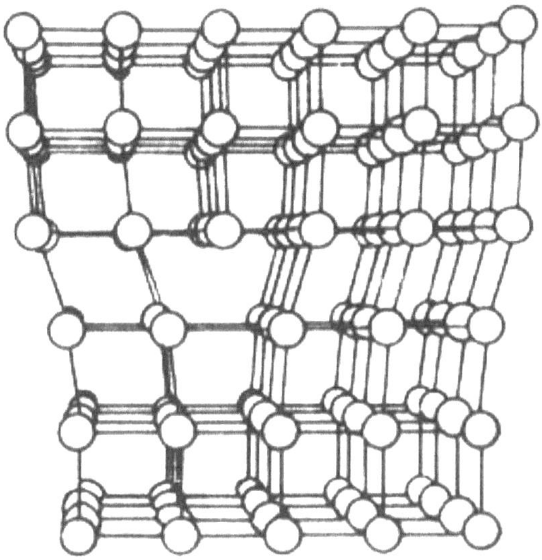

Fig. 2.3. Schematic presentation of the 3D atomic configuration corresponding to an atomic plane inserted into a perfect crystal lattice.

A two-dimensional plane of atoms of the interface layer can be analyzed in the framework of a one-dimensional FK model as an atomic chain perpendicular to the edge dislocation line. In this model, the dislocation core is modelled as a FK kink, and the inherent discreteness of the model explains the existence of an activation energy barrier, known as the Peierls-Nabarro (PN) barrier , for the dislocation motion.

The next step of modelling is to consider the dislocation line itself as (another) FK chain placed into the external periodic potential which is nothing but the primary PN relief. The motion of such a *dislocation line* is due to the creation of pairs of kinks and antikinks, when a section of the dislocation is shifted to the neighboring valley of the external PN potential. These kinks move in the *secondary* PN relief until they reach the end of the dislocation or annihilate with another kink of opposite topological charge. Note that the dislocation itself is a topological object and it cannot be broken, but it may end at the crystal surface or other defects.

At the same time, one should notice that the FK model is known to display serious disadvantages in describing the dislocation dynamics. Namely, in a one-dimensional model with a short-range coupling between particles, such as the standard FK model, the long-range interaction between the point defects (kinks) is always exponential (short-ranged). But the interaction between point defects in crystals is known to be always long-ranged. From the elastic theory, it is well known that the interaction energy follows the power law, $\propto R^{-3}$ for large distances R between point defects. This interaction emerges due to the displacements of the *substrate* atoms from their equilibrium positions, the effect omitted in the one-dimensional FK-type models. However, the values of the PN barriers which are local characteristics of the system, can be predicted correctly with the help of the FK model. In the dislocation theory, dislocations are considered as unbreakable lines interacting according to the elastic theory (in this way, the interaction is described correctly), which are artificially subjected to the PN relief calculated within the effective FK model. Another approach could be to use the FK model but with artificially assumed power-law interaction between the atoms which simulates the real elastic interaction (such an approach will be considered below in Chapter 3). Both the approaches can be coupled together if the dislocation is modelled as being surrounded by an effective tube [34, 35]. Inside the tube, the FK model can be applied, whereas the continuum elastic theory is valid outside.

2.4 Surfaces and Adsorbed Atomic Layers

One of the most important applications of the FK model is found in the surface physics where this model is used to describe the dynamics of atoms and atom layers adsorbed on crystals surfaces. In such a case, the FK substrate potential is defined by the surface atoms of a crystal, whereas the atoms adsorbed on the surface are modelled as the effective FK particles. The adsorbed atoms, simply called *adatoms*, are usually more mobile than the atoms of the substrate which vibrate around their equilibria only; as a result, for this class of problems the approximation of a rigid substrate is indeed adequate. In some cases, e.g. for absorption on the (112) plane of a bcc crystal, the surface atoms produce a furrowed potential. Then, the adatoms located inside the furrows can be considered as a one-dimensional chain. Another example is vicinal semiconductor surfaces, where the adatoms are adsorbed closely to the steps and, thus, they are organized into a structure of weakly interacting chains. In the latter case, the distance between the chains may vary depending on the angle of the vicinal surface chosen.

In the applications of the FK model in surface physics, the concentration parameter of atoms θ can vary in a wide range, from $\theta = 0$ (a single atom) to $\theta = 1$ (a commensurate layer), and even become larger when the effective size of the adatom is smaller than the period of the external potential (e.g., for Lithium atoms adsorbed on transition metal surfaces). As a result, the FK model allows to analyze many interesting physical phenomena, including transitions between different commensurate structures and the commensurate-incommensurate transitions. The parameters of the FK model can be estimated (or even evaluated from the first principles) with a good accuracy [36]. In such adsorbed layers of atoms, the FK dimensionless elastic constant g is typically of order of $0.1 \div 1$. Moreover, the adsorbed systems can be studied experimentally by direct methods such as the scanning tunnel microscopy, which provide a direct verification of many of the theoretical predictions.

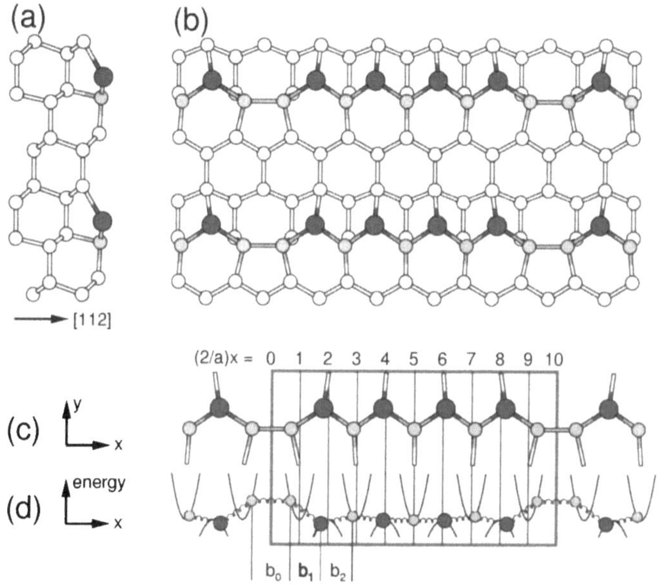

Fig. 2.4. (a) Side and (b) top views of Ga/Si(112) with vacancy period $N = 5$. The fully relaxed coordinates are from first-principles total-energy minimization. (c) Bonding chain of Ga (dark) and Si (light) atoms. (d) One-dimensional FK model representing this Ga-Si chain. The (horizontal) atomic displacements are defined relative to the ideal positions (shown by thin grid lines) of the substrate atoms; vertical displacements represent the individual substrate strain energies [38].

When a material is grown on a lattice-mismatched substrate, the resulting strain field can lead to self-organized structures with a length scale many times the atomic spacing. One well known example is the Ge/Si(001) dimerised overlayer system. The Ge film is compressively strained (by 4% relative to the bulk), and the system lowers its energy by creating dimer vacancies in the surface layer; at the vacancy sites, the exposed atoms in the second layer re-bond to eliminate their dangling bonds. The missing-dimer vacancies order into *vacancy lines* with $2 \times N$ periodicity, where the optimal N depends on the Ge coverage [37].

As was shown by several researchers, the analysis of the strain-mediated interaction of vacancy lines in a pseudomorphic adsorbate system can be performed in the framework of the FK model. For example, it was applied to the Ga/Si(112) system by extracting values for the microscopic parameters from total-energy calculations [38]. In spite of the fact that the FK model contains only nearest-neighbor harmonic interactions, it reproduces the first-principles results quite accurately, and thus allows for a particularly simple analysis of the dominant interactions.

The microscopic structure of vacancy lines on Ga/Si(112) is shown in Fig. 2.4. Since Ga is trivalent it prefers to adsorb at threefold surface sites. The bulk-terminated Si(112) substrate, which may be regarded as a sequence of double-width (111)-like terraces and single (111)-like steps, offers just such threefold sites at the step edges, as shown in Fig. 2.4(b). A single Ga vacancy leaves two step-edge Si atoms exposed, which re-bond to form a dimer. As is shown in Fig. 2.4, a Ga-Si chain (with vacancy period N) can be mapped onto a one-dimensional chain of harmonic springs connecting N Si atoms and $N-1$ Ga atoms in a unit cell of length Na, as shown in Fig. 2.4(d). This model was first proposed by Jung *et al.* [39] and subsequently confirmed by Baski *et al.* [40] and Erwin *et al.* [38] using the total-energy calculations.

Furthermore, the FK model and its generalizations can be employed for describing the dynamics of clean surfaces, provides one treats the surface atoms as atoms of the effective FK system while the atoms of the underlying layer are assumed to produce an effective substrate potential. In particular, the FK model can be used to describe the *surface reconstruction* [41, 42], the structure of *vicinal semiconductor surfaces* [43], the processes of *crystal growth* [44], and the sliding of migration islands over a crystalline surface [45].

One of the serious restrictions of the standard FK model is its one-dimensional nature. Indeed, usually quasi-one-dimensional chains of adatoms are not completely independent, and they can form a system of parallel weakly coupled chains. For example, when atoms are adsorbed on the stepped or furrowed surfaces of a crystal, they can be described as an anisotropic two-dimensional system of weakly coupled chains. Considering kinks of the primary FK chains as quasi-particles subjected to a periodic PN potential, we may analyze collective excitations of the two-dimensional system as *secondary kinks* which themselves can be described by a variant of *a secondary FK model.* A system of interacting FK chains was analyzed, for example, by Braun *et al.* [46] and Braun and Kivshar [47], and there are many papers devoted to the statistical mechanics of adsorbed layers (see, e.g., the monograph by Lyuksyutov *et al.* [48] and references therein).

More general models describing the dynamics in two-dimensional systems correspond to a vector generalization of the FK model, which is the most realistic model for two-dimensional arrays of atoms adsorbed on crystal surfaces. In this model, each atom has two degrees of freedom to move and is subjected to a two-dimensional external potential created by atoms of the surface. In fact, a variety of such models is generated by symmetry properties of various substrate potentials (isotropic models with square, triangular, hexagonal lattices or anisotropic models with, e.g., rectangular lattice, etc.). Several examples of these models will be considered in the book. Additionally, one of the ways to make the FK model more realistic for a broader class of physical applications is to include an additional degree of freedom allowing the atoms to move in the direction perpendicular to the chain. The corresponding FK model with a transverse degree of freedom was proposed by Braun and Kivshar [49]. Interesting physical effects are possible in this type of models due to the existence of nontrivial zigzag-like ground states and, correspondingly, novel types of topological kink-like excitations.

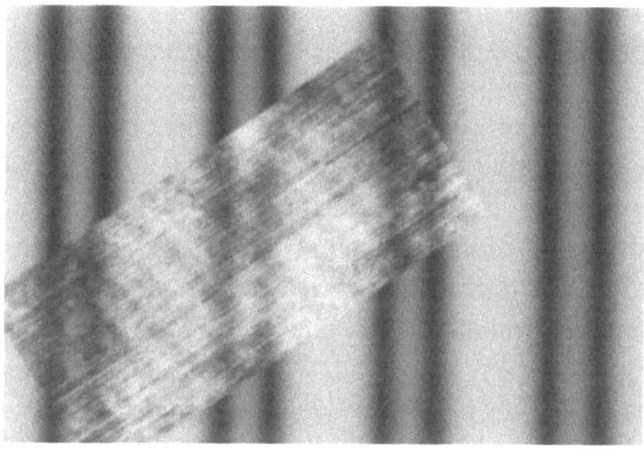

Fig. 2.5. Experimental and theoretical dislocation patterns for sub-monolayer Ag on Pt(111). The inset is a scanning tunnelling micrograph. The surrounding area shows a simulated image based on the two-dimensional FK model. The dark vertical lines are partial misfit dislocations, the light vertical lines are narrow hcp domains, the light vertical areas are wide fcc domains [50].

Many close-packed metal overlayer systems show patterns of partial dislocations which form to relieve lattice mismatch between overlayer and substrate. Indeed, the well-known herringbone reconstruction of the clean gold (111) surface is a striking example of such a dislocation pattern, formed because the lower coordinated surface gold atoms have a closer equilibrium spacing than normally coordinated bulk gold atoms [41, 51]. Similarly, bilayers of Cu on Ru(0001) form a striped dislocation pattern [52]. These patterns consist of alternating domains with fcc and hcp stacking separated by partial dislocations.

An example of the overlayer is shown in Fig. 2.5, where a scanning tunnelling micrograph of the structure for sub-monolayer Ag on Pt(111) is presented. The inset is an experimental micrograph, while the surrounding structure shows the result of the theoretical calculations. For Ag on Pt(111), the fcc domains are more than in three times wider than the hcp domains. Differences in domain widths can be correctly attributed to the fact that the fcc sites are energetically favored relative to hcp sites. However, this "monolayer stacking fault energy" is not sufficient to explain the unusually wide fcc domain widths seen for Ag on Pt(111). Hamilton *et al.* [50] demonstrated that the major difference in domain widths is caused by the immediate proximity of this system to a commensurate-incommensurate phase transition. Moreover, these authors have found that a two-dimensional FK model can reproduce the major types of dislocation structures. The prediction of the FK model is in agreement with the observation that dislocations do not form in islands until the island size is much greater than the dislocation spacing in continuous films.

2.5 Incommensurate Phases in Dielectrics

In some systems, an incommensurate (IC) phase may appear as an intermediate phase between two commensurate crystalline phases and, thus, it should play an important role in the theory of phase transitions [53]. Moreover, the study of the IC phase can be reduced in some cases to the analysis of a FK-type model. For definiteness, we consider a model that describes an array of rigid molecules joined by the elastic hinges with the elastic constant g, as shown in Fig. 2.6 (see Refs. [54, 55]). The role of elastic hinges is to keep the chain straight, because the hinge produces a moment which tends to decrease the absolute value of the rotation angle. Then, let the chain be subjected to the compression force $P > 0$ acting along its axis, which tends to destroy the horizontal arrangement of molecules. A competition between these two forces gives rise to a transverse modulational instability. The effect of a three-dimensional crystal on the linear molecular chain may be taken into account by effective bonds shown in Fig. 2.6 by vertical springs.

Under the assumption that the transversal displacements v_n of the hinges are small, $v_n \ll 1$, the Hamiltonian of the model can be written as

$$H = \frac{1}{2}\sum_n \left[\left(\frac{dv_n}{dt}\right)^2 + g(v_{n-1} - 2v_n + v_{n+1})^2 - P\,(v_{n+1} - v_n)^2 + v_n^2 + \frac{v_n^4}{2} \right],$$

and it describes the so-called Elastically Hinged Molecule (EHM) model . Using the identity $\sum_n v_n v_{n+j} = \sum_n v_{n+l} v_{n+j+l}$, one can represent the Hamiltonian in the form of that for a linear chain of point particles with a fourth-order polynomial background potential and the harmonic nearest- and next-nearest-neighbor interaction with the stiffness of the bonds between the nearest neighbors $C_1 = 4g - P$ and that between the next-nearest neighbors $C_2 = -Fg$ [56]–[59]. Introducing the coordinates $x_n = v_{n+1} - v_n$ and

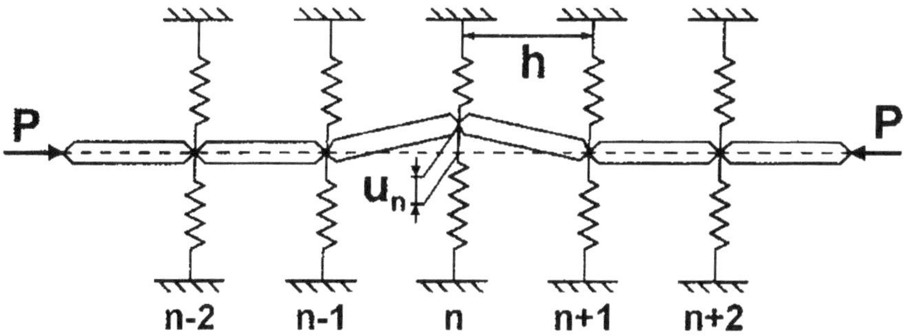

Fig. 2.6. Schematic presentation of the EHM model. Rigid molecules are joined together by the elastic hinges, and they are compressed by the force P along its axis. Each hinge has one degree of freedom, the transverse displacement u_n [55].

$y_n = v_{n+1} + v_n$, one can demonstrate that the static properties of the EHM model are identical to those of the Janssen model [60]. Notice that in the absence of the substrate potential, the IC phase can appear only if the third-neighbor interaction is taken into account.

In order to demonstrate how the EHM model and other related models can be reduced to the FK-type model, we consider the structure transformation in the EHM model when the external pressure P is fixed and the stiffness g decreases slowly. When g is very large, the only solution is a trivial state, $v_n = 0$. At a critical value of g the dispersion curve vanishes at the point $q = 2\pi\kappa$ that depends on P [55]. As a result of this second-order phase transition, the sinusoidal modulation with the wave number κ appears in the crystal. If κ is an irrational number, then the period of modulated phase is infinitely long and the resulting structure describes the IC phase. For rational wave numbers $\kappa = r/q$, where r and q are co-prime positive integers such that $2r < q$ and κ is close to $1/4$, the crystal structure can be described as the four-periodic structure with an unknown slowly varying phase $\varphi_n(t)$,

$$v_n(t) = \pm A \cos\left[\frac{\pi}{2}(n+m) + \frac{\pi}{N} + \varphi_n(t)\right], \tag{2.2}$$

where $A^2 = 2(2P-4g-1)$ is the amplitude of the commensurate four-periodic structure [55]. Averaging fast varying terms over four neighboring nodes and assuming that φ_n varies slowly, $|\varphi_n - \varphi_{n-1}| \ll 1$, we obtain the following equation of motion,

$$\frac{d^2\varphi_n}{dt^2} - 4g\left(\varphi_{n-1} - 2\varphi_n + \varphi_{n+1}\right) + \frac{A^2}{4}\sin(4\varphi_n) = 0\,, \tag{2.3}$$

which can be easily transformed into the canonical form of the FK model [55].

2.6 Crowdions and Lattice Defects

The plastic deformations in metals may result from the creation of local defects known as *crowdions* [61, 62]. A crowdion in a crystal lattice corresponds to a configuration where one extra atom is inserted into a closely packed row of atoms in a metal with an ideal crystal lattice. Additionally, some more recent studies indicate that diffusion on some strained surfaces may be mediated by newly identified adatom transport mechanism: the formation and motion of a surface crowdion [63]. The formation and structure of the surface crowdion is shown in Fig. 2.7(a-c). The saddle point configuration [Fig. 2.7(b)] is a metastable (sometimes stable) state in which the adatom is embedded into the surface, and a needlelike defect in Fig. 2.7(c) is the surface crowdion. In this case, the crowdion extends over roughly ten nearest neighbor distances, and appears only for stress states that are compressive along the axis normal to the needle direction, and tensile along the direction of the needle. The crowdion, once formed, can move along the row of atoms by relatively small changes in the positions of the atoms.

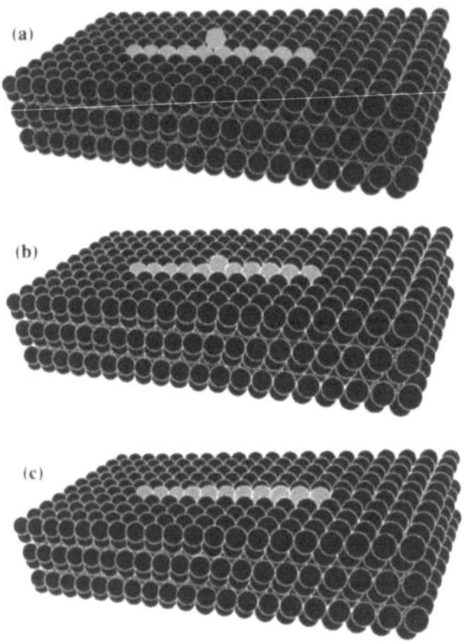

Fig. 2.7. The formation and structure of a surface crowdion: (a) initial configuration, (b) saddle point configuration, (c) surface crowdion. The crowdion extends along the [110] direction (tensile direction) for roughly 10 nearest neighbor spacings [63].

In many cases, the crystalline potential is organized in such a way that the atoms can move only along a selected direction of the row and, therefore, the inserted atom (together with the neighboring atoms) forms a one-dimensional structure which can be described as a kink of the FK model. Typically, the size of such a configuration is about ten lattice constants, so that the kink is characterized by a very small PN barrier, and it moves along the row almost freely. As a result, the crowdions plays an important role in plasticity of metals. The crowdion problem was studied, in particular, by Landau *et al.* [64] and Kovalev *et al.* [65]. The main features of this model are presented in Fig. 2.8. The whole crystal is treated as a continuous three-dimensional elastic medium with can be described in terms of the linear plasticity theory. In this medium, a cylindrical tube of the radius b along the x axis is cut out, and the discrete atomic chain is inserted inside of the created channel. The parameter b has the meaning of the interatomic separation in the direction perpendicular to the chain, while the atomic separation along it is a.

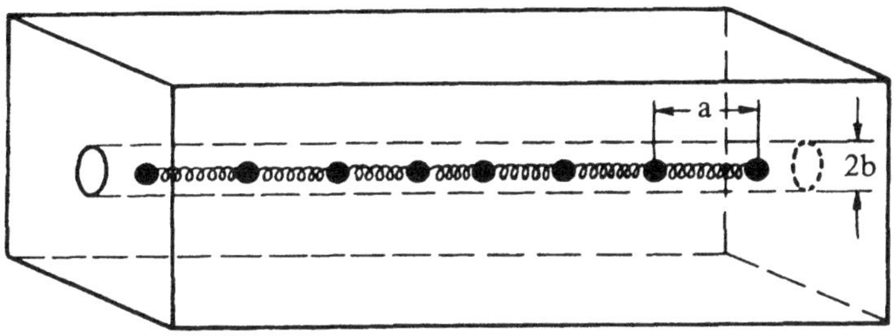

Fig. 2.8. Schematic model of a bulk crowdion in the lattice. The springs symbolize the coupling force interaction between neighboring atoms in a chain [65].

Such a model of the crowdion structure employs a simple scalar variant of the elasticity theory which considers displacements along the direction of the atomic chain only. As was shown by Kovalev *et al.* [65], this system can be described by a generalized FK model that includes the elastic properties of a crystal. In the continuous approximation, this model can be reduced to an integro-differential sin-Gordon equation that takes into account an effective long-range interaction between atoms in the chain.

2.7 Magnetic Chains

The approach similar to that used in dislocation theory and surface physics, emerges also in the description of the dynamics of plane defects such as twin boundaries [66, 67] and domain walls in felloelectrics [68]–[71], and in ferromagnetic (FM) or antiferromagnetic (AFM) chains [72]–[75].

In magnetic systems, *a domain wall* — a spatially localized region between two domains of different magnetization — is described by a kink solution that connects two equivalent ground states. The simplest one-dimensional classical model of a magnetic chain is described by the following Hamiltonian [76],

$$\mathcal{H} = -J \sum_n \mathbf{S}_n \mathbf{S}_{n+1} - D \sum_n (S_n^x)^2 . \tag{2.4}$$

The first term in Eq. (2.4) is the isotropic exchange energy and the second term describes the single-ion anisotropy. If J and D are positive, the x-axis is the preferred direction in the spin space . For the minimum energy, all spins should have maximal x-components. This leads to a degeneracy of the ground state, and two equivalent ground states are connected by a static domain wall (see Fig. 2.9). Its form represents a compromise between the counteracting effects of exchange interaction J and anisotropy D. In the classical approximation, spins are treated as vectors of length S, so that they may be described by spherical co-ordinates

$$\mathbf{S}_n = S \cdot (\sin\theta_n \cos\phi_n,\ \sin\theta_n \sin\phi_n,\ \cos\theta_n). \tag{2.5}$$

In the case of a strong planar anisotropy and a transverse magnetic field, the equation of motion for the angle ϕ_n results in the SG equation [73].

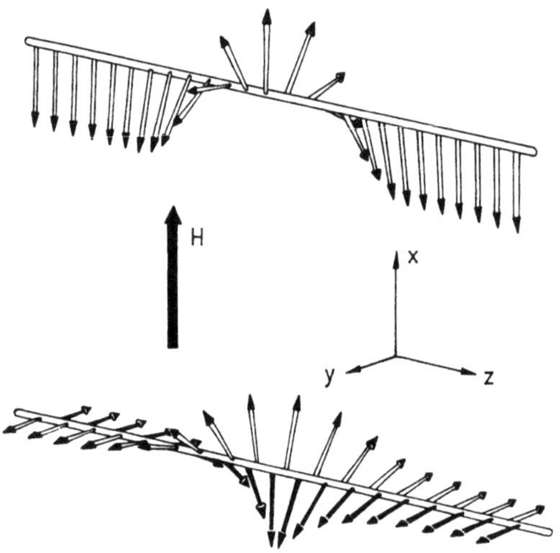

Fig. 2.9. Top: schematic structure of a FM domain wall in the presence of a magnetic field H as a 2π-twist of the ordered spin chain described by a kink of the FK model. Bottom: an AFM domain wall as a π-twist reversing both spin sublattices of the chain.

For a FM chain with the so-called easy-plane anisotropy in the direction of the transverse magnetic field H at sufficiently low temperatures, there exists only one ground state with all spins being aligned in the direction of the magnetic field. A kink soliton in this case represents a 2π-twist of the ordered spin chain which can freely move along the chain without dispersion (see Fig. 2.9, top). In an AFM chain, the ground state corresponds to a spin-flop configuration, i.e. the two sublattices, apart from a slight canting, are aligned antiparallel to each other and perpendicular to the field H. Therefore, there exist two degenerate, but topologically inequivalent ground states which differ by the sign of the spin orientation. An AFM soliton in this case represents a π-twist reversing both sublattices when passing along the chain (Fig. 2.9, bottom). The difference in topology results in different spatial and temporal behavior of the spin components and largely affects the experimental evidence. While in the FM case spin fluctuations originate only from the 2π-soliton itself, it is flipping of all spins between two neighboring π-solitons (the so-called "flipping mode") which yields the essential contribution in the AFM case. In either case, spin fluctuations in the chain direction are largely suppressed by the strong planar anisotropy.

A domain wall in a magnetic system is described by a kink soliton, and it represents a microscopic structure which cannot be directly observed but can only be probed by its influence on macroscopic properties. Different techniques can be used to study magnetic kink solitons. The most common technique is the inelastic neutron scattering [77]–[79], which allows to measure the kink's contribution to the dynamic structure factor, which is the spatial and temporal Fourier transform of the spin correlation function (see, e.g., Ref. [73]). Complementary information can be obtained by nuclear magnetic resonance [80]–[83], which probes the dynamic structure factor integrated over all wave-numbers. Two other methods, Raman scattering [84] and electron spin resonance [82, 85] were applied to probe the influence of kinks indirectly through the soliton-induced broadening of the magnon line-width. As for the physical systems, the prototype of a FM chain described by the SG equation is the well-known $CsNiF_3$ (see, e.g., Ref. [77]), whereas the prototype of an AFM chain is the TMMC compound $(CH_3)_4NMnCl_3$ [78].

2.8 Josephson Junctions

Among many physical systems, one of the closest correspondence to the SG equation has been found in the problem of fluxon dynamics in long Josephson junctions (JJs or Josephson transmission lines). A SG kink in this case is often called "fluxon" since it carries a magnetic flux quantum $\Phi_0 = h/2e = 2.07 \cdot 10^{-15}$ Wb moving between two superconducting electrodes. Several comprehensive reviews on fluxon dynamics in JJs have been published in the past [86]–[88] and some of the new topics have been addressed later [89]–[91]. This is an active area of research, especially after the discovery of high-T_c superconductivity.

A fluxon in a JJ carries a magnetic flux Φ_0 generated by a circulating supercurrent, often called Josephson vortex, which is located between two superconducting films separated by a few nanometers thin layer of insulator. As has been shown by McLaughlin and Scott [86], the fluxon corresponds to a 2π-kink of the quantum phase difference φ between the two superconducting electrodes of the junction. The perturbed sine-Gordon equation which describes the dynamics of the system reads in the normalized form as follows,

$$\frac{\partial^2\varphi}{\partial t^2} - \frac{\partial^2\varphi}{\partial x^2} + \sin\varphi = -\alpha\frac{\partial\varphi}{\partial t} + \beta\frac{\partial^3\varphi}{\partial x^2\partial t} + \gamma. \tag{2.6}$$

Here time t is measured in units of ω_0^{-1}, where ω_0 is the Josephson plasma frequency, the spatial coordinate x is measured in units of the Josephson penetration depth λ_J, α is a dissipative term due to quasi-particle tunnelling (normally assumed ohmic), β is a dissipative term due to surface resistance of the superconductors, and γ is a normalized bias current density. To account for the behavior of a real system, Eq. (2.6) must be solved together with appropriate boundary conditions which depend on the junction geometry and take into account the magnetic field applied in the plane of the junction.

A unique property of real JJs is that the parameters α, β and γ are rather small. An important solution of Eq. (2.6) with zero r.h.s. is the kink $\varphi = 4\tan^{-1} e^{\xi}$, where $\xi = (x - vt)/\sqrt{1-v^2}$. The velocity v is determined by the balance between losses and input force, $v = 1/\sqrt{1 + (4\alpha/\pi\gamma)^2}$. A fluxon behaves like a relativistic particle. Its velocity saturates at large values of γ/α and becomes close to the velocity c of linear waves in the junction, called the Swihart velocity in the JJ line.

Fluxons in weakly interacting JJs is a subject of intensive theoretical and experimental investigations. For the first time, the fluxon dynamics in two inductively coupled long JJs was considered theoretically by Mineev *et al.* [92]. The perturbation approach for small coupling has been further explored by Kivshar and Malomed [27] and Grönbech-Jensen *et al.* [93, 94]. A very important step towards quantitative comparison with real experiments was made by Sakai *et al.* [95] who derived a model for arbitrary strong coupling between the junctions. The discovery of the intrinsic Josephson effect in some high-temperature superconductors such as $Ba_2Sr_2CaCu_2O_{8+y}$ convincingly showed that these materials are essentially natural super-lattices of Josephson junctions formed on atomic scale [96]—[98]. The spatial period of such a super-lattice is only 15 Å, so the Josephson junctions are extremely densely packed. The superconducting electrodes are formed by copper oxide bilayers as thin as 3 Å and are separated by the Josephson super-lattices with many active layers.

An important application of the FK model can be found in the theory of the *Josephson junction arrays*, where a discrete chain of effective particles emerges when one considers the flux flow in discrete parallel arrays of weak links between superconductors. Strong discreteness effects appear in the case of biased arrays of small JJs, also called Josephson junction ladders. An example of such a system is shown schematically in Fig. 2.10, where a ladder consists of Nb/Al-AlO_x/Nb underdamped Josephson tunnel junctions, and each cell of it contains four small Josephson junctions. Such ladders can be fabricated with either open or periodic boundary conditions, as shown in Figs. 2.10(a,b), respectively. Both these systems have been used to study the localized modes, the *discrete breathers*.

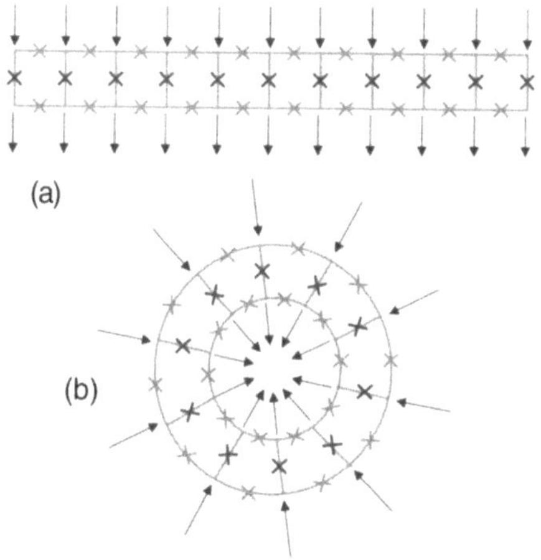

Fig. 2.10. Schematic view of (a) linear and (b) annular ladders of the Josephson junctions. Arrows indicate the direction of the externally applied bias current [99].

To describe the flux flow in discrete parallel arrays of weak links between superconductors, one can use the FK model of the form,

$$\frac{d^2\varphi_n}{dt^2} = \frac{1}{a^2}(\varphi_{n-1} - 2\varphi_n + \varphi_{n+1}) - \sin\varphi - \alpha\frac{d\varphi_n}{dt} + \gamma, \tag{2.7}$$

where $n = 1, \ldots, N$. These equations are just the Kirchhoff circuit-law equations for an array of N discrete JJ elements interconnected via a parallel resistance/inductance combination. The phases at the virtual points $n = 0$ and $n = N{+}1$ are defined through the boundary conditions. This model has been analyzed theoretically by Ustinov *et al.* [100] and their predictions have later been confirmed experimentally by van der Zant *et al.* [101, 102].

2.9 Nonlinear Models of the DNA Dynamics

The models similar to the FK model play also an important role in the interpretation of certain biological processes, such as DNA dynamics and denaturation (see,e.g., Refs. [103]–[108], the review papers by Zhou and Zhang [109], Gaeta *et al.* [110], and the book by Yakushevich [111]). Definitely, the structure of a DNA-type double-helix chain is complex. However, very general features of macromolecules such as DNA can be modelled by the Hamiltonian (1.1)–(1.5), provided we assume that all bases of the DNA chain are identical.

One of the first nonlinear models suggested for the study of the DNA dynamics took into account the rotational motion of the DNA bases [112] and, as a matter of fact, it is reduced to an equivalent mechanical model discussed above. Using the analogy between the rotational motion of bases of the DNA strands and rotational motions of pendulums, Englander *et al.* [112]

suggested that a DNA open state can be modelled by the kink solution of the FK model. To make the model more accurate, one should take into account rotational motion of bases in both DNA strands, considering the double rod-like model [111].

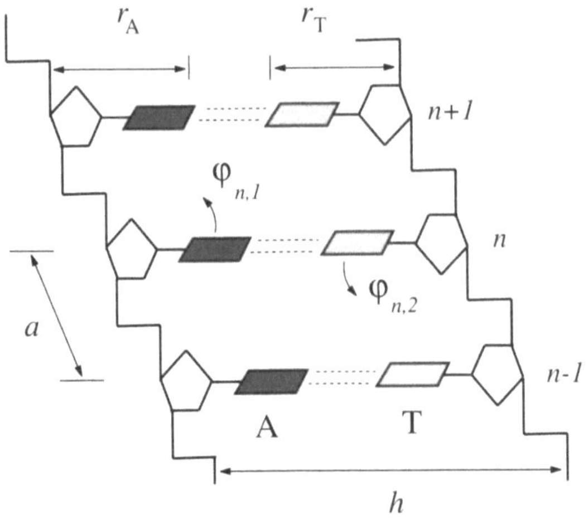

Fig. 2.11. Fragment of the DNA double chain consisting of 3 AT base pairs. Longitudinal pitch of the helix is $a = 3.4$ Å, transverse pitch is $h = 16.15$ Å.

The so-called planar model of torsional dynamics of DNA [111] has originally been developed to describe travelling kinks which are supposed to play a central role in the transcription of DNA. One of such models can be introduced for a B-form of the DNA molecule, the fragment of which is presented in Fig. 2.11. The lines in the figure correspond to the skeleton of the double helix, while black and grey rectangles correspond to bases in pairs (AT and GC). This model takes into account the rotational motions of bases around the sugar-phosphate chains in the plane perpendicular to the helix axis (positive directions of the rotations of the bases for each of the chains are shown by arrows in Fig. 2.11). If we consider the plane DNA model, where the chains of the macromolecule form two parallel straight lines separated by a distance h from each other, and additionally the bases can make only rotation motions around their own chain being all the time perpendicular to it, then the positions of the neighboring bases of different chains are described by the angles of rotation only.

In the simplest case, the equation imitating the rotational motion of the n-th base in the first chain takes the form [111]

$$mr^2 \frac{d^2\phi_{n,1}}{dt^2} = gr^2(\phi_{n+1,1} + \phi_{n-1,1} - 2\phi_{n,1}) - kr^2[2\sin\phi_{n,1} - \sin(\phi_{n,1} + \phi_{n,2})], \tag{2.8}$$

and the similar equation for the second chain. Here the variables $\phi_{n,1}$ and $\phi_{n,2}$ are the angular displacements of the n-th pendulum in the first and second chains, respectively, g is the rigidity of the horizontal thread, r and

m are the length and mass of the pendulum, and the parameter k describes the interaction strength between the chains.

The model (2.8) has at least two groups of particular solutions, $\phi_{n,1} = \pm\phi_{n,2}$. Then, the DNA dynamics can again be described by the Hamiltonian (1.1)–(1.5), where $x_n^{(i)}$ is the generalized coordinate describing the base at site n on the i-th chain of the double helix. In a general form, the model (2.8) describes two coupled FK chains.

A modification of these models taking the helicoidal structure of DNA into account has also been suggested [110]. In a double helix it happens that the bases, which are one half-wind of the helix apart, end up spatially close to each other. Introducing "helicoidal" terms to the planar models to account for this effect adds a harmonic potential to the Hamiltonian (1.1)–(1.5) for the interaction between bases with numbers n and $n+l$. In the case of the model of vibrational dynamics, these higher-order inter-base interactions modify the FK model for the common displacement $\psi_n = \frac{1}{2}(x_1 + x_2)$, making the dispersion of linear waves more complicated. From the viewpoint of physics, the novel helicoidal terms produce extremal points in the linear dispersion allowing standard breather modes, i.e. localized oscillations responsible for local openings of the double helix.

2.10 Hydrogen-Bonded Chains

More deeper background of the applications of the FK model can be found in the cases when the atoms belonging to the chain and the atoms producing the external (substrate) potential have a different physical origin. For example, in the so-called *superionic conductors* (see, e.g., Ref. [113] and also the review papers [114, 115]) an anisotropic crystalline structure forms quasi-one-dimensional channels along which ions may easily move, so that this kind of models may also be reduced to the analysis of a one-dimensional chain subjected to an effective on-site potential.

One more important model of this kind emerges in the theory of proton conductivity of hydrogen-bonded chains. Hydrogen-bonded networks are quasi-one-dimensional clusters of molecular aggregates interacting with their neighbors through hydrogen bonds. Schematically, this can be presented in the form ...X–H...X–H...X–H...X–H..., where the full line segments indicate covalent or ionic bonds, the dotted ones, hydrogen bonds, and X a negative ion. The important idea of a simple physical model for such a nonlinear chain is based on the fact that protons move in a double-well potential resulting from hydrogen bonds with the heavy-ion lattice (oxygen lattice) which is assumed to be deformable [116].

A local distortion of the oxygen lattice lowers the activation barrier for protons and, thus, promotes their motion. In order to describe this phenomenon, one- or two-component nonlinear models should include the proton sublattice which supports topological solitons (kinks), while the oxygen sublattice can be modelled as another sublattice or as an effective external potential for the proton motion. Several models of this kind have been proposed [116]–[121], and they give a simple and effective description of the proton mobility in hydrogen-bonded chains.

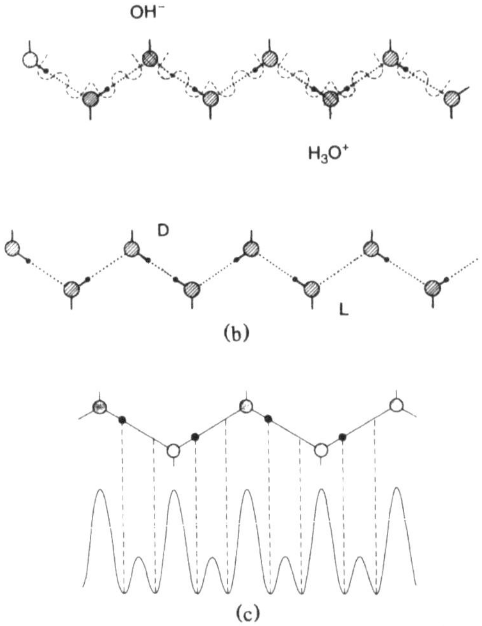

Fig. 2.12. (a) Ionic and (b) orientational (Bjerrum) defects in ice. (c) The dynamics of protons in the zigzag hydrogen-bonded diatomic chain described as the motion of protons in a doubly periodic potential [121].

In the lowest-order approximation, the oxygen atoms are assumed to have fixed positions and to produce an effective substrate potential to the mobile hydrogen atoms, for which a double-sinusoidal FK model with the effective on-site potential

$$V_{\text{sub}}(u) = 2V_0 \left[-\sin^2\left(\frac{u}{2}\right) + 2A\sin^2\left(\frac{u}{4}\right) \right] \tag{2.9}$$

can be derived (see, e.g., Ref. [121]). Then, the mechanism of proton conductivity can be explained by a migration transport of two types of defects, *ionic* and *bonding* (or Bjerrum) defects, along the chains, as shown schematically in Fig. 2.12. In the continuum limit, the defects are described by two types of kinks of the corresponding double-SG equation [121].

A similar kind of *one-component* models for hydrogen-bonded systems has been investigated in the framework of the continuum approximation [122, 123]. The concepts of the theory of kink-induced proton conductivity involve more general properties of the FK type models, e.g. the discreteness of the proton chain and thermalized kink motion [124], the effect of increased proton conductivity due to the commensurate-incommensurate phase transitions [125], a complex chain of zigzag structure [126], a mass variation along the chain [127], etc.

More rigorously, the dynamics of systems like hydrogen-bonded chains can be properly described by introducing *two interacting sublattices* for proton and oxygen atoms, respectively. In such a case, one should consider *two-component generalization* of the FK model which describes two interacting chains of particles, one of them subjected to a substrate potential which is created by the other. Several models of this type have been introduced and studied in the continuum approximation [116]–[121]. The dynamics of the two-component models has several new features in comparison with one-component models. For example, a new branch of the phonon spectrum appears in the band gap of linear excitations of the standard FK chain. In the result, the motion of kinks is stable only for low velocities which do not exceed the sound speed of acoustic phonons of the oxygen sublattice [117]. As a matter of fact, the second phonon branch plays an important role in kink scattering by local impurities [128].

Similar two-component FK models describe more realistically the dynamics of other physical objects such as dislocations, crowdions, ad-atomic chains, chains of ions in superionic conductors, etc. In all such situations the second (heavy atom) subsystem corresponds to substrate atoms, so that the whole system may be treated again as a FK chain subjected to a *deformable substrate*. A similar situation occurs for the physical models of molecular crystals and polymer chains as well as ferroelectric or ferroelastic chains where rotational and vibrational degrees of freedom are coupled [129]–[131].

2.11 Models of Interfacial Slip

There is a rapid growth of interest in the nature of friction at a microscopic level, in the so-called field of nanotribology [132]. As experimental and large-scale simulation data at the nanoscale have begun to become available, a search for models providing an interpretative framework has also begun. Several simple models have been introduced in the hope of yielding applicable analytical results. These include a many-layer model with harmonic interactions [133, 134], the Frenkel-Kontorova-Tomlinson model [135, 136], and a single-layer model with harmonic interactions and disorder [137, 138]. Other more complicated FK-type models have been simulated using molecular dynamics (MD) methods [139, 140]. This important problem can be approached by making use of large-scale two-dimensional MD simulations to justify a simple model of friction between two flat one-dimensional interfaces, and then to determine the parameters of this model. Thus, the models of intermediate complexity can be employed to justify the approximations used to derive more simple FK-type models of friction.

A simplified one-dimensional model based on the two-chain-driven Frenkel-Kontorova model with unidirectional motion was suggested in Ref. [141]. Such

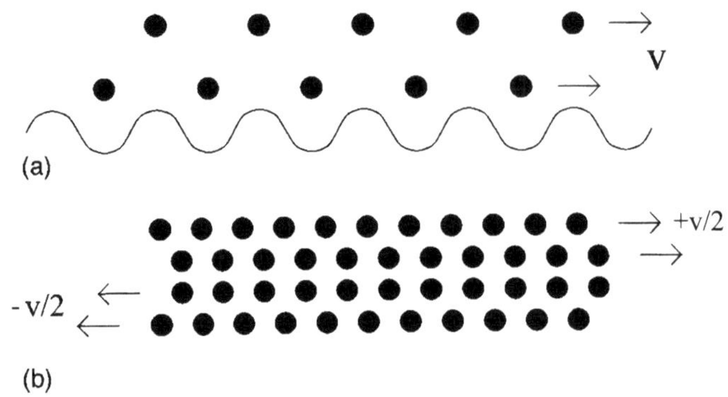

Fig. 2.13. The schematic geometry corresponding to two types of the interface slip models: (a) simplified model, (b) intermediate model [141].

a model displays many key features of the large-scale MD simulations (for example, early-time behavior at the interface, formation and motion of dislocations at the interface, the dependence of the frictional force on the velocity of slip), and at the same time it allows a more transparent and analytical approach to provide an interpretative framework.

To derive the effective FK model, Röder *et al.* [141] used the results of large-scale MD simulations and then replaced the lower half-part of the simulated system by an effective periodic sinusoidal substrate potential. Considering only the first two layers of atoms nearest to the interface and keeping the triangular lattice structure, the remainder of the upper half-part of the system is replaced by an effective viscous damping of the motion of the retained atoms. To make the model more analytically tractable, Röder *et al.* [141] restricted the motion of the atoms to be unidirectional parallel to the interface. The center of mass of the two chains moves with a constant velocity v over the substrate potential. These simplifications leads to a one-dimensional two-chain FK-type model, as shown in Fig. 2.13(a).

To further justify this model, Röder *et al.* [141] undertook a series of MD simulations of scale intermediate between the large-scale and simple models.

3 Kinks

In this chapter we discuss kinks of the FK model. One of the important properties of kinks in a discrete lattice that differ them from solitons is the existence of the discreteness-induced Peierls-Nabarro periodic potential that has a strong effect on the motion of kinks. We discuss both moving and trapped kinks as well as the dynamics of kinks in the presence of disorder. Additionally, we consider the modification of kink properties in the generalized FK models that include an on-site potential of a general form and anharmonic interactions between the particles in the chain.

3.1 The Peierls-Nabarro Potential

Kinks are fundamental topological nonlinear modes of the FK model, and they are responsible for many important physical characteristics of a chain of coupled particles on a substrate. Main properties of kinks and their existence as topological states do not depend crucially on the discreteness of the primary model, so that the approximation based on the continuous SG model is often an acceptable approach for describing the properties of kinks in the discrete FK model which allows to keep the basic features of the system dynamics. However, the very specific property of a discrete lattice is the existence of the so-called Peierls-Nabarro (PN) periodic potential $V_{PN}(X)$ (where X is the coordinate of the kink's center) that affects the kink motion. From the historical perspectives, the PN potential and its properties have been first discussed in the context of the dislocation theory in crystals [142]–[145].

To understand how the concept of the PN potential appears in the analysis the kink's motion, first we note that in the continuum limit approximation the system is invariant to any translation of the kink along the chain; such translations are possible due to the existence of the so-called Goldstone mode in the spectrum of kink's excitation. On the contrary, such an invariance is absent in discrete models; and only the kink's shifts on a lattice spacing a_s and its integer multipliers are allowed. The smallest energy barrier that kink should overcome in order to start moving in the lattice is called the PN barrier, E_{PN}. In a discrete case, the zero-frequency translational Goldstone mode is replaced by a finite-frequency localized mode known as the PN mode.

The PN barrier energy E_{PN} is a difference between two values of the kink's potential energy defined for two stationary configurations, stable and unstable (saddle) ones [see Figs. 3.1(a,b)]. The first state, shown in Fig. 3.1(a), describes a stationary configuration of coupled particles that correspond to a minimum of the energy, with a kink situated at the bottom of the potential energy relief $V_{PN}(X)$. The second state, shown in Fig. 3.1(b), corresponds to an unstable configuration when the kink is placed on one of the maxima of the PN energy relief.

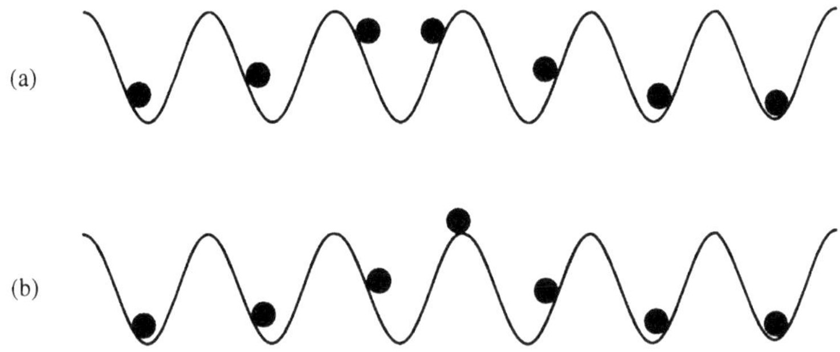

Fig. 3.1. Two stationary configurations of coupled particles in the FK model which correspond to a single kink in the chain: (a) stable, corresponding to a minimum, and (b) unstable, corresponding to a saddle point of the PN potential energy relief.

The potential energy of the chain with one kink, $U(\ldots, u_{n-1}, u_n, u_{n+1}, \ldots)$, is a function of the coordinates of all atoms of the chain. The stationary state shown in Fig. 3.1(a) corresponds to one of the minima of the function U, whereas the state shown in Fig. 3.1(b) corresponds to a saddle point of the function U, which is situated just between two nearest minima in the N-dimension coordinate (configuration) space, N being the number of atoms in the chain ($N \to \infty$). The saddle point and the neighboring minimum point can be connected in a configuration space by the so-called *adiabatic trajectory*, i.e. a curve which represents a solution of the system of coupled differential equations,

$$\frac{du_n(\tau)}{d\tau} = -\frac{\partial}{\partial u_n} U(\ldots, u_{n-1}, u_n, n_{n+1}, \ldots), \tag{3.1}$$

where τ is a parameter varying along the trajectory and characterizing it. Such a trajectory is a curve corresponding to the steepest descent, and it describes the adiabatically slow motion of a kink through the chain. Note that when the system is subjected to a low-temperature thermostat, the kink will move predominantly along the adiabatic trajectory. At finite velocities, the kink's motion will differ slightly the motion along the adiabatic trajectory.

Therefore, the PN potential of a kink moving along a discrete chain can be presented in the form,

$$V_{PN}(X) = U(\ldots, u_{n-1}, u_n, u_{n+1}, \ldots)\Big|_{x \in \text{ad.tr.}} . \tag{3.2}$$

To introduce the collective coordinate X describing the motion of the kink's center, we present the atomic coordinates as

$$u_n = f(na_s - X), \tag{3.3}$$

where the function $f(x)$ describes the kink profile. In the continuum limit approximation, i.e. when $g \gg 1$, the function $f(x)$ coincides with the function (1.19) for the SG kink. In the discrete case, the function $f(x)$ differs from the shape of the SG kink and it can be presented as the following, $f(x) = u_k^{(SG)}(x) + \Delta u_{\rm ad}(x)$, where the function $\Delta u_{\rm ad}(x)$ is called "adiabatic dressing" of the kink. Using Eq. (3.3), the coordinate of the kink's center can be defined as [146]

$$X = -\frac{\sigma}{a_s} \int x \, f'(x - X) \, dx, \tag{3.4}$$

where $f'(x) = df(x)/dx$. In numerical simulations of the system with an isolated kink, it is more convenient to define the coordinate X integrating Eq. (3.4) by parts in order to obtain $X = \sigma \sum_n u_n + C$, where the integration constant C is defined in such a way that the point $X = 0$ corresponds to the kink's position at the bottom of the PN potential (e.g., at $n = 0$).

The amplitude E_{PN} of the PN potential has been calculated in many papers, in the quasi-continuum limit [147]–[154], for the weak-bond limit [155, 156], as well as directly by means of numerical simulations [154, 155],[157]–[161]. To estimate the value of the PN potential barrier , we substitute Eq. (3.3) into the potential energy U and neglect the adiabatic dressing term for the SG kink shape, i.e. take $f(x) = u_k^{(SG)}(x)$. Then, approximating u_{n+1} as

$$u_{n+1} \approx u_n + a_s f_n', \quad f_n' = \frac{df(z)}{dz}\Big|_{z=(na_s - X)}, \tag{3.5}$$

using Poisson summation formulae and keeping only the term corresponding to the first harmonic, we finally obtain the following result [149]

$$V_{PN}(X) \approx \sum_{l=0}^{\infty} B_l \cos(lX) \approx \frac{1}{2} E_{PN}(1 - \cos X), \tag{3.6}$$

where

$$B_l = \frac{16\pi^2 \, lg}{\sinh\left(l\pi^2 \sqrt{g}\right)} \left(l^2 + \frac{1}{2\pi^2 g}\right), \quad l \geq 1, \tag{3.7}$$

so that the "bare" PN potential for the case $g \gg 1$ is given by the expression

$$E_{PN}^{(0)} = \frac{32\pi^2 g}{\sinh\left(\pi^2\sqrt{g}\right)}\left(1+\frac{1}{2\pi^2 g}\right) \approx 64\pi^2 g \exp\left(-\pi^2\sqrt{g}\right). \qquad (3.8)$$

For a kink moving with a small velocity $v \equiv dX/dt \ll c$ along the adiabatic trajectory, the kink's kinetic energy can be presented as follows,

$$K_k = \frac{1}{2}\sum_n \left(\frac{du_n}{dt}\right)^2 = \frac{m}{2}\left(\frac{dX}{dt}\right)^2, \qquad (3.9)$$

where the effective mass of the kink is defined in a simple way,

$$m(X) \equiv \sum_n \left(\frac{du_n}{dX}\right)^2 = \sum_n (f')^2. \qquad (3.10)$$

Substituting the SG kink function $f(x) = u_k^{(SG)}(x)$ into Eq. (3.10), for $g \gg 1$ we obtain the following result [152],

$$m(X) \approx m^{(SG)} + \sum_{l=1}^{\infty} A_l \cos(lX), \qquad (3.11)$$

where

$$A_l = \frac{4l}{\sinh(l\pi^2\sqrt{g})}. \qquad (3.12)$$

The effect of higher-order discreteness effects on the kink's shape was investigated numerically [150, 152, 154, 159] and also analytically [154],[160]–[162].

In the case of strong coupling, i.e. for $g \gg 1$, we can employ the method first suggested by Rosenau [7] that allows to remove singularities in the spectrum and leads to the following equation, in the first order in the discreteness parameter $\lambda = \frac{1}{12}(a_s/d)^2 \ll 1$,

$$\sin u - \frac{d^2u}{dx^2} = \lambda\left[\left(\frac{du}{dx}\right)^2 \sin u - \frac{d^2u}{dx^2}\cos u\right]. \qquad (3.13)$$

A localized solution of this equation can be easily found by the perturbation theory [27], and such a correction has the form

$$\Delta u_{\rm ad}(z) \approx -\lambda\sigma\frac{(3\tanh z - z)}{\cosh z}. \qquad (3.14)$$

Thus, as follows from this result, the discreteness reduces the kink's width, $d \to d_{\rm eff} = d(1-\lambda)$. The same result was found in numerical simulations first conducted by Currie *et al.* [159] and Willis *et al.* [152]. It is clear that the kink's width narrowing should lead to an increase in the value of the PN barrier energy E_{PN}, in comparison with the result given by Eq. (3.8) which can be estimated as follows,

$$\Delta E_{PN} \approx \frac{2d\Delta d}{a_s^2}\frac{dE_{PN}^{(0)}}{dg} = -2\lambda g \frac{dE_{PN}^{(0)}}{dg}. \tag{3.15}$$

Analytical calculations of E_{PN} in the continuum limit approximation is a subtle problem, because the kink dressing contributes to the value of E_{PN} through all orders of the perturbation in the standard perturbation theory scheme [161]. An original approach for solving this problem was proposed by Flach and Kladko [154]. It consists in rewriting the difference equation on discrete variables u_n,

$$g\left(u_{n+1} + u_{n-1} - 2u_n\right) - V'_{\text{sub}}(u_n) = 0, \tag{3.16}$$

as the differential equation on the continuous function $u(x)$,

$$g a_s^2 \frac{d^2 u}{dx^2} - V'_{\text{sub}}(u) = \left[\rho(x) - 1\right] V'_{\text{sub}}(u), \tag{3.17}$$

where

$$\rho(x) = \sum_{n=-\infty}^{\infty} \delta(x - n a_s) = a_s \left[1 + 2\sum_{k=1}^{\infty} \cos\left(\frac{2\pi k x}{a_s}\right)\right], \tag{3.18}$$

and then considering the right-hand part of Eq. (3.17) as a small perturbation of the SG equation. The result of the first-order perturbation theory is that the kink shape can be found by solving the SG equation with the actual substrate potential V_{sub} being replaced by an effective potential

$$V_{\text{eff}} = V_{\text{sub}} - \frac{1}{24g}\left(\frac{a_s}{2\pi} V'_{\text{sub}}\right)^2. \tag{3.19}$$

This approach was developed further by Kladko *et al.* [163] (see also discussion in Kevrekidis *et al.* [164]). In particular, the exponential factor in Eq. (3.8) should be corrected as follows,

$$\exp(-\pi^2\sqrt{g}) \to \exp\left(-\pi^2\sqrt{g}/\sqrt{1 - 1/12g}\right).$$

As is shown in Chap. 5 below, Eq. (3.16) for the stationary configurations can be reduced to the Taylor-Chirikov standard map for an auxiliary two-dimensional dynamical system and, therefore, static solutions of Eq. (3.16) correspond to the map manifold trajectories. In particular, the kinks correspond to two homoclinic orbits of the standard map, the stable and unstable manifolds. In the exactly integrable SG equation these two manifolds overlap; but in the discrete FK system the manifolds are different being characterized by different energies. The difference between these energies defines the PN energy barrier , and it is determined by the angle of the manifold intersection in the point closest to the middle of distance between the fixed points. This angle was calculated by Lazutkin *et al.* [153] in the $g \to \infty$ limit, and it leads to the following result for the PN energy

$$E_{PN} = Ag \exp\left(-\pi^2 \sqrt{g}\right), \tag{3.20}$$

where the numerical factor is $A = 712.26784...$ Comparing this exact result with the "bare" PN energy defined in Eq. (3.8), we find that in the limit of strong coupling the kink dressing leads to an increase of the PN energy in 1.13 times.

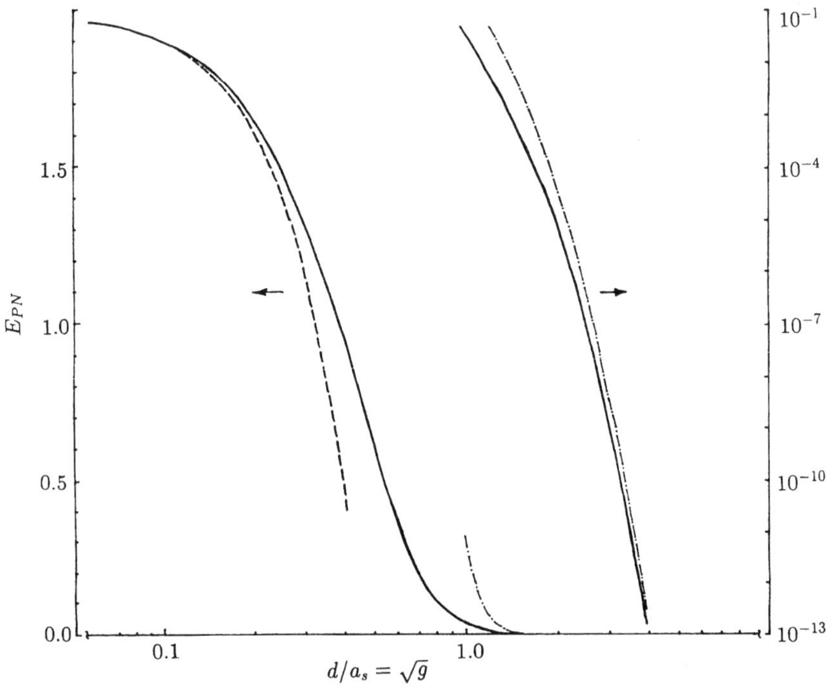

Fig. 3.2. Maximum of the PN potential relief, E_{PN}, vs. the normalized width of the kink, $d/a_s = \sqrt{g}$. Solid: numerical results of Joos [155] and Stancioff *et al.* [160]; dashed: the result of a weak-coupling approximation; dashed-dotted: the result of the continuous approximation.

In the opposite case of a weak coupling, $g \ll 1$, the kink's parameters can be calculated in the lowest approximation if we neglect the atomic displacements from the bottoms of the substrate potential wells, for all the atoms except those in the kink's core. This leads to the following expressions (see Ref. [155, 165]): $m \approx 1$, $\varepsilon_k \approx 2\pi^2 g\,(1 - 2g)$, and $E_{PN} \approx 2 - \pi^2 g$.

More accurate results may be obtained with the help of the single-active-site theory developed by Kladko *et al.* [163]. In this approach, the atoms near the kink core are treated exactly, while all other atoms are described by the linearized equation (3.16) assuming that their displacements from the minima of the substrate potential are small. Then the two solutions, one for kink-core atoms and another, for all other atoms, are matched together. This approach yields the following result for the PN force,

$$F_{PN} \approx \left[(2z - 4g - 1)^{1/2} - f(g) \cos^{-1} f(g)\right] [1 + f(g)]^{-1}, \tag{3.21}$$

where $z \equiv \sqrt{1+4g}$ and

$$f(g) = 2\left\{1 - \frac{(1+g)}{(1+2g+z)}\right\}. \tag{3.22}$$

Furuya and Ozorio de Almeida [156] have employed the approach of the standard map in order to calculate the energies ε_k and E_{PN} in the case $g < 1$. They demonstrated that the minimum energy state shown in Fig. 3.1(a), is characterized by the energy

$$\varepsilon_k \approx 2\left[1 - \cos\left(\frac{2\pi g}{1+3g}\right)\right] + \frac{g}{2}\left[2\pi\frac{(1+g)}{(1+3g)}\right]^2. \tag{3.23}$$

The saddle state shown in Fig. 3.1(b) has the energy

$$\varepsilon_{\text{saddle}} = \varepsilon_k + E_{PN} \approx 2(2 - \cos\beta) + \frac{g}{2}\left[\pi^2 + \left(\beta + \frac{1}{\beta}\sin\beta\right)^2\right], \tag{3.24}$$

where

$$\beta = \frac{2\pi g}{(1+2g) + \sqrt{1+4g}}.$$

The dependencies of E_{PN} on the parameter $d/a_s \equiv \sqrt{g}$ is shown in Figs. 3.2.

The effect of the chain discreteness on the interaction between two kinks was analyzed numerically by Joos [155]. The exponential law of the kink-kink interaction,

$$v_{\text{int}}^{(FK)}(R) = Ae^{-\gamma R/d}, \tag{3.25}$$

was shown to be valid at $R > 3a_s$ and at any value of the parameter g, although the coefficients A and γ at $g < 5$ depend on g; in particular, for $g \to 0$, $A(g) \approx 4\pi^2 g$ (see below Chap. 5). The presence of a kink in the chain changes the density of the phonon states in the system [166], this effect is important in the problems of statistical mechanics of the FK system as is discussed below in Chap. 6.

3.2 Dynamics of Kinks

3.2.1 Effective Equation of Motion

In the discrete FK model, there exist no steady-state solution for a kink moving with a constant velocity; due to the lattice discreteness a moving kink radiates linear waves (phonons). This is one of the main physical effects which explains radiative losses of dislocations calculated for various (more realistic) discrete models [167]–[170].

For the kink moving in the FK chain, the radiative effects were studied numerically by several groups. Using the extensive numerical results by Currie *et al.* [159], let us discuss the general properties of the FK kink dynamics in discrete chains. Indeed, if we start from an initial configuration with a kink-type boundary condition, i.e. we select a single-kink solution of the SG equation with (some) nonzero initial velocity, $u_{n\to-\infty} = 0$ and $u_{n\to+\infty} = -\sigma a_s$, such a configuration will never evolve into a steady state. Initially, this state will transform into radiation and a single "dressed" kink, i.e. a kink with the shape modified by the lattice discreteness. As a result, a kink will propagate through the chain not freely but with an oscillating velocity, the oscillations being caused by the lattice discreteness. Moving with a varying velocity, such a kink loses its kinetic energy emitting phonons, and eventually the kink will get trapped by the PN potential when its velocity drops below a certain critical velocity v_{PN}. The kink trapped by the PN potential oscillates near a minimum of the potential well continuously emitting phonons, and finally it reaches a stationary state corresponding to a static configuration.

Such a behavior can be easily understood in the framework of a simple physical picture of an effective particle with the mass m moving in the periodic potential, when the total particle's energy is defined as

$$E = \frac{m}{2}\left(\frac{dX}{dt}\right)^2 + V_{PN}(X). \tag{3.26}$$

When the energy E is larger than the PN energy barrier corresponding to a maximum of the potential $V_{PN}(X)$, the particle propagates along the chain and its position is described by the equation

$$X(t) = 2\,\mathrm{am}\left(\omega_{PN}t/k; k\right), \tag{3.27}$$

where $k = \sqrt{E_{PN}/E}$ is the modulus of the Jacobi elliptic function, and the particle's velocity changes periodically near its mean value with the frequency

$$\omega_{\mathrm{trav}}(E) = \frac{\omega_{PN}}{2kK(k)} = \left\langle\frac{dX}{dt}\right\rangle\left(\frac{2\pi}{a_s}\right) \tag{3.28}$$

caused by the periodic PN relief.

In the other case, when $0 < E < E_{PN}$, the effective particle (kink) stays trapped near one of the minima of the periodic potential, and the particle's coordinate oscillates according to the law

$$X(t) = 2\sin^{-1}[\widetilde{k}\,\mathrm{sn}(\omega_{PN}t;\widetilde{k})], \tag{3.29}$$

where this time the modulus of the elliptic function is $\widetilde{k} = \sqrt{E/E_{PN}} = k^{-1}$ and the frequency of the kink oscillations is given by the expression

$$\omega_{\mathrm{trap}}(E) = \frac{\pi\omega_{PN}}{2\mathrm{K}(\widetilde{k})}, \tag{3.30}$$

so that $0 < \omega_{\mathrm{trap}}(E) < \omega_{\mathrm{trap}}(0) \equiv \omega_{PN}$, where

$$\omega_{PN}^2 = \frac{1}{m}\left(\frac{d^2V_{PN}(X)}{dX^2}\right)\Bigg|_{X=0} \approx \frac{E_{PN}}{2m} \tag{3.31}$$

is the frequency of harmonic oscillations near the bottom of the PN potential, which is called the PN frequency.

For the discrete FK chain, the total kink's energy E is not a conserved quantity because of a nonlinearity-induced coupling between different modes, including phonons. This means that the kink's motion through the chain can be described with the help of an effective-particle model only approximately, and such a dynamics should be modified by additional process of excitation of phonon modes. In fact, a kink moving through the lattice with a varying velocity, or a kink trapped by the lattice discreteness and oscillating near the bottom of the PN potential, produces an effective oscillating force on the phonon subsystem with the frequencies $n\omega_{\mathrm{trav}}$ or $n\omega_{\mathrm{trap}}$ ($n = 1, 2, \ldots$), respectively. Such an oscillating force generates phonons, and the kink's energy decreases, being transformed into the energy of the excited phonon modes. As a result, the phenomenology of an effective particle introduced above should be modified by including an effective friction force,

$$F^{(\mathrm{fr})} = -m\eta\frac{dX}{dt}, \tag{3.32}$$

which causes the final trapping of the kink by the PN potential. The effective viscous friction coefficient η in Eq. (3.32) is a complex function of the system parameters, and it is discussed below in Sec. 7.2.

Adiabatic dynamics of a kink in the discrete FK chain was analyzed analytically by Ishimori and Munakata [150] by applying the soliton perturbation theory, developed earlier by McLaughlin and Scott [86], to the case when the discreteness effects are small, i.e. $\lambda = (1/12)(a_s/d)^2 \ll 1$. Ishimori and Munakata [150] calculated an effective friction produced by radiation of phonons, and they showed that the moving kink radiates phonons predominantly to the backward direction. A more careful study of the kink's dynamics in the discrete FK chain was carried out by Peyrard and Kruskal [171] and Boesch *et al.* [172] with the help of extended numerical simulations. Several results of those studies are presented in Figs. 3.3 and 3.4 below; these results show a variation of the kink's velocity in the cases of moving and trapped states.

Before discussing the features observed in numerical simulations, we introduce the density of the phonon states $\rho(\omega)$ in the one-dimensional lattice defined as follows (see, e.g., Ref. [173]),

$$\rho(\omega) = \frac{2}{\pi} \frac{\omega}{\sqrt{(\omega^2 - \omega_{\rm min}^2)(\omega_{\rm max}^2 - \omega^2)}}, \tag{3.33}$$

defined with the normalization

$$\int_{\omega_{\rm min}}^{\omega_{\rm max}} d\omega\, \rho(\omega) = 1.$$

The function $\rho(\omega)$ tends to infinity at the edges of the phonon spectrum band, $\omega_{\rm min}$ and $\omega_{\rm max}$, respectively.

Anharmonic lattice vibrations lead to the generation of higher-order harmonics in the phonon spectrum. This happens when an effective energy exchange between the kink translational or trapped motion occurs, i.e. provided the following resonance condition is satisfied,

$$k_1 \omega_k = k_2 \omega_{\rm ph}(\kappa), \tag{3.34}$$

where ω_k is replaced by $\omega_{\rm trav}$ or $\omega_{\rm trap}$ depending on the type of the kink's motion in the chain, and the integer numbers k_1 and k_2 stand for the order of the resonance. The maximum of the radiative damping produced by the resonant energy transfer to the phonon subsystem occurs for the case $k_1 = k_2 = 1$, and $\omega_{\rm ph}$ is close to one of the edges of the phonon spectrum where the phonon density takes a maximum value. In the next section, we discuss in more details the numerical simulation results for both propagating and trapped kinks.

3.2.2 Moving Kinks

We study first the general feature of the dynamics of fast kinks in discrete chains. A typical example of the kink's evolution, shown as the variation of the kink velocity, is presented in Fig. 3.3 based on the numerical results of Peyrard and Kruskal [171]. A fast kink launched at some (relatively large) velocity, $v_0 < c$, loses quickly its initial velocity up to the "critical" value v_1 during a certain time interval t_1 (see Fig. 3.3); the critical velocity v_1 being a function of the initial velocity v_0. For $t < t_1$, when $v > v_1$, the inequality

$$\omega_{\rm min} < \omega_{\rm trav} < \omega_{\rm max} \tag{3.35}$$

is valid, and this condition explains a very high rate of the energy losses observed in numerical simulations.

However, at the point $t = t_1$ the first-order resonance vanishes, and the radiation-induced damping of the kink is now caused by the second-order

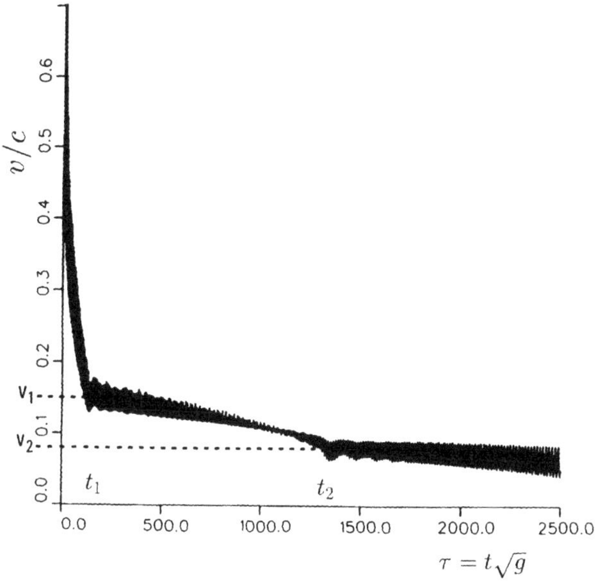

Fig. 3.3. Variation the kink's velocity in the FK chain with $\sqrt{g} = 0.95$. The initial state is a kink moving with the velocity $v_0 = 0.8\,c$. Dashed lines show the disappearance of the resonances $k_1 = k_2 = 1$ (at $t = t_1$) and $k_1 = 2$, $k_2 = 1$ (at $t = t_2$) [171].

resonance ($k_1 = 2$, $k_2 = 1$) only. The transition time t_1 is clearly observed in numerical simulations as the point where the effective friction η is abruptly lowered (see Fig. 3.3). Correspondingly, the kink radiation becomes smaller, and the mean velocity of the kink translational motion decreases much slower.

Similarly, at $t = t_2$ (i.e. when the kink's velocity drops to the value $v_k = v_2$), the second-order resonance vanishes (i.e. the resonant condition is no longer satisfied), and for $t > t_2$ the kink's velocity changes even more slowly (see Fig. 3.3). Subsequently, the kink gets trapped by the PN relief at a certain $t = t_{\rm trap}$, when the kink's kinetic energy reaches the value corresponding to the PN energy, $E = E_{PN}$. Numerical simulation results obtained by different groups show that for the case $g \gg 1$ the time interval of the kink trapping $t_{\rm trap}$ is extremely large. However, this time $t_{\rm trap}$ becomes much smaller for narrow kinks, for example, for $\sqrt{g} = 0.75$ the kink with the initial velocity $v_0 = 0.8\,c$ cannot propagate through the lattice more than for two lattice spacings and, as a result, it becomes trapped almost immediately by the lattice discreteness. Very similar behavior of a kink was observed by Combs and Yip [174] for the the kink propagation in a discrete ϕ^4 model.

3.2.3 Trapped Kinks

Evolution of trapped kinks was analyzed numerically, in particular, by Boesch *et al.* [172]. The results shown in Fig. 3.4 can be explained in a way similar to the case of a moving kink, taking into account the structure of the phonon spectrum and the density of the phonon states.

First of all, we notice that the oscillation frequency of a trapped kink, $\omega_{\rm trap}$, varies from zero to its maximum value ω_{PN}. When increasing the frequency $\omega_{\rm trap}$, the order of the resonance with the phonon modes is lowered and, for a certain harmonic, it becomes possible to cross the edge frequency of the phonon spectrum to satisfy a resonance condition, even such a condition was not satisfied initially. Consequently, the emission of phonons by a kink should increase. Typically, however, the value of the PN frequency, defined as a function of the lattice parameter g,

$$\omega_{PN}(g) \approx \left\{ \frac{2\pi^6}{3} \frac{g\sqrt{g}}{\sinh(\pi^2\sqrt{g})} \left(1 + \frac{1}{2\pi^2 g}\right)\right\}^{1/2}, \tag{3.36}$$

is much smaller than the edge frequency of the phonon band $\omega_{\rm min} = 1$, and thus the kink radiates phonons only due to the generation of higher-order harmonics, i.e. those corresponding to large numbers of the resonance condition (3.34). For example, the PN frequency calculated from Eq. (3.36) at $g = 1$ is $\omega_{PN} \approx 0.18$ (the rigorous procedure to calculate ω_{PN} was suggested by Boesch and Willis [175]; see also Ref. [176]), and it is necessary to satisfy the condition $k_2 \geq 6$ in Eq. (3.34) in order to obtain the resonant generation of phonons by the oscillating kink. Slowly changing its frequency, the kink emits suddenly large burst of radiation when its frequency (or frequencies of the higher-order harmonics) passes the edges of the phonon spectrum where the density of phonon states has a maximum. This leads to peculiarities in the temporal evolution of the kink's coordinate shown in Fig. 3.4.

Analytical results for the evaluation of the radiation-induced friction coefficient η are rather lengthy and cumbersome to be discussed here in details. However, we should mention that the first analytical calculation of the kink radiation was presented by Ishimori and Munakata [150]. This calculation employed the first-order approximation of the soliton perturbation theory, but finally the results did not reproduce well the features of the kink dynamics observed in numerical simulations. Rough estimates made by Peyrard and Kruskal [171] agreed well with the corresponding numerical results, and therefore they are more satisfactory. A rigorous procedure was proposed later by Willis *et al.* [152] (see also Ref. [177] where a projection-operator technique was developed to find the value of the effective friction η, and also the later work by Igarashi and Munakata [178]). The main idea of this approach is to look for the kink solution in a discrete case in the form $u_n(t) = f[na_s - X(t)] + q_n(t)$ and develop a Hamiltonian formalism for the kink coordinate $X(t)$ and its conjugated momentum $P(t) = m(t)(dX/dt)$

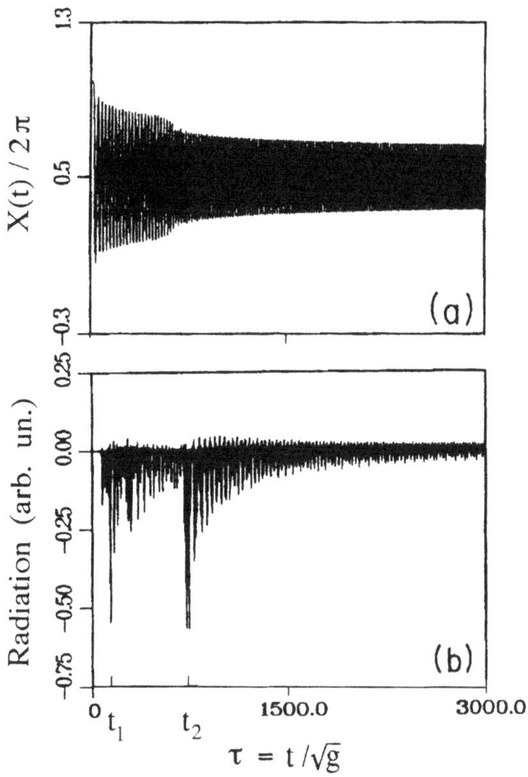

Fig. 3.4. Numerical results for (a) oscillations of the kink's coordinate $X(t)$ and (b) instantaneous Poynting's flux of the phonon radiation away from the kink trapped by the PN potential well, for the FK chain with $g = 0.791$. One can see the appearance of new resonances with $k_1 = 5$, $k_2 = 1$ (at $t = t_1$) and $k_1 = 4$, $k_2 = 1$ (at $t = t_2$) [172].

which are treated as canonical variables *extracted* from the full set of the variables of the discrete FK model. The variables $q_n(t)$ and the corresponding momenta $p_n(t) = dq_n/dt$ describe the radiation field as well as the deviation of the kink's shape from its analytical solution calculated in the continuum limit approximation. Introducing these two new canonical variables requires two constrains,

$$C_1 \equiv \sum_n f'_n q_n = 0 \quad \text{and} \quad C_2 \equiv \sum_n f'_n p_n = 0, \tag{3.37}$$

and also the modification of the Poisson brackets, $\{X, P\} = 1$ and $\{q_n, p_m\} = \delta_{n,m}$. Equations of motion are then obtained according to the Hamiltonian formalism, $d\theta/dt = \{\theta, H\}$, where θ stands for one of the canonical variables X, P, q_n, and p_n, and the PN frequency is then found by linearizing those equations for small-amplitude oscillations. The procedure described does provide an excellent agreement with numerical simulations and it takes into account an effective renormalization of the kink shape due to a strong discreteness of the FK model.

3.2.4 Multiple Kinks

For a weak interatomic interaction, e.g., $g \leq g_{4\pi} \approx 0.2025$, a repulsion between kinks is weak, and it may compete with an attractive force of the effective PN potential. Thus, two (or even more) kinks can be trapped by the lattice discreteness creating a bound state propagating as a single multikink. Such 4π-kinks, and also 6π-kinks, were discovered first numerically by Peyrard and Kruskal [171], who demonstrated that such multikink structures are rather stable to be easily detected in numerical simulations. Peyrard and Kruskal [171] have found that the velocity of such a bound state of kinks is not arbitrary, and it depends on the chain parameter g. Figure 3.5 shows the numerical results for the velocity of 4π- and 6π-kinks denoted as $v_{4\pi}$ and $v_{6\pi}$, respectively. Surprisingly, at high velocities, $v \sim c$, such multikinks remain stable even for $g > g_{2n\pi}$, where $g_{2n\pi}$ is the critical value of the elastic constant, below which n *static* 2π-kinks can be trapped together as a multikink in the same valley of the PN potential ($g_{2n\pi}$ depends of the order n of the multikink). Moreover, an effective friction coefficient for such types of multikink solutions becomes almost negligible allowing the multikink propagates in the lattice without visible radiative losses.

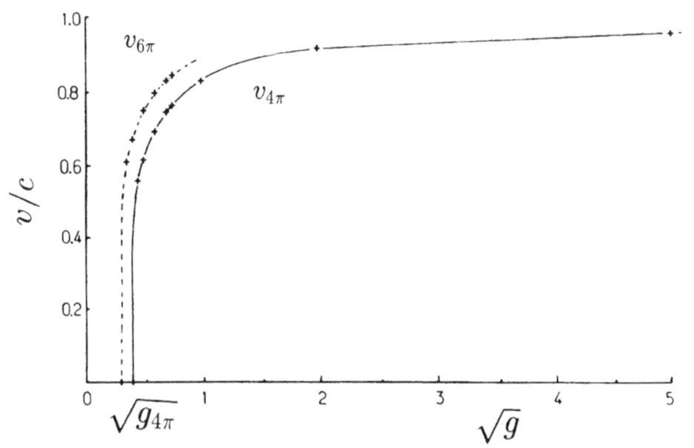

Fig. 3.5. Velocities $v_{4\pi}$ (solid curve) and $v_{6\pi}$ (dashed curve) of the 4π- and 6π-kinks as functions of the parameter $\sqrt{g} = d/a_s$ in the FK model [171].

A deep understanding of the existence of multikinks in discrete lattices at very large velocities is still a largely open problem, however, it is clear that such states become possible due to a Lorentz contraction of the kink's width at large velocities. For very narrow kinks the discreteness effects begin to play the main role. The "forward" 2π-kink of the multikink configuration emits a strong radiation behind itself, which helps the kinks that followed it to overcome the PN barriers. Thus, the stability of the multiple kink can be explained as the result of "compensation" of the waves emitted by single kinks when these waves happen to be out-of-phase. The waves suppress each other, so the composite double kink propagates almost without radiation.

The 4π-kinks were studied numerically by Savin *et al.* [179] with the help of the so-called pseudo-spectral method [180]. This method allows to find numerically all travelling-wave solutions of *stationary profile* moving with a constant velocity v. For this type of solutions one can write $u_n(t) = u(na_s - vt) \equiv u(z)$, where $z = na_s - vt$. Then the motion equation (1.9) reduces to the following differential-difference equation:

$$v^2 \frac{d^2u}{dz^2} + \sin u(z) - g\,[u(z+a_s) + u(z-a_s) - 2u(z)] = 0, \tag{3.38}$$

and one can look for its solution as the expansion in a finite Fourier series

$$u(z) \approx u_0(z) + \sum_{j=1}^{M} c_j \psi_j(z), \tag{3.39}$$

where $u_0(z)$ is some trial function with appropriate boundary conditions [for example, for the 4π-kink one can take $u_0(z) = 8\tan^{-1}\exp(z/d)$], $\psi_j(z) = \sin(2\pi jz/L)$, and $L = Na_s$ is the total length of the FK chain with the periodic boundary condition. Substitution of the expansion (3.39) into Eq. (3.38) yields the nonlinear equation of the type $F(z) = 0$, which should be satisfied for any z. Next, calculation this function for the M collocation points $z_i = (i-1)L/2(M-1)$, one obtains a system of M nonlinear algebraic equations with respect to the M unknown coefficients c_j, which may be solved, e.g., by the Newton-Raphson iteration method or the Powell hybrid method (the number M should be chosen to be large enough to acquire a given tolerance; typically it is sufficient to choose $M = 100$). Notice that the method described above automatically assumes that the kink moves freely, the radiation losses are totally forbidden.

Using this technique, Savin *et al.* [179] have found a hierarchy of the double kink states characterized by different distances between the two single kinks. Each of this bound states is dynamically stable for a certain (preferred) value of the velocity v_k for a given set of model parameters. The double-kink solutions exist only for a sufficiently strong coupling g, so that g must be bounded from below, $g > g_k$ ($g_0 \approx 0.13$, $g_1 \approx 0.31$, $g_2 \approx 0.5$, *etc.*). It is important to note that these solutions exist due to *discreteness* of the FK model, in the continuum model they are absent. The collisions of two double kinks were found to be quasi-elastic for large values of g and destructive at smaller elastic constants, so that the double kinks dissociate into separate 2π-kinks after the collision.

It is interesting that although such multikink solutions of the FK model cannot be static, in the simulation Savin *et al.* [179] observed the bound state of two standing 2π-kinks coupled through a breather. Thus, these solutions may be considered as coupled states of two nanopterons (i.e. the kinks constructed on the oscillating background) when their oscillating asymptotics annihilate each other.

As we have mentioned, there exists no clear explanation of the existence of multikinks yet, but several efforts were made to understand the origin of this kind of effects in the quasi-continuum approximation [181, 182]. Indeed, taking into account the fourth-order derivative in the SG equation that appears due to the discreteness effects of the original lattice, we can obtain the perturbed SG equation in the normalized form,

$$\frac{\partial^2 u}{\partial t^2} - \frac{\partial^2 u}{\partial x^2} - \beta\frac{\partial^4 u}{\partial x^4} + \sin u = 0, \tag{3.40}$$

where the parameter $\beta = a_s^2/12$ describes the discreteness effects.

Equation (3.40) takes into account the effect of lattice discreteness through a fourth-order dispersion term, and for $\beta = 0$ it transforms into the well-known exactly integrable SG equation that has an analytical solution for a single 2π kink moving with an arbitrary velocity $v < c$. In contrast, Champneys and Kivshar [182] demonstrated that Eq. (3.40) for $\beta \neq 0$ possess a class of localized solutions in the form of $2n\pi$ multikinks propagating with *fixed* velocities.

Following Champneys and Kivshar [182], we look for kink-type localized solutions of Eq. (3.40) that move with velocity v ($v^2 < c^2$), i.e., we assume $u(z) = u(x - vt)$. Linearizing Eq. (3.40) and taking $u(z) \propto e^{\lambda z}$, we find eigenvalues λ of the form,

$$\lambda^2 = \frac{1}{2\beta}\left[-(c^2 - v^2) \pm \sqrt{(c^2 - v^2)^2 + 4\beta}\right],$$

so that for $\beta > 0$ there always exist two real and two purely imaginary eigenvalues. Thus, the origin $u = 0$ is a saddle-center point of Eq. (3.40) and hence the kinks, which are homoclinic solutions to $u = 0$ (mod 2π), should occur for isolated values of v for a fixed value of β [182].

To find all solutions of this type, first we fix $\beta = 1/12$. Numerical shooting method applied on the ordinary differential equation for $u(z)$ using a

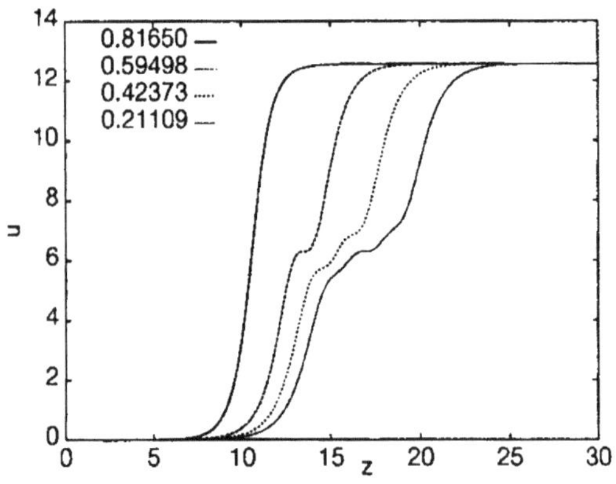

Fig. 3.6. Four 4π-kink solutions of the fourth-order SG equation (3.40) at $\beta = 1/4$. The velocity values v/c are given in the legend [182].

well-established Newton-type method for homoclinic/heteroclinic trajectories, allows to find a set of kink-like localized solutions. The first result is that there exists no 2π-kink solution at all. Instead, there exists a discrete family of 4π-kinks; specifically there exist only four such solutions at four different values of v. The first solution has an analytical form [181]

$$u(z) = 8\tan^{-1}\exp\left[(3\beta)^{1/4}z\right], \tag{3.41}$$

where

$$v^2/c^2 = 1 - 2\sqrt{\beta/3}, \tag{3.42}$$

i.e. we have $v_{4\pi}^{(1)}/c = \sqrt{2/3}$. Other values are: $v_{4\pi}^{(2)}/c \approx 0.59498$, $v_{4\pi}^{(3)}/c \approx 0.42373$, and $v_{4\pi}^{(4)}/c \approx 0.21109$. All these solutions are presented in Fig. 3.6. In addition to the 4π kinks, numerics further reveals v values at which $2n\pi$ kinks occur for all $n > 2$. The dependence (3.42) of such localized solutions resembles the numerical results of Peyrard and Kruskal [171] presented in Fig. 3.5.

Finally, we mention that the case $\beta < 0$ in Eq. (3.40) can also occur in generalized nonlinear lattices provided we take into account the next-neighbor interactions, e.g., due to the so-called helicoidal terms in nonlinear models [110]. In this case, the analysis is much simpler and, similar to the nonlocal SG equations [183], it leads to the continuous families of multikinks parameterized by v. From the mathematical point of view, in the case of $\beta < 0$ the origin $u = 0$ changes from a saddle-center point to a saddle focus, and rigorous variational principles give families of stable $2n\pi$ kinks for all $n > 1$.

3.3 Generalized On-Site Potential

The standard FK model (1.1) to (1.5) describes a chain of particles in a sinusoidal one-site potential. However, as was described above in Chapter 2, realistic physical models often include more complex types of on-site potentials that are periodic but deviate from the simple sinusoidal form. Indeed, the on-site substrate potential of the FK model is an effective potential produced by the coupling of the atoms in the chain with other degrees of freedom, e.g. with the substrate atoms, and it appears in the lowest approximation, i.e. when (i) the substrate atoms constitute a simple lattice with one atom per unit cell, and (ii) the Fourier expansion of the interaction potential is approximated by the first harmonic only. In all other physical situations, the periodic potential $V_{\text{sub}}(x)$ deviates from the sinusoidal form. For example, for the atoms adsorbed on metal surfaces the substrate potential is usually characterized by sharp bottoms and flat barriers [184]. Moreover, if the substrate is characterized by a complex unit cell, the potential $V_{\text{sub}}(x)$ has more complicated shape with several minima and/or maxima.

3.3.1 Basic Properties

First, we discuss a general case when the substrate potential $V_{\rm sub}(x)$ is a periodic function with the period $a_s = 2\pi$, and it has at least one minimum for $x \in (0, a_s)$, say at $x = x_0$ and one maximum, at $x = x_m$. Without loss of generality, we assume that the substrate potential is normalized as follows: $V_{\rm sub}(x_0) = 0$ and $V_{\rm sub}(x_m) \equiv \varepsilon_s = 2$. As above, we consider only the commensurate case when the ground state has only one atom on the period a_s of the substrate potential.

A deviation of the substrate potential from the sinusoidal form changes the parameters of both linear waves (phonons) and nonlinear modes (kinks), and this may lead to the appearance of new kink solutions and phonon branches. Nonsinusoidal substrate potential also drastically modifies breather solutions. Let us first discuss the problems related to a modification of the substrate potential from a general point of view.

Phonon modes in the FK model are characterized by the dispersion relation $\omega_{\rm ph}^2(\kappa) = \omega_{\rm min}^2 + 2g(1 - \cos\kappa)$, for $|\kappa| < \pi$, where the minimum frequency $\omega_{\rm min}$ is defined as $\omega_{\rm min}^2 = V''_{\rm sub}(x_0)$, and it corresponds to an isolated atom oscillating at the bottom of the substrate potential minimum. In the standard FK model, this normalized frequency is $\omega_{\rm min} = 1$. Thus, in the case of sharp wells, we have $\omega_{\rm min} > 1$, while for flat bottoms, we have $\omega_{\rm min} < 1$. Note that in a generalized FK model, more than one branch in the phonon spectrum may exist, provided the potential $V_{\rm sub}(x)$ has more than one minimum on the period (see below, Sec. 3.3.4).

Kinks can easily be described in the continuum approximation valid for $g \gg 1$. If the discreteness effect are negligible, the generalized SG equation becomes

$$\frac{\partial^2 u}{\partial t^2} - d^2 \frac{\partial^2 u}{\partial x^2} + V'_{\rm sub}(u) = 0. \tag{3.43}$$

Equation (3.43) is Lorentz invariant and, therefore, it always has a stationary solution $u(x,t) = \phi(y)$, $y = \gamma[x - X(t)]/d$, $\gamma = (1 - v^2/c^2)^{-1/2}$, where the kink's coordinate is defined as $X(t) = X_0 + vt$, so that the kink's velocity is $v = dX/dt$ ($|v| < c$) (recall, in the notations we use $c = d$). Equation for the function $\phi(y)$,

$$\frac{d^2\phi}{dy^2} = V'_{\rm sub}(\phi), \tag{3.44}$$

coincides with the equation of motion of an effective particle with the coordinate ϕ in the potential $U(\phi) = -V_{\rm sub}(\phi)$. Periodic oscillations of a particle near the bottom of the potential $U(\phi)$ corresponds to linear waves, the rotation of the particle corresponds to the cnoidal waves (arrays of kinks), and the separatrix trajectory generates a single-kink solution with the boundary conditions,

$$\phi(y) \to x_0 \ (\mathrm{mod}\ 2\pi) \quad \text{and} \quad \frac{d\phi(y)}{dy} \to 0 \quad \text{as} \quad y \to \pm\infty. \tag{3.45}$$

Integration of Eq. (3.44) with the use of Eq. (3.45) yields

$$\left(\frac{d\phi}{dy}\right)^2 = 2V_{\text{sub}}(\phi), \tag{3.46}$$

and the kink shape can be found from the following expression,

$$y = \mp \int_{x_m}^{\phi(y)} \frac{d\phi}{\sqrt{2V_{\text{sub}}(\phi)}}. \tag{3.47}$$

Here the upper sign corresponds to a kink solution (a local contraction of the chain), whereas the lower sign corresponds to an antikink, and the value x_m is the coordinate of a maximum of the substrate potential. Thus, the kink (antikink) solution connects two nearest neighboring minima of the substrate potential, say x_0 and $x_0 + 2\pi$. If the substrate potential has more than two minima per period, one may expect to find more than one type of kinks (see below Sec. 3.3.4).

A static kink can be characterized by the asymptotics of its tails at infinities. If $u(x) \to x_0$ for $x \to +\infty$ or $x \to -\infty$, then $|u_k(x) - x_0| \propto \exp(-\omega_{\min}|x|/d)$, for $|x| \to \infty$. The tails of the kink define the character of interaction between the kinks. Therefore, the strength of the kink-kink interaction is weaker in the case $\omega_{\min} > 1$ (i.e. for the substrate potential with sharp bottoms) than is in the case of the SG model. The "core" of the kink is determined by the expression

$$u_k(x) \approx x_m - \left(\frac{\sigma x}{d}\right) [2V_{\text{sub}}(x_m)]^{1/2} \left(1 - \frac{x^2}{6\, d_{\text{eff}}^2}\right) \tag{3.48}$$

for $|u_x(x) - x_m| \ll d$, so that the kink's effective width becomes $d_{\text{eff}} = d[-V''_{\text{sub}}(x_m)]^{-1/2}$. According to Eq. (3.10), the kink mass (at rest) can be found as

$$m = \frac{1}{4\pi^2\sqrt{g}} \int_{x_0'}^{x_0''} d\phi \sqrt{2V_{\text{sub}}(\phi)}, \tag{3.49}$$

where x_0' and x_0'' are the positions of two adjacent successive minima of the substrate potential. The energy associated with a static kink is $\varepsilon_k = mc^2$, where in our notations $c = d$, and the kink's kinetic energy K_k is defined as $K_k = mc^2(\gamma - 1) \approx \frac{1}{2}mv^2$.

In the following sections, we consider several examples of the nonsinusoidal substrate potential $V_{\text{sub}}(u)$. One of the examples is the so-called double SG potential [32, 185]

$$V_{\text{sub}}^{(DSG)}(x) = -\cos x - s\cos(2x). \tag{3.50}$$

The potential (3.50) is topologically similar to the SG model for $|s| < 1/4$, but it is characterized by flat bottoms, for $1/4 < s < 0$, or by sharp wells, for $0 < s < 1/4$. Besides, the shape of the potential (3.50) has a double well (DW) structure for $s < -1/4$ and a double-barrier (DB) form for $s > 1/4$. We would like to mention also more general on-site potentials proposed by Peyrard and Remoissenet [186, 187] which are analyzed below.

3.3.2 Kink Internal Modes

Unlike the SG kink, a kink in a discrete or continuum FK model with a nonsinusoidal one-site potential may possess additional localized degrees of freedom, the so-called "internal modes" (see, e.g., Refs. [188]–[190]). In order to understand the origin of such modes, we should linearize the motion equation (3.43) around the kink solution substituting $u(x) = u_k(x) + \Psi(x)\, e^{i\Omega t}$. The function $\Psi(x)$ satisfies the linear Schrödinger-type equation

$$-d^2 \frac{d^2\Psi}{dx^2} + W(x)\Psi(x) = \omega^2 \Psi(x), \tag{3.51}$$

where

$$W(x) = \left. \frac{d^2 V_{\rm sub}(u)}{du^2} \right|_{u=u_k(x)} . \tag{3.52}$$

Equations (3.51) and (3.52) always admit a continuum of the plane wave solutions (phonons) with the frequencies $\Omega > \omega_{\min}$, and also the so-called Goldstone mode $\Psi(x) = du_k/dx$ with $\Omega = 0$. In a discrete FK model the latter mode has a nonzero eigenvalue ω_{PN}. Additionally to these two modes, Eq. (3.51) may possess one (or more) eigenmodes with discrete frequencies in the frequency gap $(0, \omega_{\min})$ or, depending on the substrate potential, with frequencies $\Omega > \omega_{\max}$ (see details in Ref. [190]). Such modes are localized near the kink's center, and they may be treated as internal oscillations of the kink's shape. The shape modes can be excited during collisions between kinks, or due to interaction of the kinks with impurities or external periodic fields, so that they can play an important role in the kink dynamics.

The kink's internal modes appear in result of a deformation of the on-site potential, and therefore they are absent in the exactly integrable SG model. This means that a small perturbation to the SG equation may create an additional eigenvalue that split off the continuous spectrum. An analytical approach for describing a birth of internal modes of solitary waves (and, in particular, kinks) in nonintegrable nonlinear models was first suggested by Kivshar *et al.* [191], who demonstrated that a small perturbation of a proper sign to an integrable model can create a soliton internal mode bifurcating from the continuous wave spectrum (see also Refs. [190], [192]–[194] for more rigorous results).

Following Kivshar *et al.* [191], we consider kinks of the perturbed SG equation,

$$\frac{\partial^2 u}{\partial t^2} - \frac{\partial^2 u}{\partial x^2} + \sin u + \epsilon \hat{g}(u) = 0, \tag{3.53}$$

where $\hat{g}(u)$ is an operator standing for perturbation (which describes, e.g. a deformation of the sinusoidal potential). Assuming ϵ small, we look for the kink solution $u_k(x)$ of Eq. (3.53) in the form of a Taylor series, $u_k(x) = u_k^{(0)}(x) + \epsilon u_k^{(1)}(x) + O(\epsilon^2)$, where $u_k^{(0)}(x) = 4\tan^{-1} e^x$ is the kink solution of the SG equation. The spatially localized correction $u_k^{(1)}(x)$ to the kink's shape can be then found in an explicit form,

$$u_k^{(1)}(x) = \frac{1}{\cosh x} \int_0^x dx' \cosh^2 x' \int_0^{x'} dx'' \frac{\hat{g}\left(u_k^{(0)}\right)}{\cosh x''}.$$

To analyze the small-amplitude modes around the kink $u_k(x)$, we linearize Eq. (3.53) substituting $u(x,t) = u_k(x) + w(x)\, e^{i\Omega t} + w^*(x)\, e^{-i\Omega t}$, where Ω is an eigenvalue and $w(x)$ satisfies the linear equation,

$$\frac{d^2 w}{dx^2} + \left(\frac{2}{\cosh^2 x} - 1\right) w + \Omega^2 w + \epsilon \hat{f}(x) w = 0, \tag{3.54}$$

where $\hat{f}(x) \equiv u_k^{(1)}(x) \sin\left(u_k^{(0)}(x)\right) - \hat{g}'\left(u_k^{(0)}(x)\right)$ and $\hat{g}' = d\hat{g}/dz$.

In the leading order ($\epsilon = 0$), the eigenvalue problem (3.54) is described by a standard equation with a solvable potential, so that its general solution is presented through a set of eigenfunctions,

$$w(x) = \alpha_{-1} W_{-1}(x) + \int_{-\infty}^{\infty} \alpha(k) W(x,k) dk, \tag{3.55}$$

where the function $W_{-1}(x) = \operatorname{sech} x$ is the eigenmode of the discrete spectrum corresponding to the eigenvalue $\Omega^2 = 0$ (the so-called neutral mode), whereas the eigenfunction $W(x,k) = e^{ikx}(k + i\tanh x)/(k+i)$ describes the continuous wave spectrum with the infinite band of eigenvalues, $\Omega^2 = \Omega^2(k) = 1 + k^2$. We note that (i) the continuous wave spectrum bands are separated from the eigenvalues of the discrete spectrum and (ii) the eigenfunctions $W(x,k)$ include only one exponential factor in both the limits $x \to \pm\infty$ meaning that the effective potential in Eq. (3.54) is reflectionless at $\epsilon = 0$. Under the latter condition, the end point of the continuum spectrum band ($k = 0$) belongs to the spectrum and the limiting (nonoscillating) function $W(x, 0) = \tanh x$ is not secularly growing.

Now we analyze the perturbed spectral problem (3.54) in the first-order approximation in ϵ expanding the function $w(x)$ through the set of eigenfunctions, see Eq. (3.55). First, a perturbation of the effective potential in Eq. (3.54) should lead to *a deformation* of the eigenfunctions as well as to *a shift* of the eigenvalues of the discrete spectrum. Second, under the conditions (i) and (ii) the perturbation can lead to a birth of an *additional eigenvalue* of the discrete spectrum which bifurcates from the continuum spectrum band.

To find this new eigenvalue, we notice that in the first order in ϵ, a perturbation may shift the cut-off frequency $\Omega_{\min}$ of the phonon band, $\Omega_{\min}^2 =$

$1+\epsilon\,\hat{g}'(0)$. Therefore, to describe a birth of a novel discrete state, we suppose that its eigenvalue detaches from the cut-off frequency, $\Omega^2 = \Omega^2_{\rm min} - \epsilon^2\kappa^2$, where κ is the parameter which determines the location of an additional discrete-spectrum eigenvalue, $k_0 = i\epsilon\kappa$. Then, we convert Eq. (3.54) by means of Eq. (3.55) into the following integral equation,

$$a(k) = \frac{\epsilon}{2\pi} \int_{-\infty}^{+\infty} dk' \, \frac{K(k,k')\, a(k')}{k'^2 + \epsilon^2\kappa^2}, \tag{3.56}$$

where $a(k) = \alpha(k)\,(k^2 + \epsilon^2\kappa^2)$ and the integral kernel $K(k,k')$ is defined as

$$K(k,k') = \int_{-\infty}^{\infty} W^*(x,k) \left[\hat{f}(x) + \hat{g}'(0)\right] W(x,k')\, dk',$$

where $\hat{f}(x)$ and $\hat{g}(x)$ are, in general, operators. To obtain Eq. (3.56), we have neglected a nonsingular contribution of the discrete spectrum and also used the orthogonality condition,

$$\int_{-\infty}^{\infty} W^*(x,k')W(x,k)\, dx = 2\pi\delta(k'-k).$$

Because the novel eigenvalue bifurcates from the continuum spectrum band $\Omega = \Omega_{\rm min}$ at $k = 0$, we can construct an asymptotic solution to Eq. (3.56) for small k by evaluating a singular contribution of the integral, $a(k) = \mathrm{sign}(\epsilon)\,(2|\kappa|)^{-1} K(k,0)\, a(0)$. As a result, we define the parameter κ from the self-consistency condition,

$$|\kappa| = \frac{1}{2}\,\mathrm{sign}(\epsilon) \int_{-\infty}^{\infty} \tanh x \left[\hat{f}(x) + \hat{g}'(0)\right] \tanh x\, dx. \tag{3.57}$$

Therefore, the new eigenvalue $k_0 = i\epsilon\kappa$ of the discrete spectrum appears when the right-hand side of Eq. (3.57) is positive. It follows from Eq. (3.55) that the corresponding eigenfunction is exponentially localized, $w(x) \to \pm(\pi/\epsilon\kappa)\, a(0) \exp(\mp\kappa x)$ as $x \to \pm\infty$. This result confirms the observation that a thresholdless birth of the soliton internal mode from the continuum spectrum becomes possible only in those models where solitary waves generate reflectionless potentials in the associated eigenvalue problem. This property is common for many soliton bearing nonlinear models.

As an example of the application of a general result (3.57), we consider the double SG equation following from Eq. (3.53) at $\hat{g}(u) = \sin(2u)$ [see also Eq. (3.50)]. The kink's internal mode exists for $\epsilon > 0$, and it was analyzed numerically by Campbell *et al.* [195]. Applying the asymptotic method presented above, we find the first-order correction to the kink's profile,

$$u_k^{(1)}(x) = 2\left(\frac{x}{\cosh x} - \frac{\sinh x}{\cosh^2 x}\right),$$

and then, using Eq. (3.57), calculate the discrete eigenvalue, $k_0 = i(8/3)\epsilon$. Therefore, the kink's internal mode is possible only for $\epsilon > 0$ and its frequency is defined by the expansion

$$\Omega^2 = 1 + 2\epsilon - \frac{64}{9}\epsilon^2 + O(\epsilon^3). \tag{3.58}$$

At last, we would like to mention that the eigenvalues embedded into a continuous spectrum can be analyzed by means of the Evans function technique. The Evans function technique for the case of discrete systems was developed recently by Balmforth *et al.* [194].

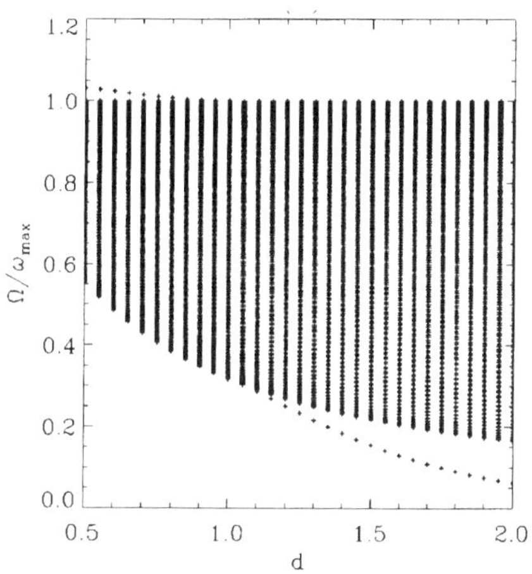

Fig. 3.7. An example of the spectrum of small-amplitude excitations around a kink in the generalized FK model as a function of the kink width d for two values of the parameter $s = -0.2$ determining the shape of the substrate potential (high discreteness corresponds to small values of d). The frequencies have been divided by ω_{max} to show more clearly the existence of a mode above the top of the phonon band for small d [190].

Braun *et al.* [190] analyzed numerically the effect of discreteness and the deformation of the on-site potential on the existence and properties of internal modes of kinks. In particular, they showed that kink's internal modes can appear not only below but also above the phonon spectrum band and, in the latter case, the localized mode describes out-of-phase oscillations of the kink's shape. Figure 3.7 shows an example of the linear spectrum of kink excitations in the generalized FK model with the potential (3.59) at $s = -0.2$, calculated numerically. To obtain these results, Braun *et al.* [190] found first the static configuration of particles, corresponding to a kink, by minimizing the energy of the chain with the corresponding boundary conditions. When all the equilibrium positions of the particles in a chain with a kink become known, the spectrum of small-amplitude oscillations around this state can be analyzed by looking for solutions of a system of linear equations. Its eigenvalues give the frequencies of the small-amplitude oscillations around the kink and the corresponding eigenvectors describe the spatial profile of each mode.

Figure 3.7 shows a deformation of the kink's spectrum for varying the kink width d. Besides the change of the phonon band, one sees clearly from these figures the growth of the frequency of the PN mode when discreteness increases. In addition, Fig. 3.7 shows how discreteness induces the formation of a novel type of high-frequency localized modes for $s < 0$. Discreteness has also more subtle effects, not visible on the figure and discussed in the work by Braun *et al.* [190].

Moreover, for the sinusoidal on-site potential, when the equation describes the classical FK model, in sharp contrast with the continuum limit, the kink's internal mode can exist in a narrow region of the discreteness parameter d [190].

3.3.3 Nonsinusoidal On-Site Potential

An important type of the on-site potential that describes many realistic situations, e.g., for the dynamics of atoms adsorbed on crystal surfaces, was suggested by Peyrard and Remoissenet [186],

$$V_{\rm sub}^{(NS)}(x) = \frac{(1+s)^2(1-\cos x)}{(1+s^2-2s\cos x)}, \quad |s| < 1. \tag{3.59}$$

The parameter s (in the original paper, the authors used the parameter $r = -s$) characterizes different shapes of the on-site potential, ranging from the case of flat bottoms to the case of flat barriers (see Fig. 3.8) .

The frequency spectrum of phonons in the model (3.59) is characterized by the minimum (gap) frequency $\omega_{\rm min} = (1+s)/(1-s)$, the kinks, by a characteristic width $d_{\rm eff} = \omega_{\rm min} d$, and the kink mass is defined as the follows:

$$m = m^{(SG)} \left(\frac{\omega_{\rm min}}{\omega_*}\right) \begin{cases} \tanh^{-1}\omega_*, & \text{if} \quad s < 0, \\ \tan^{-1}\omega_*, & \text{if} \quad s > 0, \end{cases} \tag{3.60}$$

where $\omega_* \equiv |\omega_{\rm min}^2 - 1|^{1/2}$. The kink's shape in this model was found numerically by Peyrard and Remoissenet [186]; the kink is narrow, for the case of flat bottoms ($s < 0$), and it is wide, for the opposite case of sharp wells ($s > 0$). The kink mass m vanishes for $s \to -1$, i.e. kinks can be created easier in the systems with a flat-bottom potential.

Considering the properties of a discrete chain with the substrate potential (3.59), we expect that a change of the potential shape will lead to a change of the PN barrier. This observation was confirmed in numerical simulations made by Peyrard and Remoissenet [186], and the results are shown in Fig. 3.9. The PN barrier E_{PN} for this model was estimated analytically by Ishibashi and Suzuki [151], who used the kink shape corresponding to the continuum approximation for calculating the system energy in a discrete case. The result is given by the following expressions [151],

$$\frac{E_{PN}}{E_{PN}^{(SG)}} \propto \begin{cases} \exp\left(\frac{2\pi^2\sqrt{g|s|}}{1+\sqrt{|s|}}\right), & \text{for} \quad s < 0, \\ \left|\cos\left(\frac{2\pi^2\sqrt{gs}}{1+s}\right)\right| \exp\left(\frac{2\pi^2 s\sqrt{g}}{1+s}\right), & \text{for} \quad s > 0. \end{cases} \tag{3.61}$$

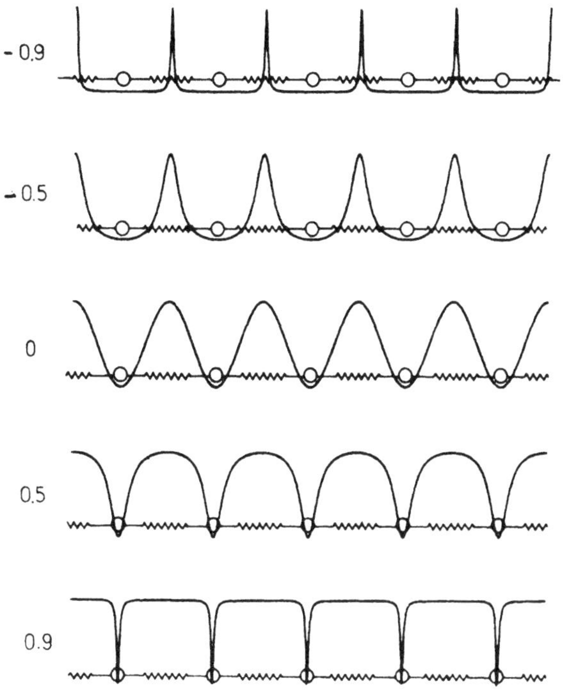

Fig. 3.8. A chain of weakly coupled particles in the periodic substrate potential (3.59). Shown are different types of the potential for marked values of s [186].

For $s > 0$ (sharp wells), the amplitude of the PN potential E_{PN} depends nonmonotonically on the elastic constant g. We should mention here that such a nonmonotonic dependence of the PN energy versus the lattice coupling had been discovered earlier for the on-site potential in the form of a periodic array of parabolas [147], [196]–[199]. In the standard FK model with the sinusoidal potential, the PN potential energy $V_{PN}(X)$ has its minimum at $X = 0$, i.e. for the structure of atoms with two central particles at the same potential well [see Fig. 3.10(a)], while a maximum of the PN potential occurs at $X = \pi$, for the atomic configuration shown in Fig. 3.10(c) when one of the atoms is at the top of one of the substrate potential maxima. Let us call this situation as the N (i.e. normal) relief. The case $s \leq 0$ in the nonsinusoidal substrate (3.59) always corresponds to the N-relief. However, the case $s > 0$ is more complicated. Apart from the N-relief, the so-called I (i.e. inverse) relief may be observed when the configuration shown in Fig. 3.10(a) corresponds to a potential maximum, and that shown in Fig. 3.10(c), to a minimum of the function $V_{PN}(X)$. Such a situation is realized for certain values of g, and the cases of the N- and I-reliefs alternate. In addition, between the regions of the N- and I-relief there exist intermediate regions where both the configurations shown in Fig. 3.10(a) and 3.10(c) correspond to two maxima of $V_{PN}(x)$ but a minimum is realized at some intermediate configuration [such as that shown in Fig. 3.10(b)] with $0 < X < \pi$. As a result, in the intermediate regions the dependence $E_{PN}(g)$ has well pronounced local minima (see Fig. 3.9). It is interesting to mention that for the substrate potential consisting of a sequence of parabolas, the value E_{PN} vanishes in the intermediate regions [198, 199]. In the limit $g \to 0$, these features of the PN potential disappear, and the PN potential approaches the function $V_{\text{sub}}(X)$.

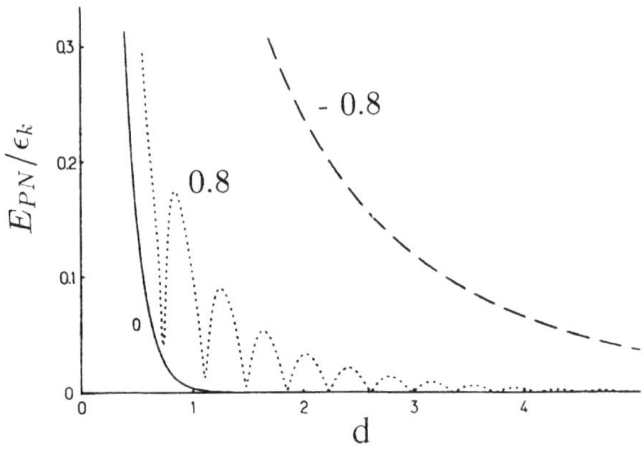

Fig. 3.9. The PN potential barrier, E_{PN}, scaled to the kink energy at rest, ε_k, vs. the width of the kind, d, shown for different values of the parameter $s = -0.8$, $s = 0$, and $s = +0.8$, respectively [186].

Dynamics of the FK model with nonsinusoidal substrate potential is qualitatively similar to that of the standard FK model described above. The motion of a FK kink is accompanied by radiation of phonons caused by the model discreteness [186]. An effective radiation-induced friction coefficient η increases with $|s|$ due to the growth of the PN barrier, E_{PN}. Collisions of kinks with phonons and other kinks are almost elastic, but the effective phase shift depends on the parameter s.

However, the FK model with a nonsinusoidal substrate potential displays at least *two novel features* in the kink dynamics, in comparison with the standard FK model. The first feature is the existence of long-living small-amplitude breathers only for a certain interval of the parameter s, namely, the breathers (described approximately by an effective NLS equation for the slowly varying wave envelope, see below) exist for the values of s in the interval $s_0 < s < 1$, where $s_0 = \sqrt{24} - 5 \approx -0.1$. The second new feature of the model is the existence of the internal modes of the kinks for $s > 0$, as was discussed above. These two main features of the nonlinear excitations in the generalized FK model with nonsinusoidal substrate significantly modify the dynamics of the kink-antikink collisions.

For the nonsinusoidal potential with $s < 0$ there exists a critical value of the kink's kinetic energy, K_{cr}, such that fast kink and antikink with the initial kinetic energy larger than K_{cr} pass through each other almost without changes of their energies. Otherwise, the collision is inelastic and, in general, the kink and antikink cannot escape from the effective (attractive) interaction potential because of a loss of a part of their energy for radiation.

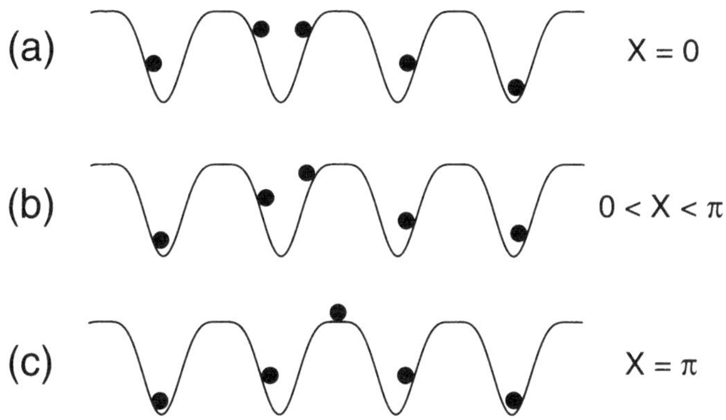

Fig. 3.10. Kink structure for the substrate potential with sharp wells.

Therefore, the kink and antikink may form a bound state as a large-amplitude (LA) breather that loses its energy transforming into a small-amplitude (SA) breather. However, the further evolution of such a bound state differs for the cases $s > s_0$ and $s < s_0$, where $s_0 = \sqrt{24} - 5 \approx -0.1$. In the former case, the LA breather transforms slowly into a SA breather (according to a power-law dependence). Otherwise, i.e. for $-1 < s < s_0$, when the LA breathers are not possible in the system, the SA breather decays more rapidly radiating two wave packets. In this latter case, the kink-antikink collisions are destructive.

In the case when the substrate potential has sharp wells ($s > 0$), the kink-antikink collisions exhibit novel phenomena caused by the kinks' internal modes. Namely, the final state of the kink-antikink collision below the threshold $K_{\rm cr}$ depends on the initial value of the relative kink velocity, so that such a collision may produce either a breather, as a final state which slowly decays, or it may result in a resonant (nondestructive) collision when the kinks do not annihilate. The resonant elastic interaction between a kink and antikink is due to the resonant energy exchange between the kink's translational mode and its internal mode, and such a type of resonances has been analyzed first for the ϕ^4 model by Campbell *et al.* [189] (see also Anninos *et al.* [200] and references therein), and later, for the potential (3.59), by Peyrard and Campbell [201]. They found that the regions where the trapping into a decaying bound state takes place (the so-called resonant velocity "windows"), and the regions characterized by almost elastic transmission of kinks, alternate. Numerical simulations showed that if the initial value of the relative kink velocity is selected in the resonant velocity "window", then kink and antikink become coupled just after the first collision and they start to oscillate. However, after a few oscillations the kinks escape to infinities. The explanation proposed by Campbell *et al.* [189] is based on the so-called resonant energy exchange mechanism. Indeed, in this case both kinks possess internal (shape) modes which are excited just after the first collision provided the relative kink velocity is not too large. When the shape modes are excited, they remove a part of the kinks' kinetic energy from the translational motion of the kinks. The kinks turn back because of a mutual attractive interaction, so that they interact with each other again. The energy stored in the kinks' shape modes may be now realized provided certain resonant conditions are satisfied,

$$\omega_B T_{12} \approx 2n\pi + \delta, \tag{3.62}$$

where ω_B is the shape mode frequency, T_{12} is the time between the first and the second kink collisions, and δ is the offset phase. The integer number n plays a role of the number of the resonance, and it determines a sequence of the resonant velocities for the kink escape below the critical value of the relative velocity for the capture. In fact, the total number of the resonances (i.e. windows) is limited by radiation. The phenomenological explanation of the resonant effects proposed by Campbell *et al.* [189] accurately describes the resonance structures in the kink collisions observed in direct numerical simulations, and the existence of such resonances has been shown for several nonlinear models, including the double SG and ϕ^4 models.

Finally, it should be mentioned that the properties of the FK model with nonsinusoidal substrate potential are rather general. For example, analogous types of the nonlinear dynamics may be observed for the double SG model [195] for $|s| < 1/4$, in particular, the LA breather modes exist provided $s > -1/16$, and the kink shape mode emerges for $s > 0$.

To conclude this section, we mention that there exist some exotic shapes of the substrate potential $V_{\text{sub}}(x)$ which produce an exactly vanishing PN potential to the kink motion, i.e., $E_{PN} = 0$. A systematic procedure for obtaining these exotic cases was developed by Speght and Ward [202] (see also Ref. [203]). In particular, above we have mentioned that such an effect may take place for the substrate potential composed of a sequence of parabolas but only for certain values of the model parameters. Another example was given by Bak [204], and this potential is defined by its first derivative, as the follows,

$$V'_{\text{sub}}(x) = 4\tan^{-1}\left(\frac{\lambda\,\sin x}{1-\lambda\,\cos x}\right), \tag{3.63}$$

where $\lambda = \tanh^2(a_s/d)$. For the potential (3.63) the *discrete* motion equation has an exact kink solution that coincides with the SG kink, $u_n(t) = 4\tan^{-1}\exp[-(na_s - X)/d]$, and it moves freely along a discrete chain, so that the kink's energy does not depend on its coordinate X. However, this model still remains nonintegrable unlike the SG equation.

3.3.4 Multiple-Well Potential

Next, we consider the FK model with the substrate potential which possesses more than one absolute minimum per its period $a_s = 2\pi$. As a consequence of this potential shape, more than one type of kinks is possible in the model, and more than one phonon spectrum branch exist. As a typical example, we consider the double-well (DW) potential proposed by Remoissenet and Peyrard [187] (see Fig. 3.11)

$$V_{\rm sub}(x) = \frac{(1-s)^4[1-\cos(2x)]}{(1+s^2+2s\sin x)^2}, \tag{3.64}$$

where $0 \le s \le +1$. This potential has two distinct minima, one at $x_{01} = 0$ and another at $x_{02} = \pi$, at which the energy values coincide, $V_{\rm sub}(x_{01}) = V_{\rm sub}(x_{02}) = 0$. The minima are separated by two barriers, at x_{m1} and x_{m2}, where the first maximum depends on the parameter s,

$$V_{\rm sub}(x_{m1}) = 2\left(\frac{1-s}{1+s}\right)^4, \tag{3.65}$$

while the second barrier has the fixed value, $V_{\rm sub}(x_{m2}) = 2$.

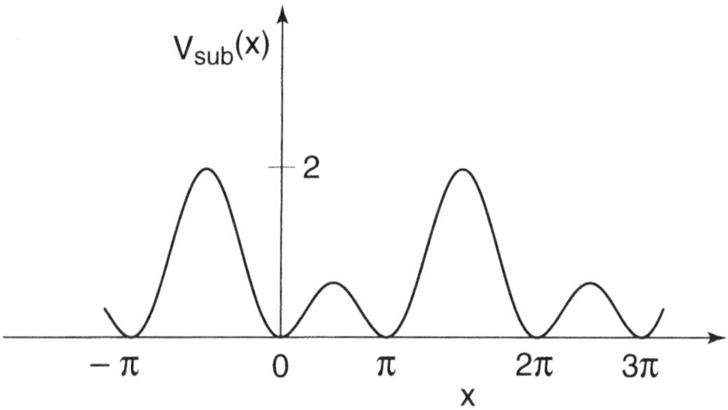

Fig. 3.11. Schematic presentation of the double-well substrate potential (3.64).

The FK model with the DW substrate potential has two types of the ground states (see Fig. 3.12). In the so-called "left ground state" (LGS) shown in Fig. 3.12(a), all the atoms in the chain occupy the left minima of the DW potential, $x_n = x_{01} + 2\pi n$ ($n = 0, \pm 1, \ldots$), and in the so-called "right ground state" (RGS) shown in Fig. 3.12(b), the atoms occupy only the right minima, i.e. $x_n = x_{02} + 2\pi n$. Both the ground states are characterized by the same phonon spectrum with $\omega_{\rm min} = 2(1-s)^2/(1+s^2)$. The standard 2π-kink, which connects two equivalent ground states, say LGS and LGS-2π, now splits into two separate sub-kinks. One sub-kink connects LGS and RGS [see Fig. 3.12(c)], and it is called *large kink* (LK) because it overcomes the largest barrier. Another sub-kink, *small kink* (SK), connects the states RGS and LGS-2π, and it overcomes the lower energy barrier [see Fig. 3.12(f)]. Analogously, large and small antikinks may be defined in the system, $\overline{\rm LK}$ and $\overline{\rm SK}$ shown in Figs. 3.12(d) and 3.12(e).

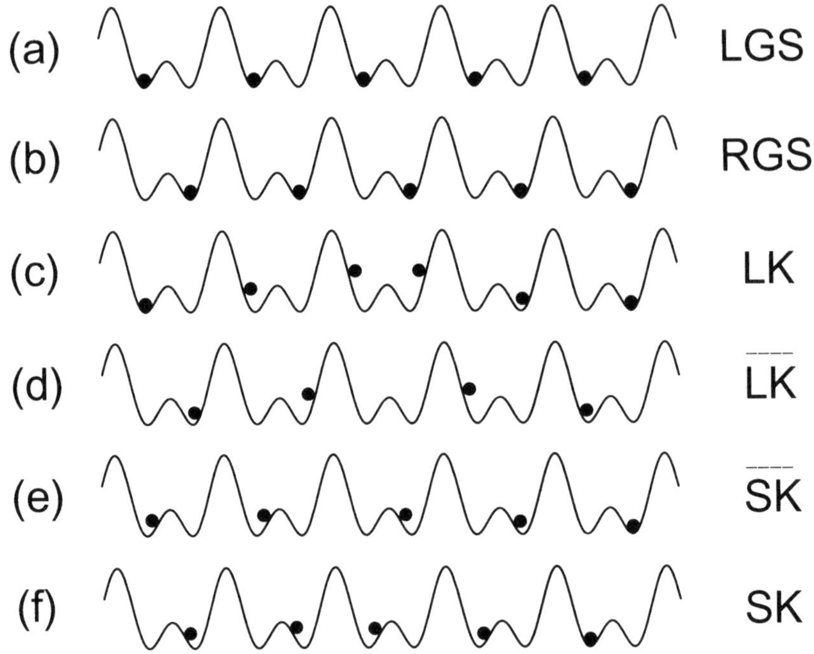

Fig. 3.12. Structure of two types of the ground state and kinks in the FK model with the double-well substrate potential: (a) "left" GS, (b) "right" GS, (c) large kink, (d) large antikink, (e) small antikink, and (f) small kink.

It is clear that the chain can support a single LK or a single SK which are independent topological excitations of the chain. The LKs and SKs have the properties similar to those of the kinks in the standard SG model, however, their parameters are different from the corresponding parameters of the SG kink (see Ref. [187]). For example, the kink masses coincide at $s = 0$ and they both vanish for $s \to +1$, but, in general, they are different, $m_{SK} < m_{LK}$. Besides, the SK has an internal mode whereas the LK has no internal mode. This makes some dynamical effects related to the kink collisions different for SKs and LKs as well.

The standard FK model allows an arbitrary sequence of the kinks and/or antikinks. In the DW model, however, some of the kink combinations are forbidden due to topological constrains. For example, in a periodic array of kinks, SKs and LKs should alternate because one "extra" atom in the chain corresponds to a pair LK+SK. In fact, the DW model allows only *four types* of the kink and antikink combinations, SK+LK, $\overline{\text{SK}}$+$\overline{\text{LK}}$, LK+$\overline{\text{LK}}$, and SK+$\overline{\text{SK}}$. Therefore, only collisions between those kinks are possible in the DW model.

SK+LK (or $\overline{\text{SK}}$+$\overline{\text{LK}}$) kink collision. As usual, two kinks of the same topological charge repel each other, and in the present case two different kinks cannot pass through each other because of topological constrains. Therefore, the kink collision should display a reflection, and such a reflection is almost elastic for $g \gg 1$, see Ref. [187].

LK+$\overline{\text{LK}}$ collision. Numerical simulations [187] have shown that this type of kink and antikink pass through each other transforming into a pair $\overline{\text{SK}}$+SK as should follow from the topological constrains. Such an effect of the kink transformation was analyzed earlier by Maki and Kumar [205] and Schiefman and Kumar [206] in the framework of the DSG equation. The difference in the kink rest energies, $\Delta\varepsilon = 2(m_{LK} - m_{SK})c^2 > 0$, is converted into the kinetic energy of the small kinks according to the energy conservation which, as has been verified, holds with a good accuracy even when the effect of radiation is not taken into account. However, in a highly discrete chain when $g \sim 1$, the energy excess $\Delta\varepsilon$ may be taken out by radiation leading to a decay of the kink-antikink pair with formation of a breather state.

SK+ $\overline{\text{SK}}$ collision. When the initial kinetic energy K_{in} of small kinks is large enough, namely $K_{\text{in}} > 2(m_{LK} - m_{SK})c^2$, they can pass through each other converting into a slowly moving $\overline{\text{LK}}$+LK pair. Otherwise, small kinks behave similar to those of the ϕ^4 model, i.e. they may be trapped into a breather state provided the initial velocity is smaller than a certain critical value.

Campbell *et al.* [195] have studied in detail the kink collisions in the DSG model (3.50), which has the DW shape for $s < -1/4$. Unlike the model with the potential (3.64) considered above, the DSG subkinks have different amplitudes, $\Delta u_{SK} < \Delta u_{LK}$, where $\Delta u = |u_k(-\infty) - u_k(+\infty)|$, so that the SK disappears when the lower barrier vanishes, i.e. for $s \to -1/4$. Besides, SK in the DSG model has an internal (shape) mode. Therefore, the SK+$\overline{\text{SK}}$ collisions at small kinetic energies exhibit resonance phenomena caused by the energy exchange between the kinks translational modes and their internal modes similar to the case described in the previous section [189, 195, 201].

Model with a more general shape of the substrate potential has been introduced in the theory of solitons in hydrogen-bonded chains, where topological solitons characterize different types of defects in such a system [122],

$$V_{\alpha,\beta}(x) \propto \left[\frac{\cos x - \alpha}{1 - \beta(\cos x - \alpha)}\right]^2 . \tag{3.66}$$

Here the parameter α ($|\alpha| < 1$) describes a relative width of two barriers as well as it controls the distance between the neighboring minima of the substrate potential. The second parameter β describes the relative height of the barriers. The potential (3.66) reduces to the sinusoidal form for the case $\alpha = \beta = 0$, and to the DSG potential, for $\beta = 0$. For the potential (3.66), a mass of the "small" kink may be larger than that of a "large" kink.

The results discussed above involved only the case of a symmetric substrate potential. However, in a general case the substrate potential may have wells with different curvatures. As a consequence, there exist more than one branch of the phonon spectrum in the model. For example, for the asymmetric double-well (ADW) potential proposed by Remoissenet and Peyrard [187] (see Fig. 3.13),

$$V_{\rm sub}(x) = \frac{(1-s^2)^2[1-\cos(2x)]}{(1+s^2+2s\cos x)^2}, \tag{3.67}$$

where $0 \leq s \leq 1$, the LGS has the minimum phonon frequency gap $\omega_{\rm min}^{(L)} = (1-s)/(1+s) < 1$, while the RGS is characterized by the gap $\omega_{\rm min}^{(R)} = (1+s)/(1-s) > 1$. In this system, there are two sub-kinks, the "left kink" (LK), which links the LGS and RGS, and the "right kink" (RK) which is just reverse, so that the kinks may be transformed to each other by a mirror transformation. The shape of a kink is asymmetric because its tails belong to the wells with different curvatures. For example, the LK has the long-range left-hand tail,

$$u_k(x \to -\infty) \propto -\exp\left(-\omega_{\rm min}^{(L)}|x|/d\right),$$

and the sharp right-hand tail,

$$u_k(x \to +\infty) \propto -\pi + \exp\left(-\omega_{\rm min}^{(R)} x/d\right).$$

The kink's mass is the same for both the kinks,

$$m = m^{(SG)} \frac{(1-s^2)}{4s} \ln\left(\frac{1+s}{1-s}\right), \tag{3.68}$$

and it varies from $\frac{1}{2}m^{(SG)}$, at $s=0$, to zero, at $s=1$.

A kink in the FK model with the ADW potential (3.67) possesses an internal mode with the frequency ω_B. Remoissenet and Peyrard [187] have found an interesting phenomenon: for $0 < s < 0.4$ the value of ω_B lies between the frequencies $\omega_{\rm min}^{(L)}$ and $\omega_{\rm min}^{(R)}$. Therefore, when the kink's internal mode is excited (e.g., during the kinks collision), it decays rapidly due to radiation of the phonons around the LGS toward a smooth kink tail. As a natural result, resonance structures may be observed only for RK+$\overline{\rm RK}$ or $\overline{\rm LK}$+LK collisions, when kink and antikink collide by their smooth tails (the "soft" collision), and the chain outside the collision region is in the RGS which is characterized by the phonon frequencies $\omega \geq \omega_{\rm min}^{(R)} > \omega_B$. Numerical simulations presented by Remoissenet and Peyrard [187] demonstrate the existence of the reflection velocity windows for this type of kink collisions.

It is clear that each the GSs (LGS or RGS) may support its own long-living breather mode in the ADW model. Remoissenet and Peyrard [187] have shown that the RGS (sharp wells) always supports a SA breathers, while the LGS (flat bottoms) supports LA breather mode provided $0 < s < s_*$, where

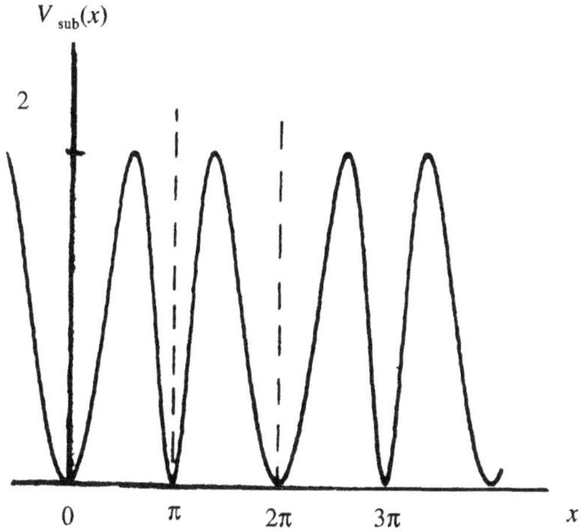

Fig. 3.13. Schematic presentation of the asymmetric substrate potential (3.67).

$s_* = 2 - \sqrt{3} \approx 0.268$. Therefore, for $s_* < s < 1$ a LA breather cannot be excited in a result of the hard-core kink-antikink collisions (i.e., in collisions LK+$\overline{\text{LK}}$ or $\overline{\text{RK}}$+RK when kinks collide from the side of their short-range tails), and the kinks are destroyed emitting phonons. Other cases of the kink collisions are similar to the standard SG-type model, and such collisions may be treated as the collisions of effective quasi-particles. Due to asymmetry of the kink shape, a kink and antikink may collide from the side of by their long-range tails ("soft" collision) and the chain outside the collision region is in the RGS, or the kink and antikink may collide from the side of by their short-range tails (the so-called "hard" collision). The soft collisions are almost perfectly elastic, while the hard collisions are inelastic and they are accompanied by strong radiation [187].

3.3.5 Multi-Barrier Potential

In some applications of the FK model (see Chapter 2), the substrate potential has a complicated structure with additional local minima, e.g. the double-barrier (DB) structure shown in Fig. 3.14. In this case, the system may be observed in a "metastable ground state" (MGS) when all atoms of the chain occupy the local minima of the substrate potential. Similar to the DW model considered above, the 2π-kink of the DB model splits and produces two subkinks connecting GS-MGS and MGS-(GS-2π). However, unlike the case of the DW potential, now the atoms in the region between the subkinks are in a metastable (excited) state (see Fig. 3.14). This leads to an effective attraction between two subkinks because the energy of the 2π-kink grows with the distance between the subkinks, i.e. with an increase of the number of atoms occupying the MGS. As a result of a competition of this attraction and the conventional repulsion of the subkinks of the same topological charge, there exists an equilibrium distance R_0 which corresponds to a minimum of the

2π-kink energy. Thus, the DB kink may be considered as a "molecule" (2π-kink) composed of two "atoms" (subkinks or π-kinks) coupled together by an interaction potential $U_{DB}(R)$ which has a minimum at $R = R_0$. It is clear that the DB kink should always possess an internal mode with the frequency ω_B, which in this case has a simple physical meaning: this mode corresponds to the relative oscillation of two "atoms" inside the "molecule".

As an example, let us consider one of the most frequently occurring substrate potentials, the double SG potential (3.50), which has the double-barrier structure provided $s > 1/4$. Introducing a new parameter r according to the relation $s = (\frac{1}{2}\sinh r)^2$, where $r > \ln(1+\sqrt{2}) \approx 0.881$, the DSG potential can be presented as

$$V_r(x) = \frac{\sinh^2 r}{\cosh^4 r}\left\{4(1-\cos x) + \sinh^2 r[1-\cos(2x)]\right\}. \tag{3.69}$$

The potential (3.69) has a minimum, at $x_0 = 0$, and two maxima, at $x_m = \pi \pm \cos^{-1}(\cosh^2 r)$, $V_r(x_m) = 2$, and a relative minimum at $x_b = \pi$, $V_r = 8\sinh^2 r/\cosh^4 r$, per one period $a_s = 2\pi$.

Fig. 3.14. Kink in the FK model with the double-barrier substrate potential.

In the continuum limit approximation, when $g \gg 1$, the static DB kink (antikink) has a simple form,

$$u^{(DB)}(x) = \mp 2\tan^{-1}\left(\frac{\sinh y}{\cosh r}\right), \tag{3.70}$$

where

$$y = 2(x - X)/d. \tag{3.71}$$

The mass of the kink is defined as

$$m^{(DB)}(r) = m^{(SG)}\frac{\sinh r}{\cosh r}\left[1 + \frac{2r}{\sinh(2r)}\right] > m^{(SG)}. \tag{3.72}$$

It is interesting to mention that the static kink (3.70), (3.71) of the DSG equation can be exactly expressed as a sum of two single kinks of the SG model [195, 207],

$$u^{(DB)}(x) = \mp\left[u_\pi(\widetilde{R} + y) - u_\pi(\widetilde{R} - y)\right], \tag{3.73}$$

where $u_\pi(y) = 2\tan^{-1}\exp(y)$, $\widetilde{R} = r$, and y is determined by Eq. (3.71). Thus, the parameter X can be considered as the coordinate of the center of mass of two kinks, and the parameter $R = \widetilde{R}d$, as the distance between two subkinks of the DB kink.

Willis *et al.* [208] have analyzed a complete Hamiltonian dynamics of a DB kink in the DSG model introducing two collective variables $X(t)$ and $R(t)$, and the corresponding conjugate momenta as canonical variables. Analogously to the Hamiltonian formalism for the SG model, in the present case one should add one more degree of freedom (internal oscillations) and to modify the Poisson brackets. After some lengthy calculations, Willis *et al.* [208] have proved that the energy of the DSG kink in such an approach may be presented in the form,

$$H_{DB} = \frac{m^{(DB)}}{2}\left(\frac{dX}{dt}\right)^2 + \frac{m^{(R)}}{2}\left(\frac{dR}{dt}\right)^2 + U_{DB}(R), \tag{3.74}$$

where the effective mass $m^{(DB)}$ is determined by Eq. (3.72),

$$m^{(R)}(r) = \int_{-\infty}^{\infty}\frac{dx}{2\pi}\left(\frac{\partial u^{(DB)}}{\partial R}\right)^2 = \frac{1}{4}m^{(SG)}\frac{\sinh r}{\cosh r}\left[1 - \frac{2r}{\sinh(2r)}\right], \tag{3.75}$$

and

$$U_{DB}(R) = m^{(SG)}c^2\frac{\sinh r}{2\cosh r}\left\{1 + \frac{\tanh^2 r}{\tanh^2 \widetilde{R}} + \right.$$
$$\left. 2\widetilde{R}\left[\frac{1}{\sinh(2\widetilde{R})} + \frac{\coth \widetilde{R}}{\cosh^2 r} - \frac{\tanh^2 r \coth \widetilde{R}}{2\sinh^2 \widetilde{R}}\right]\right\}, \tag{3.76}$$

where $\widetilde{R} \equiv R/d$. For the small-amplitude internal oscillations, the energy $U_{DB}(R)$ may be presented in the form

$$U_{DB}(R) \approx m^{(DB)}c^2 + \frac{1}{2}m^{(R)}\omega_B^2(r)(R - R_0)^2, \tag{3.77}$$

where $R_0 = rd$, and $\omega_B(r)$ is the frequency of the internal oscillation,

$$\omega_B^2(r) - \frac{1}{m^{(R)}(r)}\left(\frac{d^2U_{DB}}{dR^2}\right)\Bigg|_{R=R_0}. \tag{3.78}$$

The frequency ω_B can be found from Eqs. (3.75), (3.76), and (3.78) as well as it can be determined as an eigenvalue of the Schrödinger equation (3.51), (3.52). The latter method leads to the result [209] $\omega_B \approx \omega_{\min}\sqrt{1-\beta^2}$, where

$$\beta = \frac{1}{2}\left[\left(1 + \frac{8}{\alpha^2}\tanh^2 r\right)^{1/2} - 1\right], \tag{3.79}$$

and

$$\alpha = \frac{\tanh^2 r \, \sinh(2r)}{[\sinh(2r) - 2r]} . \tag{3.80}$$

The potential $U_{DB}(R)$ defined by Eq. (3.76) is anharmonic. Therefore, when the amplitude of the "molecule" oscillations grows, the oscillations become nonlinear and the anharmonicity of the potential $U_{DB}(R)$ becomes important, and it may result in the generation of higher-order harmonic and the phonon emission. However, a coupling of the kink's internal oscillations to the phonons is extremely weak because it is caused only by higher-order resonances. Numerical simulations carried out by Burdick *et al.* [210] have shown that an effective damping of the internal oscillations is negligible even when the oscillation become strongly nonlinear.

In the limit $r \to \infty$, the DSG potential (3.69) reduces to the SG one with the period $a = \pi$, the DB kink (3.73) splits into two separate SG π-kinks, and $\omega_B \to 0$. We note also that in the discrete case the Hamiltonian (3.74) should include the PN potential for the subkinks.

Campbell *et al.* [195] have studied numerically the DSG kink-antikink collisions, and they demonstrated the existence of the "resonance windows" due to an energy exchange mechanism between the kinks translational and internal modes. The windows of the resonance velocities correspond to the situation when the kinks collide inelastically and form a breather, otherwise the kinks scatter not changing their identities. For $r > 1/4$, a qualitatively new effect occurs, namely, two counter-propagating breathers may emerge as the final result of the kink-antikink collision. The resonant energy exchange can be studied in the framework of an effective collective-coordinate model that treats the scattering process as a collision of two "molecules", each consisting of the DB kink and DB antikink.

3.4 Disordered Substrates

For realistic physical models, interaction of nonlinear excitations (kinks or breathers) with impurities should play an important role in the transport properties because kinks (or breathers) can be trapped or reflected by local impurities. Additionally, a breather captured by an impurity excites *a nonlinear impurity mode*, and this observation makes a link between the theory of nonlinear chains and the theory of harmonic lattices with defects (see, e.g., Refs. [211]–[214]).

Many features of the soliton-impurity interactions have been already discussed in review papers by Kivshar and Malomed [27] and Gredeskul and Kivshar [215] in the framework of the SG model with local or extended inhomogeneities. For the discrete FK model, two new features of the soliton-impurity interactions appear and they should be discussed. First, in a discrete chain, a kink moves in the presence of an effective PN potential whose amplitude is always lower than the amplitude of the substrate potential. Thus, the kink parameters are varying periodically and this simple mechanism generates phonons leading to the subsequent pinning of the kink by the lattice discreteness. As a result, the discreteness effects which are absent in the SG model may significantly modify the adiabatic kink scattering [49]. Second, the important feature of the kink scattering by impurities in a discrete chain is the possible excitation of impurity modes during the scattering. In fact, such an effect is also possible for continuous models provided one considers strong disorder, but the discreteness modifies the impurity mode frequency making the process of its excitation more easier [216].

A simple generalization of the FK model to include defects of different kind was discussed by Braun and Kivshar [49]. The FK model with disorder in a general case may be described by the following Hamiltonian [cf. Eqs. (1.1) – (1.5)]

$$\mathcal{H} = \sum_j \left\{ \frac{m_j}{2} \left(\frac{dx_j}{dt} \right)^2 + \frac{g_j}{2} (x_{j+1} - x_j - a_0)^2 + U(x_j) \right\},$$

with the potential

$$U(x_j) = \varepsilon_j \left[1 - \cos\left(\frac{2\pi x_j}{a_s} \right) \right] + v(x_j).$$

where impurities are taken into account through the parameters m_j (change of the particle mass), g_j (change of the interparticle interaction), ε_j (local distortion of the substrate potential), and $v(x_j)$ (an additional change of on-site potential created by impurities). The motion equation for the atomic displacements $u_j = x_j - ja_s$ takes the form [we consider here the simplest case when $a_0 = a_s$]

$$m_j \frac{d^2 u_j}{dt^2} + g_j(u_j - u_{j+1}) + g_{j-1}(u_j - u_{j-1}) \\ + \varepsilon_j \sin u_j + v'(ja_s + u_j) = 0. \quad (3.81)$$

When one of the atoms of the chain, say at $j = 0$, has properties which are different from those of the lattice atoms, it may be characterized by a local change of the parameters (in dimensionless units adopted in Chapter 1), $\varepsilon_0 = 1 + \Delta\varepsilon$, $m_0 = 1 + \Delta m$, and $g_0 = g_{-1} = g + \Delta g$, so that the perturbation-induced correction $\delta\mathcal{H}$ to the Hamiltonian of the FK chain is written as

$$\delta\mathcal{H} = \frac{\Delta m}{2} \left(\frac{du_0}{dt} \right)^2 + \frac{\Delta g}{2} [(u_1 - u_0)^2 + (u_0 - u_{-1})^2] + \Delta\varepsilon\, (1 - \cos u_0). \quad (3.82)$$

In the continuum limit approximation such an impurity is introduced by the changes like $\varepsilon_j \to \varepsilon(x) = 1 + \Delta\varepsilon\, a_s\, \delta(x)$, and so on.

First of all, the combined effect of nonlinearity and disorder can modify the kinks properties even in the static case. This problem is easier to be analyzed for the SG model, i.e. for the continuous version of the FK chain. In fact, the SG model with defects was introduced by Baeriswyl and Bishop [217] who analyzed the linear properties of that model. For the case of the delta-like impurities, a number of exact results to the SG model can be obtained for defect stationary states, nonlinear static structures created by the effect of kink's pinning due to impurities. Several cases where such stationary structures may be treated analytically have been considered for both the linear coupling between the defect and the wave field, i.e. when $v(x) \sim \lambda\,\delta(x - x_0)$ (see, e.g., Ref. [218]), and for nonlinear coupling (when, e.g., $\varepsilon(x) \sim \lambda\,\delta(x - x_0)$, see [219]. The derivative mismatch introduced by such a δ-function allows to get (for isolated defects) the nonlinear stationary conditions which can be solved analytically. With the help of those exact results, the correlation function in the presence of defects can be calculated, as well as the free energy of the various possible configurations. This program can be realized not only for one or two impurities but also for a random distribution of defects in the limit of small concentration [218].

More complicated behavior is observed in a generalized FK model where, e.g., an extension of the model beyond the limits of the harmonic approximation for the interatomic potential leads to some qualitatively novel results such as the existence of distortion chain configurations [220] or formation of cracks when the tensile strength of the chain exceeds a certain critical value [221, 222]. When local impurities are inserted into the chain, they may act as traps in both pinning the antikinks and increasing the threshold for a chain breakup. Such an effect was analyzed for the FK chain with nonconvex interaction between neighboring atoms by Malomed and Milchev [223] who showed that the breakup threshold for an antikink pinned by an inhomogeneity which locally decreases the substrate potential is higher than for a free antikink, the effect they related to the observed formation of cracks out of misfit dislocations in III-V hetero-structures [44].

3.4.1 Effective Equation of Motion

First we consider the continuum limit approximation of the FK model described by the SG equation with inhomogeneous parameters. In this case, the effective equation for the kink's coordinate can be derived by a simplified version of the collective-coordinate approach [27, 224, 225]. As an example, we consider the simplest case of the inhomogeneous SG model,

$$\frac{\partial^2 u}{\partial t^2} - \frac{\partial^2 u}{\partial x^2} + \sin u = \epsilon f(x) \sin u, \tag{3.83}$$

when the impurity is modelled by introducing the external potential

$$U_{\text{ext}}(x) = \int^x dx\, f(x).$$

Analyzing the kink dynamics in the framework of the collective-coordinate approach, we can obtain, in a simple way, an effective equation of motion for the kink's coordinate (see, e.g., Refs. [86, 159]). To derive such an equation, we note that the unperturbed SG system has an infinite number of quantities (system invariants) that are conserved during the evolution, among which there is the momentum,

$$P \equiv - \int_{-\infty}^{\infty} dx \, \frac{\partial u}{\partial t} \frac{\partial u}{\partial x}. \tag{3.84}$$

For the SG kink, Eq. (3.84) takes the form of the well-known relativistic expression $P = mV/\sqrt{1 - V^2}$, V being the kink velocity. In the presence of perturbations, the momentum is no longer conserved; using Eq. (3.83) it is possible to show that it varies according to the equation

$$\frac{dP}{dt} = \epsilon \int_{-\infty}^{\infty} dx \, f(x) \, \frac{\partial}{\partial x}(\cos u).$$

provided the boundary conditions $u \to 0$ (2π) at $x \to \pm\infty$ holds. The adiabatic approach is now defined by the assumption that, for ϵ small enough, the kink shape is not affected and only the kink's coordinate X becomes a slowly varying function of time. Within this hypothesis it can be shown that, in the non-relativistic limit, the kink center obeys the following equation of motion, $m \, d^2X/dt^2 = -U'(X)$, where

$$U(X) \equiv -2\epsilon \int_{-\infty}^{\infty} dx \, \frac{f(x)}{\cosh^2(x - X)}, \tag{3.85}$$

and we have used the approximate expression $P \approx m\,(dX/dt)$, valid for small velocities. Thus, in the framework of such an adiabatic approach, the motion of the SG kink can be thought of as that of a particle with (kink) mass m in the external potential $U(X)$ defined by Eq. (3.85). The similar properties can be shown for relativistic kinks [146].

The following two cases arise naturally from Eq. (3.85). If $f(x)$ changes rapidly over distances of the order of the kink length, then ϵ has to be small for our approximation to hold. For example, in the case $f(x) = \delta(x)$, we have [86] $U(X) = -2\epsilon \operatorname{sech}^2 X$. On the other hand, if $f(x)$ changes slowly, i.e., its characteristic length (say L) is much larger than the kink width, it is not necessary for ϵ to be small, because all the parameters of the perturbation theory are of the order of L^{-1}, and we are left with $U(X) \approx 4\epsilon \, f(X/L)$.

The approximation involved in the derivation presented above is based on the assumption that the kink moves slow through the region of inhomogeneity. In this case, the kink's width does not change much and its variation can be neglected. This corresponds to the so-called "nonrelativistic" interaction of the kink with an impurity. However, relativistic effects can be taken into account by introducing one more collective coordinate associated with kink's width [226, 227]. A more detailed analysis of this effect was presented by Woafo and Kofaneé [228] who observed that the kink is shortened in the attractive potential and extended in the repulsive potential of the impurity.

The adiabatic theory presented above becomes not valid in the case when a localized impurity can support an impurity mode, an oscillating linear mode at the impurity site. In this latter case, the kink's position and the impurity mode amplitude are two effective collective coordinates as discussed in detail below. Different types of the so-called resonant interactions of solitons and kinks with impurities have been reviewed by Belova and Kudryavtsev [229], and the simplest example of such an interaction can be found below in Sect. 4.3.2.

In the discrete FK lattice, the motion equation for the kink's coordinate is modified by the PN relief. One of the ways to derive the effective equation of motion for the kink has been already mentioned in Sect. 3.2.1 and it is based on the projection-technique approach developed by the group of Willis. Another approach is based on the Lagrangian formalism which we will apply here just to mention the example how such a method really works (see, e.g., Refs. [49, 230, 231]).

Let us start from the Lagrangian of the inhomogeneous FK chain

$$L = \sum_j \left\{ \frac{m_j}{2} \left(\frac{du_j}{dt} \right)^2 - \frac{g_j}{2}(u_{j+1} - u_j)^2 - \varepsilon_j \left[1 - \cos\left(\frac{2\pi u_j}{a_s} \right) \right] \right\}. \tag{3.86}$$

Considering now the simplest case of a single-point defect at the site $n = 0$ (the case of several impurities can be treated in a similar manner), we put $\varepsilon_j = \varepsilon_s + \Delta\varepsilon\, \delta_{j0}$, $m_j = m_a + \Delta m\, \delta_{j0}$, and $g_j = g + \Delta g\, \delta_{j0}$. Introducing the dimensionless variables, $\tau = (c/a_s)\,t$ and $\phi_j = (2\pi/a_s)\,u_j$ and setting $\mu = a_s/l$, $c^2 = g a_s^2/m_a$, $l = c/\omega_0$, where $\omega_0^2 = 2\pi^2\varepsilon_s/m_a a_s^2$, the Lagrangian (3.86) becomes

$$L = A \sum_j \left\{ \frac{1}{2} \left(\frac{d\phi_j}{dt} \right)^2 \left(1 + \frac{\Delta m}{m_a}\, \delta_{j0} \right) - \frac{1}{2}(\phi_{j+1} - \phi_j)^2 \left(1 + \frac{\Delta g}{g}\, \delta_{j0} \right) \right.$$
$$\left. - \mu^2 (1 - \cos\phi_j) \left(1 + \frac{\Delta\varepsilon}{\varepsilon_s}\, \delta_{j0} \right) \right\},$$

where $A = m_a (c/2\pi)^2$. In the notations adopted above, the parameter μ has the meaning of a ratio of the lattice spacing to the kink's width. We now assume that the value μ is small, so that distorted kink in the discrete chain may be approximated by the SG kink ansatz

$$\phi_j(\tau) = 4 \tan^{-1} e^{\mu \xi_j}, \tag{3.87}$$

where $\xi_j = j - Y(\tau)$, $Y(\tau)$ being a collective coordinate of the kink. Substituting Eq. (3.87) into the system Lagrangian and evaluating the sums with the help of the Poisson sum formula,

$$\sum_{n=-\infty}^{\infty} h\, f(nh) = \int_{-\infty}^{\infty} dx\, f(x) \left[1 + 2\sum_{s=1}^{\infty} \cos\left(\frac{2\pi s x}{h}\right)\right],$$

we obtain the effective Lagrangian in the following reduced form [49]

$$L/A = 4\mu \left\{ \left(\frac{dY}{dt}\right)^2 - \frac{4\pi^2}{\sinh(\pi^2/\mu)} \cos(2\pi Y) \right\}$$
$$+ \frac{2\mu^2}{\cosh^2(\mu Y)} \left\{ \left(\frac{dY}{dt}\right)^2 \left(\frac{\Delta m}{m_s}\right) - \left(\frac{\Delta g}{g} - \frac{\Delta \varepsilon}{\varepsilon_s}\right) \right\}. \tag{3.88}$$

The equation of motion for the kink's coordinate $X = \mu Y$ can be obtained from Eq. (3.88) in a straightforward manner. The simple analysis shows that the discreteness yields an additional potential field associated with the PN relief so that the kink may be treated as an effective particle of a variable mass moving in an effective potential $U_{\text{eff}}(X) = U_{\text{PN}}(X) + U_{\text{im}}(X)$, where

$$U_{\text{PN}}(X) = \frac{2\pi^2 \mu}{\sinh(\pi^2/\mu)} \cos\left(\frac{2\pi X}{\mu}\right) \tag{3.89}$$

and

$$U_{\text{im}}(X) = \frac{1}{4}\left(\frac{\Delta g}{g} + \frac{\Delta \varepsilon}{\varepsilon_s}\right) \frac{\mu^3}{\cosh^2 X}. \tag{3.90}$$

The analysis of the kink motion in the vicinity of the impurity can be found in the paper by Braun and Kivshar [49], but a qualitative physical picture of such an interaction is rather simple: the kink's motion is affected by the potentials of two kinds, localized, from the impurity, and nonlocalized from the periodic PN relief. In particular, if the kink is pinned by the discreteness not far from the impurity, its PN frequency is renormalized to be

$$\omega_j^2 = \omega_{PN}^2 - \frac{\mu^3}{2}\left(\frac{\Delta g}{g} + \frac{\Delta \varepsilon}{\varepsilon}\right) \frac{(1 - 2\sinh 2X_j)}{\cosh^4 X_j}, \tag{3.91}$$

where $\omega_{PN}^2 = [8\pi^4/\mu \sinh(\pi^2/\mu)]$ is the PN frequency, $X_j = (j + \frac{1}{2})\,\mu$, and ja_0 is the distance from the impurity. Note that the local impurity potential U_{im} acts on a kink and antikink in the same way.

The similar technique can be applied to the problem of the DNA promoters to explain its role as dynamical activators of transport processes of the RNA polymerase along DNA macromolecules. By introducing an effective potential for the kink in a disordered FK model, suggested as a simplest model of the DNA chain [103]–[106],[112],[232]–[234]. In particular, Salerno and Kivshar [231] demonstrated the existence of a dynamically "active" region inside of a DNA promoter, in a qualitative agreement with experimental data [235].

Several interesting effects can be observed for a multi-kink dynamics when two or more kinks interact near the impurity. In particular, when a moving kink collides with a kink trapped by at an impurity site, three different outcomes of the collision are possible: depinning, capture, and exchange [236]. For the kinks of different polarities, such collisions may result also in annihilation of the kinks at the impurity. The scattering of a SG breather can be treated as that of a coupled pair of kinks, at least in the case of low-frequency breathers. A complex dynamics of the breather-impurity scattering has been demonstrated by Kenfack and Kofané [237] and Zhang [238].

The adiabatic effects discussed above are based on the approximation when radiative losses are negligible. However, during the scattering by impurities, the kink radiates phonons which may change the general pictures of the scattering by introducing an effective radiative losses [27]. In fact, radiative losses in the inhomogeneous FK model may be of three types:

(i) *The first type* of radiation losses is due to existence of PN relief, and they have been discussed above in detail.

(ii) *The second type* of the resonant effects is related to the change of the kink's velocity caused by impurities. This kind of the kink's emission can be calculated in the lowest order as emission of the SG kink. A number of such problems was mentioned in a review paper on the soliton perturbation theory [27]. The importance of this type of radiative effects has been demonstrated for the case of the kink scattering by two impurities by Kivshar *et al.* [239], who demonstrated that, for low kink's velocities, the reflection coefficient of the kink depends oscillatory on the distance between the impurities, the effect caused by an interference of the radiation emitted by the kink.

(iii) *The third type* of inelastic effects which are not taken into account by the adiabatic approach is the excitation of impurity localized modes by the scattering kink, and in the limit of the SG model such a problem will be discussed below.

3.4.2 Point Defects

When the function $f(x)$ in Eq. (3.83) describes random impurities, we may consider the simplest case of delta-like inhomogeneities, $\epsilon f(x) = \sum_n \epsilon_n \delta(x - a_n)$, where the numbers ϵ_n and a_n are chosen to be random, and it is assumed that the distances $b_n = a_{n+1} - a_n$ are identically distributed random numbers with the probability density $p(b) = b_0^{-1} e^{-b/b_0}$. Then, the equation for the kink coordinate X takes the form $m\, d^2X/dt^2 = -U'(X)$, where

$$U(X) = \sum_n u_n(X), \quad u_n(X) \equiv u(X - a_n), \quad u(X) \equiv -\frac{2\varepsilon}{\cosh^2 X}. \tag{3.92}$$

Here, as above, we have approximated $P \simeq mV \simeq m\, dX/dt$. Thus, in the collective coordinate framework, the motion of the SG kink can be interpreted as the motion of a nonrelativistic particle with the unit mass m in an effective random potential defined by Eq. (3.92).

In the paper by Gredeskul *et al.* [240] the kink scattering was analyzed for the case when disorder appears as randomly distributed point impurities with equal intensities, i.e. for $\epsilon_n = \epsilon$. The general methods usually used for time-dependent random perturbations (see, e.g., Refs. [241]–[243], and also the review papers [215, 244, 245]) cannot be directly applied to this problem because for randomly distributed spatial impurities we cannot derive the Fokker-Planck equation [notice that the potential (3.92) is not Markovian]. Gredeskul *et al.* [240] developed a statistical procedure to compute the mean characteristics of the kink propagation, e.g. the kink's mean velocity, assuming the velocity is rather large and the strength of impurities small.

One of the important problems related to the FK chain dynamics is the influence of disorder on the dislocation motion in crystals with a high PN potential. The basic concepts of the dislocation dynamics in crystals with a high PN relief were formulated by Lothe and Hirth [246] and Kazantsev and Pokrovsky [247]. Transverse displacement of the dislocation line is treated as creation of kink-antikink pairs by thermal fluctuations under the action of the applied constant force. Experimental data show that the dislocation mobility is also significantly affected by point defects, e.g. doping of crystals can give rise to an increase in the dislocation speed [248, 249]. This effect can be explained in the framework of the FK model as a consequence of local lowering of the PN barrier due to the interaction between the dislocation and impurities [250, 251] and we would like to mention that such conclusions were confirmed by extensive numerical simulations which showed that point defects in crystal lattice have an extremely profound effect on dislocation mobility [252, 253].

3.4.3 External Inhomogeneous Force

The concept of a kink as an effective point-like particle is valid, rigorously speaking, when the external field or effective force produced by impurities do not change rapidly on the spatial scale of order of the kink's width. Otherwise, one should expect some differences between, for example, stability conditions for a kink treated as a point particle and for a kink studied as an extended object. From the physical point of view, such differences come from interactions of kink's tails with zeros of an effective force. In particular, in the presence of a static force there exist static solutions for kinks whose centers are connected (or coincide) with position of zeroes of this force. This result follows from the effective-particle approach described above.

However, the condition of the kink stability is generally not equivalent to the stability condition that follows from simple energetic considerations treating the kink as a point particle. This was first noticed by Gonzaléz and Holyst [254] (see also Refs. [255, 256]) who considered a very special form of the static force $F(x)$ in the ϕ^4 model where the exact solution for a static kink, $u_k(x) = A\tanh(Bx)$, exists with the parameters A and B defined by the force $F(x)$. Linear eigenvalue problem for small-amplitude oscillations around the kink $u_k(x)$, due to the special choice of $F(x)$, coincides with the linear Schrödinger equation with the Pösch-Teller potential, and it can be solved exactly to describe the stability of the kink as an extended object. As a result, when the external force $F(x)$ has one stable ($x = 0$) and two unstable ($x = x_{1,2}$) extremum points, the kink located at the local minimum $x = 0$ becomes *unstable* provided the two maximum points at $x = x_{1,2}$ are closer to the kink center at $x = 0$, and the interactions of kink's tails with these zeroes are sufficiently strong to make the stable kink unstable [254].

The similar results were obtained later by Holyst [256] for the kink in the Klein-Gordon model with a double-quadratic substrate potential. To be more specific, the model studied by Holyst [256] has the form,

$$\frac{\partial^2 u}{\partial t^2} - \frac{\partial^2 u}{\partial x^2} + (|u| - 1)\,\mathrm{sgn}(u) = F(x), \tag{3.93}$$

where the force $F(x)$ was selected in the form

$$F(x) = \left\{A - 1 + A\,(B^2 - 1)\,e^{-B|x|}\right\}\mathrm{sgn}(Bx) \tag{3.94}$$

in order to have the exact solution for the kink of the shape

$$u_k(x) = A\left(1 - e^{-B|x|}\right)\mathrm{sgn}(Bx). \tag{3.95}$$

Treating the kink as a point particle, we obtain the stability condition

$$\left(\frac{dF(x)}{dx}\right)_{x=0} > 0,$$

i.e. $AB^2 > 1$. However, the linear eigenvalue problem for the kink (3.95) requires for the kink stability the condition $\Omega^2 > 0$, that yields a different condition $AB > 1$. A simple analysis of these two conditions and the extremum points of the external force $F(x)$ gives the following results [256]: (i) when the force (3.94) has *a single zero*, the both conditions coincide, (ii) when the force (3.94) has *three zeros*, the presence of additional zeros can *destabilize* the kink when the point $x = 0$ is stable, or *stabilize* it, otherwise.

These results were shown to modify the kink tunnelling with sub-barrier kinetic energy. Gonzaléz *et al.* [257] investigated, theoretically and numerically, the dynamics of a kink moving in an asymmetrical potential well with a finite barrier. For large values of the width of the well, the width of the barrier, and/or the height of the barrier, the kink behaves classically. On the other hand, they obtained the conditions for the existence of soliton tunnelling with sub-barrier kinetic energies; the condition is linked to the effect of the stability change in the presence of many zeroes of the external force, as discussed above. Indeed, when the kink behaves as a classical particle, the minimum of the potential corresponds to a stable kink sitting at the bottom of the potential well [see Fig. 3.15(a)], and it will not move to the right of

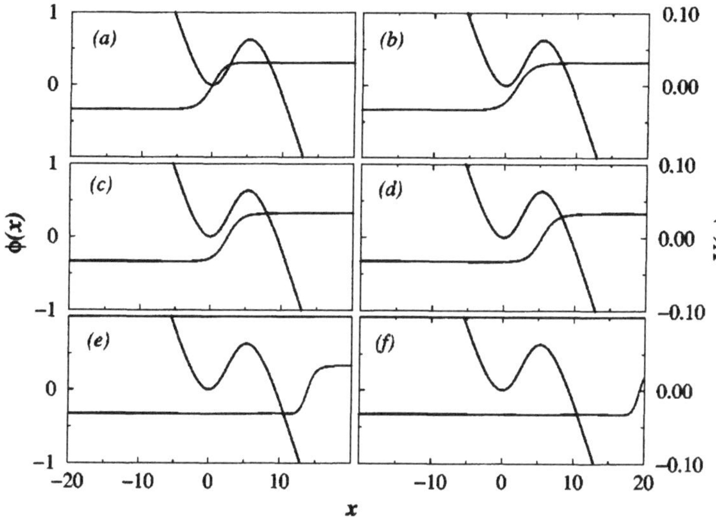

Fig. 3.15. Numerical simulation of the kink tunnelling with sub-barrier kinetic energy. The pale curve is the potential and the bold curve is the soliton. (a) shows the initial configuration at $t = 0$, while (b) to (f) describe the dynamics in successive time instants [257].

point $x = 0$. On the other hand, when the kink at the point $x = 0$ becomes unstable, it will move to the right, crossing the barrier even if its center of mass is placed in the minimum of the potential and its initial velocity is zero [see Figs. 3.15(b)-(f)]. In this case the kink performs tunnelling with sub-barrier kinetic energy.

Gonzaléz *et al.* [257] applied these results to the study of kink propagation in disordered systems. They considered the force $F(x)$ defined in such a way that it possesses many zeroes, maxima, and minima. Such a system describes an array of inhomogeneities, which can be analyzed as a series of elements with two zeroes and a maximum. If for each element the tunnelling condition is satisfied, then the kink can cross the whole inhomogeneous zone, even if the array is completely disordered. Thus, if the tunnelling condition is fulfilled, there is no localization for the kink motion.

3.5 Anharmonic Interatomic Interaction

For describing realistic physical systems, different types of anharmonic interatomic potentials should be considered. Such generalized FK models describe qualitatively new physical effects such as breaking of the kink-antikink symmetry, new types of dynamical solitons (supersonic waves), a breakup of the antikink solitons followed by a rupture of the chain, a change of the ground state of the model, a change of the interaction between the kinks, etc.

There are several ways to introduce anharmonic interparticle interaction into the FK model for modelling realistic physical systems. As an example, we briefly discuss here several mechanisms of interaction between atoms adsorbed on a crystal surface (for details, see, e.g., Refs. [36, 258, 259]).

When adatoms are charged, then the Coulomb repulsion $V_{\text{int}}(x) \approx e^2/x$ (e is the adatomic charge) acts between them for distances $x < a^*$ (a^* being the screening radius which is equal to the Debye screening radius, for a semiconductor substrate, and to the inverse Thomas-Fermi momentum, for a metallic substrate) [260]. For a semiconductor substrate, the value a^* is large enough, and the main contribution into the interaction potential is power-law, $V_{\text{int}}(x) \propto x^{-1}$. For adsorption on a metallic substrate the value of the screening radius a^* is of order of the lattice constant; for $x > a^*$ the interaction of adatoms has a dipole-dipole character: $V_{\text{int}}(x) \approx 2p_A^2/x^3$, p_A being the dipole moment of an adatom [261]. If adatoms are neutral, then the overlap of their electronic shells gives rise in a direct interaction, which decreases exponentially with the distance, $V_{\text{int}}(x) \propto \exp(-\beta x)$ [258]. More complex interaction laws are possible as well, such as for the so-called "indirect" mechanism of adatom interaction [36, 258, 259] which may be approximated by the following generalized law,

$$V_{\text{int}}(x) \sim x^{-n} \sin(2k_F x + \phi), \tag{3.96}$$

where n may have values from 1 to 5 depending on the electronic structure of the substrate, ϕ is a constant phase, and k_F is the Fermi momentum of the substrate electrons. In the latter case, an attraction (or "effective" attraction) can emerge for adatoms at some distances.

To cover a larger class of physically important systems, here we consider the following interaction potentials:

(i) *Exponential*

$$V_{\text{int}}(x) = V_0 \exp[-\beta(x - a_s)], \tag{3.97}$$

where V_0 is the energy of interaction between adatoms occupying the nearest neighboring minima of the substrate potential, and the parameter β characterizes anharmonicity of the potential;

(ii) *Power-law*

$$V_{\text{int}}(x) = V_0 \left(\frac{a_s}{x}\right)^n, \tag{3.98}$$

where n is an integer number ($n \geq 1$);

(iii) *Morse potential*

$$V_{\text{int}}(x) = V_m \left[e^{-2\beta(x-a_0)} - 2e^{-\beta(x-a_0)}\right], \tag{3.99}$$

where a_0 is the equilibrium distance, and V_m is the depth of the potential well, and, at last,

(iv) *Double-well potential*

$$V_{\text{int}} = V_m \left[\frac{1}{2}\beta^4 (x - a_*)^4 - \beta^2 (x - a_*)^2 \right]. \tag{3.100}$$

The latter potential has two minima at $x = a_* \pm \beta^{-1}$ and it can approximate qualitatively the oscillating potential (3.96) which appears for the "indirect" interaction of adatoms.

The potentials (3.97) and (3.98) are repulsive. Of course, one can add also an attractive branch of the form $V_{\text{int}} \propto x$ so that the resulting potential will have a minimum at some distance x_0. Such a modification will produce changes only for the case of the finite chain. In the present Section, however, we consider an infinite FK chain.

The potentials (3.97) and (3.98) are convex, i.e. $V''_{\text{int}}(x) > 0$ for all $x > 0$, while the potentials (3.99) and (3.100) are nonconvex. The potential (3.99) is concave for $x > a_0 + \beta^{-1} \ln 2$, and the potential (3.100) is concave for the region $a_* - \beta^{-1}/\sqrt{3} < x < a_* + \beta^{-1}/\sqrt{3}$. As a result, at some values of the parameters a_0 (or a_*), β, and V_m the ground state of the system becomes nontrivial, and its excitation spectrum is changed. The FK model with nonconvex potentials (3.99) and (3.100) is considered in Sect. 3.5.2. The exponential potential (3.97) and, especially, the power-law potential (3.98) are long-range potentials. Therefore, the interaction of more neighbors than nearest neighbors should be taken into account. This kind of problems is discussed in Sects. 3.5.3 and 3.5.4, where we show that the interaction of all neighbors changes the system parameters for the exponential interaction (3.97) while in the case of the power-law interaction (3.98) the motion equation of the system becomes nonlocal even in the continuum limit approximation.

3.5.1 Short-Range Interaction

For small anharmonicity of the interatomic interaction, the potential can be expanded into a Taylor series to yield the following motion equation [165]

$$\frac{d^2 u_n}{dt^2} + \sin u_n = V''_{\text{int}}(a_s)\,(u_{n+1} + u_{n-1} - 2u_n) \left[1 + \frac{V'''_{\text{int}}(a_s)}{2V''_{\text{int}}(a_s)} (u_{n+1} - u_{n-1}) \right].$$

Neglecting the discreteness effects, i.e. using the continuum limit approximation, this equation can be reduced to the form,

$$\frac{\partial^2 u}{\partial t^2} - d^2 \left[1 + \alpha d \left(\frac{\partial u}{\partial x} \right) \right] \frac{\partial^2 u}{\partial x^2} + \sin u = 0, \tag{3.101}$$

where we have introduced the parameter d defined as $d = a_s \sqrt{g}$, $g = V''_{\text{int}}(a_s)$, and the dimensionless anharmonicity parameter α,

$$\alpha = \left(\frac{a_s}{d} \right) \frac{V'''_{\text{int}}(a_s)}{V''_{\text{int}}(a_s)}. \tag{3.102}$$

Anharmonicity does not change the spectrum of linear excitations (phonons) of the chain. However, a kink solution of Eq. (3.101) differs from that in the harmonic FK chain. At small α a stationary kink solution can be found by the perturbation theory [27],

$$u_k(z) = u_k^{(SG)}(z) + \alpha u_\alpha(z), \tag{3.103}$$

where

$$u_\alpha(z) = -\frac{4}{3}\tan^{-1}(\sinh z)\,\mathrm{sech} z \tag{3.104}$$

with $z = x/d$. A simple analysis shows that anharmonicity of interatomic interaction destroys the symmetry between a kink and an antikink because according to Eq. (3.103), the correction u_α is independent on the kink topological charge σ. This means that the effective kink width changes by an amount of $\sigma\alpha(\pi d/3)$, i.e.

$$d_{\text{eff}} = d\,(1 - \pi\sigma\alpha/3)\,. \tag{3.105}$$

This leads to the corresponding changed in other parameters characterizing the properties of the kink and antikink, e.g., the effective mass, $m_\sigma \approx m^{(SG)}\,(1 + \pi\sigma\alpha/6)$, and the amplitude of the PN potential which may be estimated as

$$E_{PN} \approx E_{PN}^{(SG)}(g) - \frac{2\pi}{3}\sigma\alpha g\,\frac{dE_{PN}^{(SG)}(g)}{dg}, \tag{3.106}$$

where the function $E_{PN}^{(SG)}(g)$ is defined by Eq. (3.8). We note also that the energy of kink-antikink pair creation is given by

$$\varepsilon_{\text{pair}} = \varepsilon_{\text{pair}}^{(SG)}\left(1 - \frac{4}{27}\alpha^2\right), \quad \varepsilon_{\text{pair}}^{(SG)} = 2\varepsilon_k^{(SG)}. \tag{3.107}$$

Such a symmetry breaking between the kink and antikink was found by Milchev and Markov [262] (see also Refs. [263, 264]). The change of the kink shape may result in the appearance of low-frequency (or high-frequency) shape mode of the kink, as has been shown by Zhang [265].

For the exponential interatomic potential (3.97) the parameters g and α are calculated to be $g = V_0\beta^2$, $\alpha = -\beta a_s/d$. If the anharmonicity parameter α is negative, the effective width of the kink ($\sigma = +1$) is larger, while that for an antikink ($\sigma = -1$) is lower than for the SG kink. This phenomenon has a simple physical interpretation. Indeed, effective interaction forces for a kink (i.e. in the region of local contraction of a chain) exceed those for an antikink (in the region of local extension of a chain). Because of that, at the same value of the system parameters, V_0 and β, a kink, as compared with an antikink, is characterized by lower values of the effective mass and Peierls-Nabarro barrier. These qualitative consideration is substantiated by Fig. 3.16 adopted from the paper by Braun *et al.* [165] which presents the results of calculation of the dependencies $E_{PN}(g)$ and $m(g)$ for the FK model when the interaction between only nearest neighboring atoms is taken into account, but, in contrast to the standard FK model, the interaction potential is exponential as in Eq. (3.97). It can be seen that a "splitting" of the curves in Fig. 3.16 is larger for lager values of β.

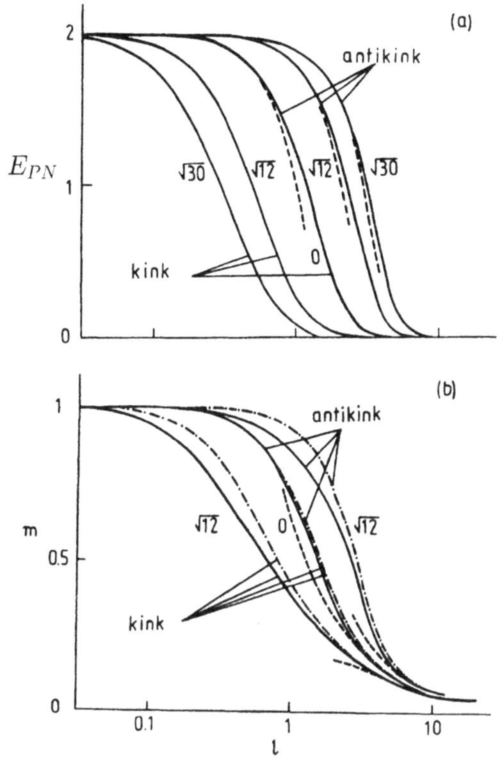

Fig. 3.16. (a) Amplitude of the PN potential and (b) effective mass for kink and antikink as functions of the parameter $l = \pi\sqrt{g}$ for the case when the nearest neighbors interact via exponential forces (3.97) at various values of the anharmonicity parameter β: $\beta = 0$ (the classical FK model), $\beta = \sqrt{12}/2\pi$, and $\beta = \sqrt{30}/2\pi$. Dashed curves show analytical asymptotics [165].

Equation (3.105) indicates that the width of an antikink vanishes with increasing of the anharmonicity parameter β. In order to analyze such an effect in more details, Milchev and Markov [262] applied the operator relation, $u_{n\pm1} = \exp(\pm a_s \nabla) u_n$ to the discrete version of the FK model with the exponential interaction and they obtained the operator equation

$$2g \exp\{-\beta[\sinh(a_s \nabla) u\} \sin\{\beta[\cosh(a_s \nabla) - 1]u\} = \beta \sin u. \qquad (3.108)$$

Keeping the lowest-order derivatives in Eq. (3.108), it can be reduced to [221]

$$d^2 \frac{d^2 u}{dx^2} \exp\left(\alpha d \frac{du}{dx}\right) = \sin u. \qquad (3.109)$$

Integration of Eq. (3.109) yields

$$1-\left(1-\alpha d\frac{du}{dx}\right)\exp\left(\alpha d\frac{du}{dx}\right)=\frac{\beta^2}{g}(C-\cos u), \tag{3.110}$$

where C is a constant. The value $C = 1$ corresponds to a separatrix curve on the phase plane (u_x, u), which connects the points $u = 0$ and $u_x = 0$. This homoclinic trajectory corresponds to a kink (antikink) solution of Eq. (3.109). Substituting $u = \pi$ into Eq. (3.110), we find

$$\left(1-\alpha d\frac{du}{dx}\right)\exp\left(\alpha d\frac{du}{dx}\right)=1-\frac{2\beta^2}{g}, \tag{3.111}$$

which determines the value u_x at the kink center. It is easy to see that Eq. (3.111) has no solution for antikink ($u_x > 0$) for $\beta > \beta_{\rm cr} = \sqrt{g/2}$, i.e. if $\alpha < \alpha_{\rm cr} = -1/\sqrt{2}$. In this case the separatrix corresponding to an antikink (an extra hole) is discontinuous (i.e., $u_x \to \infty$). The latter means that the exponential potential (3.97) cannot withstand the chain extension, and the chain should break into two disconnected (semi-infinite) parts [221]. The analysis made by Milchev and Mazzucehelli [266] has shown that the antikink's effective width tends to zero as $d_{\rm eff} = d\sqrt{1-2\alpha^2}$ when $\alpha \to \alpha_{\rm cr}$. The energy of the antikink-antikink repulsion, $v_{k\bar{k}}(R) \propto \exp(-R/d_{\rm eff})$, also vanishes for $\alpha < \alpha_{\rm cr}$. Thus, for large enough anharmonicity of the interatomic potential, antikinks may come closely to each other creating a cluster of extra holes in the chain, and this explains the effect of the chain "rupture". Such an effect may be observed in collisions of kinks and antikinks as well as in kink interactions with inhomogeneities (such as interfaces) [267].

Of course, the continuum limit approximation used above breaks down at $\alpha \to \alpha_{\rm cr}$ even in the case of $g \to \infty$. The predicted "rupture" of the atomic chain is an artifact caused by nonapplicability of the continuum limit approximation; such a "rupture" indicates only that the effective width of an antikink becomes *smaller* than the lattice spacing a_s, and the energy of disorder of a regular chain of antikinks is rather small. The real rupture of the chain is possible only for nonconvex interatomic potentials such as the Morse potential (3.99) (see Chap. 5 below).

In the case when $\alpha \sim \alpha_{\rm cr}$, the antikink's parameters can be calculated with the help of a weak-bond approximation [155, 165]. For $\alpha < \alpha_{\rm cr}$ the amplitude of the PN potential for the antikink tends to the value of the substrate potential amplitude, $\varepsilon_s = 2$ (see Fig. 3.16), so that an antikink cannot move freely along the chain and it is strongly pinned at a PN potential well. Otherwise, a kink (a local contraction of the chain) propagates along the anharmonic FK chain more freely than along the harmonic one, because $d_{\rm eff}(\sigma = +1) > d$ and $E_{PN}(\sigma = +1) < E_{PN}^{(SG)}$.

Apart the kinks and antikinks, an anharmonic chain supports the so-called *supersonic shock waves*. To show this, let us neglect the substrate potential, then the chain of atoms interacting via the exponential forces coincides with the well-known Toda lattice [268]–[270]. The Toda soliton has the following shape,

$$u_n(t) = m a_s + \frac{1}{\beta} \ln \left[\frac{1 + \exp(-2\kappa)\phi_n(t)}{1 + \phi_n(t)} \right], \tag{3.112}$$

where

$$\phi_n(t) = \exp(z_n/d_{\text{eff}}), \quad z_n = n a_s - vt. \tag{3.113}$$

Soliton in the Toda lattice propagates with the velocity $v > c$, and it is characterized by the effective width, $d_{\text{eff}} = a_s/2\kappa$, mass, $m = 1/\beta d_{\text{eff}}$, momentum, $p = mv$, and energy, $\varepsilon_{\text{Toda}} = 2V_0 (\sinh\kappa \cosh\kappa - \kappa)$. The parameter $\kappa = \kappa(v)$ used above is determined by the equation $(\sinh\kappa)/\kappa = v/c$ (recall that we use the units where $c = d = a_s\sqrt{g}$). The Toda soliton is a kink-like excitation which carries a jump of the atomic displacements equal to $\Delta u = -m a_s$ propagating along the chain. Such an excitation is *dynamical* and it cannot be static similar to the topological kink of the FK model. However, one may suppose that a Toda-like soliton can propagate in the FK chain for some finite time. Owing to the periodic substrate potential, a travelling soliton will lose its kinetic energy decreasing its velocity. When the value of v will approach the sound speed c, the Toda soliton should decay into FK kinks and radiation. The total number of the FK kinks may be estimated from the viewpoint of topological constrains, $n_k = [m - 1/2]$, where $[\ldots]$ stands for an integer part.

Indeed, the supersonic motion of topological solitons has been described by Savin [271] for the ϕ^4-model with anharmonic interatomic interaction. Savin has found that for certain kink velocities, when the jump in the atomic displacements Δu matches exactly the width of the double-well potential, $\Delta u = -a_s$, the supersonic kink moves almost without radiation of phonons. This supersonic kink may be considered as n acoustic Boussinesq's solitons bounded together by the topological constrain due to the external substrate potential, when the sum of their amplitudes exactly coincides with the width of the barrier of the ϕ^4 potential. The maximum number $n_{\max}$ of acoustic solitons that can be coupled together in one kink, depends on the anharmonicity parameter of the interaction and increases with it, so that at low anharmonicity there is only one (supersonic) kink velocity ($n_{\max} = 1$), but at larger anharmonicity the number of the allowed discrete values for the velocity increases.

Later similar results were obtained by Zolotaryuk *et al.* [272] with the help of the pseudo-spectral method (see Sect. 3.2.4) for the FK chain subjected to the sinusoidal substrate potential, when the atoms interact via the exponential potential (3.97). Again, it was found that the supersonic 2π-kink has a hierarchy of states, $n = 1, 2, \ldots, n_{\max}$ (each state corresponds to the kink travelling with its own preferred velocity only), which may be considered as n acoustic Toda solitons bounded together, when the sum of their amplitudes coincides with the period of the substrate potential a_s. Moreover, multiple supersonic $2Q\pi$-kinks with the topological charge $Q > 1$ were also found, and they may exhibit a hierarchy of states as well. All these solutions were found to be asymptotically unstable, i.e. if one modifies, e.g., the distance between acoustic sub-solitons in the n-state of the kink, then the kink on the oscillating background appears. However, starting from the kink shape obtained by the pseudo-spectral method and then carrying out molecular dynamics simulation, the authors showed that these kinks are dynamically stable. Finally, let us emphasis that all these solutions exist for kinks (local compression of the chain) only.

It is interesting to note that for a special form of the interatomic potential,

$$V_{\text{int}}(x) = \frac{g}{2}(x - a_s)^2 \left[1 + \gamma (x - a_s)^2\right], \tag{3.114}$$

where the anharmonicity parameter is of certain value, $\gamma = 1/48$, the motion equation of the FK model in the continuum limit approximation has an exact kink solution of the standard form,

$$u_k(x,t) = 4 \tan^{-1} \exp\left\{-\frac{(x - vt)}{d_{\text{eff}}}\right\}, \tag{3.115}$$

which can propagate with an *arbitrary* velocity v [273, 274]. The effective width of the kink (3.115) is given by the formula

$$d_{\text{eff}} = \frac{a_s}{\sqrt{6}} \left\{\left[\left(\frac{v}{c}\right)^2 - 1\right] + \sqrt{\left[\left(\frac{v}{c}\right)^2 - 1\right]^2 + \frac{1}{3}\left(\frac{a_s}{d}\right)^2}\right\}^{-1/2}, \tag{3.116}$$

so that in the limit $|v| \ll c$ the kink's width approaches the value given by the standard SG model, $d_{\text{eff}} = d\sqrt{1-(v/c)^2}$, while in the case of the supersonic motion, when $|v| \gg c$, the kink width is given by the expression $d_{\text{eff}} \approx a_s c/\sqrt{12(v^2 - c^2)}$, which looks like the corresponding width of a dynamical Toda soliton.

3.5.2 Nonconvex Interatomic Potentials

In the sections above we have assumed that the interatomic interaction in the chain is described by a convex function, i.e. $V''_{\text{int}}(x) > 0$ for all $x > 0$. The opposite case of the concave potential, i.e. when $V''_{\text{int}}(x) < 0$, is less interesting from the physical point of view because, according to the inequality

$$V_{\text{int}}(a - \Delta a) + V_{\text{int}}(a + \Delta a) - 2V_{\text{int}}(a) < 0, \tag{3.117}$$

all the atoms will come together to one well of the substrate potential.

In the present section we will consider the nonconvex interatomic interaction potential which has an inflection point a_i defined by the equation $V''_{\text{int}}(a_i) = 0$. The FK model with such a potential exhibits complicated properties and a rich nonlinear dynamics due to the existence of two competing length scales, the period of the substrate potential, $a_s = 2\pi$, and the scale which is given by the inflection point a_i at which the strength of the interatomic bond reaches its maximum. For definiteness, let us suppose that the potential $V_{\rm int}$ is convex beyond the inflection point, i.e. $V''_{\rm int} > 0$ for $x < a_i$, and it is concave at larger distances similar to the Morse potential (3.99). The opposite case is reduced to that mentioned if kinks are replaced by antikinks.

The nonconvex potential $V_{\rm int}(x)$ is obviously anharmonic. Thus, when the anharmonicity is large enough, antikinks in the chain may lead to the chain rupture. In contrast to the case of the exponential potential analyzed above, now the rupture is a real breaking of the chain into two independent semi-infinite chains, because of attractive interaction between antikinks [222]. More precisely, a coupling of antikinks can lead to a creation of a cluster which consists of n antikinks and makes the system energy *lower* for $n > n_{\rm cr}$. The critical size of the cluster, $n_{\rm cr}$, can be estimated from the inequality

$$V_{\rm int}(n_{\rm cr}a_s) + V_{\rm int}(2a_s) \geq V_{\rm int}(n_{\rm cr}a_s + a_s) + V_{\rm int}(a_s), \tag{3.118}$$

so that $n_{\rm cr} \approx a_i/a_s$. Thus, the chain's rupture due to increasing of the number of antikinks in the chain has the nucleation character similar to the first-order phase transitions.

Another feature of the FK model with the nonconvex interatomic interaction is the instability of the trivial ground state (GS). Looking at Eq. (3.117), we may expect that the trivial GS becomes unstable provided $a_i < a_s$ when $V''_{\rm int}(a_s) < 0$. Indeed, it is easy to show that for $V''_{\rm int}(a_s) \leq -1/4$ the trivial GS [see Fig. 3.17(a)] becomes unstable, and the chain will be dimerized so that short and long bonds alternate as shown in Fig. 3.17(b) for the FK chain with the Morse interatomic interaction [220, 275]. This phenomenon is due to the fact that the average energy of one long and one short bonds is smaller than the energy of a bond with an intermediate length. Decreasing further the value $V''_{\rm int}(a_s)$, the ground state of the chain may be trimerized for $V''_{\rm int}(a_s) \leq -1/3$ [Fig. 3.17(c)], tetramerized [Fig. 3.17(d)], pentamerized [Fig. 3.17(e)], and so on. The simple linear analysis shows that for

$$V''_{\rm int}(a_s) \leq -\frac{1}{2}\left[1 - \cos\left(\frac{2\pi}{q}\right)\right], \tag{3.119}$$

the trivial GS becomes unstable with respect to creation of a superstructure with the period $a = qa_s$ [Eq. (3.119) is valid for $q > 2$ only].

To find all GS configurations for the FK model with a nonconvex interatomic interaction is a difficult problem because the system of stationary equations, $\partial U/\partial u_n = 0$ (where U is the total potential energy), has usually many solutions, only one of them is the ground state while others correspond to metastable and unstable configurations. Griffits and Chou [276] proposed an algorithm focused directly on the GS and valid for arbitrary potentials $V_{\rm sub}(x)$ and $V_{\rm int}(x)$. According to these authors, the GS of an infinite chain may be found as a solution of the functional eigenvalue equation ($u, u' \in [-a_s/2, +a_s/2]$),

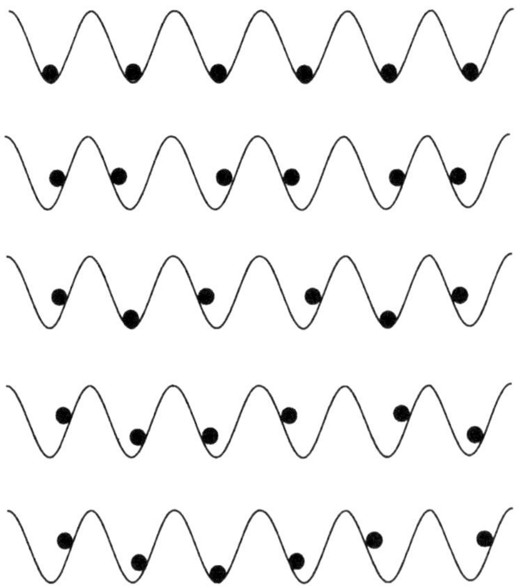

Fig. 3.17. Ground states of the FK model with the Morse interatomic potential (3.99): (a) undistorted chain, (b) dimerized GS, (c) trimerized GS, (d) tetramerized GS, and (e) pentamerized GS [220].

$$\epsilon_0 + \widetilde{V}(u') = V_{\rm sub}(u') + \min_u [V_{\rm int}(u' - u - a_s) + \widetilde{V}(u)]. \qquad (3.120)$$

The function $\widetilde{V}(u)$ is called an effective potential [it has the same period as the primary substrate potential, $\widetilde{V}(u + a_s) = \widetilde{V}(u)$] and the value ϵ_0 is the average energy per particle in the GS. It was shown [276, 277] that the function $\widetilde{V}(u)$ always exists, and the corresponding value ϵ_0 is unique to be given by a solution of Eq. (3.120). One can construct the map

$$u = M(u'), \qquad (3.121)$$

obtained by looking for u which, at a given u', minimizes the r.h.s. of Eq. (3.120). The attraction point of this map, $u_{n+1} = M(u_n)$, generates the corresponding GS configuration, $x_n = na_s + u_n$.

In order to get some physical interpretation of Eqs. (3.120) and (3.121), let us consider a semi-infinite chain of atoms with the edge atom fixed at the position u_0 [278]. Suppose that we let the rest part of the chain to relax freely reaching a minimum energy configuration corresponding to the boundary condition. Then the value of the derivative, $d\widetilde{V}(u_0)/du_0$, gives the value of the force which should be applied to hold the edge atoms at the position at $u = u_0$. Then the location of the n-th atom is given by the function $M^n(u_0)$.

Usually the functional equation (3.120) is solved numerically by using a grid of a hundred (or more) equally spaced points in an interval around the point $u = 0$ and applying the r.h.s. of Eq. (3.120) to the functions defined at these points. The sequence of the iterations $\widetilde{V}^{(n)}$ is stopped when $\widetilde{V}^{(n+1)}$ and $\widetilde{V}^{(n)}$ differ only by a constant ϵ_0 within a chosen accuracy. Note that for a hard-core interatomic potential, Eq. (3.120) has an analytic solution [279].

As a typical example, let us consider, following to Marchand *et al.* [280], the simplified FK model where the sinusoidal on-site potential, $V_{\rm sub}(x)$, is replaced by a sequence of parabolas, $(1-\cos x_n) \to \frac{1}{2}u_n^2$. For the double-well interatomic potential (3.100) with $\beta = 1$, $a_* = a_s + \delta$, and $V_m = \frac{1}{2}K$, the Hamiltonian of the model takes the form

$$H = \sum_n \left[\frac{1}{2}\left(\frac{du_n}{dt}\right)^2 + \frac{1}{2}u_n^2 + \frac{1}{4K}(u_{n+1} - u_n - \delta)^4 - \frac{1}{2K}(u_{n+1} - u_n - \delta)^2 \right]. \tag{3.122}$$

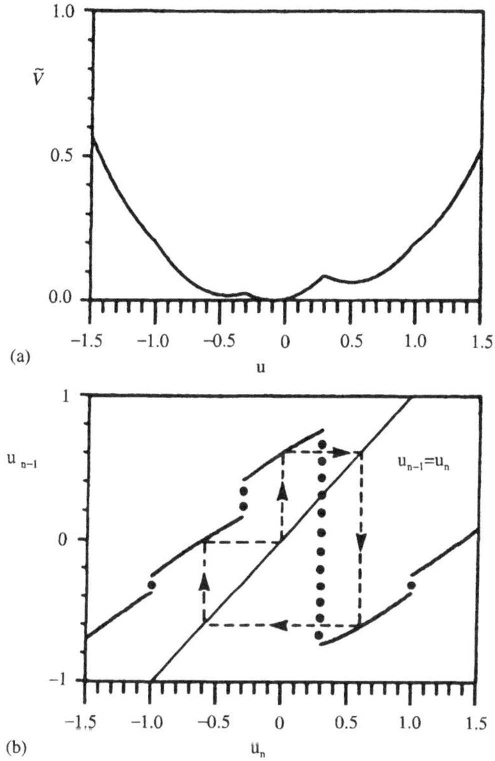

Fig. 3.18. (a) Effective potential $\tilde{V}(u)$ and (b) associated map $u_{n+1} = M(u_n)$ of the model (3.122) with the double-well interatomic potential for $K = 0.5$ and $\delta = 0.33$. Also shown in (b) are discontinuities (dotted lines), the line $u_{n+1} = u_n$, and the limit cycle of the period $q = 3$ [280].

Figure 3.18 shows an example of the effective potential $\widetilde{V}(u)$ and the associated map $M(u)$ obtained for this model [280]. Note that $\widetilde{V}(u)$ is continuous but it has a discontinuous first derivative at the same point where M is discontinuous. This corresponds to a situation when the ground state is "pinned" to the substrate potential, whereas for incommensurate GS configurations, when the chain of atoms can "slide" under zero force, one can expect the functions $\widetilde{V}$ and M to be smooth.

Phase diagram obtained by Marchand *et al.* [280] for the model (3.122) is shown in Fig. 3.19. The different GS configurations are marked by the ratio of two integers, p/q, where q characterizes the period of the modulated structure, $a = qa_s$, and p is the number of long bonds per one elementary cell, $p = \sum_1^q \Theta(u_{n+1} - u_n)$ (here Θ is the Heaviside function, $\Theta = +1$ for $x \geq 0$, and $\Theta = 0$ for $x < 0$). Numerical results suggest that the model (3.122) exhibits a complete Devil's staircase even through a rigorous proof of this statement is not possible within the framework of the effective potential algorithm. First, all the configurations are structurally stable. Second, the phase characterized by a ratio $\nu = (p+r)/(q+s)$ is always found to be between p/q and r/s phases for sufficiently small values of K. Hence, there is an infinite number of phases between any two given phases (including the incommensurate phases characterized by irrational values of ν).

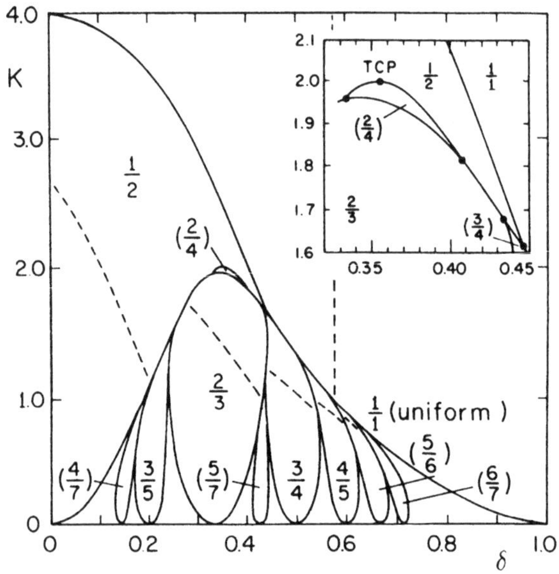

Fig. 3.19. Phase diagram for the model (3.122). The numbers p/q indicates the type and structure of modulated phases. The unlabelled regions contain additional commensurate phases. Inset shows the tricritical point [280].

When the system parameters, i.e. δ or K in Eq. (3.122), are adiabatically varying, the phase transitions between different phases should take place. Numerical simulations [278, 280] show that the transition between the homogeneous (1/1) and any modulated (p/q with $q > 1$) phases is usually a continuous (second-order) phonon-driven transition, while transitions between the modulated ground states, such as $\frac{1}{2} \to \frac{2}{3}$, $\frac{1}{2} \to \frac{3}{4}$, $\frac{2}{3} \to \frac{3}{4}$, etc., are typically first-order transitions, and they take place via creation of kink-type defects with subsequent nucleation of the defects.

It is worth to mention the work of Byrne and Miller [279] who considered the FK model with nonconvex Lennard-Jones and double-well interatomic potentials, and the studies by Takeno and Homma [281] and Yokoi *et al.* [282] where a sinusoidal interatomic potential was analyzed. Marianer and Bishop [283] investigated the FK model for which, in addition to the double-well interparticle interaction with $a_* = a_s$, the strain gradients are taken into account via next-nearest neighbor interactions, so that the system Hamiltonian is taken to be

$$H = \sum_n \left[\frac{1}{2} \left(\frac{du_n}{dt} \right)^2 + (1 - \cos u_n) + \frac{1}{2} V_m \beta^4 (u_{n+1} - u_n)^4 \right.$$
$$\left. - V_m \beta^2 (u_{n+1} - u_n)^2 + \frac{\gamma}{2} (u_{n+1} + u_{n-1} - 2u_n)^2 \right]. \quad (3.123)$$

The model (3.123) can be useful in describing twinning in martensite materials [284, 285]. To apply the effective potential method, Marianer and Floria [278] transformed the Hamiltonian (3.123) into that with only nearest neighbor interactions but with vector variables defined as $v_n \equiv \{u_{2n}, u_{2n+1}\}$. As a result, the effective potential $\widetilde{V}(v_n)$ becomes two-dimensional. The calculated phase diagram consists of various modulated commensurate and incommensurate GS structures.

Let us now briefly discuss an excitation spectrum of the modulated GS. First, we should note that the GS with a complex elementary cell (i.e. $q > 1$) may have more than one phonon branch. Second, the modulated GS with the period $a = qa_s$ is q-times degenerated because the shift of all atoms in the chain on the distance which is integer multiplier of the substrate period, $\Delta x = ja_s$, $j = 1, \ldots, q-1$, will transform a GS to a nonequivalent one. Thus, the situation is quite similar to that which emerges in the case of a multiple-well substrate potential (see Sect. 3.3.4 above). A standard 2π-kink splits into q independent subkinks undergoing repulsive interactions. One of those subkinks (in fact, the largest one) is a SG-like kink which describes a configuration where the atoms occupy neighboring minima of the substrate potential. The other $(q-1)$ subkinks are confined to be in an elementary cell of the substrate; sometimes they are called "interface kinks" or "domain walls". Note, however, that all types of subkinks are topologically stable. Of course, the subkinks as well as their interactions should satisfy some topological constrains.

It is clear that the dynamics of kinks for the modulated GS is much more complicated than that of the original FK model. As a simplest example, let us consider here the FK model with a double-well interatomic potential (3.100) (with $a_* = a_s$ or $\delta = 0$) following the paper by Marianer *et al.* [286]. As can be seen from Fig. 3.19, the GS of the chain is dimerized if $V_m \beta^2 > 1/8$, i.e. if $V''_{\rm int}(a_s) \leq -1/4$. This GS is two-times degenerated, and the first GS describes the "short-long" spring length configuration with the atomic coordinates $x_n = na_s + u_n$, $u_n = (-1)^n b$, where $b \approx (1/2\beta)(1 - 1/8V_m\beta^2)^{1/2}$ for $\beta \gg 1$. The second GS corresponds to the "long-short" length configuration with $u_n = (-1)^{n+1} b$. To consider a subkink ("interface") which links these two GSs, we introduce a dimensionless variable $v_n = (-1)^n u_n / b$ and use the continuum limit approximation, $v_n \to v(x)$ and $v_{n\pm1} \to v \pm a_s v_x$, so that the Hamiltonian is reduced to the form [286],

$$H \approx b^2 \int \frac{dx}{a_s} \left\{ \frac{1}{2} \left(\frac{\partial v}{\partial t} \right)^2 + \frac{1}{2} C(v) \left(\frac{\partial v}{\partial x} \right)^2 + A \left(v^2 - 1 \right)^2 \right\}, \tag{3.124}$$

where

$$A = 2 \left(V_m \beta^2 - \frac{1}{8} \right) \quad \text{and} \quad C(v) = 2 V_m \beta^2 \left[1 - 3v^2 \left(1 - \frac{1}{8 V_m \beta^2} \right) \right].$$

Hamiltonian (3.124) corresponds to the ϕ^4-model with an effective spring constant $C(v)$ which depends on the variable $v(x,t)$. A kink of the ϕ^4-model (3.124), $v(x) \propto \tanh(x/d_{\text{eff}})$, has an effective width

$$d_{\text{eff}} = b\sqrt{2} \left(\frac{1 - 12 b^2 \beta^2}{1 - 1/8 V_m \beta^2} \right)^{1/2}. \tag{3.125}$$

The kink's width becomes infinite at $V_m \beta^2 \to 1/8$ (when the dimerized GS disappears) and it vanishes when $b \to 1/2\beta\sqrt{3}$, or for $V_m \beta^2 \to \frac{1}{8} + \frac{1}{16} = \frac{3}{16}$. The latter case is similar to the case of an antikink in the FK model with exponential interatomic interaction when the nonlinearity parameter α is less than the critical value α_{cr} (see Sect. 3.5.1). Analogously to this, continuum limit approximation breaks down and for $V_m \beta^2 \geq 3/16$ the subkink becomes pinned by the substrate potential.

General method to analyze the kinks excited on a modulated GS requires straightforward but rather lengthy calculations. Therefore, we outline here only the main idea of this approach not going into specific details. Atomic coordinates are given by the relation

$$x_n = n a_s + X_n + u_n, \tag{3.126}$$

where X_n corresponds to the kink coordinate (for an "interface" we take $X_n = 0$) and u_n ($|u_n| < a_s/2$) describes the modulation of the GS. Displacements u_n are expanded into a Fourier series,

$$u_n(t) = v_n(t) e^{iQn} + v_n^*(t) e^{-iQn} + \text{h.h.}, \tag{3.127}$$

h.h. stands for higher harmonics, with some wavenumber $Q = 2\pi p/q$ (p and q are integers) characterizing the modulated GS. (Note that if we restrict ourselves only by the first harmonic terms in Eq. (3.127), the approximate ground state may be infinitely degenerated for $q > 2$, and associated kinks will not be topologically stable). Then the expressions (3.126) and (3.127) are substituted into the Hamiltonian of the model, the periodic substrate potential is changed to be

$$V_{\text{sub}}(x_n) = 1 - \cos u_n \cos X_n + \sin u_n \sin X_n, \tag{3.128}$$

and the functions $\cos u_n$ and $\sin u_n$ are expanded into Taylor series in small u_n. The resulting Hamiltonian can be then considered in the continuum limit approximation in a straightforward way using, for example, the methods described in details by Slot and Janssen [57, 287] for the frustrated ϕ^4-model. Namely, the variables X_n and v_n are assumed to be slowly varying on the scale of order of the lattice spacing a_s, the latter assumption allows us to use the continuum limit expansions, $na_s \to x$, $S_n \to S(x,t)$, $S_{n\pm1} \to S \pm a_s S_x$, where $S \sim \mathcal{O}(1)$, $S_x = \partial S/\partial x \sim \mathcal{O}(\epsilon)$, $S_x^2, S_{xx} \sim \mathcal{O}(\epsilon^2)$, etc., with $\epsilon \ll 1$, and S_n stands for X_n or v_n. Substituting these expansions into the Hamiltonian, neglecting fast varying terms, and making some transformations, we can derive an approximate Hamiltonian which yields an effective motion equation which has to be solved together with appropriate boundary conditions. However, the procedure described above is rather lengthy, so that direct numerical simulations with a discrete FK model is usually more straightforward.

To conclude this section, we would like to mention that the FK model with Morse or Lennard-Jones interatomic potentials has in fact *three* characteristic lengths, a_s, a_i, and a_0. The additional spatial scale, a_0, corresponds to a minimum of the interaction potential. For the boundary conditions used above (i.e. the chain's ends are *fixed* at infinities) this fact does not change the results provided $a_0 \geq a_{FM}$, where the value a_{FM} ($a_{FM} < a_s$) introduced by Frank and van der Merwe [31, 32] describes the situation when the ground state of the chain with *free* ends contains kinks with a finite density. For $a_0 < a_{FM}$ the infinite chain (with fixed ends) will rupture into two semi-infinite chains because this effect leads to a lower system energy in the case when $V_{\text{int}}(x) \to 0$ at $x \to \infty$. However, such a rupture is not connected with "extra" antikinks as in the case analyzed above.

3.5.3 Kac-Baker Interaction

Now we extend the classical FK model, assuming that not only nearest neighboring atoms interact in the chain. We consider the interaction potential $V_{\text{int}}(x)$ which remains convex and falls fast enough for $|x| \to \infty$ (e.g., as in the case of exponentially decaying potential). In fact, the dynamics of the FK model in this case is similar to that for the model when the nearest neighbors interact only, but it is characterized by the *renormalized* coupling parameter,

$$g \to g_{\text{eff}} = \sum_{j=1}^{\infty} j^2 V''_{\text{int}}(ja_s). \tag{3.129}$$

As an example, we take the exponential interaction (3.97), for which Eq. (3.129) yields

$$g_{\text{eff}} = g\frac{(1+S)}{(1-S)^3}, \tag{3.130}$$

where $g = V_0\beta^2$ is defined above, and

$$S = e^{-\beta a_s}. \tag{3.131}$$

For the long-range interatomic potential, when $\beta a_s \ll 1$, from Eq. (3.130) it follows

$$g_{\text{eff}} \approx \frac{2g}{(\beta a_s)^3} \gg g. \tag{3.132}$$

For an exponential interatomic interaction the results mentioned above may be simply proved with the help of the method firstly proposed by Sarker and Krumhansl [288] (see also Refs. [165],[289]–[291]). Following this procedure, we expand the interaction potential (3.97) into a Taylor series keeping the cubic terms, for interaction of the nearest-neighbors, and quadratic terms, for interaction of other atoms. In this case the interaction energy is

$$H_{\text{int}} = \frac{1}{2}\sum_{i\neq j} V_{\text{int}}(x_i - x_j) \approx \frac{A}{6}\sum_i (u_i - u_{i-1})^3 + J\frac{(1-S)}{4S}\sum_{i\neq j} S^{|i-j|}(u_i - u_j)^2, \tag{3.133}$$

where we have introduced the following notations, $A = \alpha(d/a_s)^3$, $J = (d/a_s)^2/(1-S)$, and the parameters $d = a_s\sqrt{g}$ and $\alpha = -\beta a_s/d$ are defined above. Thus, Eq. (3.133) describes a one-dimensional chain of atoms interacting via a pair potential of the Kac-Baker form [292, 293]. The equations of motion which correspond to the Hamiltonian (3.133), is

$$\frac{d^2 u_i}{dt^2} + \sin u_i + \frac{1}{2}A\left[(u_i - u_{i-1})^2 - (u_{i+1} - u_i)^2\right] + 2Ju_i = L_i, \tag{3.134}$$

where the auxiliary quantity

$$L_i = J\,\frac{(1-S)}{S}\sum_{j=-\infty (j\neq 0)}^{+\infty} S^{|j|} u_{i+j} \tag{3.135}$$

satisfies the following recurrence relation [288],

$$\left(S + \frac{1}{S}\right) L_i = L_{i+1} + L_{i-1} + J\,\frac{(1-S)}{S}\,(u_{i+1} + u_{i-1} - 2Su_i), \tag{3.136}$$

which allows to reduce Eqs. (3.134) to (3.136) to an effective problem which includes only the interaction of the nearest-neighbor atoms.

In the continuum limit approximation, Eqs. (3.134) to (3.136) can be presented in the form,

$$\frac{\partial^2 u}{\partial t^2} - d_{\text{eff}}^2\,\frac{\partial^2 u}{\partial x^2} + \sin u - \alpha d^3\,\frac{\partial u}{\partial x}\frac{\partial^2 u}{\partial x^2} = \frac{a_s^2 S}{(1-S)^2} f(u), \tag{3.137}$$

where

$$f(u) = \frac{\partial^4 u}{\partial^2 x\,\partial t^2} - \left(\frac{\partial u}{\partial x}\right)^2 \sin u - \frac{\partial^2 u}{\partial x^2}(1 - \cos u), \tag{3.138}$$

and

$$d_{\text{eff}}^2 \equiv d^2 \frac{(1+S+S/J)}{(1-S)^3}. \tag{3.139}$$

Using the dimensionless coordinate, $x \to x/d_{\text{eff}}$, we derive the equation

$$\frac{\partial^2 u}{\partial t^2} - \left(1 + \alpha_{\text{eff}} \frac{\partial u}{\partial x}\right) \frac{\partial^2 u}{\partial x^2} + \sin u = \epsilon f(u), \tag{3.140}$$

where

$$\alpha_{\text{eff}} \equiv \alpha \left(\frac{d}{d_{\text{eff}}}\right)^3 \quad \text{and} \quad \epsilon = \frac{S}{S + J(1+S)}. \tag{3.141}$$

In the case $d \gg a_s$ we have $J \gg 1$ and $\epsilon \ll 1$; therefore, the perturbation $\epsilon f(u)$ in Eq. (3.140) can be neglected. Consequently, a long-range exponential character of the atomic interaction, as compared with the considered-above short-range interactions, results only in an effective renormalization of the kink parameters, e.g. the kink's width increases ($d \to d_{\text{eff}} > d$), while its nonlinearity parameter decreases ($\alpha \to \alpha_{\text{eff}} < \alpha$). We would like to note also that interaction between two kinks is always more extended than that for the direct interaction of two extra atoms via the potential (3.97) because $d_{\text{eff}}^{-1} \approx \beta\,(\beta_s/2g)^{1/2} \ll \beta$ for $\beta a_s \ll 1$ and $g \gg 1$.

When the interatomic interaction extends over the whole chain, $\beta \to 0$ so that $S \to 1$, the effective kink width increases indefinitely according to Eq. (3.130). Thus, the Peierls-Nabarro barrier should vanish in this limit. However, Mingaleev *et al.* [294] have shown that this is true in the case of $J > 1/2$ only. Using the technique of pseudo-differential operators (e.g., see Eqs. (3.108)–(3.111) above in Sect. 3.5.1; in fact, this technique is equivalent to Rosenau's approach [7] described in detail in Sect. 1.1), Mingaleev *et al.* [294] studied the model (3.133) for the $A = 0$ case, i.e. when the anharmonicity of the interaction is ignored, and found the implicit analytical form of the kink as well as its energy in the continuum limit approximation. They showed that at small values of the parameter J, $J < J_c \equiv S/(S+1)$, the kink becomes "S-shaped" (multivalued), i.e. the slope of the kink shape at its center becomes vertical at $J = J_c$ and takes positive values at lower J. This indicates that the continuum limit approximation breaks down at $J \leq J_c$. Analogously to the case of anharmonic interaction described above in Sect. 3.5, in this case the kink width becomes smaller than the lattice constant [see Eq. (3.139)], so that such a kink may be called as "intrinsically localized" one. The PN barrier, however, remains finite even in the limit $\beta \to 0$ if $J < J_c$. It is interesting that if one ignores the nonphysical part of the S-kink and replace it with a vertical slope, then the resulting form of the kink agrees well with the shape obtained numerically for the discrete chain, provided β is small enough. Indeed, in this case the kink is almost a point-like object and consist of its tails only, which can be well described with the help of the continuum limit approximation.

Another interesting effect found by Mingaleev *et al.* [294], is that the long-range interatomic interaction strongly enhances the creation of kink's shape modes (see Sect. 3.3.2) at small values of β. For example, while the standard ($\beta = \infty$) FK model demonstrates the existence of only one shape mode for a narrow interval of values of the parameter J around $J \approx 0.5$, the long-range interaction with, for example, the exponent $\beta = 0.2\,\pi$ gives seven internal localized modes, and their number grows indefinitely in the limit $\alpha \to 0$. Finally, Mingaleev *et al.* [294] pointed out that due to break of the Lorentz invariance in the model with nonharmonic interaction, the *moving* kink should radiate phonons with the wavevector proportional to kink's velocity even in the continuum limit approximation, when the discreteness effects are ignored and the PN potential is absent.

At last, from Eqs. (3.137) to (3.139) it follows that the long-range character of the interatomic interaction changes the dispersion relation for phonons. Indeed, for the wave numbers $|\kappa| \ll \pi$, the dispersion relation can be obtained in the following form,

$$\omega_{\rm ph}^2(\kappa) = \frac{\omega_{\rm min}^2 + g_{\rm eff}\kappa^2}{1 + S(1-S)^{-2}\kappa^2}, \tag{3.142}$$

where $g_{\rm eff} = (d_{\rm eff}/a_s)^2$ and $\omega_{\rm min} = 1$. We would like to mention also that the double SG model with the Kac-Baker interactions was considered by Croitoru [290] with qualitatively similar conclusions.

3.5.4 Long-Range Interaction

The interaction potentials discussed up to now allow a reduction of the motion equation in the continuum limit approximation to a SG-type equation with local interaction. In contrast to that case, the motion equation for the FK model with a power-law interatomic interaction,

$$V_{\rm int}(x) = V_0 \left(\frac{a_s}{x}\right)^n, \tag{3.143}$$

can be reduced to a *nonlocal integro-differential* SG equation [34, 165, 295]. To derive such an equation, let we use continuum limit approximation, $j \to y = ja_s$, $\sum_j \to \int dy/a_s$, and change the variable, $y \to x = y + u(y)$, so that approximately, $dx = (1 + u_y)\,dy \approx (1 + u_x)\,dy$ and $dy \approx (1 - u_x)\,dx$. Then the interaction energy takes the form

$$H_{\rm int} = \frac{1}{a_s^2} \iint dx dx' \frac{\partial u(x)}{\partial x} \frac{\partial u(x')}{\partial x'} V_{\rm int}(x - x'). \tag{3.144}$$

The result (3.144) has a simple physical meaning, since the value $\rho(x) \equiv -u_x(x)/a_s$ is the density of the atomic excess (with respect to the initial commensurable structure). For a local potential of the atomic interactions,

$$V_{\rm int}(x) = a\,\delta(x)\,d^2, \tag{3.145}$$

Eq. (3.144) takes the form of the standard SG equation. For the nonlocal potential (3.143) the integral (3.144) diverges provided $(x - x') \to 0$, and, as a result, one should make a cut of the integration interval at some distance $a^* \approx a_s$ [165]. Introducing dimensionless variables, we obtain the Hamiltonian for a nonlocal SG model,

$$H = H_{\text{local}} + \frac{A}{2\delta} \int dx \frac{\partial u}{\partial x} \int_{\delta}^{\infty} \frac{dx'}{(x')^n} \left[\frac{\partial u}{\partial x}(x + x') + \frac{\partial u}{\partial x}(x - x') \right], \quad (3.146)$$

where $\delta = a_s/d$, $A = V_0\, \delta^{n+1}/(2\pi)^2$. If the potential $V_{\text{int}}(x)$ is short-range, then, without the last term in Eq. (3.146), the expression for the energy should take the form corresponding to the SG model, for which $d^2 = a_s^2 V''_{\text{int}}(a_s) = V_0 n(n+1)$. We use this relation to reduce the number of independent parameters and express V_0 of the potential (3.143) in terms of the parameter d. As a result, we obtain $A = V_0 \delta^{n-1}/n(n+1)$ and the Hamiltonian (3.146) is a function of only two parameters, δ and n. The motion equation corresponding to the Hamiltonian (3.146) has the form

$$\frac{\partial^2 u}{\partial t^2} - \frac{\partial^2 u}{\partial x^2} + V'_{\text{sub}}(u) = A \frac{\partial}{\partial x} \int_{\delta}^{\infty} \frac{dx'}{(x')^n} [u_x(x + x') + u_x(x - x')], \quad (3.147)$$

and it describes the dynamics of a chain with a nonlocal interaction.

From Eq. (3.147), one can see that the core structure of a kink (i.e. its shape at $|x - X| < d$) is determined mainly by local terms of the motion equation [295]. Therefore, "local" characteristics of a kink, such as its effective mass or amplitude of the PN potential, will not differ significantly from those calculated for the local FK model with the renormalized elastic constant $g_{\text{eff}} = \sum_{j=1}^{\infty} V''_{\text{int}}(j a_s) = g S_{n+2}$, $g = n(n+1)V_0/a_s^2$, where $S_m = \sum_{j=1}^{\infty} j^{-m}$ (e.g., $S_3 \approx 1.202$, $S_5 \approx 1.037$), and the anharmonicity of the interaction is determined by the parameter

$$\alpha_{\text{eff}} = \alpha S_{n+2}^{-3/2}, \quad \alpha = -(n+2)/d. \quad (3.148)$$

Indeed, the dependencies $E_{PN}(l)$ and $m(l)$ (where $l \equiv \pi\sqrt{g}$) calculated by Braun *et al.* [165] for the Coulomb ($n = 1$) and dipole-dipole ($n = 3$) atomic interactions shown in Fig. 3.20 are qualitatively similar to those in Fig. 3.16 obtained for the local anharmonic interaction in the FK model. The difference between the parameters E_{PN} and m for a kink and an antikink at the same value of the parameter g for the dipole interaction, is much larger than for the Coulomb interaction, which is accounted for by larger anharmonicity of the dipole potential, according to Eq. (3.148). We note that the amplitude of the PN potential for the FK model with the Coulomb interaction ($n = 1$) at some particular values of the system parameters was calculated by Wang and Pickett [296]. Braun *et al.* [165] have calculated also the kink's parameters for the power-law FK model with nonsinusoidal substrate potential. The dependencies (see Fig. 3.21) are similar to those for the local FK model described in Sect. 3.3.2.

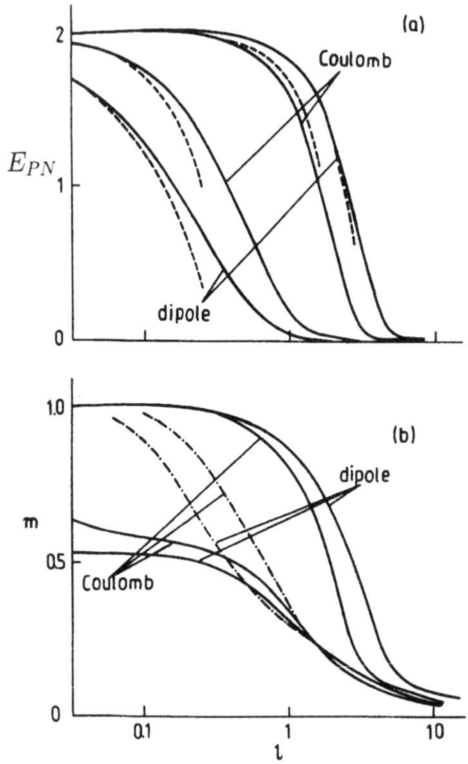

Fig. 3.20. (a) The PN energy $E_{PN}(l)$ and (b) effective mass $m(l)$ for a kink and antikink in the presence of the Coulomb and dipole interatomic interaction. The parameter $l = \frac{1}{2}\sqrt{V_0 n(n+1)}$ for $n = 1$ or 3 is defined by the kink's width d. Dashed curves show analytical asymptotics [165].

In spite of the fact that local characteristics of the kink are similar to those for the classical FK model, asymptotics of the kink of Eq. (3.147), which are determined by the last term, are very different from those of the SG kink, and they are power-like [34, 295]. Indeed, linearizing Eq. (3.147) near the asymptotic value $u(\infty)$ and integrating by parts, we obtain

$$|u(x) - u(\infty)| \approx \frac{2\pi n A}{\omega_{\min}^2 |x|^{n+1}}, \quad x \to \pm\infty, \tag{3.149}$$

where $\omega_{\min}^2 = V''_{\rm sub}(0)$. It is clear that for the power-law interatomic forces, the interaction between kinks (i.e., between "extra" atoms or holes in the chain) should also be power-like. This has been shown by Kosevich and Kovalev [34], for crowdions in a bulk of a crystal, by Gordon and Villain [297], Lyuksyutov [298] and Talapov [299], for elastic interaction of atoms adsorbed on a crystal surface, and by Haldane and Villain [300] and Pokrovsky and Virosztek [295], for dipole-dipole interaction of adatoms. To show this directly, let us consider the chain with two kinks of topological charges σ_1 and σ_2, respectively, which are separated by some distance x_0. In the zero-order approximation, the solution of Eq. (3.147) can be presented as a superposition of two SG kinks,

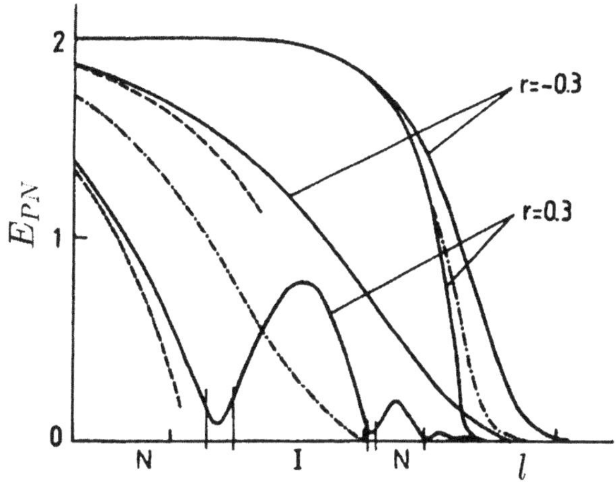

Fig. 3.21. The PN energy $E_{PN}(l)$, where $l = \sqrt{3V_0}$, for the dipole repulsion of atoms in the case of the nonsinusoidal substrate potential (3.59) for $s = -r = \pm 0.3$. For comparison, dash-dotted curves show the kink characteristics for the sinusoidal potential. Dashed lines show the results of the weak-bond approximation. Regions of the normal and inverse PN relief are indicated by the letters N and I, respectively [165].

$$u(x,t) = u_k^{SG}\left(x - \frac{1}{2}x_0\right) + u_k^{SG}\left(x + \frac{1}{2}x_0\right). \tag{3.150}$$

Within the framework of the adiabatic perturbation theory [27] the change of the relative coordinate x_0 due to the kink interaction is given (for large values of x_0) by the following equation [165]

$$\frac{d^2x_0}{dt^2} = (2\pi)^2 nA\,\frac{\sigma_1\sigma_2}{x_0^{n+1}}, \tag{3.151}$$

which reduces the problem to a motion of an effective particle in the potential

$$V_{\rm int}^{(0)}(x_0) = \frac{(2\pi)^2 A\,\sigma_1\sigma_2}{x_0^n}. \tag{3.152}$$

It is interesting to note that after introducing again the dimensional variables we find that the interaction law obtained above is nothing but the interaction of two "extra" atoms (or holes) in the chain. Such a contribution to the kink interaction is absent in the standard FK model where only nearest-neighboring forces are taken into account, i.e. for the classical FK model we always have $V_{\rm int}^{(0)} \equiv 0$. In the local FK model, however, the kink interaction is caused by an overlapping of their tails, and such an overlapping gives the interaction energy $V_{\rm int}^{(\rm loc)}(x) \propto u_x(x)$, which is proportional to the density of the "excess" atoms. Of course, the same effect will give a contribution to the kink interaction energy for the power-law forces as well. However, in that case this contribution is *smaller* in comparison with the main interaction described by Eq. (3.152), i.e. $V_{\rm int}^{(\rm loc)} \propto x^{-(n+2)}$. Indeed, as follows from the numerical simulations of Braun *et al.* [165], in the case of the dipole interaction the result $V_{\rm int}^{(\rm loc)}(x) \propto x^{-5}$ is in a good agreement with numerical data.

The phonon spectrum of the nonlocal FK model is described by the expression

$$\omega_{\rm ph}^2(k) = \omega_{\rm min}^2 + 2g \sum_{j=1}^{\infty} \frac{1-\cos(\kappa j)}{j^{n+2}}, \tag{3.153}$$

which is similar to the dispersion relation of a local FK model. However, parameters of the breather excitations differ remarkably from those for the local FK model as will be shown in Chapter 4.

Recently, the nonlocal SG equation of the form,

$$\frac{\partial^2 u}{\partial t^2} + \sin u = \frac{\partial}{\partial x} \int_{-\infty}^{+\infty} dx'\, G(x-x')\, \frac{\partial u}{\partial x'}(x',t), \tag{3.154}$$

with the exponential kernel, $G(x) = (2\lambda)^{-1}\exp(-|x|/\lambda)$, or the McDonald kernel, $G(x) = (\pi\lambda)^{-1}\mathrm{K}_0(|x|/\lambda)$, has been derived to describe nonlocal effects in the electrodynamics of long Josephson junctions [301]. For this type of nonlocal models, it has been shown that the nonlocal SG equation (3.154) with the exponential interaction does not support any *moving* 2π-kinks, but instead, it allows the moving 4π-, 6π-, etc. kinks. However, these complex kinks can propagate only with certain velocities [183, 302].

3.5.5 Compacton Kinks

When the coupling in the FK chain is purely *nonlinear* and linear coupling vanishes, the kinks can be localized on a finite interval being characterized by the absence of the exponentially decaying tails, the effect due to solely *nonlinear dispersion*. Such a type of solitary waves with a compact support are usually called *compactons*, they have been first discovered for a generalized Korteweg-de Vries equation with purely nonlinear dispersion [303].

Kivshar [305] reported that intrinsic localized modes in purely anharmonic lattices may exhibit compacton-like properties. Later, Dusuel *et al.* [30] demonstrated that the same phenomenology can also appear in nonlinear Klein-Gordon systems with anharmonic coupling, then obtained the experimental evidence of the existence of a static compacton in a real physical system, made up by identical pendulums connected by springs. Very recently, dark compacton solutions have been found in a model of Frenkel excitons [304].

To describe a kink on a compact support, we consider the anharmonic interaction in the FK model written in the standard form,

$$\frac{d^2u_n}{dt^2} - [V'_{\rm int}(u_{n+1} - u_n) - V'_{\rm int}(u_{n-1} - u_n)] + V'_{\rm sub}(u_n) = 0, \qquad (3.155)$$

and assume that the inter-particle interaction energy takes the form,

$$V_{\rm int}(u_{n+1} - u_n) = \frac{c_{\rm l}}{2}(u_{n+1} - u_n)^2 + \frac{c_{\rm nl}}{4}(u_{n+1} - u_n)^4. \qquad (3.156)$$

Here, the parameters $c_{\rm l}$ and $c_{\rm nl}$ control the strength of the linear and nonlinear couplings, respectively.

In the continuum limit, the corresponding equation takes the form,

$$\frac{\partial^2 u}{\partial t^2} - \left[c_{\rm l} + 3c_{\rm nl}\left(\frac{\partial u}{\partial x}\right)^2\right]\frac{\partial^2 u}{\partial x^2} - V'_{\rm sub}(u) = 0, \qquad (3.157)$$

where we neglected the higher-order derivative terms, assuming $c_{\rm nl}(u_x)^2 u_{xx} \gg c_{\rm l} u_{xxxx}$. Looking for steady-state localized solutions of this equation propagating with a velocity v, we find that the effective linear dispersion vanishes in two cases, i.e. when (i) $c_{\rm l} = 0$ and $v = 0$, and (ii) $c_{\rm l} \neq 0$ and $v = \pm\sqrt{c_{\rm l}}$, which correspond to a kink with a compact support. In the case of the normalized sinusoidal substrate potential, $V_{\rm sub}(u) = (1/2)[1+\cos(\pi u)]$, the static compacton kink can be found in an analytical form as a solution consisting of four pieces,

$$u_k(x) = \pm\frac{2}{\pi}\cos^{-1}\left\{{\rm cn}^2\left[\frac{\pi}{2\alpha^{1/4}}(x - x_0), k\right]\right\}, \qquad (3.158)$$

when $|x - x_0| < 1$, and $u = \pm 1$, otherwise. In Eq. (3.158) $\alpha = 3c_{\rm nl}$, the signs "+" and "−" correspond to the semi-axes $x \geq x_0$ and $x \leq x_0$, respectively, and cn is a Jacobi elliptic function with the modulus k. The same solution can be found for a compacton moving with the fixed velocity when $c_{\rm l} \neq 0$. Such a solution can be obtained from the static compacton by the change of the variable, $x \to (x \pm \sqrt{c_{\rm l}}\, t)$.

Solution (3.158) describes a topological compacton, a generalization of the original concept suggested for the Korteweg-de Vries equation with nonlinear dispersion; its shape consists of two pieces of an elliptic function which connect two straight lines, $u = \pm 1$. More importantly, the concept of the kink compacton can be generalized to any type of the Klein-Gordon model and the substrate potential $V_{\rm sub}(u)$, and exact analytical solutions can be found as well. Dusuel *et al.* [30] presented the exact solutions for the ϕ^4, double quadratic, and SG models. Figure 3.22 shows two example of a kink compacton for the ϕ^4 and sinusoidal potentials, respectively.

Dusuel *et al.* [30] investigated numerically the stability of the kink compactons to the propagation and noise. They found that the static compacton kink is stable, whereas the compacton moving with the velocity $v = \sqrt{c_{\rm l}}$ is unstable. They also constructed a mechanical analog of a nonlinear chain with a double-well potential, an experimental lattice of coupled pendulums. Adjusting the tension of a metallic stripe, Dusuel *et al.* [30] were able to make

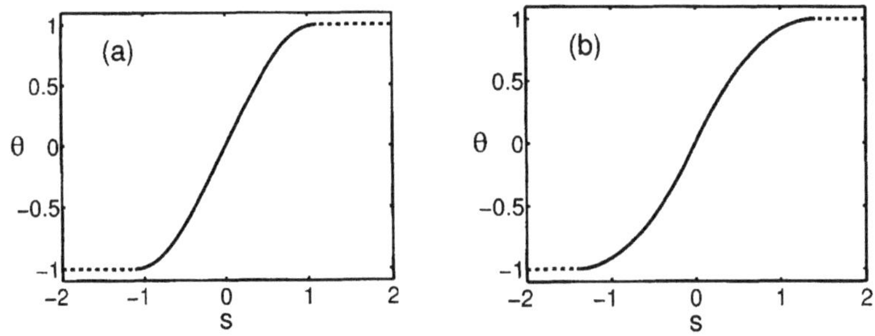

Fig. 3.22. Normalized compactons for two kinds of on-site potential: (a) ϕ^4 potential and (b) sinusoidal potential. For each potential the S-shaped wave form represents the compact part which connects the two constant parts, shown by horizontal dotted lines of the solution [30].

the linear coupling almost vanishing to observe a shape closely resembling the form of a static compacton kink. As for a moving compacton formally existing in the model with both linear and nonlinear dispersion, it was found in numerical simulations that it evolves into a conventional kink after emitting some radiation.

Dinda *et al.* [306] analyzed the ability of a compacton-like kink to execute a stable ballistic propagation in a discrete chain with an on-site ϕ^4 potential and anharmonic coupling. They demonstrated that the effects of lattice discreteness, and the presence of a linear coupling between lattice sites, are detrimental to a stable ballistic propagation of the compacton, because of the particular structure of the small-oscillation frequency spectrum of the compacton in which the lower-frequency internal modes enter in direct resonance with phonon modes. These studies revealed the parameter regions for obtaining a stable ballistic propagation of a compacton-like kink and the character of interactions between compacton-like kinks. It comes out from those simulations that collisions between the compacton and anti-compacton kinks that travel at low incoming velocities always end in a long-lived bound state, similar to the collisions of the ϕ^4 kink and antikink. However, Dinda *et al.* [306] found out that collisions between compactons never end in reflection for any incoming velocities, including the case when two compactons are launched with different velocities, or the case of collisions of compactons of different widths. Thus, all the translation energy of compactons is transferred to their internal modes during the collision.

4 Breathers

In this chapter we present the properties of breathers in the FK model. First, we discuss the breathers in a weakly discrete SG model where the discreteness is treated as a small perturbation. One of the most remarkable features in this case is associated with multi-soliton effects and fractal soliton scattering. Then we discuss nonlinear modes localized at impurities–nonlinear impurity modes. Last, we describe the case of very localized discrete modes – discrete breathers.

4.1 Perturbed Sine-Gordon Breathers

Even a weak discreteness does not allow oscillating breathers to exist as dynamical nonlinear modes of the chain; a weak discreteness acts as an external perturbation breaking the exact integrability of the SG model. As a result, the breathers radiate linear waves and slowly decay. However, for the chains with a strong coupling between particles, the breather dynamics can be considered in the framework of the perturbed SG equation which takes into account the discreteness effects. Then, the breather dynamics and the corresponding lifetime depend of the input energy, and are different for the breathers of small and large amplitudes [307]. As a matter of fact, in many nonintegrable models breathers are long-lived nonlinear excitations which play an important role in many physical processes such as the nonequilibrium dynamics of nonlinear systems [308], the energy transmission in the forbidden gap of the spectrum [309], being precursors to strong first-order structural phase transitions [310].

4.1.1 Large-Amplitude Breathers

For $g \lesssim 1$, a low-frequency breather may survive as two separate kink and antikink trapped in the corresponding wells of the PN potential. The effect of discreteness of the breather dynamics and the calculation of the effective PN potential can be found in a paper by Boesch and Peyrard [311]. We will outline some essential ideas here, but the reader is referred to the original work [311] for the calculational details.

To describe a breather in a discrete chain, we use the ansatz based on the exact breather solution of the SG model,

$$u_n^{(\mathrm{br})} = 4\tan^{-1}\left\{\frac{\sinh(k_B Z)}{\cosh[k_B(n-X)]}\right\}, \tag{4.1}$$

where X is the center of the breather (here we use $a_s = 1$), which is treated as an independent parameter, $2Z(t)$ represents the distance between two subkinks (in fact, a kink and anti-kink) which form the breather, and in this notation the breather's frequency is defined as $\omega_{\mathrm{br}} = \sqrt{1-k_B^2}$. The ansatz (4.1) is used to calculate the Hamiltonian of the FK chain [311] in terms of the collective coordinates X, Z, and k_B. It consists of two parts describing, independently, internal and translational dynamics of the breather. If we choose the initial breather profile (4.1) at the time $t=0$ when the subkinks are at their maximum separation, i.e. $Z(t)|_{t=0} = Z_0$, this yields the initial profile condition

$$\tanh(k_B Z_0) = k_B. \tag{4.2}$$

Then, the breather's total energy can be simplified keeping the first two terms in the Fourier series [since the coefficients decay exponentially as $1/\sinh(\pi^2/k_B)$], so that the total energy becomes

$$E_{\mathrm{br}}(Z_0, X) = 16\,k_B\left[1 + \frac{2\pi^2/k_B}{\sinh(\pi^2/k_B)}\cos(2\pi Z_0)\cos(2\pi X)\right]. \tag{4.3}$$

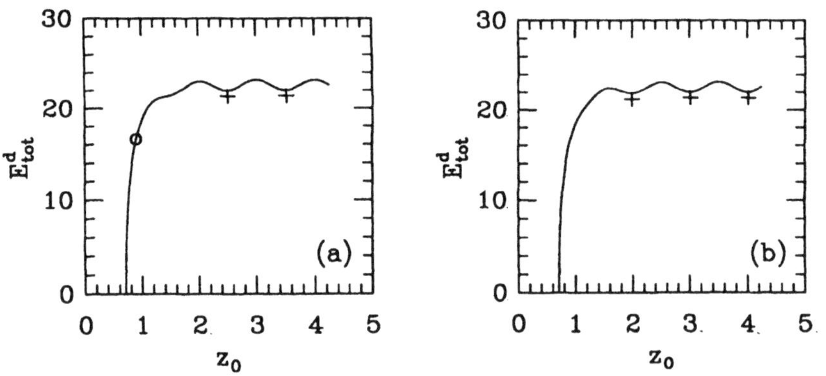

Fig. 4.1. Total energy of the breather vs. Z_0 for (a) $X=0$ and (b) $X=1/2$. The solid line corresponds to Eq. (4.3), the pluses mark the energies of a static chain configuration, and the circle is a numerical result [311].

Equation (4.3) must be considered simultaneously with the condition (4.2) which changes the value of k_B for each Z_0. Notice that for high-frequency (i.e., small-amplitude) breathers when discreteness effects are small, $k_B \to 0$, we recover the well known continuum expression for the SG breather energy, $E^{\mathrm{br}} = 16\, k_B$. For large-amplitude breathers, Eq. (4.1) reveals the existence of two PN potentials, one for X and the other for Z_0, and it defines the positions where the initial breather profile will be trapped if started from rest.

Figure 4.1 shows the total breather energy calculated by different methods for two values of the breather coordinate X. Solid line is a result found from Eq. (4.3), and the symbols mark the energies calculated from numerical simulations. Since Z_0 is measured with respect to X, the panels (a) and (b) in Fig. 4.1 each indicate that the minima of the PN wells are located at the kinks' positions between the particles.

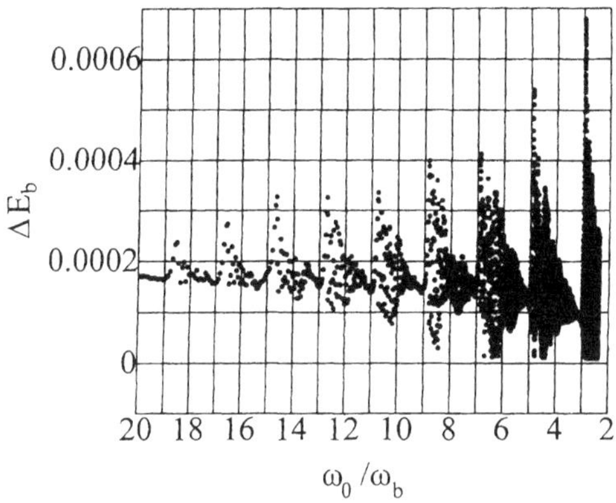

Fig. 4.2. Energy lost by a breather, ΔE_{br}, as a function of $\omega_0/\omega_{\mathrm{br}}$ [312].

Using molecular-dynamics and Fourier-transform numerical techniques, Boesch and Peyrard [311] demonstrated that a breather in the FK model spontaneously makes remarkably sharp transitions from a short lifetime to a long lifetime; this type of the breather dynamics can be explained by the structure of radiation spectrum emitted by the breather, which differ by more than four orders of magnitude on each side of the transition.

For a small degree of discreteness, phonon emission from a large-amplitude discrete breather in the FK model was studied more recently by Dmitriev *et al.* [312]. In contrast to the case of highly discrete system investigated by Boesch and Peyrard [311], it was found that in the case $g \gg 1$ the resonance between the breather's oscillation and the phonons of the lower phonon band edge (with the wavevector $k = 0$) takes place. In particular, Dmitriev *et al.* [312] established numerically that in a weakly discrete FK model [i.e., for the lattice spacing $0.1 \leq h \leq 0.5$, where $h^{-1} = \sqrt{g}$], the resonance between large-amplitude breather and phonons of the lower phonon band edge takes place when $(2m + 1)\, \omega_{\mathrm{br}} = \omega_0$ with a positive integer m. Under this condition the breather starts to emit comparatively large bursts of radiation in the form of small-amplitude wave packets. The most prominent resonance takes place for $m = 1$. For a highly discrete system, the resonance effect does not manifest itself, because the breather passes through the resonances very quickly. Thus, the resonance effect is prominent for a moderate discreteness and it is noticeable for a weak discreteness.

In Fig. 4.2, the variation of the energy $\Delta E_{\rm br}$ lost by breather in the l-th half-period of oscillations as a function of $\omega_0/\omega_{\rm br}$ is presented. One dot corresponds to one half-oscillation of the breather or, in other words, to a certain oscillation period l. As Fig. 4.2 suggests, the energy emitted in a half-oscillation, on the average, sharply increases at $(2m+1)\,\omega_{\rm br} = \omega_0$. The smaller the m, the sharper the increase, so the most prominent resonance takes place when $3\omega_{\rm br}$ becomes equal to ω_0.

4.1.2 Small-Amplitude Breathers

In the small-amplitude limit, when the breather width is much larger than the lattice spacing, the effects of discreteness are almost negligible, and the radiation-induced losses of the breather energy can be readily small. These losses are usually *exponentially small* in the parameter defined as a ratio between the lattice spacing and the breather's width, and it is "beyond of all order". Neglecting radiation, we can find approximate periodic solutions for the breathers by means of the multi-scale asymptotic expansion. The rigorous procedure to find such high-frequency breather modes for general models with arbitrary substrate potential was suggested by several authors (see, e.g., Refs. [313]–[315], and references therein).

We look for a solution of the nonlinear equation in the limit of small amplitudes using the asymptotic expansion,

$$u_{\rm br}(x,t) = \mu\Phi(x,t)e^{i\Omega t} + \text{c.c.}, \tag{4.4}$$

where c.c. stands for complex conjugate and $\mu = (\omega_{\min}^2 - \Omega^2)^{1/2}$ is a small parameter of the asymptotic procedure. Substituting Eq. (4.4) into the discrete motion equation and expanding the substrate potential for small u_n, $V'_{\rm sub}(u) \approx \omega_{\min}^2 u + \beta u^2 + \widetilde{\beta} u^3$, we may derive an effective evolution equation for the wave envelope Φ assuming that the latter is changing slowly on the scales of order of the lattice spacing (see details, e.g., in Ref. [314]),

$$2i\Phi_t - Q\Phi_{zz} + G\,|\Phi|^2\Phi = 0, \tag{4.5}$$

where the variable z is connected with a reference frame moving with the breather, and the parameters Q and G are functions of the parameters of the effective potential [314]. Equation (4.5) is the nonlinear Schrödinger (NLS) equation and it has a localized soliton solution provided $QG > 0$, the latter condition is that for the small-amplitude breathers to exist.

The effective NLS equation (4.5) describes a breather for any type of the substrate potential. However, due to the presence of higher-order harmonics, only the SG model supports nonradiating breathers because the integrability of the SG model implies a cancellation of this kind of higher-order effects. For other type of the substrate potential, it has been rigorously shown that exact breather solutions do not exist [316, 317] and a breather, being excited in a chain, radiates slowly energy. This kind of long-lived radiation process has been estimated for the breather of the SG model [311, 312] and ϕ^4 model [318].

In more general cases, e.g. for the case of long-range interaction between the particles in a chain, the asymptotic expansion results cannot be applied, and some other techniques to describe the small-amplitude breathers should be used (see, e.g., Ref. [165]). In this case the breather parameters differ remarkably from those for the local FK model [319].

4.2 Breather Collisions

As was mentioned above, for $g \gg 1$ the FK model is described with a high accuracy by the SG equation. However, this statement is not always valid for many-soliton collisions which may become highly inelastic even for the case of a weak discreteness.

Soliton collisions in a weakly perturbed SG equation were studied first theoretically by Kivshar and Malomed [27]. In the case of two-soliton collisions, the discreteness leads to a break-up of a large-amplitude (i.e. low-frequency) breather into a kink-antikink pair due to its collision with another kink or breather, and also to a radiation-induced fusion of a kink-antikink pair into a breather. Recently, Kevrekidis *et al.* [320] used the exact solution of the SG equation describing an array of breathers, and studied the asymptotic interaction between two breathers. They identified the exponential dependence of the interaction on the breather separation as well as its power-law dependence on the frequency.

More interesting effects are possible for many-soliton collisions. In particular, collisions of three kinks in a weakly perturbed SG equation were studied theoretically by Kivshar and Malomed [27] who found that the radiationless energy and momentum exchange between the colliding kinks is possible when all kinks collide almost at one point.

More recently, many-particle effects in the soliton collisions have been studied extensively both numerically and analytically [321]–[325] in a weakly discrete FK model. As a result of these studies the possibility of comparatively strong, radiationless energy exchange between kinks and breathers in the FK model with a very weak degree of discreteness was demonstrated numerically confirming the earlier theoretical predictions. In many cases, resonant energy exchange between colliding solitons can be explained through the excitation of soliton internal modes [326]. However, in the case of many-soliton collisions, the energy exchange between colliding solitons can be noticeable, while the role of the soliton internal mode is negligible [191], so that one should find qualitatively different explanation.

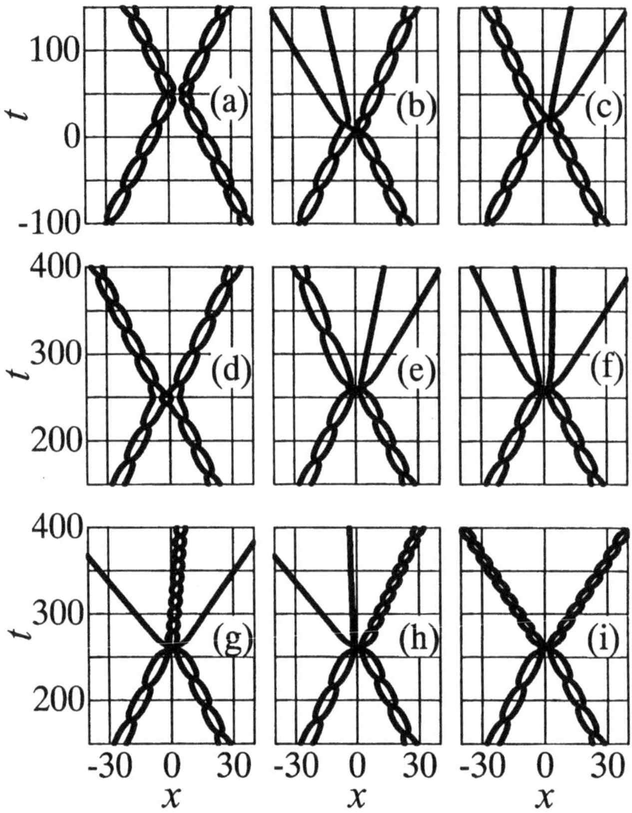

Fig. 4.3. Examples of the breather scattering in the FK model with $g = 25$ for the initial velocities $v_{B2} = -v_{B1} = 0.2$, and the frequencies $\omega_{B1} = 0.1$ and $\omega_{B2} = 0.8\,\omega_{B1}$. The initial distance between the colliding breathers is (a) $D^0 = 20.0$, (b) $D^0 = 4.3$, (c) $D^0 = 8.1$, (d) $D^0 = 100.0$, (e) $D^0 = 104.1$, (f) $D^0 = 104.13$, (g) $D^0 = 104.26$, (h) $D^0 = 104.285$, and (i) $D^0 = 104.33$ [324].

Many-particle effects become possible only if at least three solitons participate in the interaction and collide almost at one point. For the case of three kinks, the probability of this event is very small. However, if one of the colliding solitons is a breather that consists of two coupled kinks which are always near each other, inelastic collisions with a kink or another breather become more typical.

To illustrate this, in Fig. 4.3 we show some typical results of a collision between two breathers in a weakly discrete FK chain. The discreteness parameter is set to $h = 0.2$ (we recall that $h^{-1} = \sqrt{g}$), so that the perturbation due to discreteness is indeed very small. Nine panels in Fig. 4.3 correspond to different values of the initial distance between the colliding breathers D^0. In the cases (a) and (d), the collisions are practically elastic, i.e., there is no energy and momentum exchange between the colliding breathers. In all other cases shown in Fig. 4.3, the breather collisions are inelastic. In the case (a), at the moment of collision, the breathers oscillate nearly out of phase, and that is why they repel each other. In the case (d), the colliding breathers oscillate nearly in phase, but the collision remains elastic because it occurs without involving the three- or four-kink interaction effects. In contrast, in the cases (b) and (c), the three-kink effects can be seen and, in the cases (e–i), all four subkinks participate in the collisions. As a possible result of the collisions, one of the colliding breathers (or even both of them) can break-up into free kinks. The break-up process takes place only for the breathers with sufficiently small internal frequencies; otherwise, the inelasticity of collisions manifests itself in the energy and momentum exchange between the breathers, as is shown for the case (i). In all the cases, the radiation losses are very small or negligible.

4.2.1 Many-Soliton Effects

The origin of radiationless energy exchange between solitons in the weakly discrete FK chain can be understood from the analysis of the *unperturbed* SG equation. In this section we demonstrate that if the parameters of colliding particles are properly chosen, then two different solutions of the SG equation can be very close to each other at a particular time moment. Therefore, if a very weak perturbation is added to the system, then one solution can be easily transformed to another. This may explain the inelastic collision of breathers in a nearly integrable system. Moreover, since this effect is predicted from the analysis of the unperturbed SG equation, it does not depend on a particular type of the perturbation.

Let us consider the kink-breather collision described by the three-soliton solution of the SG equation of the form $u = v + w$, where the function v describes a kink, $v = 4\tan e^{B}$, with $B = \delta_k(x - x_k - v_k t)$. Here $\delta_k^{-1} = \sqrt{1 - v_k^2}$ is the kink's width, $0 \le v_k < 1$ is the kink's velocity, and the parameter x_k defines the kink's position at $t = 0$. The second part of the solution is $w = 4\tan[(\eta X)/(\omega Y)]$ with

$$X = 2\omega\,(\sinh D - \cos C \sinh B) + 2\delta_k\delta_B\,(v_k - v_B)\sin C\cosh B$$

$$Y = 2\eta\,(\cos C + \sinh D \sinh B) - 2\delta_k\delta_B\,(1 - v_k v_B)\cosh D\cosh B,$$

where $C = -\omega\delta_B\,[t - v_B(x - x_B)] + 2\pi m$ with an integer m, $D = \eta\delta_B\,(x - x_B - v_B t)$, $\delta_B^{-1} = \sqrt{1 - v_B^2}$, $0 \le v_B < 1$ is the velocity of the breather, $\eta = \sqrt{1 - \omega^2}$, the breather frequency ω $(0 \le \omega < 1)$ defines the amplitude of the breather (or its rest energy), and x_B defines the position of the breather at the time $t = 0$.

The coordinate and time of the point of the kink-breather collision are defined as follows,

$$x_c = \frac{v_k x_B - v_B x_k}{v_k - v_B}, \qquad t_c = \frac{x_B - x_k}{v_k - v_B}. \tag{4.6}$$

Let us consider two different solutions of this form, u and u^*, for which we take the soliton velocities with the opposite signs, $v_k = -v_k^*$ and $v_B = -v_B^*$, and the soliton positions at the time $t = t_c$ in the solution u^*

$$x_k^* = \frac{x_k(v_k + v_B) - 2v_k x_B}{v_B - v_k}, \quad x_B^* = \frac{2v_B x_k - x_B(v_k + v_B)}{v_B - v_k} \tag{4.7}$$

are chosen in such a way that $x_c = x_c^*$ and $t_c = t_c^*$. The magnitudes of all other parameters in the solutions u and u^* remain the same. One can see that if

$$\omega\,(x_B - x_k) = 2\pi m \delta_B\,(v_k - v_B)\,, \tag{4.8}$$

where m is an integer, then at the time $t = t_c$ we have $u(x,t_c) \equiv u^*(x,t_c)$, while for the derivatives we have $u_t(x,t_c) = -u_t^*(x,t_c)$. Under the additional conditions $v_k \to 0$ and $v_B \to 0$, we have $u_t(x,t_c) \to 0$ and $u_t^*(x,t_c) \to 0$ for any x, or $u_t(x,t_c) \to u_t^*(x,t_c)$ for any x.

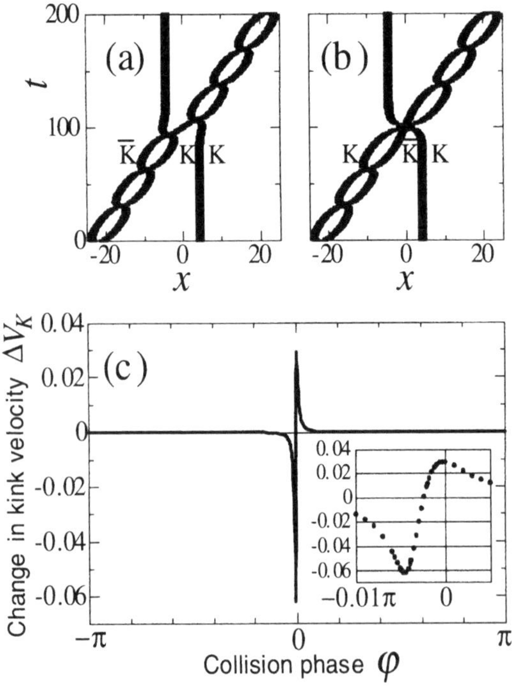

Fig. 4.4. Kink-breather collisions described by the SG equation. Depending on the phase, two– [(a)] or three–[(b)] soliton collisions are observed. (c) Change of the kink velocity vs. the collision phase in the presence of weak discreteness ($h = 0.1$). The insert shows an enlarged domain of the plot.

Thus, if those conditions are nearly fulfilled, then at a certain time, $t = t_c$, the two qualitatively different solutions, u and u^*, as well as their derivatives with respect to time, u_t, u_t^*, become close to each other. From the physical point of view, the condition (4.8) means that all three solitons meet at one point. When this happens, the kink-breather collision become strongly inelastic even in a weakly perturbed system. The condition (4.8) can also be written through the breather wavelength $\lambda = 2\pi\delta_B|v_B|/\omega$ as $x_B/\lambda = m$, where m is integer.

Numerical results presented in Fig. 4.4 demonstrate the exceptionally high sensitivity of the kink-breather collisions to a variation of the relative soliton phase. In order to change the relative phase, the breather's initial position x_B was varied while all other parameters were kept fixed: $x_k = 0$, $v_k = 0$ for the kink, and $\omega = 0.1$, $v_B = -0.2$ for the breather. Before the collision with a breather, the kink was at rest, so that the change of kink's velocity after the collision, ΔV_k can be used as a measure of inelasticity of the kink-breather interaction. The inelasticity of the kink-breather collision increases in the vicinity of the point $\phi = 0$ by three orders of magnitude.

4.2.2 Fractal Scattering

One of the most intriguing properties of soliton interactions in nonintegrable models is the observation of the fractal nature of their scattering, first discussed for kink-antikink collisions in the ϕ^4 model [200]. The main features of fractal soliton scattering are usually explained by the excitation of the soliton internal mode, which is an important property of solitary waves of many nonintegrable soliton-bearing models [191], and the physics of fractal soliton scattering can be understood as a resonant energy exchange between the soliton translational motion and its internal mode [189, 195]. A similar mechanism was revealed for the interaction of a kink with a localized impurity [326].

Recently, Dmitriev *et al.* [324] described a different physical mechanism of fractal soliton scattering in a weakly discrete FK model. In this case, the role of the soliton internal mode is negligible [191], and the fractal structures observed in soliton scattering should be explained by a qualitatively different mechanism that can be understood as manifestations of multi-particle effects in the soliton collisions, due to resonant coupling between the "atomic" and "molecular" degrees of freedom of the colliding composite solitons.

In order to demonstrate this effect, we consider collisions of two breathers in a weakly discrete FK model, when two weakly interacting breathers break up into two separate breathers flying away. After breaking up, the two independent breathers move in opposite directions. The absolute values of their velocities are nearly equal because in a weakly discrete chain the momentum conservation is nearly fulfilled and the breathers were taken to have nearly the same energies. The breather velocity after splitting v_B^* is shown in Fig. 4.5 as a function of the initial distance between the colliding breathers D^0.

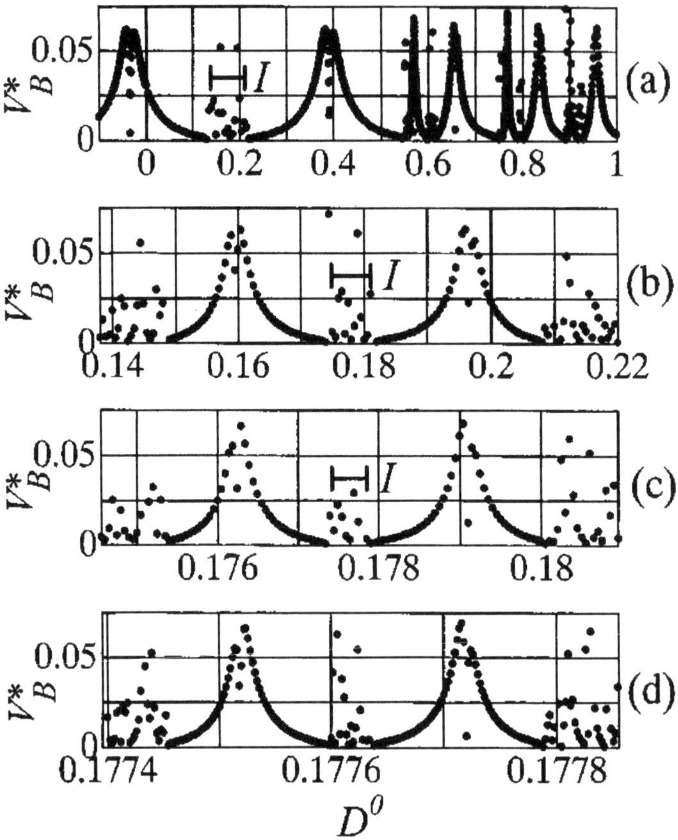

Fig. 4.5. Fractal structure (four scales are shown) of the breather collisions shown as the function $v_B^*(D^0)$ [324].

The function $v_B^*(D^0)$ shows the property of self-similarity at different scales usually associated with fractal scattering. Four levels of such similarity are presented in Figs. 4.5(a)–(d), where each succeeding figure is shown for the interval expanded from a smaller region marked by I in the preceding one. The expansion coefficient is about 13.5 for each step. At each scale, the function $v_B^*(D^0)$ looks like an alternation of smooth and chaotic domains. However, at larger magnification, each chaotic domain again contains chaotic regions and smooth peaks. Thus, the output velocity $v_B^*(D^0)$ is actually a set of smooth peaks of different scales. In some regions, the width of the peaks vanishes, so that the density of the peaks goes to infinity, while the height of the peaks remains the same at each scale.

The fractal structure of the function $v_B^*(D^0)$ proves the chaotic character of breather scattering in a weakly discrete case. The fractal nature of breather collisions can have a simple physical explanation. As was shown above, in a weakly discrete (and, therefore, weakly perturbed) system the breathers attract each other with a weak force. As can be seen from Fig. 4.5, the chaotic regions appear where the extrapolation of the smooth peaks gives nearly zero velocity $v_B^*(D^0)$. In these regions, the breathers gain a very small velocity after interaction and subsequent splitting. With such a small initial velocity, the breathers cannot overcome their mutual attraction and collide again. In the second collision, due to momentum exchange, the breathers can acquire an amount of kinetic energy sufficient to escape each other, but there exists a finite probability of gaining kinetic energy below the escape limit. In the latter case, the breathers will collide for a third time, and so on. Thus, a series of collisions leads to a resonant energy exchange between the "atomic" (kink's translational) and "molecular" (relative oscillatory) breather degrees of freedom, and to fractal scattering.

4.2.3 Soliton Cold Gas

As was described above, the kinks, breathers and phonons in the SG equation are considered as a gas of free quasi-particles which do not interact in the framework of the exactly integrable model. As a result, the total number of the quasi-particles of each type is conserved. It is natural to assume that such a picture will remain valid in the case of a weakly perturbed SG equation. However, this conjecture is only valid in the absence of multi-particle effects which, according to the results presented above, may lead to a completely different behavior [325].

As an illustration of the importance of the multi-particle effects, we consider the dynamics of the so-called "cold soliton gas" consisting of kinks and antikinks which are initially at rest. Figure 4.6 shows the time evolution of a gas of 800 kinks and antikinks (only a small part of the whole system is shown) in the FK system with a weak discreteness $h = 0.2$ (so that $g = 25$). Initial conditions were set with the help of a periodic solution of the SG equation,

$$u(x,t) = 2\sin^{-1}\left\{\mathrm{dn}\left[\frac{x-vt}{\sqrt{1-v^2}}, \kappa\right]\right\}, \tag{4.9}$$

with two independent parameters, v and κ, in the limit when the velocity vanishes, $v = 0$. The parameter κ defines a spatial period of the solution (which is 16 in the example presented below).

When the solution period is sufficiently large, a pair of kinks on each period of the solution (4.9) can be regarded either as a kink-antikink pair or as a large-amplitude breather, where the subkinks inside the pair are coupled. Due to the mutual attraction between the kinks of different polarity, the periodic chain of alternating kinks and antikinks is unstable. Even without any perturbation (except discreteness), the kinks start to move after a time

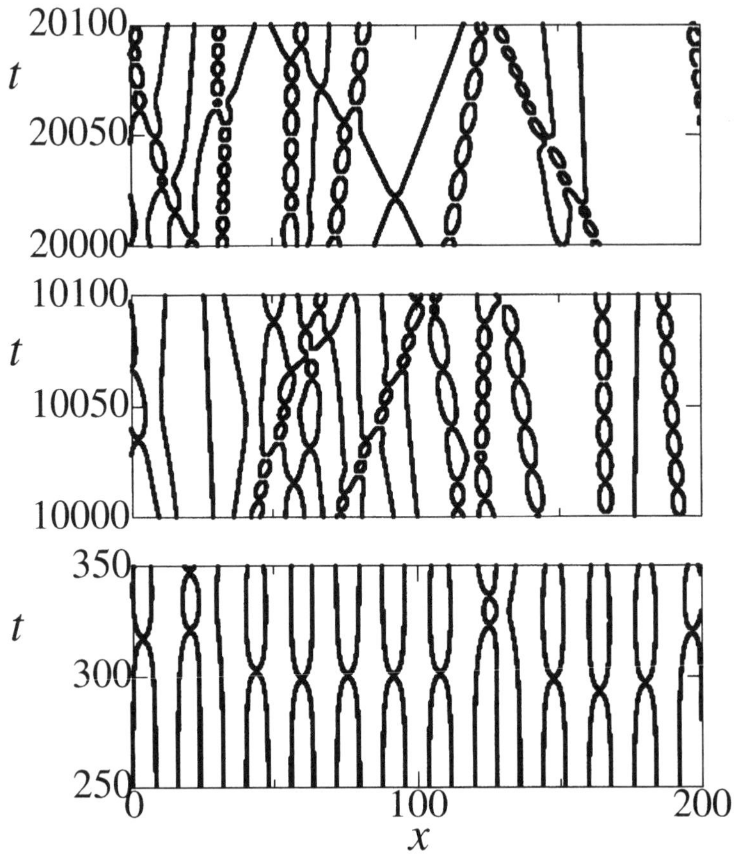

Fig. 4.6. Time evolution of a gas of 800 kinks of alternating polarity (only a small part of the chain is shown) in a weakly discrete FK chain [327].

$t \sim 300$ as shown in the bottom part of Fig. 4.6. When kinks interact in pairs, they collide practically elastically. However, many-soliton collisions become more and more frequent and probability of the inelastic collisions gradually increases.

In Fig. 4.7 we show the long-term evolution of the number of kinks, breathers and, in order to identify the origin of the inelastic processes, the frequency of the corresponding three-soliton states in the system. For $t < 300$, before the onset of instability of the kink array, there are no breathers in the system. For $t > 300$ the number of kink-antikink pairs decreases gradually, and the number of breather and three-soliton states increases. This happens due to two major mechanisms. The first mechanism is a fusion of a kink and an antikink moving with a small relative velocity into a breather. The second mechanism is the formation of breathers due to inelastic many-soliton colli-

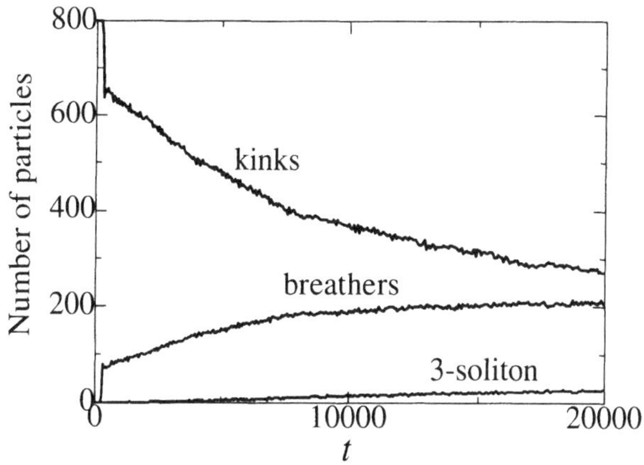

Fig. 4.7. Number of kinks, breathers and three-soliton states as functions of time [327].

sions. Indeed, when a breather collides with a kink or with another breather, the probability of the three-kink collision is not small being proportional to the breather density. Thus, the probability of inelastic collisions increases with a growth of the number of breathers in the gas (see Fig. 4.7, where the fraction of three-soliton states shows actually the number of the solitons colliding inelastically).

From Fig. 4.7 one can also note that the total number of solitons calculated as $N_{\text{tot}} = N_1 + 2N_2 + 3N_3 + \ldots$, where N_n is the number of n-soliton states, is conserved. However, the number of kinks and breathers themselves is not conserved. Notice also that because the discreteness is very weak in this example, the radiative losses are also very small (the estimates show that at $t = 20,000$ the relative losses to radiation are less that 0.5%).

4.3 Impurity Modes

Before analyzing discrete breathers, we introduce the concept of nonlinear impurity modes as spatially localized modes supported by defects in the lattice. In the low-amplitude limit, such modes are well known in the linear theory of lattice vibrations, and they can also be found in nonlinear lattices. Both linear and nonlinear impurity modes provide a link between the linear theory of lattice vibrations and intrinsic localized modes (discrete breathers) supported by nonlinearity in a perfect chain.

4.3.1 Structure and Stability

Localized modes supported by impurities (e.g., by point defects in crystals) are well known in the theory of lattice vibrations (see, e.g., Ref. [214]). They are also possible in many types of lattice models, including the FK chain. To investigate such kind of localized modes in the framework of the linearized

FK model, we consider small-amplitude oscillations of the atoms making the expansion $\sin u_j \approx u_j$. Then, spatially localized modes of the linear lattice at the impurity site can be found analytically with the help of the Green-function technique [173, 214], and for the case of the linearized FK model this analysis was carried out by Braun and Kivshar [49].

The Green function of the chain with impurities satisfies the Dyson equation, and a simple analysis of its solutions gives the conditions for the impurity modes to exist [49]. As a result, the impurity-induced localized modes become possible with the frequencies lying either *above* or *below* the phonon frequency band defined as $(\omega_{\min}, \omega_{\max})$, where $\omega_{\min} = 1$ and $\omega_{\max} = \sqrt{1+4g}$ for the standard FK model. For example, in the case of an isotopic impurity (i.e. a defect atom with a mass that differs from the mass of the lattice atoms) first analyzed by Lifshitz [211], we should take $\Delta m \neq 0$ but $\Delta g = \Delta\epsilon = 0$. As a result, the localized impurity mode exists for $\Delta m > 0$ (i.e. for a heavy-mass defect) below the frequency band, $0 < \omega_l < \omega_{\min}$, and for $\Delta m < 0$ (i.e. a light-mass defect), above the frequency band, $\omega_l > \omega_{\max}$. The mode frequency ω_l is defined by the expression,

$$\omega_l^2 = \frac{(1+2g) \mp \sqrt{4g^2 + \Delta m^2(1+4g)}}{(1-\Delta m^2)}, \tag{4.10}$$

where the sign $\pm$ defines the case $\Delta m > 0$ or $\Delta m < 0$, respectively.

The impurity-induced localized modes can be easily described in the nonlinear FK chain in the long-wavelength limit when the effective SG equation becomes valid. In such a case, the solution for a nonlinear impurity mode can be obtained as a breather mode captured by an impurity [49, 328, 329]. Analogously to the linear approximation discussed above, the nonlinear impurity modes may exist with the frequencies lying either below or above the phonon spectrum band. However, the shape of the nonlinear mode is modified by nonlinearity giving rise several new features. In particular, the nonlinearity itself may extend the condition for the nonlinear modes to exist (see, e.g., Refs. [49, 328]–[331]), however, the stability analysis shows that such nonlinear localized modes excited near local impurities are stable only in the regions of existence of corresponding linear modes [49, 330].

As an example, let us consider the case when only $\Delta\epsilon \neq 0$ so that the corresponding continuous version of the FK model is described by the perturbed SG equation

$$u_{tt} - u_{zz} + \sin u = -\epsilon_1(z)\sin u, \tag{4.11}$$

where $z = x/d$, $d = a_s\sqrt{g}$, and $\epsilon_1 = \Delta\epsilon/2\sqrt{g}$. In the linear approximation, the impurity mode for Eq. (4.11) is given by the expression $u = A\exp(\epsilon|z|/2)\cos(\omega_l t)$, where $\omega_l = \sqrt{1-\epsilon^2/4}$, and such a mode exist only provided $\epsilon < 0$. To analyze the impurity mode in the nonlinear case, it is convenient to derive an effective envelope equation instead of Eq. (4.11) making the transformation $u = \Psi e^{it} + \Psi^* e^{-it}$ (the asterisk stands for the complex conjugation), where the envelope function Ψ is assumed to be slowly varying and small enough to consider the nonlinearity in the lowest order. Then, it is possible to reduce the problem to the effective NLS equation

$$2i\Psi_t - \Psi_{zz} - \frac{1}{2}|\Psi|^2\Psi = 0 \tag{4.12}$$

with the matching condition at $z = 0$

$$\Psi_z|_{0+} - \Psi_z|_{0-} = \epsilon_1 \Psi(0). \tag{4.13}$$

Matching two soliton solutions of the NLS equation (4.12)–(4.13), we may find the approximate solution for the nonlinear impurity mode,

$$u(z,t) = 4\beta \frac{\cos(\Omega_l t)}{\cosh[\beta(|\mathrm{z}| + \mathrm{z}_0)]}, \tag{4.14}$$

where the impurity mode frequency Ω_l is determined by the relation $\Omega_l = \sqrt{1 - \beta^2/2}$ and, unlike the linear case, now it depends on the mode amplitude β. The equation which follows from the matching condition (4.13) takes the form

$$\tanh(\beta z_0) = -\frac{\epsilon_1}{2\beta}, \tag{4.15}$$

and it determines the structure of the nonlinear impurity mode which has different shapes for different signs of ϵ_1. For $\epsilon_1 < 0$, Eq. (4.15) yields $z_0 > 0$, and the impurity mode has a shape similar to the harmonic case [see Fig. 4.8(a)], and, moreover, for $z_0 \to \infty$ it recovers the linear case. In the case of $\epsilon_1 > 0$, Eq. (4.15) leads to the solution with $z_0 < 0$, and the impurity mode has two maxima [see Fig. 4.8(b)]. The latter case shows that, in principle, impurity localized mode may be supported by nonlinearity in the cases when it is not possible in the linear limit. However, this new kind of localized impurity modes does not give stable solutions [49, 330].

Analogously, we may find the impurity modes with the frequencies lying above the cut-off frequency of the linear lattice $\omega_{\max}$, using again the approximation of the slowly varying mode envelope. In this case we start from the FK model looking for a solution which describes out-of-phase oscillations of the atoms. The approximate solution is found to be [49]

$$u_j = 4(-1)^j \beta \frac{\cos(\Omega_l t)}{\sinh[\beta(|z| + z_0)]}, \tag{4.16}$$

and it is shown in Fig. 4.9. The matching parameter z_0 is determined by the equation

$$\coth(\beta z_0) = \frac{\epsilon_1}{2\beta}, \tag{4.17}$$

and the mode frequency,

$$\Omega_l = \omega_{\max} + \frac{\beta^2}{2\omega_{\max}}, \tag{4.18}$$

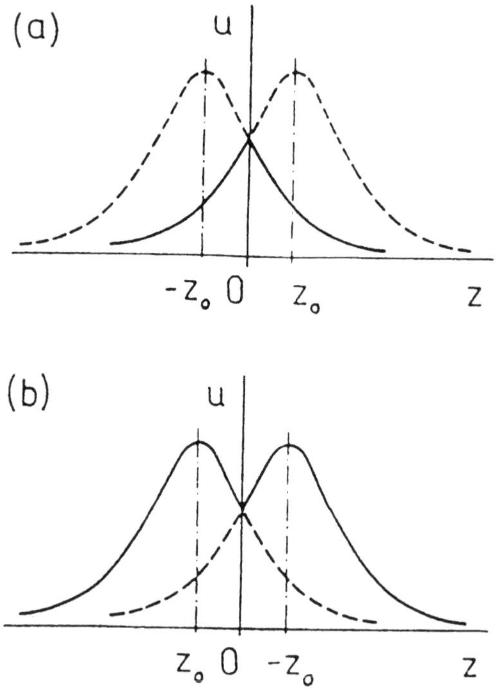

Fig. 4.8. Profiles of a low-frequency nonlinear mode for (a) $\epsilon_1 < 0$ (stable), and (b) $\epsilon_1 > 0$ (unstable). The parameter z_0 is defined in Eq. (4.15).

lyes above the upper (cut-off) frequency of the spectrum. Such a mode exists only for the case $\epsilon_1 > 0$.

Similar types of impurity modes can be found in other models. Wattis *et al.* [332] analyzed a short section of a DNA chain with a defect, with the aim of understanding how the frequency, amplitude, and localization of breathing modes depend on the strength of the bonds between base pairs, both along the chain and between the chains. The results show that the presence of a defect in the chain permits the existence of a localized breather mode. Parameter values for the interaction energy of a base with its nearest neighbors were found to be in a good agreement with both the amplitude and the number of base pairs affected by defect-induced breathing motion.

As has been mentioned above, one of the main problems for the nonlinear impurity modes to exist is their stability. Although such impurity modes may exist even for the conditions when the linear impurity modes are forbidden, in most of the cases these new modes are in fact unstable. One of the simplest ways to carry out the stability analysis for the nonlinear modes discussed above is to introduce a small mismatch between two parts of the composed solution (4.14). Then the small-amplitude oscillations around the stationary solution are characterized by the frequency [49] $\widetilde{\omega}^2 = -3\epsilon_1(\beta + \epsilon/2)$ which clearly shows that the nonlinear mode is unstable for $\epsilon_1 > 0$, i.e. just for the condition when the linear problem does not have spatially localized solutions. Thus, even supporting stationary localized solutions of a new form, nonlinearity itself does not extend the conditions for the impurity localized modes to exist.

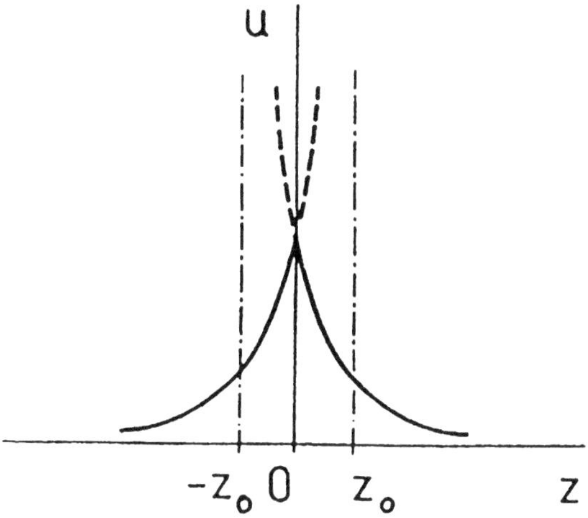

Fig. 4.9. Profile of a high-frequency nonlinear impurity mode for $\epsilon_1 > 0$. The parameter z_0 is defined in Eq. (4.17).

Stability of nonlinear impurity modes was discussed by Bogdan *et al.* [333] for the case of the NLS equation. They employed the analysis similar to that developed by Vakhitov and Kolokolov [334] for solitary waves of the generalized NLS equation in a homogeneous medium and formulated the stability in terms of the invariant

$$N(\omega_l) = \int_{-\infty}^{\infty} |\Psi(x;\omega_l)|^2 dx.$$

As a result, they confirmed that the nonlinear mode for $\epsilon_1 > 0$ is unstable, the instability corresponds to the condition $dN(\omega_l)/d\omega_l < 0$. Such an instability manifests itself in an exponential growth of antisymmetric perturbations which shift the soliton to one side from the impurity mode that finally repel the localized mode due to the repulsive effective interaction with it, as consistent with the prediction based on an effective potential. More general stability analysis for the impurity modes was developed by Sukhorukov *et al.* [335], who considered the nonlinear localized modes supported by a nonlinear impurity in the generalized NLS equation and described *three types of nonlinear impurity modes*, one- and two-hump symmetric localized modes (as discussed above) and asymmetric localized modes (possible for a nonlinear impurity). They derived a *general analytical stability criterion* for the nonlinear localized modes and considered the case of a power-law nonlinearity in detail. Sukhorukov *et al.* [335] discussed also several scenarios of the instability-induced dynamics of the nonlinear impurity modes, including the mode decay or switching to a new stable state, and collapse at the impurity site.

In a discrete chain, the potential of the interaction between the localized mode and an impurity is modified by an effective periodic Peierls-Nabarro potential. Due to a complex structure of extremum points of a total effective potential, the mode can be shifted from the impurity site, creating an *asymmetric nonlinear impurity mode*. Such a kind of (high-frequency) nonlinear impurity mode has been analyzed for a lattice without a substrate potential but with nonlinear interatomic coupling [331, 336], and it has been shown that it may exist even for a heavy-mass impurity (i.e. for $\epsilon_1 > 0$, see above). Even being expected, such modes are not investigated yet for a lattice with on-site potential.

Another very important problem related to the theory of nonlinear impurity modes is the radiative damping of the mode oscillations. For the low-frequency impurity modes such an effective decay is usually power-law [49, 307] while for the high-frequency nonlinear modes the mode lifetime may be much shorter (see Ref. [49] and references therein).

4.3.2 Soliton Interactions with Impurities

In a general case, the kink-impurity interaction may be described by a simple picture where a local inhomogeneity gives rise to an effective potential to the kink (see Chap. 3 above). However, the model of a classical particle is valid only in the case when the impurity does not support an *impurity mode*, a local oscillating state at the impurity site. Such an impurity mode can be *excited* due to the kink scattering and it may change the result for the kink transmission. The importance of the impurity modes in the kink-impurity interactions has been pointed out in the papers by Fraggis *et al.* [337], Kivshar *et al.* [326], Zhang *et al.* [338]–[341], Malomed *et al.* [342], and Belova and Kudryavtsev [343]. A comprehensive overview of different types of the kink-impurity interactions can be found in a review paper [229].

An important effect revealed in numerical simulations [326] is that a kink may be totally *reflected by an attractive impurity* due to a resonance energy exchange between the kink translational mode and the impurity mode. This resonant phenomenon is quite similar to the resonances observed in the kink-antikink collisions in nonlinear Klein-Gordon equations [189, 195, 200, 201].

To demonstrate the origin of the resonant kink-impurity interactions, we start from the SG model (3.83) which includes a local point-like impurity, i.e. $f(x) = \delta(x)$. When the impurity is absent, the model (3.83) displays the kink propagation without perturbations. In the presence of the δ-like impurity, the potential (3.85) becomes $U(X) = -2\epsilon \operatorname{sech}^2 X$, i.e. for $\epsilon > 0$ the impurity attracts the kink.

Kivshar *et al.* [326] (see also Zhang *et al.* [338]–[340]) have studied the kink-impurity interactions for $\epsilon > 0$ by numerical simulations. They found that there are *three* different regions of the initial kink velocity, namely, region of pass, of capture, and of reflection; and a critical velocity v_c (e.g., $v_c \approx 0.2678$ for $\epsilon = 0.7$) exists such that if the incoming velocity of the kink is larger than v_c, the kink will pass the impurity inelastically and escape without change of the propagation direction, losing a part of its kinetic energy through radiation and exciting an impurity mode. In this case, there is a linear relationship between the squares of the kink initial velocity v_i and its final velocity v_f: $v_f^2 = \alpha(v_i^2 - v_c^2)$, $\alpha \approx 0.887$ being constant. If the incoming velocity of the kink is smaller than v_c, the kink cannot escape to infinity from the impurity after the first collision, hence it will stop at a certain distance and return back (due to an attracting force acting on the kink from the impurity site) to interact with the impurity again. For most of the velocities, the kink will lose energy again in the second interaction, and finally it gets trapped by the impurity. However, for some special incoming velocities, the kink may escape to the direction *opposite* to the incident one after the second collision, i.e., the kink may be totally *reflected* by the impurity. The reflection is possible only if the kink initial velocity is taken from certain resonance windows (see Fig. 4.10). By numerical simulation, Kivshar *et al.* [326] have found a number of such windows. Using the idea of the resonant energy exchange between the kink translational mode and the impurity mode, it is possible to predict analytically the positions of the resonance windows [326]

$$v_n^2 \approx v_c^2 - \frac{11.0153}{(n\, t_{\rm im} + 0.3)^2}, \quad n = 2, 3, \ldots, \tag{4.19}$$

where $t_{\rm im}$ is the period of the impurity mode oscillation, and v_c is the critical velocity. This formula has been shown to provide a very good prediction (see the corresponding data in the table presented by Kivshar *et al.* [326]).

In order to provide an explanation of the resonance structures observed in the kink-impurity interactions, first we notice that the nonlinear system (3.83) supports a localized mode. By linearizing Eq. (3.83) for small u, the shape of the impurity mode can be found analytically to be

$$u_{\rm im}(x,t) = a(t)\, e^{-\epsilon|x|/2}, \tag{4.20}$$

where $a(t) = a_0 \cos(\Omega t + \theta_0)$, Ω is the frequency of the impurity mode, $\Omega = \sqrt{1 - \epsilon^2/4}$, and θ_0 is an initial phase. As a matter of fact, the impurity mode (4.20) can be considered as a small-amplitude oscillating mode trapped by the impurity, with energy $E_{\rm im} = \Omega^2 a_0^2/\epsilon$.

Now we may analyze the kink-impurity interactions through the collective-coordinate method taking into account two dynamical variables, namely the kink coordinate $X(t)$ and the amplitude of the impurity mode oscillation $a(t)$ [see Eq. (4.20)]. Substituting the ansatz

$$u = u_k + u_{\rm im} = 4 \tan^{-1} e^{[x - X(t)]} + a(t)\, e^{-\epsilon|x|/2} \tag{4.21}$$

into the Lagrangian of the system, and assuming that a and ϵ are small enough so that the higher-order terms can be neglected, it is possible to

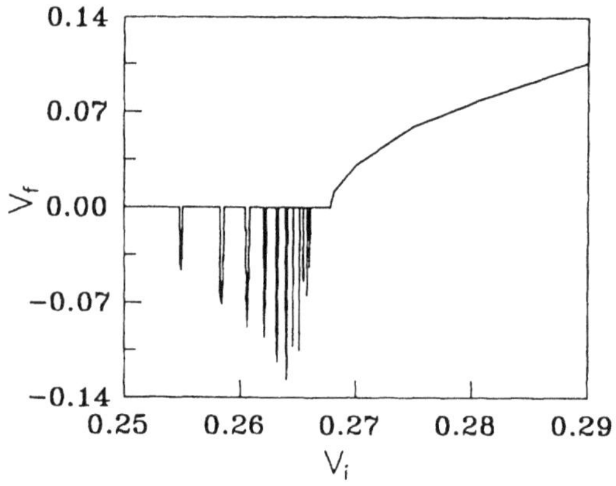

Fig. 4.10. Final kink velocity v_f as a function of the initial kink velocity v_i at $\epsilon = 0.7$. Zero final velocity means that the kink is captured by the impurity [326].

derive the following (reduced) effective Lagrangian

$$L_{\text{eff}} = \frac{1}{2}m\left(\frac{dX}{dt}\right)^2 + \frac{1}{\epsilon}\left[\left(\frac{da}{dt}\right)^2 - \Omega^2 a^2\right] - U(X) - aF(X), \tag{4.22}$$

where $U(X)$ is given above, and $F(X) = -2\epsilon \tanh X \, \text{sech} X$. The equations of motion for the two dynamical variables become

$$\begin{aligned} m\,\frac{d^2X}{dt^2} + U'(X) + aF'(X) &= 0, \\ \frac{d^2a}{dt^2} + \Omega^2 a + \frac{\epsilon}{2}F(X) &= 0. \end{aligned} \tag{4.23}$$

The system (4.23) describes a particle (kink) with coordinate $X(t)$ and effective mass m placed in an attractive potential $U(X)$ ($\epsilon > 0$), and "weakly" coupled with a harmonic oscillator $a(t)$ (the impurity mode). Here we say "weakly" because the coupling term $aF(X)$ is of order $O(\epsilon)$ and it falls off exponentially. The system (4.23) is a generalization of the well-known equation, $m_k d^2X/dt^2 = -U'(X)$, describing the kink-impurity interactions in the adiabatic approximation.

The dynamical system (4.23) can describe all features of the kink-impurity interactions. First, it may be used to calculate the threshold velocity of kink capture, which is given by the equation [339, 340],

$$v_{\text{thr}} = \frac{\pi\epsilon}{\sqrt{2}}\,\frac{\sinh[\Omega Z(v_{\text{thr}})/2v_{\text{thr}}]}{\cosh(\Omega\pi/2v_{\text{thr}})}, \tag{4.24}$$

where $Z(v) = \cos^{-1}[(2v^2 - \epsilon)/(2v^2 + \epsilon)]$. Comparing the analytical results with the direct numerical simulations of Eq. (3.83), Zhang *et al.* [338] found that the perturbation theory is valid only for very small ϵ, ($\epsilon \le 0.05$), while formula (4.24) gives good estimations of $v_{\text{thr}}(\epsilon)$ for ϵ over the region (0.2, 0.7).

As was pointed out by Kivshar *et al.* [326] and Zhang *et al.* [339], Eqs. (4.23) can be used as a qualitative model to explain the mechanism of resonant energy exchange between a classical particle and an oscillator. The resonant reflection of a particle by a potential well corresponds to the reflection of the kink by an attractive impurity. Therefore, the collective-coordinate approach can give a qualitative explanation of the resonance effects in the kink-impurity interactions. At the same time, the collective-coordinate model (4.23) is conservative, so that it cannot explain the inelastic effects such as the subsequent kink trapping by the impurity. Such effects can be explained only by introducing other degrees of freedom of the system, for example, an effective coupling to phonons or another subsystem [342].

It is important to note that similar resonance phenomena have been observed in the kink-impurity interactions in the ϕ^4 model [340, 343]. However, the resonant structures in the ϕ^4 kink-impurity interactions are more complicated than in the SG model because the ϕ^4 kink has an internal (shape) mode which also can be considered as an effective oscillator. Zhang *et al.* [340] have developed a collective-coordinate approach taking into account three dynamical variables, and they have found that due to the joint effect of the impurity and the kink internal mode oscillation, some resonance windows may disappear.

The resonant interactions described above have been analyzed numerically and analytically for a single impurity, and we can say that the physical mechanism of this effect has been well understood. It is clear that in the case of several impurities the well-defined energy-exchange process will be more difficult to observe (for the case of two impurities, see Zhang *et al.* [341]). However, increasing the number of impurities, it is likely to expect that the fine structure of resonances will be destroyed, especially for the case of a random lattice. However, the possibility of exciting impurity modes during the kink propagation will lead to an additional and, as we have seen for a single impurity, an efficient source of the energy loss during the kink propagation. So, a possible mechanism of the kink damping in disordered media is the excitation of localized mode vibrations due to impurities but not radiation of small-amplitude waves. This mechanism was mentioned in the earlier paper by Tsurui [344] who discussed the soliton propagation in a nonlinear lattice with isotopic disorder (see also the paper by Malomed [345] for the case of the ϕ^4 kink).

When a discrete breather (or nonlinear localized mode) interacts with an impurity, the resonant effects similar to those described above can be observed [216]. In particular, due to an overlapping of the breather and impurity mode frequencies, a local change of only $5 - 10\%$ of the particular mass is already sufficient to trap the breather in a lattice. Forinash *et al.* [216] also observed that the lattice discreteness enforce a stronger interaction between the localized breather mode and impurity mode, although they were not able to describe this effect quantitatively within the collective coordinate

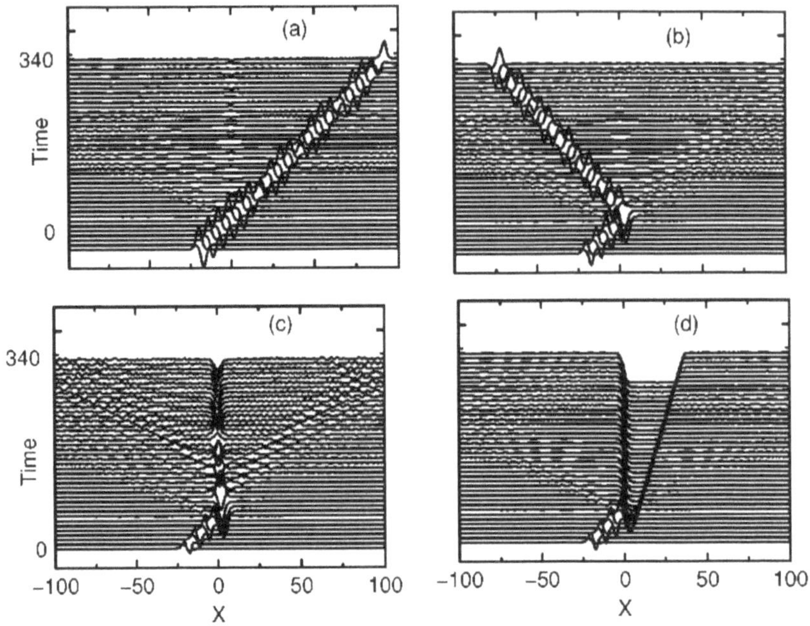

Fig. 4.11. Evolution of the field $u(x,t)$ for a breather with initial frequency $\omega_{\mathrm{br}} = 0.4$, velocity $v_{\mathrm{br}} = 0.35$, and different initial phases, scattered by an attractive impurity. The breather (a) passes the impurity for phase $\theta = \pi/2$, (b) is reflected for $\theta = 2\pi/15$, (c) is trapped for $\theta = \pi/30$, and (d) decays into a kink-antikink pair for $\theta = \pi/15$. In the latter case, the kink is trapped at the impurity and the antikink moves forward [238].

approach because the form of a breather moving in a discrete lattice is not known even numerically. A number of interesting effects was described by Forinash *et al.* [216] for the case of the interaction of breather with an excited impurity. In particular, they noticed that the disorder associated with impurities can act as a catalyst for nonlinear energy localization because it can cause the fusion of two nonlinear excitations into a single larger one.

Later, Zhang [238] presented more detailed results on the breather-impurity scattering in the SG model, and demonstrated how the outcome of the scattering depends on the breather's initial velocity, internal oscillation frequency, and phase. In particular, for an attractive impurity a breather can either be trapped by the impurity, pass the impurity, be totally reflected, or break into a kink-antikink pair. Some examples of such interactions are presented in Fig. 4.11 (a–d), for the variation of the input phase of the incoming breather.

As was found by Zhang [238], for a repulsive impurity, a small-amplitude breather behaves like a rigid particle, and the scattering result is mainly determined by the breather's initial velocity. In contrast, the scattering of a large-amplitude breather depends on both the breather's initial velocity and phase, and the breather can either pass the impurity or be reflected for a fixed initial velocity but different phases, as shown in Fig. 4.11(a)–(d). Existence of these complex and interesting phenomena is due to the interplay between the breather's internal and its translational degrees of freedom, which become strongly coupled when the breather is near the impurity region.

Chen *et al.* [346] investigated experimentally the interactions between the impurities and breathers in nonlinearly coupled pendulum chains subject to vertical vibrations. This system is described by a driven FK model. The experiments show that a defect pendulum attracts or repels the solitons including breather and kink, supplying an evidence to numerical results. The characteristic of the interactions, attraction or repulsion, is not only dependent on the polarity of the impurities, but also on the driving frequency and topology of the solitons. These observations testify to some perfect symmetries between long and short defects, breather and kink solitons in impurity-soliton interactions. The intensity of the interaction is related to the defect intensity and driving amplitude.

4.4 Discrete Breathers

4.4.1 General Remarks

In solid-state physics, the phenomenon of localization is usually perceived as arising from extrinsic disorder that breaks the discrete translational invariance of the perfect crystal lattice. Familiar examples include the localized vibrational modes around impurities in crystals [347] and Anderson localization of electrons in disordered media [348]. The typical perception is that in perfect lattices, free of extrinsic defects, phonons and electrons exist only in extended, plane wave states. This firmly entrenched perception was severely jolted in the late 1980's by the discovery of the so-called *intrinsic localized modes*, or *discrete breathers* as typical excitations in perfect but strongly nonlinear, spatially extended discrete systems.

For an anharmonic lattice without on-site potential, such localized breather-like states exist in a form of the so-called *intrinsic localized modes* [349, 350] for which the energy is localized on a few sites only, due to nonlinear interaction between the particles in the chain. Different properties of the intrinsic localized modes were discussed in a number of publications (see, e.g., Refs. [161],[351]–[370], and they were reviewed by Takeno *et al.* [351], Sievers and Page [371], Kiselev *et al.* [372], and Flach and Willis [373]. An important theoretical result was a rigorous mathematical proof of the existence of nonlinear localized modes [374] based on the analysis of a system of weakly coupled nonlinear oscillators (in the so-called anticontinuous limit, see Ref. [375, 376]).

The original model first studied for describing the discrete breathers is a chain of particles with anharmonic interatomic interaction; it describes a one-dimensional lattice without a substrate potential in which each atom interacts only with its nearest neighbors via a symmetric nonlinear potential [349, 350]. This model supports two types of localized modes (see Fig. 4.12 below). The localized Sievers-Takeno (ST) mode has a symmetric pattern [350] with its maximum amplitude localized at a lattice site, whereas the Page (P) mode has the asymmetric pattern being localized between the neighboring sites [352]. Both these modes can propagate through the lattice [377, 378] but their motion is strongly affected by the lattice discreteness, similar to the case of the FK kinks described above. Adiabatic motion of the localized mode can be also viewed as a sequential change between ST and P modes. Due to the lattice discreteness, these two states have different energies [379], and can viewed as two states of the same mode moving adiabatically trough an effective PN periodic potential. In such a picture, the P-mode corresponds to a minimum of the effective PN potential, whereas the ST-mode corresponds to a maximum and it displays a dynamical instability.

Nonlinear localized mode are known to exist in the models where nonlinearity appears through an on-site substrate potential, similar to the simplest type of the FK chain. The most typical model is the small-amplitude expansion of the FK model with a quartic on-site potential, discussed by several authors [161, 351, 357, 367, 380], where discrete breathers with the frequencies *below* the minimum frequency of the linear spectrum were found.

The existence of an effective PN potential for moving localized mode, similar to that for the FK kink, was demonstrated analytically and numerically (see, e.g., Refs. [381, 382]). Additionally, the spectrum of a linear discrete FK chain is limited from above by an upper cut-off frequency existing due to discreteness, so that one naturally expects to find localized modes with the frequencies *above* the cut-off frequency similar to the odd-parity ST modes or even-parity P modes in a chain with anharmonic interatomic interactions. The physically important problem related to these localized nonlinear modes is to prove that they are long-lived excitations which can contribute to many properties of nonlinear discrete systems.

4.4.2 Existence and Stability

The proof of existence of discrete breathers essentially requires two ingredients: (i) First, the system is truly nonlinear, i.e., the frequency of a mode depends on its amplitude (or, equivalently, its action); (ii) Second, the system is discrete so that the linear phonon spectrum exhibits gaps and does not extend up to infinite frequencies. With these conditions, local modes may be found with the frequency and its higher-order harmonics outside the linear phonon spectrum. Then, it can be understood intuitively that since this local mode cannot emit any radiation by linear phonons (at least to leading order), the local mode energy remains constant. This local mode may persist over very long time as a quasi-steady solution. Actually, a stronger result under

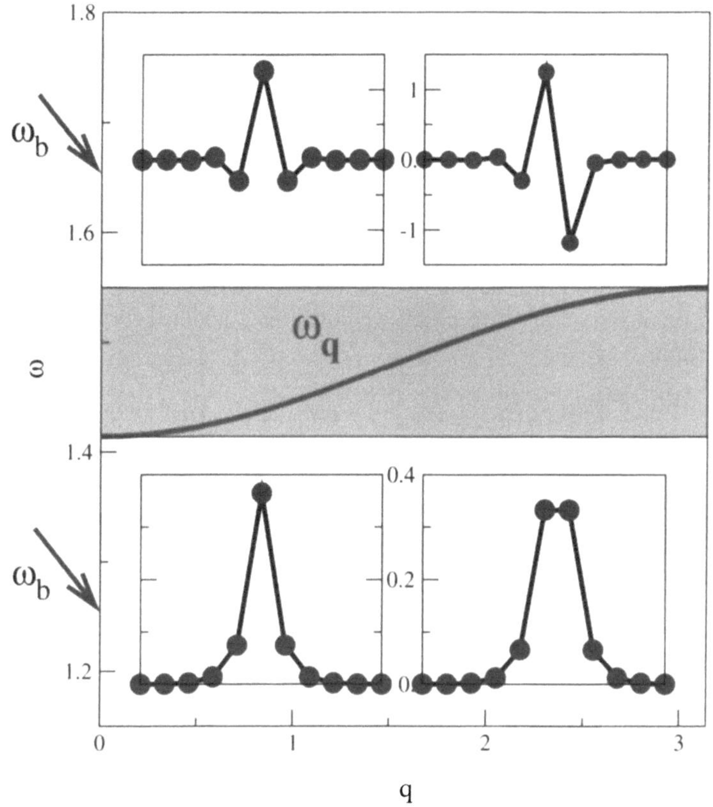

Fig. 4.12. Linear spectrum $\omega(q)$ and the structure of four types of discrete breathers (insets) in the chain with an on-site interaction. The arrows indicate the exact breather frequencies for the two bottom and top structures, respectively.

these assumptions can be proved, i.e., the existence of exact solutions with infinite lifetime [374, 375].

These arguments make clear why the exact breather solutions in the SG model are non generic. There are always breather harmonics in the linear phonon spectrum which extends up to infinity, and the absence of linear radiation can be viewed as a highly exceptional phenomenon. Although the discrete NLS equation is discrete, it is also particular, since for ensuring the existence of an exact solution it suffices that the fundamental frequency of the discrete breather (there are no harmonics) does not belong to the linear phonon spectrum. The same arguments also suggest that the existence of quasi-periodic discrete breathers in Hamiltonian systems with extended linear phonons, is also non-generic. The reason is that the time Fourier spectrum of such a solution is dense on the real axis and necessarily overlaps with the linear phonon bands. As a result, it should radiate energy by phonon emission till the solution either becomes time-periodic or vanishes. The initial method by MacKay and Aubry [374] for proving the existence of discrete breathers works close enough to a limit called anticontinuous, where the system decouples into an array of uncoupled anharmonic oscillators and where the existence of local modes is trivial. Then, the implicit function theorem can be used for proving that this exact solution persists when the anharmonic oscillators are coupled. There are several ways for determining an anticontinuous limit for a given model. These models can be classified in two classes. The first class involves only optical modes with a phonon gap. The second class, which is more realistic for real materials, involves acoustic phonons.

As an example of the first class, we consider a general form of the Klein-Gordon chain

$$\frac{d^2u_n}{dt^2} - g\left(u_{n+1} + u_{n-1} - 2u_n\right) + V'_{\text{sub}}(u_n) = 0, \tag{4.25}$$

where the atoms with scalar coordinate u_n and unit mass are submitted to an on-site potential $V_{\text{sub}}(u)$ with zero minimum at $u = 0$ and are elastically coupled to their nearest neighbors with coupling constant g.

This local potential expands for small u as $V_{\text{sub}}(u) = \frac{1}{2}u^2 + \ldots$, and the linear phonon frequency at wave vector q is $\omega(q) = \sqrt{1 + 4g\sin^2(q/2)}$ which yields that the phonon spectrum occupies the interval $[1, \sqrt{1+4g}]$ and exhibits a gap $[0, 1]$, where we assume the normalized gap frequency $\omega_0 = 1$. An anticontinuous limit is obtained for this model at $g = 0$, when the anharmonic oscillators are uncoupled. Then, the motion of each oscillator is periodic and its frequency $\omega(I)$ depends on its amplitude or equivalently on its action I [which is the area of the closed loop in the phase space $(u, \dot{u})$]. Thus, we generally have $d\omega(I)/dI \neq 0$.

There are trivial breather solutions at the anticontinuous limit corresponding, for example, to a single oscillator oscillating at frequency ω_{br} while the other oscillators are immobile. Weak coupling between the neighboring particles leads to the mode extension over a few sites (see Fig. 4.12). Depending on its sign, the nonlinearity may support spatially localized time-periodic states (breathers) with frequencies outside the spectrum of linear oscillations, as shown in Fig. 4.12 for the case of the ϕ^4-model with the potential $V_{\text{sub}}(u) = -\frac{1}{2}u^2 + \frac{1}{4}u^4$, where the gap frequency is $\omega_0 = \sqrt{2}$. Imagine setting up a discrete breather just below in the gap, $\omega_{\text{br}} < \sqrt{2}$. This can be achieved, for instance, by setting just one of the nonlinear oscillators in motion at a fairly small amplitude, so its frequency is just smaller than the smallest allowed linear frequency. Then, if the coupling between the sites is weak enough, not only will the fundamental frequency ω_{br} of this mode be below the gap but also all harmonics of ω_{br} will be above the gap. Hence, there is no possibility of a (linear) coupling to the extended modes, even in the limit of an infinite system when the spectrum $\omega(q)$ becomes dense. This means that the breather cannot decay by emitting linear waves (phonons) and is hence (linearly) stable.

In the case when $n\,\omega_{\rm br} \neq 1$ for any integer n, i.e., the breather frequency and its harmonics are not equal to the degenerate phonon frequency $\omega(q)$ at $q = 0$, the implicit function theorem can be used for proving, in a rigorous mathematical sense, that this solution persists up to some non-zero coupling g as a breather solution at frequency $\omega_{\rm br}$ [374]. More generally, even in complex models with an anticontinous limit, the existence of discrete breathers can be easily proven not too far from this limit [383]–[385].

However, models of the second class with acoustic phonons create problems because the phonon spectrum contains the frequency $\omega(q) = 0$. Then, $n\,\omega_{\rm br}$ always belongs to this phonon spectrum for $n = 0$. However, it was shown by Aubry [386] that in molecular crystal models with harmonic acoustic phonons and non-vanishing sound velocity, the resonant coupling between the breather harmonics at $n = 0$ and the acoustic phonons becomes harmless and that the breather persists in the coupled system.

4.4.3 The Discrete NLS Equation

Discrete Model

In spite of the fact that the discrete nonlinear Schrödinger (NLS) equation does not describe all the properties of the discrete breathers in nonlinear lattices with on-site potentials, it serves as a fundamental model that has many applications and allow to study many of the fundamental problems associated with the dynamics of discrete breathers.

To derive the discrete model, we consider the dynamics of a one-dimensional FK chain with an on-site potential of a general shape. Taking into account relatively small amplitudes of the modes, we expand the substrate potential keeping lower-order nonlinear terms. For the standard FK model, the coefficient α in front of the cubic term vanishes due to the symmetry of the potential, but the case considered below is more general (and it allows to treat, for example, the situation of a small dc field applied to the FK chain). Denoting by $u_n(t)$ the displacement of the n-th particle, its equation of motion is

$$\frac{d^2u_n}{dt^2} - g\left(u_{n+1} + u_{n-1} - 2u_n\right) + \omega_0^2 u_n + \alpha u_n^2 + \beta u_n^3 = 0, \tag{4.26}$$

where g is the coupling constant, ω_0 is the frequency of small-amplitude on-site vibrations in the substrate potential, α and β are the anharmonicity parameters of the potential. Linear waves of the frequency ω and wavenumber k are described by the dispersion relation

$$\omega^2 = \omega_0^2 + 4g\sin^2\left(\frac{ka_s}{2}\right), \tag{4.27}$$

a_s being the lattice spacing. As shown by Eq. (4.27), the linear spectrum has a gap ω_0 and it is limited by the cut-off frequency $\omega_{\rm max} = (\omega_0^2 + 4g)^{1/2}$ due to discreteness.

Analyzing slow temporal variations of the wave envelope, we will try to keep the discreteness of the primary model *completely*. In fact, this is possible only under the condition $\omega_0^2 \gg g$, i.e. when a coupling force between the neighboring particles is weak. Looking for a solution in the form

$$u_n = \phi_n + \psi_n e^{-i\omega_0 t} + \psi_n^* e^{i\omega_0 t} + \xi_n e^{-2i\omega_0 t} + \xi_n^* e^{2i\omega_0 t} + \dots, \tag{4.28}$$

we assume the following relations (similar to the continuum case, see, e.g., Ref. [314]): $\phi_n \sim \epsilon^2$, $\xi_n \sim \epsilon^2$, $\psi_n \sim \epsilon$, and also the following relations between the model parameters, $g \sim \epsilon^2$, $\omega_0^2, \alpha, \beta \sim 1$, $(d/dt) \sim \epsilon^2$. It is clear that this choice of the parameters corresponds to large values of ω_0^2 (we may simply divide all the terms by the frequency gap value).

Substituting Eq. (4.28) into Eq. (4.26) and keeping only the lowest order terms in ϵ, we obtain the equation for ψ_n,

$$2i\omega_0 \frac{d\psi_n}{dt} + g\left(\psi_{n+1} + \psi_{n-1} - 2\psi_n\right) - 2\alpha\left(\phi_n\psi_n + \psi_n^*\xi_n\right) - 3\beta\,|\psi_n|^2\psi_n = 0, \tag{4.29}$$

and two algebraic relations for ϕ_n and ξ_n,

$$\phi_n \approx -\frac{2\alpha}{\omega_0^2}\,|\psi_n|^2, \quad \xi_n \approx \frac{\alpha}{3\omega_0^2}\,\psi_n^2. \tag{4.30}$$

The results (4.29) and (4.30) are generalizations of the well known ones for the continuum case [314]. Thus, the final discrete NLS (or DNLS) equation stands,

$$i\frac{d\psi_n}{dt} + K(\psi_{n+1} + \psi_{n-1} - 2\psi_n) + \lambda|\psi_n|^2\psi_n = 0, \tag{4.31}$$

where $K = g/2\omega_0$, $\lambda = [(10\alpha^2/3\omega_0^2) - 3\beta]/2\omega_0$. Equation (4.31) is used below to analyze different types of localized modes in the FK chain. In fact, the discrete NLS equation (4.31), also known as the discrete self-trapping equation [387, 388], is rather known to have numerous physical applications, and it describes the self-trapping phenomenon in a variety of coupled-field theories, from the self-trapping of vibron modes in natural and synthetic biomolecules [387, 389] to the dynamics of a linear array of vortices, being a special limit of the discrete Ginzburg-Landau equation [390]. A generalized version of the discrete NLS equation with an arbitrary degree of the nonlinearity has also been considered to study the influence of the nonlinearity on the structure and stability of localized modes (see, e.g., Refs. [391]–[395]). We would like to point out ones more that in the present context Eq. (4.31) emerges as an approximate equation under the assumption of slow (temporal) variation of the envelopes as well as the neglecting of higher-order harmonics, and the latter means that we assume the gap frequency ω_0 large with respect to the other frequencies in the system, i.e. $\omega_0^2 \gg 4g$, and $\omega_0^2 \gg \alpha u_0$, βu_0^2, where u_0 is the wave amplitude. The former inequality is valid in a weakly dispersive system where ω_0 is close to $\omega_{\max}$, while the latter one means that the nonlinearity of the substrate potential is not large. These are usual conditions to get the NLS equation. In the lattice, however, the condition $\omega_0^2 \gg 4g$ means also that discreteness effects are considered strong pointing out the interest to the discrete modes localized on a few particles.

Modulational Instability

The discrete NLS equation (4.31) allows to analyze modulational instability of the constant-amplitude modes. As is well known, nonlinear physical systems may exhibit an instability that leads to a self-induced modulation of the steady state as a result of an interplay between nonlinear and dispersive effects. This phenomenon, referred to as modulational instability, has been studied in continuum models (see, e.g., Refs. [396]–[399]) as well as in discrete models [400]. As has been pointed out, modulational instability is responsible for energy localization and formation of localized pulses.

For the DNLS equation (4.31), derived in the single-frequency approximation, modulational instability in the lattice can be easily analyzed. Equation (4.31) has the exact continuous wave solution

$$\psi_n(t) = \psi_0\, e^{i\theta_n} \quad \text{with} \quad \theta_n = k a_s n - \omega t, \tag{4.32}$$

where the frequency ω obeys the *nonlinear* dispersion relation

$$\omega = 4K \sin^2\left(\frac{ka_s}{2}\right) - \lambda\psi_0^2 \,. \tag{4.33}$$

The linear stability of the wave (4.32), (4.33) can be investigated by looking for the perturbed solution of the form $\psi_n(t) = (\psi_0 + b_n)\exp(i\theta_n + i\chi_n)$, where $b_n = b_n(t)$ and $\chi_n = \chi_n(t)$ are assumed to be small in comparison with the parameters of the carrier wave. In the linear approximation two coupled equations for these functions yield the dispersion relation

$$\begin{gathered}[\Omega - 2K\sin(Qa_s)\sin(ka_s)]^2 = \\ 4K\sin^2(Qa_s/2)\cos(ka_s)\left[4K\sin^2(Qa_s/2)\cos(ka_s) - 2\lambda\psi_0^2\right]\end{gathered} \tag{4.34}$$

for the wavenumber Q and frequency Ω of the linear modulation waves. In the long wavelength limit, when $Qa_s \ll 1$ and $ka_s \ll 1$, Eq. (4.34) reduces to the usual expression obtained for the continuous NLS equation [397].

Equation (4.34) determines the condition for the stability of a plane wave with the wave-number k in the lattice. Contrary to what would be found in the continuum limit, now the stability depends on k. An instability region appears only if [400]

$$\lambda\cos(ka_s) > 0. \tag{4.35}$$

For positive λ and a given k, i.e. $k < \pi/2a_s$, a plane wave will be *unstable* to modulations in all this region provided $\psi_0^2 > 2K/\lambda$. The stability of the plane wave solutions to modulations of the wave parameters allows to conclude on possible types of nonlinear localized modes which may exist in the chain.

Modulational instability was shown to be an effective mean to generate localized modes in discrete lattices, because the lattice discreteness modifies drastically the stability condition. Daumont *et al.* [401] have analyzed, following the technique suggested by Kivshar and Peyrard [400], the modulational instability of a linear wave in the presence of noise in a lattice with cubic and quintic on-site potential, and they have demonstrated that the modulational instability is the first step towards energy localization (see also Ref. [402] for the case of the deformable discrete NLS model).

Spatially Localized Modes

As has been mentioned above, one of the main effects of modulational instability is the creation of localized pulses (see, e.g., Ref. [403]). In the present case, this means that for $\lambda > 0$ the region of small k is unstable and, therefore, the nonlinearity can lead to a generation of localized modes *below* the smallest frequency of the nonlinear spectrum band (4.33). Such a localized mode can be found in an explicit form from the DNLS equation (4.31) following the method by Page [352]. Looking for the stationary solutions of Eq. (4.31) in the form $\psi_n(t) = Af_n e^{-i\omega t}$, we obtain a set of coupled algebraic equations for the real functions f_n,

$$\omega f_n + K(f_{n+1} + f_{n-1} - 2f_n) + \lambda A^2 f_n^3 = 0. \tag{4.36}$$

We seek now two kinds of strongly localized solutions of Eq. (4.36), which are centered *at* and *between* the particle sites. First, let us assume that the mode is centered at the site $n = 0$ and take $f_0 = 1$, $f_{-n} = f_n$, $|f_n| \ll f_1$ for $|n| > 1$. Simple calculations yield the pattern of the so-called "A-modes" (left low mode in Fig. 4.12),

$$\psi_n^{(A)}(t) = A\,(\ldots, 0, \xi_1, 1, \xi_1, 0, \ldots)\, e^{-i\omega t}, \tag{4.37}$$

where the parameter $\xi_1 = K/\lambda A^2$ is assumed to be small (i.e., terms of order of ξ_1^2 are neglected). The frequency ω in Eq. (4.37) is $\omega = -\lambda A^2$, and it indeed lies below the lowest band frequency.

The second type of the localized modes, the "B-modes", may be found assuming that the mode oscillation is centered symmetrically between two neighboring particles (right low mode in Fig. 4.12),

$$\psi_n^{(B)}(t) = B\,(\ldots, 0, \xi_2, 1, 1, \xi_2, 0, \ldots)\, e^{-i\omega t}, \tag{4.38}$$

where the values ω and ξ_2 are defined as $\xi_2 = K/\lambda B^2$ and $\omega = -\lambda B^2$.

The calculation of an effective PN potential for the localized mode is much more difficult task than that for the kinks, because the localized modes possess more parameters and the PN potential cannot be defined rigorously [404].

However, the existence of a kind of PN potential affecting the motion of a nonlinear localized mode through the lattice can be easily demonstrated. First, following the paper by Kivshar and Campbell [381], we can imagine a localized wave *of a fixed shape* being translated rigidly through the lattice. Then, it is clear that when the peak is centered on a lattice site, the symmetry is of the "A" form, whereas when the peak is centered halfway between the sites, the symmetry is of the "B" form. This observation motivates a comparison between the energies of these two modes provided, for the case of the NLS equation, the integral of motion $N = \sum_n |\psi_n|^2$ is conserved. Such a comparison of the integrals N calculated for A- and B-modes gives the relations between the amplitudes A and B in the lowest order in the small parameter ξ_1 and ξ_2, $A^2 = 2B^2$. With this condition on A and B, we can interpret now the two modes as stationary states of *the same* localized mode, calculating a difference in the energy between these two stationary states, $\Delta E_{AB} = E_A - E_B = -\frac{1}{2}\lambda A^4 + \lambda B^4 = -\frac{1}{4}\lambda A^4$.

From this simple estimate, it follows an important conclusion that there exists an effective energy barrier (the height of the effective PN potential) between these two stationary states of the discrete NLS equation, also meaning that any translation of the nonlinear localized modes through the lattice will be affected by a periodic energy relief. In particular, a localized mode may be *captured* by the potential (i.e. trapped by the lattice discreteness). A simple way to observe this effect is "to push" the localized mode to move through the lattice by variation of the mode initial phase [405]. The B-mode, corresponding to a maximum of the potential, starts to move almost immediately, whereas it is clear that a certain energy barrier must be overcome to move through the lattice [405]. This is an indirect manifestation of the effective barrier due to the lattice discreteness, in spite of the fact it cannot be defined in a rigorous way.

Importantly, the analysis presented above is somewhat related to the stability properties of the nonlinear localized modes: the stationary localized mode corresponding to a local *maximum* of the PN potential should display an instability whereas the mode corresponding to a *minimum* should be stable. This qualitative observation is in an agreement with the work by Sandusky *et al.* [360] who have shown numerically and analytically (using other arguments and not referring to the PN potential) that for the case of interatomic quartic anharmonicity the ST localized mode (left upper mode in Fig. 4.12) is in fact unstable, but the P mode (right upper mode in Fig. 4.12) is extremely stable.

A simple way to calculate the shape of the PN potential in the case of the NLS equation is to use the integrable version of the lattice equation, i.e. the Ablowitz-Ladik (AL) model [406], and to take the difference between these two models as a perturbation (see also Refs. [234, 382, 407], where the perturbed AL chain has been considered as a novel physically important model). To do so, we present the primary discrete NLS equation (4.31) in the form

$$i\frac{d\psi_n}{dt} + K(\psi_{n+1} + \psi_{n-1} - 2\psi_n) + \frac{1}{2}\lambda(\psi_{n+1} + \psi_{n-1})|\psi_n|^2 = R(\psi_n) , \quad (4.39)$$

where

$$R(\psi_n) = \frac{1}{2}\lambda \mid \psi_n \mid^2 (\psi_{n+1} + \psi_{n-1} - 2\psi_n) . \quad (4.40)$$

We start from the exact solution of the AL model (Ablowitz and Ladik, 1976) for the unperturbed case ($R = 0$) which we present in the form

$$\psi_n(t) = \frac{\sinh\mu \, \exp\left[i\nu(n - x_0) + i\alpha\right]}{\cosh\left[\mu(n - x_0)\right]} , \quad (4.41)$$

where $d\mu/dt = 0$, $d\nu/dt = 0$, $dx_0/dt = (2/\mu)\sinh\mu\,\sin\nu$, and $d\alpha/dt = 2[\cosh\mu\,\cos\nu - 1]$. In Eq. (4.41) and the subsequent calculations related to Eqs. (4.39), (4.40) we use the normalized variables, $t \to t/K$ and $|\psi_n|^2 \to (2K/\lambda)|\psi_n|^2$.

Considering now the right-hand-side of Eq. (4.39) as a perturbation (that is certainly valid for not too strongly localized modes), we may use the perturbation theory based on the inverse scattering transform [27]. For the case of the AL model, the corresponding version of the soliton perturbation theory was elaborated by Vakhnenko and Gaididei [408]. According to this approach, the parameters of the localized solution (4.41), i.e. μ, ν, α and x_0, are assumed to be slowly varying in time. The equations describing their evolution in the presence of perturbations were obtained by Vakhnenko and Gaididei [408]. Substituting Eq. (4.40) into those equations and applying the Poisson formula to evaluate the sums appearing as a result of discreteness of the primary AL model, we obtain two coupled equations for the soliton parameters ν and x_0:

$$\frac{dx_0}{dt} = \frac{2}{\mu}\sinh\mu\,\sin\nu , \quad (4.42)$$

$$\frac{d\nu}{dt} = -\frac{2\pi^3\sinh^2\mu\,\sin(2\pi x_0)}{\mu^3\sinh(\pi^2/\mu)} , \quad (4.43)$$

and also $d\mu/dt = 0$. In Eq. (4.43) we took into account only the contribution of the first harmonic because the higher ones, of the order of s, will always appear with the additional factor $\sim \exp(-\pi^2 s/\mu)$ which is assumed to be small.

The system (4.42)-(4.43) is a Hamiltonian one, and the corresponding Hamiltonian is given by the expression

$$H = -\frac{2}{\mu}\sinh\mu\,\cos\nu - \frac{\pi^2\sinh^2\mu}{\mu^3}\frac{\cos(2\pi x_0)}{\sinh(\pi^2/\mu)} , \quad (4.44)$$

where the parameters x_0 and ν have the sense of the generalized coordinate and momentum, respectively. The first term is the kinetic energy of the effective particle, the second one is the periodic potential, which is, as a matter of fact, the effective periodic PN relief. In the approach assuming the difference between the two model small, i.e. the parameter μ small, the amplitude of the PN potential defined as

$$U_{\max} = \frac{\pi^2 \sinh^2 \mu}{\mu^3 \sinh(\pi^2/\mu)}, \tag{4.45}$$

is exponentially small in the parameter μ^{-1}. As we can see, the dependence (4.45) and the periodic potential $U_{\max} \cos(2\pi x_0)$ are similar to those in the problem for the topological kink in the FK model discussed above. As a result, all types of motion of the effective particle remain the same as in the case of the kink, in particular, the nonlinear mode may be trapped by discreteness similar to a trapping of a kink.

The approach based on the perturbations of the integrable AL lattice model have been examined by Bang and Peyrard [315] who found only qualitative agreement with numerical results on nonlinear localized modes in the generalized FK models. One of the possible reason for this is the strong assumption in the derivation of the adiabatic equations (4.42) and (4.43) which is valid only in the case of a very small perturbation. In a general case, all higher-order harmonics in the Poisson sums give a contribution of the same order and the quantitative agreement should be no longer expected. However, the theory based on the AL model may serve as a simple example explaining how the effective periodic dependence of the mode parameters appears due to lattice discreteness when the lattice model becomes nonintegrable.

4.4.4 Dark Breathers

As is well known, in the continuum limit approximation the DNLS equation (4.31) supports two different kinds of soliton solutions, *bright* and *dark* solitons. The bright solitons are similar to spatially localized modes discussed above, whereas localized structures which are similar to dark solitons are less discussed in literature. It is the purpose of this section to present two types of these structures in the FK type lattice following the paper by Kivshar [409] (see also Refs. [410]–[412]). This type of structures has been observed experimentally by Denardo *et al.* [413] in an array of parametrically driven pendulums (see also Ref. [414] for the case of a diatomic lattice). Similar types of dark-soliton localized modes for a chain with anharmonic interatomic coupling has been analyzed by Bortolani *et al.* [415].

First, we notice that for positive λ, the continuous wave solution is stable only for $k > \pi/2a_s$, so that dark-profile structures are possible, for example, near the cut-off frequency $\omega_m = 4K$. Substituting $\psi_n = (-1)^n \Psi(x,t)\, e^{i\omega_m t}$ into Eq. (4.31), where the slowly varying envelope Ψ is found to be a solution of the continuous NLS equation, it is easy to obtain the dark soliton solution in the continuum approximation,

$$\psi_n = (-1)^n A \tanh(Ax)\, e^{-i\Omega t}, \tag{4.46}$$

where $\Omega = 4K - \lambda A^2$, and $x = n a_s \sqrt{K}$ is considered as a continuous variable.

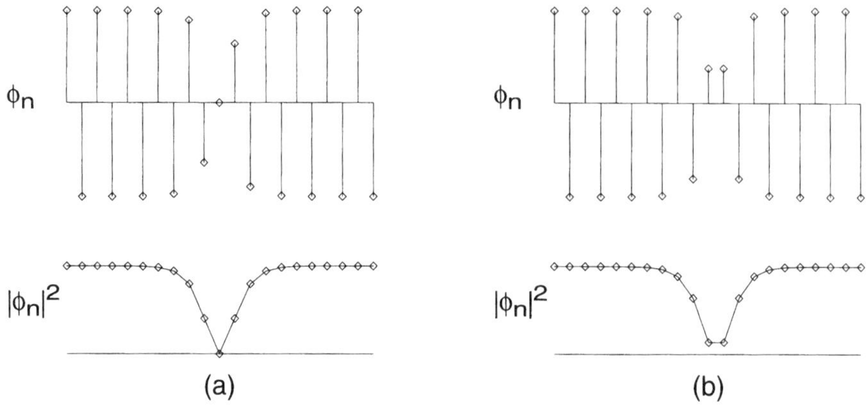

Fig. 4.13. Two types of "staggered" dark-soliton modes in a lattice. Shown are the oscillation amplitude at each site (upper row) and the value $|\psi_n|^2$ (lower row) [416].

Looking now for the similar structures in the discrete NLS equation (4.31), we find that they are possible, for example, in the form

$$\psi_n = A\,e^{-i\Omega t}(\ldots, 1, -1, 1, -\xi_1, 0, \xi_1, -1, 1, \ldots), \tag{4.47}$$

where $\Omega = 4K - \lambda A^2$, and $\xi_1 = 1 - \Delta_1$, $\Delta_1 = K/\lambda A^2 \ll 1$. The structure (4.47) is a solution of the NLS equation with the accuracy better for smaller Δ_1, and it is a phase-kink excitation with the width localized on a few particles in the lattice [see Fig. 4.13(a)]. Because the frequency Ω coincides with the cut-off frequency of the nonlinear spectrum, we call these solutions "cutoff kinks".

The dark-soliton modes presented in Figs. 4.13(a) and 4.13(b) describe two types of stationary "black" solitons in a discrete lattice, the on-site mode (A-mode) centered with zero intensity at a lattice site, and the inter-site mode (B-mode) centered between two sites. These two modes can be uniquely followed from the continuous limit ($g \to \infty$) to the anticontinuous limit ($g \to 0$). At $g \to 0$, the A-mode describes a single "hole" in a background wave with constant amplitude and a π phase shift across the hole. Similarly, the B-mode describes the lattice oscillation mode with a π phase shift between two neighboring sites and no hole.

Johansson and Kivshar [416] studied the dark breathers and their stability in the discrete NLS equation, and described the oscillatory instability of dark solitons. They found the critical value of the lattice coupling constant g, $g_{\rm cr} \approx 0.07647$ that defines the stability region. At $g = g_{\rm cr}$, a Hopf-type bifurcation occurs, as two complex conjugated pairs of eigenvalues leave the imaginary axis and go out in the complex plane. Thus, an oscillatory instability occurs for the A-mode when $g > g_{\rm cr}$, with the instability growth rate given by the real part of the unstable eigenvalue. For the B-mode, all eigenvalues lie at zero when $g = 0$. As soon as g is increased, one pair goes out on the real axis and stays there for all $g > 0$. Thus, the B-mode is always unstable. This instability appears due to inherent discreteness of a nonlinear lattice model, and the universality of the instability scenario suggests that it should be also observed in other nonlinear models supporting dark solitons. The instability observed here may be regarded as an extension of the instabilities existing for short arrays [387, 388].

Another type of dark-profile nonlinear localized structures which may be also described analytically in continuum as well as discrete models, is realized in the case when the mode frequency is just at the middle of the spectrum band, i.e. the wavenumber is equal to $\pi/2a_s$. In this case we may separate the particles in the chain into two subsets, odd and even ones, and describe their dynamics separately, introducing the new variables, i.e. $\psi_n = v_n$, for $n = 2l$, and $\psi_n = w_n$, for $n = 2l + 1$. The main idea of such an approach is to use the continuum approximation for *two* envelopes, v_n and w_n [409].

Looking now for solutions in the vicinity of the point $k = \pi/2a_s$, we may use the following ansatz,

$$v_{2l} = (-1)^l\, V(2l, t)\, e^{-i\omega_1 t}, \quad w_{2l+1} = (-1)^l\, W(2l+1, t)\, e^{-i\omega_1 t}, \tag{4.48}$$

where $\omega_1 = 2K$ is the frequency of the wavelength-four linear mode, and we assume that the functions $V(2l, t)$ and $W(2l+1, t)$ are slowly varying in space. Substituting Eqs. (4.48) into Eq. (4.31), we finally get the system of two coupled equations,

$$i\frac{\partial V}{\partial t} + 2a_s K\, \frac{\partial W}{\partial x} + \lambda |V|^2 V = 0, \tag{4.49}$$

$$i\frac{\partial W}{\partial t} - 2a_s K\, \frac{\partial V}{\partial x} + \lambda |W|^2 W = 0, \tag{4.50}$$

where the variable x is treated as continuous one. Analyzing localized structures, we look for stationary solutions of Eqs. (4.49) and (4.50) in the form $(V, W) \propto (f_1, f_2)e^{i\Omega t}$ assuming, for simplicity, the functions f_1 and f_2 to be real. Then, the stationary solutions of Eqs. (4.49) and (4.50) are described by the system of two ordinary differential equations of the first order,

$$\frac{df_1}{dz} = -\Omega f_2 + \lambda f_2^3, \tag{4.51}$$

$$\frac{df_2}{dz} = \Omega f_1 - \lambda f_1^3, \tag{4.52}$$

where $z = x/2a_s K$. Equations (4.51), (4.52) represent the dynamics of a Hamiltonian system with one degree of freedom and the conserved energy, $E = -\frac{1}{2}\Omega\,(f_1^2 + f_2^2) + \frac{1}{4}\lambda\,(f_1^4 + f_2^4)$, and they may be easily integrated with the help of the auxiliary function $\phi = (f_1/f_2)$, for which the following equation is valid,

$$\left(\frac{d\phi}{dz}\right)^2 = \omega_1^2 \Omega^2 (1 + \phi^2)^2 + 4\lambda E\,(1 + \phi^4). \tag{4.53}$$

Different kinds of solutions of Eq. (4.53) may be characterized by different values of the energy E [409]. On the phase plane (f_1, f_2), the soliton solutions correspond to the separatrix curves connecting a pair of the neighboring saddle points $(0, f_0)$, $(0, -f_0)$, $(f_0, 0)$, or $(-f_0, 0)$, where $f_0^2 = \Omega/\lambda$. Calculating the value of E for these separatrix solutions, $E = -\Omega^2/4\lambda$, it is possible to integrate Eq. (4.53) in elementary functions and to find the soliton solutions,

$$\phi(z) = \exp(\pm\sqrt{2}\,\Omega z), \tag{4.54}$$

$$f_2^2 = \frac{\Omega e^{\mp\sqrt{2}\,\Omega z}[2\cosh(\sqrt{2}\,\Omega z) \pm \sqrt{2}]}{2\lambda\cosh(2\sqrt{2}\,\Omega z)}, \qquad f_1 = \phi f_2. \tag{4.55}$$

The solutions (4.54), (4.55), but for negative Ω, exist also for defocusing nonlinearity when $\lambda < 0$.

These results give the shapes of the localized structures in the discrete nonlinear lattice. The whole localized structure represents two kinks in the odd and even oscillating modes which are composed to have opposite polarities.

Highly localized nonlinear structures in the lattice corresponding to the solutions (4.54), (4.55) may be also found, and one of these structures has the following form, $\psi_n = Ae^{-i\Omega t}(\ldots, 1, 0, -1, 0, \xi_2, \xi_2, 0. -1, 0, 1, \ldots)$, where $\Omega = 2K - \lambda A^2$ is the frequency at the middle of the nonlinear spectrum, and $\xi_2 = 1 - \Delta_2$, $\Delta_2 = K/2\lambda A^2 \ll 1$. The approximation is better for smaller values of the parameter Δ_2.

4.4.5 Rotobreathers

Oscillatory modes are not the only nonlinear localized modes in the nonlinear discrete model. For lattices of coupled rotators, Takeno and Peyrard [369] demonstrated the existence of the so-called *rotating modes* in which a central rotator performs a monotone increasing rotation while its neighbors oscillate around their equilibrium positions. In the phase space of the system, the motion of the central site and the motion of the neighbors lie on opposite sides of a separatrix. As a result the rotating modes are *intrinsically discrete.*

However, due to the specificity of the rotating modes, getting an analytical solution turns out to be much more difficult than for oscillatory modes because there is a qualitative change between the motion of the central site and the oscillations of the others. This precludes any continuum limit, and different functions must be used to describe the dynamics of different sites. Takeno and Peyrard [369] showed how an approximate solution could be derived. Numerical checks have indicated that this solution was rather good because it treated intrinsically the discreteness of the lattice. Later, they presented a solution based on lattice Green's function [417] The numerical solution that can be deduced from the analytical expressions given by the

Green's functions is slightly better than the previous one, but, more importantly, the method is not restricted to one-dimensional lattices and can be formally extended to two or three dimensions.

The model of coupled rotators can be written in the following form,

$$\ddot{u}_n + g \sin u_n = J[\sin(u_{n+1} - u_n) - \sin(u_n - u_{n-1})], \tag{4.56}$$

where u_n is the field variable associated to the n-th site of the lattice, and J is constant. This set of equations describes the dynamics of a chain of coupled pendulums or molecules in a plastic crystal.

We look for a particular solution of Eq. (4.56) in which a single rotator (say, $n = 0$) undergoes a rotational motion while the others oscillate. This means that $|u_0| \gg |u_n|$ for $n \geq 1$, and we can assume that the large amplitude motion of the central site at $n = 0$ is not perturbed by the small vibrations of the neighboring sites. The central rotator is described by the equation

$$\ddot{u}_0 + \omega_0^2 \sin u_0 = 0, \tag{4.57}$$

where $\omega_0 = \sqrt{g + 2J}$, and its solution rotating with the angular velocity Ω has the form, $u_0(t) = 2\mathrm{am}[(\Omega/2)t, k]$, where k is the modulus of the Jacobi elliptic function, $k < 2\omega_0/\Omega < 1$. Coupling this rotation to the neighboring site allows to construct a mode in which a central rotator performs a monotone increasing rotation while its neighbors oscillate around their equilibrium positions. The numerical simulations show therefore that a rather accurate expression of the rotating modes can be obtained with the lattice Green's function method [417]. The starting point is to linearize the equations outside of the site that performs a continuous rotation. This approximation is justified by the qualitative difference between the motion of the center and that of the other sites, as well as by the small amplitude of the vibrations of the off-center sites, even in the case of a rather strong coupling.

There is no smooth way to go from a full rotation to an oscillation. The theorem of MacKay and Aubry [374] can be extended to show that rotating modes can be exact solutions of the coupled-rotator equations of motions, and numerical investigations of the thermalization of the rotator lattice [369] show that, like the breathers, the rotating modes can be thermally excited. Moreover, while it is very easy to thermally excite rotating modes that involve one or a few lattice sites, nonlocalized rotating modes are not observed unless one reaches very high temperatures because they have a very large energy. This is a sharp contrast with oscillatory modes for which the nonlocalized counterparts, i.e., oscillatory waves, or phonons, are easily found.

4.5 Two-Dimensional Breathers

Although most of this methodology and associated theoretical techniques were developed for the case of one spatial dimension, there have been some studies of higher-dimensional systems. Most importantly, a rigorous proof of the existence of breathers in higher-dimensional nonlinear lattices applies for the case of arbitrary dimensions. Also, the results of numerical studies have been published for several simple nonlinear lattices as, for example, two-dimensional Fermi-Pasta-Ulam chains [418]–[420], Klein-Gordon chains [421]–[422], and also for discrete NLS systems [423]–[425]. Systems of higher dimensionality have also been investigated [426].

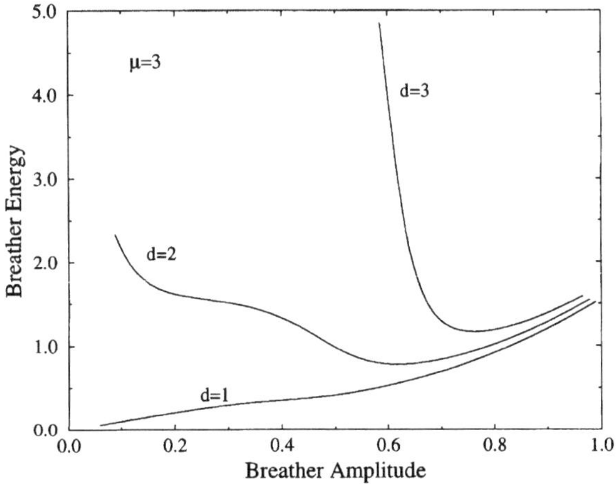

Fig. 4.14. Breather energy versus its amplitude for the d-dimensional discrete NLS system for different lattice dimensions, marked on the plot. System sizes for $d = 1, 2, 3$ are $N = 100$, $N = 25^2$, $N = 31^3$, respectively [424].

Discrete breathers are time-periodic, spatially localized solutions of equations of motion for classical degrees of freedom interacting on a lattice. In spite of the fact that their existence and the physical origin do not depend crucially on the lattice dimension, their properties such as the excitation energy and stability, are different in the systems of different dimensions. Flach *et al.* [424] were first to report on studies of energy properties of (one-parameter) breather families in one-, two-, and three-dimensional lattices. Taking the example of the d-dimensional discrete NLS model, they demonstrated that breather energies have a positive lower bound if the lattice dimension of a given nonlinear lattice is greater than or equal to a certain critical value. In particular, they estimated the breather energy as $E_{\rm br} \sim A_0^{(4-zd)/2}$, where A_0 is the breather amplitude, d is the lattice dimension, and z is a power of nonlinearity. Thus, for $d > d_c = 4/z$, the breather energy diverges for small amplitudes, so that there exists a threshold of the breather energy in the system (see Fig. 4.14). These findings could be important for the experimental detection of discrete breathers.

Kevrekidis *et al.* [425] developed a methodology for the construction of two-dimensional discrete breathers. Application to the discrete NLS equation on a square lattice reveals *three different types of breathers*. Considering an elementary plaquette, the most unstable mode is centered on the plaquette, the most stable mode is centered on its vertices, while the intermediate (but also unstable) mode is centered at the middle of one of the edges. Thus, interesting differences between the cases of one and two dimensions arise through the existence of a hybrid mode, as well as from a much richer stability scenario. This picture has been characterized using bifurcation theory tools. Kevrekidis *et al.* [425] studied the bifurcations of the two-dimensional breather modes in a frequency-power phase diagram; the bifurcation diagram was found to be consistent with the numerical stability analysis.

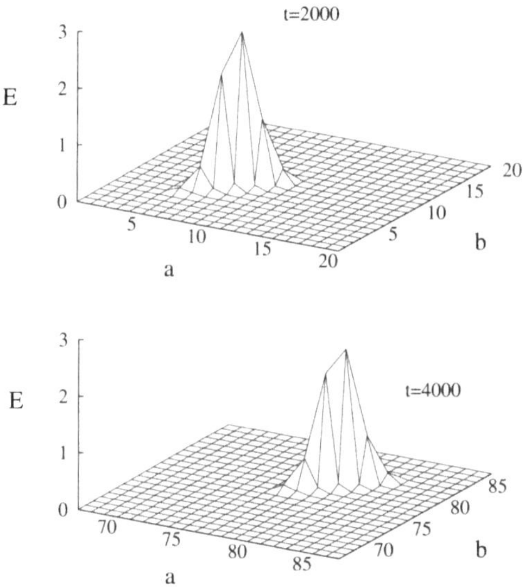

Fig. 4.15. A breather travelling along the direction of the CuO_2 plane of atoms in some high-T_c materials. The structure has fcc layers of fixed atoms sitting symmetrically above and below, that create an on-site potential for the atoms in the (a, b) plane [429].

The studies of multi-dimensional breathers are often motivated by their expected applications for describing the properties of realistic physical systems. For example, it has been reported that single layers of cuprate superconducting material can have a threshold temperature of the superconducting transition T_c similar to that of the bulk material [427]. This suggests that the underlying pair-bonding mechanism in high-T_c superconductors can be active in a two-dimensional plane of the crystal. In addition, other studies support the idea that the lattice structural and dynamic properties play a critical role in the mechanisms for superconductivity [428]. Being motivated by a previous studies that the superconducting properties of cuprates were correlated to the existence of linear chains of atoms, Marin *et al.* [429] provided a realization of these waves in the form of mobile discrete breathers found earlier in the hexagonal lattice case [430]. They describe numerical simulations which suggest that lattice nonlinearities allow the transport of strongly localized and robust packets of vibrational energy (discrete breathers) along these chains. The results support previous studies which correlated these particular structural properties with superconductivity in these cuprates.

Marin *et al.* [429] modelled the two-dimensional copper-oxide layer of a typical YBCO compound by means of lattice dynamics applied to a two-dimensional generalized FK model. They used pair potentials between the atoms of the plane, and simulated the three-dimensional environment of the crystal via a layer of fixed atoms sitting above and below the plane. Figure 4.15 shows a typical simulation of a breather envelope, shown for the energy of the breather E in arbitrary units. In between the two snapshots, the excitation has travelled about 100 lattice cells, losing only some 10% of the energy, but remaining as an exponentially localized excitation. Moving breathers in general are not expected to be infinitely lived, due to the non-integrability of these lattice models, but two-dimensional moving breathers, like their one-dimensional counterparts, were found to exhibit remarkably long lifetimes.

4.6 Physical Systems and Applications

Localized solutions describing the breather modes were predicted for many nonlinear models, including classical spin lattices [431]–[434]. More recently, one of the first attempts of the experimental observation of discrete breathers in antiferromagnets has been made by Schwarz *et al.* [435].

However, one of the most visually impressive recent observation of discrete breathers (in fact, rotobreathers) was reported in the field of superconductivity. Spatially localized excitations in the form of discrete breathers were observed experimentally by two independent groups in ladders of small Josephson junctions [436, 437]. The localized excitations correspond to spatially localized voltage drops in the presence of a spatially homogeneous dc bias (current) threading the ladder. A few junctions are in the resistive state, while the other junctions are superconducting. Notably the superconducting junctions generate ac voltages due to their coupling to the resistive junctions. These dynamical effects cause varieties of resonances and related hysteresis loops in current-voltage characteristics. The breather states of the ladder are visualized using a low temperature scanning laser microscopy, and were compared with the numerical results for the discrete breathers solutions in the corresponding model equations. The stability analysis of these solutions was used to interpret the measured patterns in the current-voltage characteristics. Various states shown in Fig. 4.16 account for different branches in the current-voltage plane. Each resistive configuration is found to be stable along

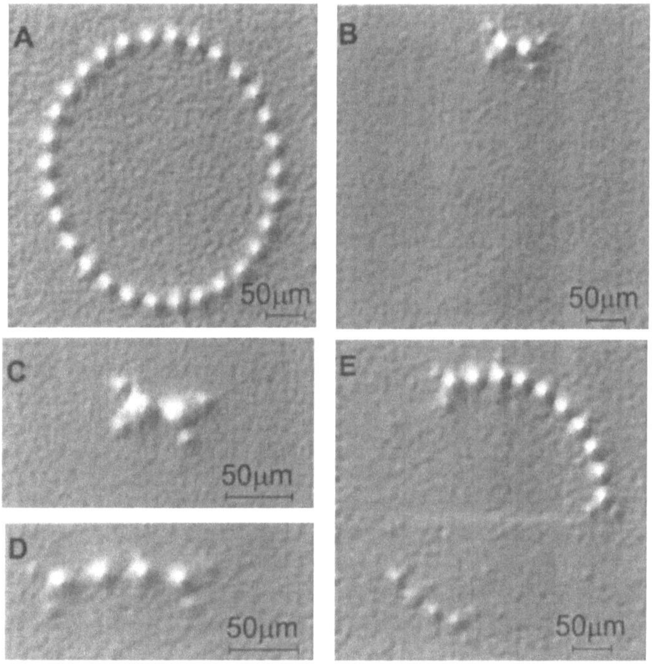

Fig. 4.16. Whirling states measured in the annular ladder using the low temperature scanning laser microscope: (A) spatially homogeneous whirling state, (B – E) various localized states corresponding to discrete breathers [436].

its particular branch. The transitions between the branches are discontinuous in voltage, and all branches of localized states lose their stability at a voltage of about 1.4 mV. The occurrence of discrete breathers is inherent to these systems.

One of the first experimental observations of discrete breathers was reported in guided-wave optics, for the case of discrete spatial solitons excited in waveguide arrays [438] (see also a comprehensive review paper [439] and the book [440]). Spatial optical solitons are self-trapped optical beams that can propagate in slab waveguides or a bulk medium due to self focussing in nonlinear Kerr media. All experiments to date have been performed in waveguide arrays made of AlGaAs. Each waveguide is 4 μm wide, and has a length of a few mm. Arrays contain typically 40–60 waveguides. The strength of the coupling between neighboring guides was controlled through the spacing between the waveguides, between 2 μm to 7 μm. The light source in these experiments was a synchronously pumped laser. Light was launched into a single waveguide on the input side of a 6 mm long sample, and the light distribution at the output end was recorded. At low light power, the propagation is linear, and the light expands over all waveguides at the output. As the input power is increased above a threshold value, the width of the output distribution shrinks. At a power more than 200 W, light is confined to about 5 waveguides around the input waveguide and a highly localized, nonlinear mode is created. Eisenberg *et al.* [438] have also shown that the theoretical description of this system is reduced to a discrete NLS equation. While the discrete cells are due to the discrete set of waveguides, the continuous variable (time) in the equation corresponds to the spatial coordinate along the propagation direction along the waveguide.

Russell *et al.* [441] described an experimental model consisting of an anharmonic chain of magnetic pendulums acting under gravity. This nonlinear model seems to be the first experimental example which visually demonstrates highly mobile and strongly localized dynamically stable oscillating states in the form of the breather modes. This type of nonlinear excitations is different from the well-known supersonic Toda-like solitons which describe the propagation of a lattice deformation of a constant profile. Moreover, the model suggested by Russell *et al.* [441] has interest in its own right as an excellent pedagogical tool in the study of nonlinear lattices.

Many physical applications of discrete breathers have been suggested. One of them is understanding slow relaxation properties of glassy materials. Another application concerns energy localization and transport in biological molecules by targeted energy transfer of breathers. A similar theory could be used for describing targeted charge transfer of nonlinear electrons (polarons) and, more generally, for targeted transfer of several nonlinear excitations (e.g., Davydov solitons).

5 Ground State

So far, we discussed only the case when the mean distance between the atoms, a_A, coincides with the period of the substrate potential, a_s. In a more general case we should consider $a_A \neq a_s$. Then the FK model possesses two competing length scales and its ground state may become nontrivial. In this chapter we discuss a number of novel effects associated with the existence of nontrivial ground states in the FK chain.

5.1 Basic Properties

The FK chain is described by a one-dimensional model of classical mechanics, so that we may assume that its ground state (GS), i.e. the atom configuration with the lowest potential energy, should be quite simple. This is indeed the case when the mean distance between the atoms a_A is equal to the period of the substrate potential a_s. This case corresponds to a trivial GS when all atoms occupy the minima of the substrate potential, except the case of the nonconvex potential discussed above in Sect. 3.5.2.

However, in a more general case of $a_A \neq a_s$, the FK model possesses two competing length scales, and the GS problem becomes less trivial. Indeed, the interatomic interaction favors an equidistant separation between the atoms, while the interaction with the substrate tends to force the atoms into a configuration in which they are regularly spaced at the distance a_s. As a result of this competition, the system may exhibit two distinct types of phases. The first phase is *commensurate* (C-phase, it is known also as *periodic* or *crystalline* phase) when the atom coordinates x_l form an arithmetic series of increment a_s (or multiple of a_s). The second phase is *incommensurate* (IC-phase) when the sequence x_l does not form an arithmetic series. The concept of incommensurability is a very interesting topic from both mathematical and physical viewpoints. Besides, the equations for stationary configurations reduce to the so-called *standard map*, which is one of the classical models of the theory of chaos.

In this chapter we deal with an infinite FK chain and consider the case when both the length L and the number of atoms N tend to infinity, but the linear concentration of atoms,

$$n = \lim_{N,L\to\infty} N/L, \tag{5.1}$$

remains finite. Thus, the mean spacing between the atoms in the chain is equal to $a_A = L/N = n^{-1}$, and the dimensionless concentration (the so-called *coverage parameter*) θ is determined by

$$\theta = \frac{N}{M} = \frac{a_s}{a_A}, \tag{5.2}$$

where we have used the relationship $L = Na_A = Ma_s$. It is convenient to introduce the so-called *window number* w defined by the equation

$$w = a_A \,(\mathrm{mod}\, a_s) \equiv \left\{\frac{a_A}{a_s}\right\}, \tag{5.3}$$

where $\{x\} = x - \mathrm{int}(x)$, and $\mathrm{int}(x)$ is the integer part of x, i.e. the largest integer smaller than or equal to x. Besides, when the interatomic potential $V_{\mathrm{int}}(x)$ has the absolute minimum at some distance $a_{\min}$, it is useful to introduce the *misfit parameter* P defined as

$$P = \frac{a_{\min} - q_0 a_s}{a_s}, \tag{5.4}$$

where $q_0 = \mathrm{int}\left(\frac{a_{\min}}{a_s} + \frac{1}{2}\right)$ is the integer reducing P to the interval $\left[-\frac{1}{2}, \frac{1}{2}\right]$.

It is important to emphasize that there are two distinct situations, the "*fixed-density*" FK chain and the "*free-end*" chain, which we discuss below in more details.

In the "fixed-density" case (sometimes called as the "fixed length" or the "fixed boundary conditions" case) we assume that the values L and N are fixed externally, so that the parameter n (or θ, or w) is the parameter of the model. Since the potentials V_{sub} and V_{int} do not depend on the atomic index l, the model is homogeneous and a relabelling of the atoms does not change the physical properties of a configuration. Because of the periodicity of V_{sub}, the transformation $\sigma_{i,j}$ defined as

$$\sigma_{i,j}\{x_l\} = \{x_{l+i} + ja_s\} \tag{5.5}$$

(i and j are arbitrary integers) produces another configuration with the same potential energy. Given two configurations $\{x_l\}$ and $\{y_l\}$ one says that the one is *less than* the other, $\{x_l\} < \{y_l\}$, if $x_l < y_l$ for all l. A configuration $\{x_l\}$ is called *rotationally ordered* [442], if for any symmetry transformation $\sigma_{i,j}$ the transformed configuration $\{y_l\} = \sigma_{i,j}\{x_l\}$ is either less than $\{x_l\}$, or the opposite, unless both coincide. For the rotationally ordered configuration we have

$$\begin{aligned} \sigma_{i,j}\{x_l\} < \{x_l\} \text{ if } iw + j < 0, \\ \sigma_{i,j}\{x_l\} > \{x_l\} \text{ if } iw + j > 0. \end{aligned} \tag{5.6}$$

Note that the GS configuration, as well as minimum-energy configurations (see below), are always rotationally ordered.

If the value of θ is rational,

$$\theta = \frac{s}{q} \tag{5.7}$$

with s and q being relative prime, so that

$$w = q \,(\mathrm{mod}\, s) \equiv \left\{\frac{q}{s}\right\} = \frac{r}{s}, \tag{5.8}$$

where $r = q - s\,\mathrm{int}(q/s)$, then the GS configuration is the C-phase with the period

$$a = qa_s = sa_A, \tag{5.9}$$

and the elementary cell consists of s atoms. Otherwise, for an irrational value of θ the GS is an incommensurate phase. The structure of the IC phase is sufficiently determined by the value of the parameter g of the FK model. Namely, with increasing of g above of some critical value $g_{\mathrm{Aubry}}(\theta)$ the "transition by breaking of analyticity" occurs abruptly, and the "sliding mode" appears simultaneously (see below Sect. 5.2.2). Aubry and Le Daeron [443] have proved that the SG configuration is uniquely determined by the value θ, if the interaction potential is strictly convex, i.e. if there exists a constant C such that $V_{\mathrm{int}}''(x) \geq C > 0$ for all x, $0 < x < \infty$. For a nonconvex interatomic potential (such as the Morse potential or the sinusoidal one, see Sect. 5.4.3) the situation is more complicated; for example, in the case of $a_{\mathrm{min}} < a_A$ the chain may break into two independent semi-infinite parts with free ends.

Another situation emerges when the chain has one or two free ends so that its length L is not fixed (the "free-end" case or the "open-end boundary condition"). Of course, the "free-end" chain may exist only if the potential $V_{\mathrm{int}}(x)$ has an attractive branch, i.e. a minimum at some distance a_{min}. However, if we suppose that some external force ("pressure" Π) is applied to the end atoms of the chain, the "free-end" situation takes place for interatomic repulsion as well. Analogous situation emerges when the FK system is "in contact" with a "vapor phase" having a nonzero chemical potential μ so that the number of atoms N is not fixed. In these cases one of the parameters, a_{min} (or P), or Π, or μ, is the model parameter which determines the value θ or w and, therefore, defines the GS configuration. Namely, for small values of the misfit P, the atoms of the GS configuration are uniformly spaced at bottoms of the substrate potential wells (the trivial GS). With increasing of $|P|$, at a critical value P_{FM} the GS goes continuously to a state characterized by a finite density of kinks (the "kink" will be defined rigorously below in Sect. 5.2.1). As $|P|$ increases beyond P_{FM}, the average spacing a_A rises through an infinite number of steps of varying width, each step corresponding to a higher-order commensurate phase. The curve $a_A(P)$ (or similar ones, see the dependence $w(P)$ in Fig. 5.1) is known as *Devil's staircase* due to B.

Mandelbrot [444] (in the mathematical literature such a function is known as a "Cantor function"). If g is large, $g \gg 1$, the total width of the steps is very small, and separating each step are IC phases with irrational values of w so that the Devil's staircase is "incomplete". However, when the magnitude of g is reduced beyond a critical value g_{Aubry} ($g_{\text{Aubry}} \approx 1$), the transition by breaking of analyticity manifests itself by the disappearance of the IC ground states; IC-phases are no longer present as the GS because a nearly, possibly high-order C-phase becomes always a more favorable lowest-energy configuration. In this case the curve $a_A(P)$ is described by the "complete" Devil's staircase.

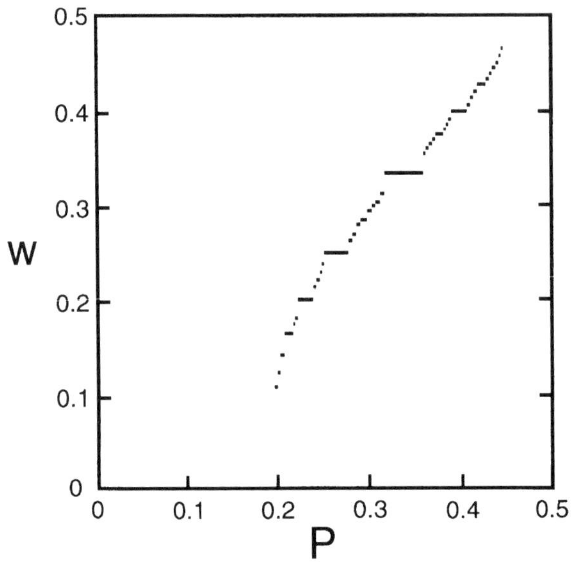

Fig. 5.1. Window number w vs. misfit parameter P for the FK chain with free ends ($g = 1$) [445].

Investigation of the questions mentioned above had been started in the works of Ying [446], Sokoloff [447], Aubry [448]–[454], Bulaevsky and Chomsky [455], Theodorou and Rice [456], Pokrovsky and Talapov [457], Sacco and Sokoloff [458], Bak [459] and continues up to now (see also review papers [277, 460, 461]).

Recently, with scanning tunnelling microscopy Hupalo *et al.* [462] found that ordered phases in Pb/Si(111) are one of the best examples of the Devil's staircase phase diagram. Phases within a narrow coverage range ($1.2 < \theta < 1.3$ monolayers) are constructed with the rules similar to the ones found in theoretical models.

Each configuration $\{x_l\}$ is characterized by its potential energy

$$E_N(\{x_l\}) = \sum_{l=1}^{N} V_{\text{sub}}(x_l) + \sum_{l=2}^{N} V_{\text{int}}(x_l - x_{l-1}), \qquad (5.10)$$

where we have assumed the interaction of nearest neighboring atoms only. For an infinite chain of atoms we can define the energy per atom $\varepsilon(\{x_l\})$,

$$\varepsilon(\{x_l\}) = \frac{1}{N} \lim_{N \to \infty} E_N(\{x_l\}) . \tag{5.11}$$

A stationary configuration is defined in such a way that the force acting on each atom vanishes,

$$\frac{\partial E_N(\{x_l\})}{\partial x_l} = 0 \quad \text{for all } l. \tag{5.12}$$

The stationary configuration is unstable if a small displacement of a finite number of atoms will lower the potential energy of the system. Of course, most of the unstable configurations are the saddle ones, for which the energy decreases in one directions in the N-dimensional configurational space and increases in others. The stable stationary configurations correspond to local minima of the function $\varepsilon(\{x_l\})$, and most of them describe metastable states of the system. In order to find the ground state of the FK chain, we have to look for the stable configuration which satisfy the following three conditions:

(a) It should satisfy appropriate boundary conditions which are different for "fixed density" and "free-end" cases;
(b) It should correspond to the absolute minimum of the potential energy;
(c) It should be "homogeneous", or "nondefective", or "recurrent" according to Aubry [460]. This condition is important for the infinite chain of atoms, because a *finite* number of local defects (kinks) inserted into the GS configuration will not change the energy $\varepsilon(\{x_l\})$.

Explicit rigorous calculation of the GS configuration can be carried out only on particular but pathological models, such as ones with the substrate potential $V_{\rm sub}(x)$ being replaced by a piecewise parabola periodic potential [448]. For a general case the set of equations (5.12) is nonintegrable, therefore numerical calculations have to be performed. One of approaches used by Peyrard and Aubry [463] and Braun [176], the so-called "gradient method", is to start with an appropriate initial configuration and then to solve the set of equations

$$\frac{dx_l}{d\tau} = -\frac{\partial E_N(\{x_l\})}{\partial x_l} . \tag{5.13}$$

The solution of Eqs. (5.13) gives the system trajectory in the overdamped limit, $\eta \to \infty$ (the so-called "adiabatic" trajectory, see Sect. 5.2.1), and it always tends to a stable configuration. In order to avoid metastable configurations, the Gauss random force may be added to the r.h.s. of Eq. (5.13). During the simulation the amplitude of noise should be decreased to zero with a small speed γ. In the $g \ll 1$ case, however, the required simulation time t_s drastically increases, $t_s \propto \ln\ln 1/\gamma$, and a special procedure of introducing of "numerical enzymes" has to be used [464]. Note also that τ in Eq.

(5.13) is not the time but a variable along the trajectory. The main problem of this approach is to guess a "good" initial configuration which will lead to the GS, and then to distinguish the GS final configuration from metastable ones.

An another approach is to solve directly the equations (5.12) by Newton's method, the superconvergence of which greatly reduces the computing time. This method, however, requires an extremely good guess of the correct initial condition [465]–[467].

The most powerful numerical method which guarantees to yield the GS (or at least a GS in cases of degeneracy) was proposed by Griffiths and Chou [276]. This method is called the "minimization eigenvalue approach" or "effective potential method"; if was discussed shortly in Chapter 3.5.2, see also the original papers [277, 445]. Note that numerical methods always deal with commensurate phases because the number N is finite in computer simulation. However, an appropriate choice of N and M such as provided by the Fibbonacci sequence, leads to window numbers which are practically undistinguished from irrational ones. Note also that in the "free-end" case the ends of the chain have to be let free, while for the "fixed-density" case the periodic boundary conditions,

$$x_{N+1} = x_1 + N\, a_A, \tag{5.14}$$

should be used.

Besides numerical calculation, important results on the structure of stationary FK configurations can be obtained by reducing Eq. (5.12) to the "symplectic map". For example, for the standard FK model with sinusoidal substrate potential,

$$V_{\rm sub}(x) = 1 - \cos x, \tag{5.15}$$

and the harmonic interaction of nearest neighbors only,

$$V_{\rm int}(x) = \frac{1}{2} g\, (x - a_{\rm min})^2, \tag{5.16}$$

stationary configurations should satisfy the set of difference equations

$$g(x_{l+1} + x_{l-1} - 2x_l) = \sin x_l. \tag{5.17}$$

Aubry [448] was the first who showed that Eq. (5.17) may be rewritten as the two-dimensional nonlinear area-preserving twist map,

$$\begin{aligned} x_{l+1} &= x_l + p_{l+1}, \\ p_{l+1} &= p_l + K\, \sin x_l, \end{aligned} \tag{5.18}$$

where $K \equiv g^{-1}$. Using the substitution $x_l = \widetilde{x}_l + q_0 l a_s$, $p_l = \widetilde{p}_l + q_0 a_s$ with an appropriate integer q_0, the map (5.18) may be rewritten as

$$Y_{l+1} = T\, Y_l, \tag{5.19}$$

where the point $Y_l = (\widetilde{x}_l, \widetilde{p}_l)$ is defined in the tour $(0, a_s) \times (0, a_s)$. Thus, the map (5.19) allows one to represent any stationary configuration of the FK model by a trajectory of the "artificial" nonlinear dynamical system with the discrete "time" l and the "evolution" operator T.

Area-preserving twist maps have been extensively studied (see, e.g., Ref. [468]) as examples of Hamiltonian dynamical systems which exhibit the full range of possible dynamics from regular (integrable) to chaotic motion, and occur commonly in a variety of physical contexts, from accelerator to chemical and fluid physics. Equation (5.18) known as the Taylor-Chirikov standard map, is the best-studied example of an area-preserving map. It was firstly considered by Chirikov [469], and then it was extensively studied in a number of works [470]–[473], because it describes time evolution of a sinusoidally driven pendulum which is one of classical models exhibiting a transition from regular to chaotic motion.

Starting from any point Y_0 in the tour, successive application of T or T^{-1} produces an orbit $\{Y_l\}$ which may be either regular, i.e. periodic if $T^s Y_l = Y_l$ or quasi-periodic if all points Y_l lie on the continuous curve in the tour, or chaotic (see Fig. 5.2). Not going into details, we recall only main aspects of behavior of such systems (the complete theory may be found in, e.g., Refs. [473, 474]).

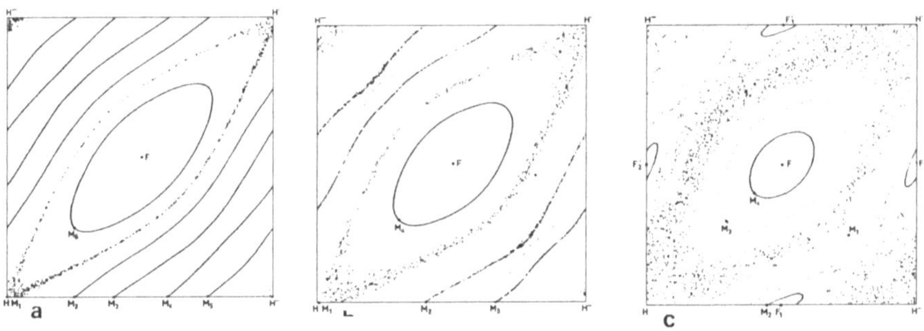

Fig. 5.2. Map of the transformation $(x_{l+2}, x_{l+1}) = T'(x_{l+1}, x_l)$ showing the trajectories of the initial points M_i plotted on the figures for the standard FK model for (a) $g = 13.333$, (b) $g = 10$, and (c) $g = 8$ [460].

At $K = 0$ (or $g = \infty$) the map (5.19) is integrable, and all the orbits are regular. They may be classified by the window number w: all points Y_l of a given orbit $\{Y_l\}$ lie on a circle $\widetilde{p} = wa_s = \text{const}$ in the tour and turns the angle $2\pi w$ per iteration of the map. The corresponding FK configurations have a constant separation between the atoms. All the orbits (including a countable set of periodic orbits with rational values of w) are sliding (not pinned), i.e. they can be rotated as a whole by any angle. This means that the translation of a FK configuration by any Δx leads to an equivalent configuration with the same energy. Irrational window numbers produce an uncountable infinite set of quasiperiodic orbits which correspond to the IC phases of the FK model. Thus, at $K = 0$ the sliding orbits cover the whole tour and form a set with the measure $\nu_s = 1$.

Regular orbits remain to exist at any $K \neq 0$. First, for any rational $w = r/s$ there exists the periodic orbit with the period s, $T^s Y_l = Y_l$. However, at $K \neq 0$ the periodic orbit is always pinned, i.e. it consists only of a finite number of fixed points, and the corresponding FK configuration is locked by the substrate potential. (Note that for exotic substrate potentials such as the piecewise parabolic $V_{\rm sub}(x)$ with continuous first derivative the map is integrable and, therefore, the unpinned (sliding) configurations with rational w exist even at $K \neq 0$ [445].) Second, for any irrational value of w the regular orbits exist too. The well known Kolmogorov-Arnold-Moser theorem guarantees that these orbits (called as the "invariant" or "KAM-tori" orbits) survive small perturbations and have a finite measure ν_s in the tour, $0 < \nu_s < 1$, so that the sliding configurations of the FK model continue to exist at least for small $K \neq 0$. The circles with irrational w which lie between the regular ("invariant") ones, are "stochastically destroyed", so that the corresponding orbits are "chaotic", and the FK configurations are locked. According to general stochastic theory, the "area" ν_c occupied by chaotic orbits should increase with increasing of K. The destruction of regular orbits begins at "resonance" circles which are "close" to "simple" periodic orbits (i.e., to the orbits with $w = 0, \frac{1}{2}, \frac{1}{3}$, etc.), and is finished by the circle with the "most irrational" number, the golden mean $w_{\rm gm} = (\sqrt{5} - 1)/2$ (or $w'_{\rm gm} = 1 - w_{\rm gm}$) at $K = K_c$. Beyond the chaotic threshold K_c there are no more unpinned orbits, the majority of the tour being taken up by chaotic orbits, and $\nu_c = 1$. An uncountable set of the regular orbits with irrational w still exists, but now they form a Cantor set which takes zero measure in the tour, $\nu_s = 0$. Thus, at $K \geq K_c$ all the stationary configurations of the FK model are locked. The magnitude of the chaotic threshold $K_c \equiv g^{-1}_{\rm Aubry}(w_{\rm gm})$ may be found with the help of Greene's residue criterion [471, 472, 475], and it is estimated as $K_c \approx 0.971635406$ (see, e.g., Refs. [460, 473, 476]). (Note that the irrational golden-mean value $w_{\rm gm} = (\sqrt{5} - 1)/2$ can be approximated by rational approximants which form the Fibonacci sequence $w_n = s_{n-1}/s_n$ with $s_n = s_{n-1} + s_{n-2}$ and $s_0 = 1$, $s_1 = 2$, so that $w_5 = 8/13$, $w_6 = 13/21$, $w_7 = 21/34$, $w_8 = 34/55$, $w_9 = 55/89$, $w_{10} = 89/144$, etc., and $|w_{10} - w_{\rm gm}| \approx 2\,10^{-5}$, $|w_{14} - w_{\rm gm}| \approx 5\,10^{-7}$, etc.)

Each orbit of the map (5.19) corresponds to a stationary configuration of the FK model. Note that although the mapping (5.19) does not depend on the magnitude of the model parameter $a_{\rm min}$, the associated FK configurations are characterized by their potential energies $\varepsilon(\{x_l\})$ which do depend on $a_{\rm min}$. Most of the orbits correspond to unstable configurations of the FK model. (Note that physical stability of the FK configuration must not be confused with the stability in the map of the associated orbit. For example, the trivial FK configuration corresponds to the "dynamically unstable" fixed point at the origin in the standard map). A small fraction of the orbits corresponds to metastable configurations and only the one describes the GS. Recall that the GS configuration:

(a) must satisfy the boundary conditions (for the "free-end" case we have to consider all orbits, while for the "fixed-density" case we have to extract the orbits with a given window number w);
(b) must correspond to the lowest potential energy;
(c) must be "recurrent" (according to Aubry [460], the recurrent orbit returns to any neighborhood of any point of the orbit. Intuitively, the idea is that any part of the FK configuration will appear again within an arbitrary close approximation, later in the same configuration).

Aubry and Le Daeron [443] have proved that the GS configuration always is associated with a regular orbit. At rational w the GS orbit is periodic, at irrational w it is regular either on the continuous circle (smooth "KAM tour") or on a discontinuous Cantor set. The Aubry "transition by breaking of analyticity" which occurs at $g = g_{\mathrm{Aubry}}(w)$ for the "fixed-density" case corresponds to the stochastic destroying of the continuous circle with the given w. Note ones more that the GS orbit remains regular (not chaotic) despite it is embedded in the chaotic region of the map, i.e. it just belongs to the zero measure Cantor set of orbits which are not chaotic. The transition from "incomplete" to "complete" Devil's staircase for the "free-end" model occurs at $K_c = g_c^{-1}$, where $g_c = \min_w g_{\mathrm{Aubry}}(w) = g_{\mathrm{Aubry}}(w_{\mathrm{gm}})$, when all "invariant" circles are destroyed.

Metastable FK configurations describe excited states of the model. In particular, a fraction of the "chaotic" orbits corresponds to metastable configurations with chaotically pinned kinks. Thus, the FK model may be used in investigation of amorphous solids (see Sect. 6.3). Another physically important solution of the model is the homoclinic orbit, which tends to different periodic orbits for $l \to \pm\infty$. This orbit corresponds to kink excitation of the FK model. The homoclinic points are intersections of the stable manifold, a continuum of points which tends to the periodic orbit for $l \to +\infty$, and the unstable manifold, similarly defined for $l \to -\infty$. If these manifolds intersect transversely, the kinks are pinned and the standard map may be used for calculation of the pinning energy [149, 156]. Otherwise, when the manifolds merge into a unique continuous curve, the PN barrier vanishes.

5.2 Fixed-Density Chain

5.2.1 Commensurate Configurations

A configuration $\{x_l\}$ is commensurate if there are two irreducible positive integers s and q such that

$$x_{l+s} = x_l + qa_s \tag{5.20}$$

for all l. The number of atoms s per a unit cell of the C-structure is known as the "order" of the C-phase [446]. For $s = 1$ the C-phase is trivial, or registered. To determine the atomic coordinates in a nontrivial C-phase (i.e. for $s \geq 2$), the system of $s - 1$ transcendental equations is to be solved [446, 456].

A phonon spectrum of the system describes small oscillations around the stable stationary positions of atoms. The phonon spectrum of C-phases is optical, i.e. it always starts at a finite frequency $\omega_{\min} > 0$ [446, 456]. For the trivial GS with $s = 1$ it has the form

$$\omega^2(\kappa) = \omega_0^2 + 2c^2(1 - \cos\kappa), \quad |\kappa| \leq \pi, \tag{5.21}$$

where $\omega_0 = 1$ and $c = \sqrt{g}$ for the standard FK model. For higher-order C-phases with $s \geq 2$ the excitation spectrum can be determined just in the same manner as in the study of phonon spectrum of crystalline lattice with a basis. The normal modes are organized into s branches (possibly separated by forbidden gaps), and the minimum phonon frequency $\omega_{\min}$ decreases with increasing of the order s of the structure.

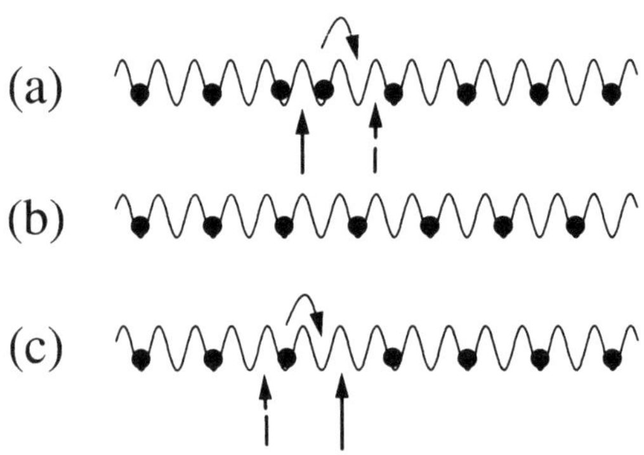

Fig. 5.3. (a,b) Kink and antikink in a commensurate structure $\theta_0 = 1/2$ ($s = 1$, $q = 2$, $w = 0$). Solid (dashed) arrow shows the kink before (after) displacement of an atom to the right.

Besides phonons, the FK model has essentially nonlinear excitations called kinks. Their existence follows from the fact that due to periodicity of the substrate potential $V_{\text{sub}}(x)$, any FK configuration is infinitely degenerated. Namely, equivalent configurations can be obtained from a given one by translation and/or relabelling, i.e. by transformation $x_l \to x_{l+i} + ja_s$ with arbitrary integers i and j. Kink configurations describe the configurations which "link" two commensurate ground states of the infinite FK chain. Owing to boundary conditions, such excitations are topologically stable. It has to be emphasized that kinks exit only for commensurate structures, for IC phases a kink excitation does not exist. Aubry and Le Daeron [443] have proved rigorously the theorem which states that there always exist a couple of commensurate ground states $\{x_l^-\}$ and $\{x_l^+\}$ such that there are no ground states $\{x_l\}$ which satisfy for all l the inequality

$$x_l^- < x_l < x_l^+ . \tag{5.22}$$

However, there exists a *minimum-energy (m.e.) configuration* (i.e., a stationary configuration which satisfies the boundary conditions and corresponds to the lowest potential energy but is not recurrent) $\{y_l\}$ such that for all l

$$x_l^- < y_l < x_l^+ \tag{5.23}$$

and

$$\lim_{l\to-\infty}(x_l^+ - y_l) = 0, \quad \lim_{l\to+\infty}(y_l - x_l^-) = 0. \tag{5.24}$$

Analogously, it also exists a m.e.-configuration which satisfy the conditions

$$\lim_{l\to-\infty}(y_l - x_l^-) = 0, \quad \lim_{l\to+\infty}(x_l^+ - y_l) = 0. \tag{5.25}$$

Equations (5.24) [or (5.25)] define the kink (or antikink) configuration which is an excited state of the reference C-phase. Notice that Aubry used for kink (antikink) a more rigorous but too long name "delayed (or advanced) elementary decommensuration".

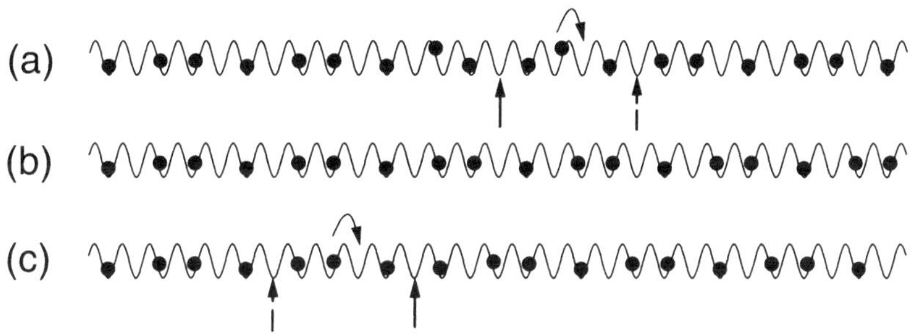

Fig. 5.4. The same as Fig. 5.3 but for $\theta_0 = 3/5$ ($s = 3$, $q = 5$, $w = 2/3$).

Thus, every time over a few sites, the sequence x_l loses or gains an amount a_s from the nearest C-phase, a kink or antikink is present in the chain. Otherwise, a kink (antikink) with a topological charge $\sigma = +1$ ($\sigma = -1$) may be defined as the minimally possible contraction (extension) of the C structure when at infinity, i.e. for $l \to \pm\infty$, the arrangement of atoms relative to minima of the substrate potential coincides with their arrangement in the GS. Examples of kink structures for different C-phases are shown in Figs. 5.3–5.6.

The configuration $\{y_l'\} \equiv \{y_{l-s} + qa_s\}$ is also a configuration which satisfies the same conditions (5.23) to (5.25). It corresponds to the kink translated by $a = qa_s$, i.e. by the elementary cell of the commensurate GS. The

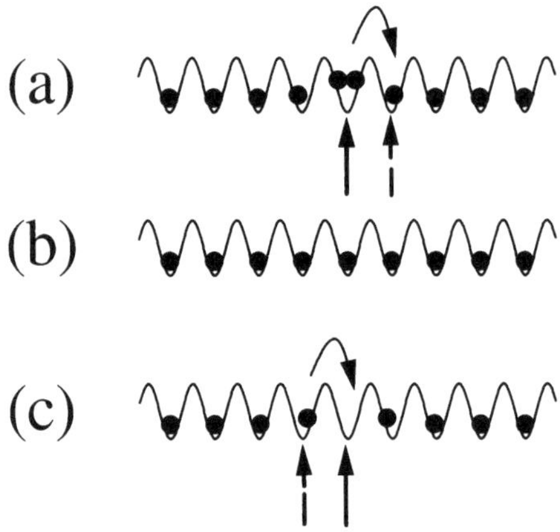

Fig. 5.5. The same as Fig. 5.3 but for $\theta_0 = 1$ ($s = 1$, $q = 1$, $w = 0$).

Peierls-Nabarro (PN) barrier ε_{PN} [142, 143] is defined as the lowest energy barrier which must be passed for a continuous translation of the configurations $\{y_l\} \to \{y_l'\}$ (for a precise definition of the PN barrier see Ref. [477]). Thus, kink is a localized quasi-particle which can move along the chain and may be characterized by its coordinate, mass, energy, etc.

For the concentration $\theta_0 = s$ the kink structure is trivial (an excess of an atom corresponds to the kink, and a vacancy, to the antikink, see Figs. 5.5 and 5.6), but at arbitrary concentration $\theta_0 = s/q$ with $q \neq 1$ it is more complex. (A general method of construction of kink excitations for a complex reference structure is discussed in Sect. 11.3.3). As an example, Fig. 5.3(a) [Fig. 5.3(c)] shows the kink (antikink) on the background of the commensurate structure with concentration $\theta_0 = \frac{1}{2}$ shown in Fig. 5.3(b). The arrow in Fig. 5.3(a) [Fig. 5.3(c)] indicates the atom, whose displacement to the right into the nearest neighboring minimum of the substrate potential results in a displacement of the kink to the right (antikink, to the left) by the structure period a. Similarly, a displacement to the right of an atom in the commensurate structure of Fig. 5.3(b) results in the creation of a kink-antikink pair in the system. Figs. 5.4 to 5.6 show also kink structures for $\theta_0 = 3/5$, 1 and 2. Note that, as compared with the initial commensurate structure, a kink contains σ/q excess atoms.

The kink shape is conveniently characterized by the quantities

$$u_l = y_l - x_l^- . \tag{5.26}$$

For a C-structure with a complex unit cell it is convenient to split the index l into two subindexes, $l = (i, j)$, where $j = 0, \pm 1, \ldots$ corresponds to a number of the unit cell, while $i = 1, \ldots, s$ numerates the atoms in the cell. Then, for an isolated kink the displacements u_l satisfy the sum rule

$$\sum_{i=1}^{s} (u_{i,j=+\infty} - u_{i,j=-\infty}) = -\sigma a_s . \tag{5.27}$$

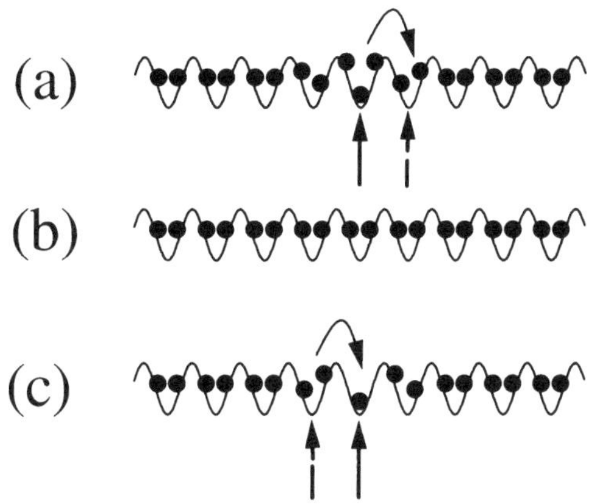

Fig. 5.6. The same as Fig. 5.3 but for $\theta_0 = 2$ at $s = 2$, $q = 1$, and $w = 1/2$.

Following Bergman *et al.* [146], a coordinate of an isolated kink may be defined as the center of mass of the system,

$$X = q \sum_{l} u_l + C, \tag{5.28}$$

where the additive constant C is to be chosen so that a maximum deviation from the reference configuration occurs at the atom $l_0 = X/a$ (see arrows in Figs. 5.3–5.6). The kink rest energy as measured with respect to the reference GS with the same number of atoms, is defined as

$$\varepsilon_k = \lim_{N \to \infty} \left[E_N(\{y_l\}) - E_N\left(\{x_l^-\}\right)\right], \tag{5.29}$$

or, more rigorously for a complex reference structure [478], as

$$\varepsilon_k = \lim_{i,p \to \infty} \frac{1}{s} \sum_{j=p+1}^{s} \sum_{l=-i+1}^{j} \left[V(y_l, y_{l-1}) - V\left(x_l^-, x_{l-1}^-\right)\right], \tag{5.30}$$

where

$$V(x, x') = V_{\text{int}}(x - x') + V_{\text{sub}}(x). \tag{5.31}$$

For the harmonic potential, $V_{\text{int}}(x) = \frac{1}{2} g (x - a_{\min})^2$, the kink energy can be represented as

$$\varepsilon_k = -g a_s P \left(x_{\text{c.m.}}^+ - x_{\text{c.m.}}^-\right) + E_k, \tag{5.32}$$

where E_k is independent on the misfit P, and the center-of-mass coordinate of a unit cell is defined by

$$x_{\text{c.m.}}^{\pm} = \frac{1}{s} \sum_{l=-(s-l)/2}^{(s-1)/2} x_l^{\pm} \quad (\text{odd } s), \tag{5.33}$$

or

$$x^{\pm}_{\rm c.m.} = \frac{1}{s} \sum_{l=-s/2+1}^{s/2} x_l^{\pm} \quad (\text{even } s). \tag{5.34}$$

When the FK chain contains more that one kink, the mutual influence of kinks on each other changes their shapes and leads to a contribution to the system energy, which can be interpreted as an interaction between the kinks. Of course, this interaction is not pairwise in a general case. Besides, as kinks contain an excess number of atoms (equal to σ/q), the interaction of these atoms leads to an additional contribution to the energy of kink's interaction if the interatomic potential $V_{\rm int}(x)$ is long ranged. Usually kinks with the same topological charges repel, while a kink and an antikink attract each other (see Chap. 3).

When a kink moves along the chain, we may write formally $dx_l/dt = (\delta x_l/\delta X)(dX/dt)$, and the system kinetic energy takes the form

$$K = \frac{m}{2}\left(\frac{dX}{dt}\right)^2, \tag{5.35}$$

where m is the effective mass of the kink defined as

$$m = m_a \sum_l \left(\frac{\delta u_l}{\delta X}\right)^2. \tag{5.36}$$

Recall that kink's motion is carried out in the PN potential $V_{PN}(X)$ which has the amplitude ε_{PN} and the period a. Both the characteristics, the effective mass and the shape of the PN potential, depend on the trajectory of kink's motion. In what follows we will assume that kink moves along the "adiabatic" trajectory defined as the curve in the N-dimensional configuration space of the system, which links the two m.e. configurations $\{y_l\}$ and $\{y_l'\}$, passes through the nearest saddle configuration with the lowest potential energy and is defined by the differential equations (5.13). Physically this means that kink moves adiabatically slow. It is to be noted that all kink's parameters depend on the structure of the reference commensurate GS, i.e. on the model parameter θ_0. Because the FK model is nonintegrable, these parameters should be found numerically. In computer simulation, instead of the motion of an isolated kink in an infinite chain, it is more convenient to consider a simultaneous motion of kinks in an infinite periodic structure of kinks (the cnoidal wave). Let the period of this structure be Ma_s. Then it is sufficient to place a chain of N atoms into the periodic substrate potential with M minima and to impose periodic boundary conditions. To have a single kink over the length Ma_s on the background of a commensurate structure $\theta_0 = s/q$, the N and M values should satisfy the integer equation [165]

$$qN = sM + \sigma. \tag{5.37}$$

The resulting commensurate structure of kinks is characterized by the average concentration

$$\theta = N/M = \theta_0 + \sigma/Mq, \tag{5.38}$$

and the distance between kinks is equal to Ma_s. Emphasize that this method gives the kink characteristics for the reference GS with the concentration θ_0 and not with θ. However, a mutual interaction of kinks in their periodic structure may disturb the results, so the number M should be large enough. It has to be mentioned also the numerical method of Griffiths and collaborators [277] which gives not only the GS but kink's configurations as well.

Besides computer simulation, approximate results can be obtained in two limiting cases considered below, the cases of weak and strong interatomic interactions.

Weak-Bond Approximation

First, we consider the case of weak interaction between the atoms, $V_{\rm int}(a_A) \ll \varepsilon_s$, when all atoms are situated near the corresponding minima of the substrate potential. For instance, at a concentration $\theta_0 < 1$, when not more than one atom is in one potential well, we may neglect by small displacements of atoms from their positions corresponding to minima of $V_{\rm sub}(x)$. Then for the motion of kinks or antikinks on the background of the concentration $\theta_0 = s/q$ lying within the interval $(1+p)^{-1} < \theta_0 < p^{-1}$ (including the kink on the background of the concentration $\theta_0 = (1+p)^{-1}$ and the antikink on the background of $\theta_0 = p^{-1}$), where $p \equiv \text{int}\left(\theta_0^{-1}\right)$ is an integer, from a simple geometric consideration [see Figs. 5.3, 5.4, and 5.5(c)] we obtain for the amplitude ε_{PN} of the PN potential the following expression [165],

$$\varepsilon_{PN} \approx \varepsilon_s + 2V_{\rm int}\left(pa_s + \frac{1}{2}a_s\right) - V_{\rm int}(pa_s) - V_{\rm int}(pa_s + a_s). \tag{5.39}$$

It is important to note that the kink mass m, the kink and antikink energies ε_k and $\varepsilon_{\bar{k}}$, and the difference in the PN potential amplitudes for the kink and antikink, $\delta\varepsilon_{PN}$, depend only on the interaction of unit cells on the distance $a = qa_s$ and not on the interaction of neighboring atoms within the cell on the distance $a_A = a/s$:

$$m \approx 1/q^2, \tag{5.40}$$

$$\varepsilon_k \approx V_{\rm int}(a - a_s) - V_{\rm int}(a), \tag{5.41}$$

$$\varepsilon_{\bar{k}} \approx V_{\rm int}(a + a_s) - V_{\rm int}(a), \tag{5.42}$$

$$\begin{aligned}\delta\varepsilon_{PN} \approx &\left[2V_{\rm int}\left(a + \frac{1}{2}a_s\right) - V_{\rm int}(a) - V_{\rm int}(a + a_s)\right] \\ &- \left[2V_{\rm int}\left(a - \frac{1}{2}a_s\right) - V_{\rm int}(a - a_s) - V_{\rm int}(a)\right].\end{aligned} \tag{5.43}$$

From Eqs. (5.39) to (5.43), it follows that

$$\varepsilon_k^\sigma = -\sigma a_s V_{\rm int}'(a) + E_k, \quad E_k \approx \pi^2 g_a \varepsilon_s, \tag{5.44}$$

$$\varepsilon_{PN} \approx \varepsilon_s \left(1 - \frac{1}{2}\pi^2 g_p\right), \tag{5.45}$$

and

$$\delta\varepsilon_{PN} \simeq \frac{1}{2}\pi^2 \alpha g_a \varepsilon_s, \tag{5.46}$$

where

$$g_p = \frac{a_s^2}{2\pi^2 \varepsilon_s} V_{\rm int}'' \left(p a_s + \frac{1}{2} a_s\right), \quad g_a = \frac{a_s^2}{2\pi^2 \varepsilon_s} V_{\rm int}''(a), \tag{5.47}$$

and α is the dimensionless parameter of anharmonicity of the potential $V_{\rm int}(x)$,

$$\alpha = -\frac{a_s V_{\rm int}'''(a)}{V_{\rm int}''(a)}. \tag{5.48}$$

The case of a concentration $\theta_0 > 1$ is somewhat more complex. For example, for a kink on the background of the $\theta_0 = 1$ structure, two atoms with coordinates x_1 and x_2 are situated in the same well of the substrate potential [see Fig. 5.5(a)], and their coordinates are $x_{1,2} = \pm x_0$, where the value $x_0 \ll 1$ is determined by the stationary solution of the motion equation

$$m_a \omega_0^2 x_0 + V_{\rm int}'(2x_0) - V_{\rm int}'(a_s - x_0) = 0,$$

where we put $V_{\rm sub}(x) \approx \frac{1}{2} m_a \omega_0^2 x_0^2$ near the well's bottom. Then,

$$\varepsilon_k \approx 2V_{\rm sub}(x_0) + V_{\rm int}(2x_0) + 2V_{\rm int}(a_s - x_0) - 3V_{\rm int}(a_s) \tag{5.49}$$

and

$$\varepsilon_{PN} \approx \left[\varepsilon_s + 2V_{\rm int}\left(\frac{1}{2}a_s\right) + V_{\rm int}(a_s)\right] - \left[2V_{\rm sub}(x_0) + V_{\rm int}(2x_0) + 2V_{\rm int}(a_s - x_0)\right]. \tag{5.50}$$

In the lowest order of the weak-bond approximation the amplitude of PN potential is the same for kinks and antikinks on the background of any concentration θ_0 within the range of $1 < \theta_0 < \frac{3}{2}$. Note also that for the standard FK model with $V_{\rm int} = \frac{1}{2} g (x - a_{\rm min})^2$ the value $\varepsilon_{PN} \approx 2 - \pi^2 g$ is the same for all C-structures ($0 < \theta_0 < \infty$) in the first order in g. More accurate results in the weak-bond case may be obtained by the standard map technique [156].

Continuum Approximation

In the opposite case of strong interaction between the atoms, $V_{\rm int}(a_A) \gg \varepsilon_s$, relative displacements of equivalent atoms in the nearest neighboring cells are small, $|u_{i,j+1} - u_{i,j}| \ll a$ (while relative displacements of nearest neighboring atoms in the same cell may be not small if $s \geq 2$), we can use a continuum approximation,

$$j \to x = ja, \quad u_{i,j}(t) \to u_i(x,t), \quad i = 1, \ldots, s, \quad \sum_j \to \int \frac{dx}{a}, \tag{5.51}$$

and the motion equations reduces to a set of s coupled differential equations on functions $u_i(x,t)$ [165, 282, 457]. Methods of solution of this system were shortly discussed in Sect. 3.5.2. Note only that one has to keep the interaction between s nearest neighbors at least. Below we consider the simplest case of the trivial GS ($s = 1$) with the concentration $\theta_0 = 1/q$ and interaction of nearest neighbors only according to the harmonic law

$$V_{\rm int}(x) \simeq V_{\rm int}(a) + V'_{\rm int}(a)(x-a) + \frac{1}{2} V''_{\rm int}(a)(x-a)^2 \tag{5.52}$$

[in what follows the constant term $V_{\rm int}(a)$ is omitted]. In the continuum limit approximation the Hamiltonian of the FK chain is reduced to

$$H[u] = \varepsilon_0[u] + \int \frac{dx}{a} \left\{ \frac{1}{2} \left(\frac{\partial u}{\partial t} \right)^2 + \frac{d^2}{2} \left(\frac{\partial u}{\partial x} \right)^2 + V_{\rm sub}[u] \right\} \tag{5.53}$$

(recall $a_s = 2\pi$ and $\varepsilon_s = 2$), where

$$\varepsilon_0[u] = V'_{\rm int}(a) \int dx \, \frac{\partial u}{\partial x} \tag{5.54}$$

and

$$d = a\sqrt{g_a}. \tag{5.55}$$

For the standard FK model, $V_{\rm sub}(u) = 1 - \cos u$, the Hamiltonian (5.53) leads to the SG equation. The SG kink is characterized by the following parameters (see Sects. 1.2 and 3.1),

$$m = 2/q^2\pi^2\sqrt{g_a}, \tag{5.56}$$

$$\varepsilon_k^\sigma = -\sigma a_s V'_{\rm int}(a) + E_k, \quad E_k = 4\varepsilon_s\sqrt{g_a}, \tag{5.57}$$

$$\varepsilon_{PN} \simeq (8\pi^4/3)\varepsilon_s g_a \exp(-\pi^2\sqrt{g_a}), \tag{5.58}$$

where $g_a = a_s^2 V''_{\rm int}(a)/2\pi^2\varepsilon_s$. Kink's shape may be characterized by the density of excess atoms

$$\rho(x) = -\frac{1}{a}\frac{\partial u}{\partial x}, \tag{5.59}$$

so that

$$\int \rho(x) dx = \sigma/q. \tag{5.60}$$

When the FK chain contains two kinks with topological charges σ_1 and σ_2 separated by a distance R, they interact with the energy

$$v_{\rm int}(R) \simeq 16\sigma_1\sigma_2\varepsilon_s\sqrt{g_a}\exp(-R/d). \tag{5.61}$$

Equations (5.56), (5.57), (5.58), and (5.61) are valid provided $g_a \gg 1$. They may be compared with those obtained in the weak-coupling approximation ($g_a \ll 1$) for the standard FK model:

$$m = 1/q^2, \tag{5.62}$$

$$\varepsilon_k^\sigma = -\sigma a_s V'_{\rm int}(a) + E_k, \quad E_k \simeq \pi^2 g_a(1-2g_a)\varepsilon_s, \tag{5.63}$$

$$\varepsilon_{PN} \simeq \varepsilon_s\left(1 - \frac{1}{2}\pi^2 g_a\right), \tag{5.64}$$

$$v_{\rm int}(R) \simeq 2\pi^2\sigma_1\sigma_2\varepsilon_s\exp(-\xi R/d), \tag{5.65}$$

where the numerical factor $\xi(g_a)$ ($\xi < 1$) was tabulated by Joos [155]. As will be shown in Sect. 6.3.1, $\xi(g) \simeq -\sqrt{g}\ln g$ at $g \ll 1$. Note that for the standard FK model the expressions (5.62), (5.63), (5.64), and (5.65) are valid for kinks at $\theta_0 = 1$ as well.

Besides a single kink solution, the SG equation admits the cnoidal wave stationary solutions which describe a regular equidistant sequence of kinks. Repulsion between the kinks leads to narrowing of their widths,

$$d \to d^* = kd, \tag{5.66}$$

where the "modulus" k, $0 < k \leq 1$, is defined by the equation

$$Q = g_a^{1/2}2k\mathbf{K}(k), \tag{5.67}$$

and $\mathbf{K}(k)$ is the complete elliptic integral of the first kind, while Q is the number of atoms per one period of the cnoidal wave, so that a distance between the kinks is equal to

$$R = Qa = Qqa_s. \tag{5.68}$$

Recalling that kinks are quasi-particles interacting via potential $v_{\rm int}(R)$ and subjected to the periodic PN potential $V_{PN}(X)$, we come to renormalization ideas considered in the next subsection.

Renormalization Approach

Let us suppose that we know the parameters of a single kink for a simple commensurate structure C_0 with a concentration $\theta_0 = s_0/q_0$. This structure can be modelled by N_0 atoms distributed on a length $L_0 = M_0 a_s$ with N_0 and M_0 satisfying the integer equation $N_0 q_0 = M_0 s_0$. Then, let us add or subtract ΔN atoms so that the new commensurate structure C will have $N = N_0 + \Delta N$ atoms on the same length L_0, and the concentration will be equal to $\theta = s/q$ with $Nq = M_0 s$. Each additional atom (or vacancy) will split into q_0 kinks, the topological charge of the kink being equal to $\sigma_k = \mathrm{sign}(\theta - \theta_0)$. Thus, the structure C may be considered as a regular lattice of kinks for the reference structure C_0, the distance between the kinks being equal to

$$R = \frac{L_0}{|N - N_0| q_0} = \frac{a_s}{|\theta - \theta_0| q_0}. \tag{5.69}$$

Considering kinks as quasiparticles subjected to the PN potential with the period $a_0 = q_0 a_s$ and interacting via a potential $v_{\mathrm{int}}(R)$, the structure C can be viewed as a new effective FK chain of kinks instead of atoms with the concentration

$$\theta_k = a_0/R = q_0^2 |\theta - \theta_0|. \tag{5.70}$$

We denote this structure as C_k.

The renormalization procedure $C \to C_k$ was proposed by Joos *et al.* [479] (see also Ref. [165]), and it is summarized in Table 5.1. Note that in a rigorous approach one should take into account that the kink parameters for the kink lattice are not equal to those of an isolated kink. As was mentioned above, the kink-kink repulsion leads to narrowing of their widths, $d \to d^* = kd < d$. As a result, local kink's characteristics such as the effective mass m^* and the amplitude of PN potential ε_{PN}^*, will be larger than the isolated kink values m and ε_{PN}. The parameters m^* and ε_{PN}^* may be calculated with the help of Eqs. (5.56) to (5.65) with $g \to g^* = k^2 g < g$.

Thus, we can "renormalize" a complex structure C to a new structure C_k of kinks defined on the background of some simple commensurate structure C_0. Then, a kink excitation of the structure C is equivalent to the kink ("superkink") excitation of the structure C_k (the lattice of kinks on the reference structure C_0). So, choosing an appropriate reference structure C_0 such that $|\theta - \theta_0| \ll \theta_0$, we can essentially simplify the calculation of kink's parameters of a complex structure C. The renormalization procedure can be repeated by a required number of times if necessary. Note, however, that the shape of the substrate potential $V_{PN}(X)$ as well as the interaction law $v_{\mathrm{int}}(X)$ for the renormalized FK chain may deviate from those of the initial FK chain

5.2.2 Incommensurate Configurations

The renormalization approach helps also to analyze the incommensurate structures [479]. Namely, a GS with an irrational θ can be considered as a kink lattice on the nearest commensurate phase C_0. Now, however, the renormalized FK chain will have the structure C_k which is an IC structure again. Therefore, the renormalization procedure is to be repeated an infinite number of times. In a result, we obtain an infinite sequence of dimensionless elastic constants $g^{(i)}$ defined as

$$g^{(i)} = \frac{1}{2\pi^2} \frac{\left(a_s^{(i)}\right)^2}{\varepsilon_s^{(i)}} \left(\frac{d^2}{dx^2} V_{\rm int}^{(i)}(x) \right)_{x=a_A^{(i)}} . \tag{5.71}$$

The renormalization procedure has two stable fixed points, $g^{(\infty)} = 0$ and $g^{(\infty)} = \infty$, and one unstable fixed point at $g = g_{\rm Aubry}(\theta)$. If the starting value $g^{(0)} = g$ is lower then the critical one, $g < g_{\rm Aubry}(\theta)$, the sequence $g^{(i)}$ tends to the limiting point $g^{(\infty)} = 0$. This indicates that the initial incommensurate GS is locked, or pinned by the substrate potential. The pinning energy can be estimated by approximating the GS by a nearest C-configuration. In particular, for the standard FK model the pinning energy is equal to $\varepsilon_{PN} \simeq 2 - \pi^2 g$ at $g \ll 1$ (recall that in this case ε_{PN} approximately does not depend on the structure of the GS). Evidently that the phonon spectrum of the locked GS starts from a nonzero value $\omega_{\rm min} > 0$. The propagation and localization of phonon modes in the pinned IC GS have been studied by Burkov *et al.* [480], Ketoja and Satija [481] (see also Refs. [482, 483], where the heat conduction along the chain is discussed).

Otherwise, if we start from the initial value $g^{(0)} = g > g_{\rm Aubry}(\theta)$, the sequence $g^{(i)}$ tends to the point $g^{(\infty)} = \infty$, i.e., the final renormalized FK chain corresponds to the continuum (SG) limit where the PN barriers are vanished. So, in this case the incommensurate GS is "sliding", or "truly IC" state, it can be accelerated freely when an arbitrary small force is applied to

Table 5.1. Reduction of a complex structure C of atoms (the "primary" FK model) to the structure C_k of kinks (the "secondary" FK model). The reference structure C_0 is characterized by the dimensionless concentration $\theta_0 = s_0/q_0$ and the period $a_0 = q_0 a_s^{(0)}$, so that the dimensionless elastic constant is $g_a^{(0)} = [a_s^{(0)}]^2 V''_{\rm int}(a_0)/2\pi^2 \varepsilon_s^{(0)}$, and the dimensionless anharmonicity parameter is equal to $\tilde{\beta}^{(0)} = -a_s^{(0)} V'''_{\rm int}(a_0)/V''_{\rm int}(a_0)$.

parameter	C structure	C_k structure
particle's mass	$m_a^{(0)} = m_a$	$m_a^{(1)} = m^*$
coordinates	$x_l^{(0)} = x_l$	$x_l^{(1)} = X_l$
interaction	$V_{\rm int}^{(0)} = V_{\rm int}(x_{l+1} - x_l)$	$V_{\rm int}^{(1)} = v_{\rm int}(X_{l+1} - X_l)$
substrate	$V_{\rm sub}^{(0)} = V_{\rm sub}(x)$	$V_{\rm sub}^{(1)} = V_{PN}(X)$
period	$a_s^{(0)} = a_s$	$a_s^{(1)} = a_0 = q_0 a_s^{(0)}$
height	$\varepsilon_s^{(0)} = \varepsilon_s$	$\varepsilon_s^{(1)} = \varepsilon_{PN}^*$
coverage	$\theta^{(0)} = \theta = s/q$	$\theta^{(1)} = \theta_k = s_k/q_k$
order	$s^{(0)} = s$	$s^{(1)} = s_k = q_0 \lvert s q_0 - s_0 q \rvert$
	$q^{(0)} = q$	$q^{(1)} = q_k = q^{(0)}$
period	$a^{(0)} = a = q a_s$	$a^{(1)} = a_k = q a_0 = q q_0 a_s$
mean distance	$a_A^{(0)} = a_A = a/s$	$a_A^{(1)} = R = a_k/s_k$
linear concentration	$n^{(0)} = n = a_A^{-1} = \theta/a_s$	$n^{(1)} = n_k = R^{-1}$
elastic constant	$g^{(0)} = g = a_s^2 V''_{\rm int}(a_A)/2\pi^2 \varepsilon_s$	$g^{(1)} = g_k = a_0^2 v''_{\rm int}(R)/2\pi^2 \varepsilon_{PN}^*$

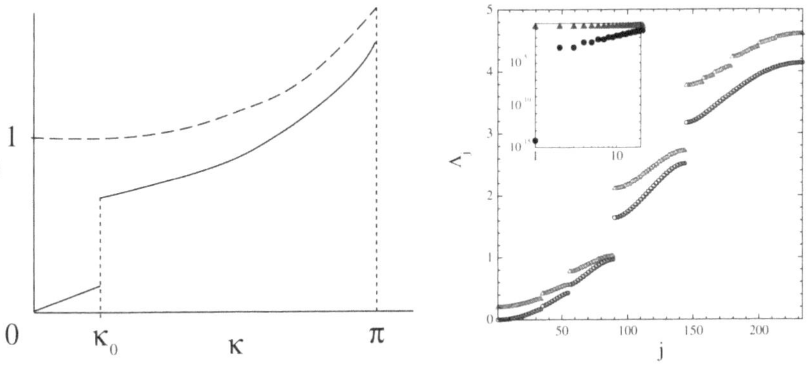

Fig. 5.7. Left panel: phonon spectrum of IC phase in continuum-limit approximation (schematically). Dashed curve describes the optical phonon spectrum of the trivial GS. $\kappa_0 \approx \pi a_s/R$, R being a distance between kinks in the cnoidal wave corresponded to IC phase. Right panel: eigenvalues spectrum of the linear stability matrix ($\Lambda_j = \omega_j^2$) for the "incommensurate" GS with the window number $w = 144/233$ at both sides of the Aubry transition: triangles for $K \equiv g^{-1} = 1.5 > K_c$ (below the Aubry transition, the pinned state) and circles for $K = 0.7 < K_c$ (above the transition, the sliding state). Inset shows the Goldstone mode and the gap at κ_0 [487].

each atom [458]. The "sliding mode" of the "truly IC" state is not a usual zero-wavevector acoustic mode in which the chain of atoms or a lattice of kinks moves rigidly. To explain the sliding mode, it is convenient to consider the IC phase as the limit of a sequence of C-phases as the size of the unit cell goes to infinity. Then, the sliding mode of the IC phase corresponds to the motion of a single kink in these C-phases as the kink width tends to infinity and the amplitude of the PN potential goes to zero.

To investigate the phonon spectrum of the GS, we have to substitute the functions

$$x_l(t) = x_l^{\mathrm{GS}} + \chi_l(t), \quad \chi_l(t) \propto \exp(i\omega t - i\kappa l) \tag{5.72}$$

into the motion equation and then linearize the obtained equation in small displacements $\chi_l(t)$. In the continuum limit approximation, where an incommensurate ground state x_l^{GS} is represented by a cnoidal wave with a distance R between the kinks, the equation for $\chi(x,t)$ reduces to the Lame equation

$$\frac{\partial^2 \chi}{\partial t^2} - \frac{\partial^2 \chi}{\partial x^2} - \cos[u^{\mathrm{GS}}(x)]\,\chi = 0. \tag{5.73}$$

Equation (5.73) is exactly solvable through Jacobian elliptic functions [484]. It follows that the phonon spectrum consists of two branches (see Fig. 5.7, left panel) [447, 455, 457, 485, 486]. At low wave vectors, $\kappa \ll \kappa_0$, where $\kappa_0 \simeq \pi a_s/R$, there is an acoustic branch with the dispersion law

$$\omega(\kappa) = c_s \kappa, \tag{5.74}$$

where the sound velocity c_s is equal to

$$c_s = g^{1/2}(1 - k^2)^{1/2}\mathbf{K}(k)/\mathbf{E}(k), \tag{5.75}$$

$\mathbf{K}(k)$ and $\mathbf{E}(k)$ being the complete elliptic functions of the first and second types respectively, and k is the modulus determined by Eq. (5.67). This mode represents the collective motion of kinks in the cnoidal wave, and at $\kappa \to 0$ it corresponds to the sliding mode. At higher wave vectors, $\kappa \geq \kappa_0$, the phonon spectrum is optical analogously to the spectrum of pieces of the nearest C-phase separated by kinks in the cnoidal wave. The acoustic and optical branches are separated by a forbidden gap at a wave vector κ_0 (in the truly IC GS $\kappa_0 \to 0$). Note that for the discrete FK chain the optical branch is much more complicated because it consists of an infinite number of subbranches separated by forbidden gaps (see Fig. 5.7, right panel).

Using the KAM theory and topological arguments, Aubry *et al.* [56, 443, 460, 463] have proved that for the discrete FK chain with strictly convex potential $V_{\rm int}(x)$ an incommensurate GS can be parameterized by one or two hull functions $h(x)$,

$$x_l = h(la_A + \beta), \tag{5.76}$$

where β is an arbitrary phase, so that the potential energy $\varepsilon(\{x_l\})$ is independent of β. The hull function $h(x)$ obviously depends on the concentration θ (or the window number w). It describes a structure of atoms at distance a_A which is modulated by the function $\phi(x)$,

$$\phi(x) = h(x) - x, \tag{5.77}$$

which is periodic with the same period a_s as the substrate potential $V_{\rm sub}(x)$. Above the critical value, $g > g_{\rm Aubry}(\theta)$, the hull function $h(x)$ is continuous (analytical). However, when the parameter g is smaller than $g_{\rm Aubry}(\theta)$, the system undergoes abruptly the transition by breaking of analyticity, and the function $h(x)$ becomes discontinuous on a dense set of points which form a Cantor set (see Fig. 5.8). Below the transition the hull function $h(x)$ has two possible determinations, $h^+(x)$ which is right continuous, and $h^-(x)$ which is left continuous:

$$\lim_{\delta \to 0^+} h^-(x+\delta) = h^+(x), \quad \lim_{\delta \to 0^-} h^+(x+\delta) = h^-(x). \tag{5.78}$$

Both functions $h^{\pm}(x)$ determine the same GS. They can be written as a sum of Heaviside step functions,

$$h^{\pm}(x) = \sum_l h_l \,\Theta^{\pm}(x - x_l), \tag{5.79}$$

where h_l is the amplitude of the step function located at x_l [by definition $\Theta^{\pm}(x) = 0$ for $x < 0$, $\Theta^{\pm}(x) = 1$ for $x > 0$, and $\Theta^+(0) = 1$, $\Theta^-(0) = 0$].

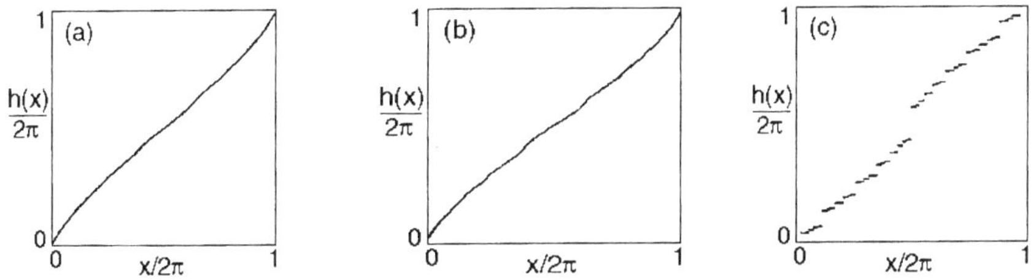

Fig. 5.8. Hull function $h(x)$ in (a) unpinned, $g > g_{\rm Aubry}$, (b) critical, $g = g_{\rm Aubry}$, and (c) pinned phase, $g < g_{\rm Aubry}$, respectively, for the golden-mean w [488]. At $g = g_{\rm Aubry}$, $h(x)$ has a fractal structure.

The transition by breaking of analyticity can be interpreted in the following intuitive way [463, 489, 490]. It is clear that in the limit $g \to 0$ all atoms in the GS will lie close to minima of the substrate potential $V_{\rm sub}(x)$. Therefore, in order for the chain to be moved by a discrete amount to a new position with the same (lowest) potential energy, at least one atom must jump from one valley of the $V_{\rm sub}(x)$ to another by going over the maxima (tops) in the $V_{\rm sub}(x)$. As a result, the GS is pinned, the lowest phonon frequency $\omega_{\rm min}$ is nonzero, and it must be applied a nonzero force F_{PN} (known as the depinning force) to each atom in order for the chain to be moved. Otherwise, a GS will be unpinned (as it should be at least in the SG limit, $g \to \infty$) if and only if there are atoms in the GS arbitrary close to a top of the potential $V_{\rm sub}(x)$. Thus, it is convenient to introduce a "disorder" parameter ψ defined as the minimum distance of any atom from the nearest top of the substrate potential,

$$\psi = \min_{l,j} \left| x_l - \left(j + \frac{1}{2} \right) a_s \right| . \tag{5.80}$$

Computer simulations show that near the critical point the transition from pinned to sliding ground states occurs according to a power law,

$$\psi,\ d,\ \omega_{\rm min},\ E_{PN},\ F_{PN} \propto (g_{\rm Aubry} - g)^{\chi}, \tag{5.81}$$

where the threshold $g_{\rm Aubry}$ and the critical super-exponents $\chi_{\dots}$ depend on the concentration θ (or the window number w). In particular, for the golden mean window number, $w = w_{\rm gm} \equiv (\sqrt{5} - 1)/2$, when $K_{\rm Aubry} = g_{\rm Aubry}^{-1} = 0.971635406$ (see Ref. [476]), it was found that $\chi_\psi \approx 0.7120835$ for the disorder parameter [the largest gap in the hull function (5.76)], $\chi_d \approx -0.9874624$ for the coherent length, $\chi_{\omega_{\rm min}} \approx 1.0268803$ for the phonon gap $\omega_{\rm min}$, and $\chi_{PN} \approx 3.0117222$ for the PN energy barrier and the depinning force (see also Refs. [463, 467, 476, 489], [490]–[493]). Note that this transition exhibits a scaling behavior as usually in critical phenomena.

The exponents above are called super-critical because they are zero in the sliding ground state, $g > g_{\rm Aubry}$. On the other hand, one may introduce sub-critical exponents for the sliding state above the Aubry transition. One of them is "compressibility" $C = (\partial w/\partial P)_g$ which scales as $C(g) \propto (g - g_{\text{Aubry}})^{\chi_c}$ with $\chi_c \approx 0.049328$ for the golden-mean window number. In the SG limit ($K \to 0$ or $g \to \infty$) we have $w(P) = P$ so that $C = 1$. In the truly IC GS, $g > g_{\text{Aubry}}$, C decreases with g and vanishes in the pinned GS, where the Devil's staircase is complete and $w(P)$ has zero slope everywhere. Another important subcritical quantity is the effective viscosity $\Gamma = \lim_{F\to 0} F/v$ which describes the steady-state average velocity $v = \langle \dot{x}_l \rangle$ in response to an infitisemally small dc force F applied to all atoms (to avoid infinite acceleration, an external damping η should be included in the motion equation). Γ is zero in the SG limit ($g \to \infty$, or $K \to 0$) and diverges at the Aubry transition. For the golden-mean window number, it scales as

$$\Gamma(g) \propto (g - g_{\text{Aubry}})^{-\chi_\Gamma} \tag{5.82}$$

with $\chi_\Gamma \approx 0.029500$. The critical exponents are coupled by the scaling relations [476, 491, 493]

$$2\chi_{\omega_{\min}} + \chi_d = \chi_\Gamma + \chi_{PN}, \tag{5.83}$$

$$\chi_c = \chi_{PN} - 3\chi_d. \tag{5.84}$$

Biham and Mukamel [494] have studied numerically the threshold g_{Aubry} and the critical exponents $\chi_{\dots}$ as functions of the window number w. They found that the critical curve $g_{\text{Aubry}}(w)$ has a fractal nature and is characterized by the Hausdorff dimension 0.87 ± 0.02.

The truly IC GS, $g > g_{\text{Aubry}}$, is often called "frictionless", because the chain begins to slide freely when an arbitrary small dc force is applied to all atoms. This "frictionless" motion (also called "superlubrication") exists, however, only in the SG limit ($g \to \infty$) and, moreover, only for an infitisemally small velocity of the center of mass (c.m.), $v_{\text{c.m.}} \to 0$, where $v_{\text{c.m.}} = \dot{X}$, $X = N^{-1} \sum_l x_l$. As one can see from Fig. 5.9, the center-of-mass velocity is lower than the maximum value $v_f = F/m_a\eta$ even in the $F \to 0$ limit in agreement with Eq. (5.82). At any finite velocity, $v_{\text{c.m.}} > 0$, the kinetic energy of the c.m. motion decays due to excitation of phonons, although with very long time scales in some velocity windows (note that in simulation of a short chain, when the phonon spectrum is discrete, the superlubrication may be observed for some velocity windows). The mechanism of this damping is, however, a rather subtle one [495]. In the truly IC state the configuration (5.76) corresponds to the equidistant arrangement of atoms (with the spacing a_A) modulated with the wave vector $\kappa_s = 2\pi/a_s$ (and its higher harmonics $n\kappa_s$) due to the substrate potential. In the $g \gg 1$ case, the amplitudes of the modulation scales as K^n, where $K \equiv g^{-1} \ll 1$, and may be considered as the "frozen-in" phonons with the frequencies $\omega_{n\kappa_s}$, where $\omega_q = 2\sqrt{g}\,\sin(qa_A/2)$ is the phonon spectrum of the elastic chain without the substrate potential. Now, if such a configuration will rigidly slide with the c.m. velocity $v_{\text{c.m.}}$, its atoms pass over the maxima of the substrate potential with the washboard

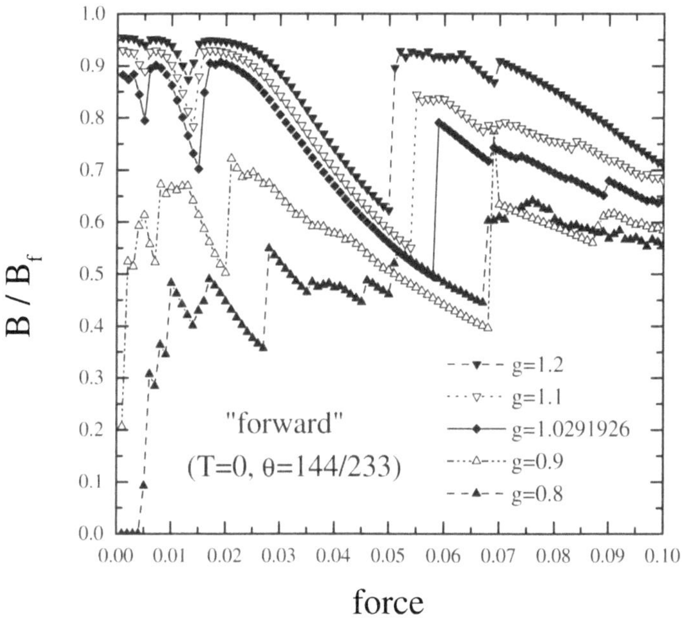

Fig. 5.9. Mobility $B = v_{\mathrm{c.m.}}/F$ normalized on the free-motion value $B_f = (m_a \eta)^{-1}$ as a function of the dc force F for the classical FK model with the "golden-mean" concentration $\theta = 144/233$ for different values of the elastic constant g below and above the Aubry threshold $g_{\mathrm{Aubry}} \approx 1.0291926$. The motion equation included an external viscous damping with the friction coefficient $\eta = 0.1$.

frequency $\omega_{\mathrm{wash}} = (2\pi/a_s)\, v_{\mathrm{c.m.}}$. The following decay of the c.m. velocity occurs in two steps. On the onset of motion, the nonlinear coupling of the c.m. to the "frozen-in" mode ω_{κ_s} (and its higher harmonics) leads to long wavelength oscillations (emerging due to Mathieu *parametric* resonance) with the frequency $\Omega \sim |\omega_{\mathrm{wash}} - \omega_{\kappa_s}|$ (but not with the washboard frequency ω_{wash} as one could suppose). Then, on the second step, these long wave-length oscillations drive a complex parametric resonance involving *several* resonant modes (see details in Ref. [495]).

5.3 Free-End Chain

While the GS of the "fixed-density" FK chain is determined by two parameters g and θ, the "free-end" GS is determined by the parameters g and P [the misfit P is defined in Eq. (5.4)], and corresponds to the absolute minimum of the potential energy

$$E(P, N) = U_{\mathrm{sub}} + U_{\mathrm{int}},$$

$$U_{\text{sub}} = \sum_{l=1}^{N} V_{\text{sub}}(x_l), \tag{5.85}$$
$$U_{\text{int}} = \sum_{l=2}^{N} V_{\text{int}}(x_l - x_{l-1}),$$

where one or both ends of the chain are let free, and the limit $N \to \infty$ is implied. Clearly that the free-end chain can exit only if the interatomic potential $V_{\text{int}}(x)$ has an attractive branch. To avoid this restriction, let us suppose that the chain is subjected to one-dimensional external "pressure" Π (which is minus the "stress" or "tension") which allows one to change the distance between neighboring atoms. In this case we have to look for a minimum of the entalpy

$$W(\Pi, N) = E + \Pi L, \tag{5.86}$$

where L is the total length of the chain,

$$L = \sum_{l=2}^{N} (x_l - x_{l-1}) = x_N - x_1, \tag{5.87}$$

so that Π is the thermodynamic conjugate to the extensive variable L. (Recall that at zero system temperature the potential energy E coincides with the Helmholtz free energy F, and the entalpy W, with the Gibbs free energy G). For the standard FK model with $V_{\text{int}}(x) = \frac{1}{2}g(x - a_{\min})^2$ it follows

$$U_{\text{int}} + \Pi L = \sum_{l=2}^{N} \left\{ \frac{1}{2} g(x_l - x_{l-1} - a_{\min})^2 + \Pi(x_l - x_{l-1}) \right\}$$
$$= \sum_{l=2}^{N} \left\{ \frac{1}{2} g(x_l - x_{l-1} - a'_{\min})^2 + \Pi'(x_l - x_{l-1}) \right\} + C, \tag{5.88}$$

where C is the constant, $C = \frac{1}{2}g(N-1)\left[a_{\min}^2 - (a'_{\min})^2\right]$, and the old and new pressure and the corresponding misfit parameters are connected by the relationship

$$\Pi - g a_s P = \Pi' - g a_s P'. \tag{5.89}$$

Thus, a chain with $P = 0$ subjected to the external pressure Π, is equivalent to the chain with $P = -\Pi / g a_s$ subjected to zero external pressure. Therefore, a description in which g and P are the major variables and that where g and Π are the major variables are equivalent. Note also that when the FK model describes a chain of pendulums [496, 497], x_l means the angle of rotation, and L is the total average phase shift generated by the "torque" Π.

There is also an alternative description where one is looking for a GS for a chain which is in equilibrium with a particle reservoir having some chemical potential μ. In this approach we have to minimize the function

$$J(L,\mu) = E - \mu N, \tag{5.90}$$

and the total number of atoms is determined as $N = \partial J/\partial\mu$.

Thus, the GS problem for the "free-end" FK chain reduces to construction of a phase diagram of the system in one of phase planes (g, P), (g, Π), or (g, μ). In this subsection we discuss the standard FK model. Because the behavior of this model is the same for any value of the integer q_0 in the definition of P [see Eq. (5.4)], it is enough to consider the case of $q_0 = 1$. Moreover, the standard FK model is symmetric on the sign of the misfit P (the kink-antikink symmetry), so the variation of P can be restricted by the interval $0 \le P \le \frac{1}{2}$.

5.3.1 Frank-van-der-Merwe Transition

It is evident that when the value $a_{\min}$ is equal or close to the period of the external potential a_s, so that $P \simeq 0$, the GS of the "free-end" FK chain is trivial. Let us now insert a kink with a topological charge $\sigma = -\text{sign}P$ into the trivial state through a free end of the chain. As a result, the system energy is changed on the value [see Eqs. (5.57) and (5.63)]

$$\varepsilon_k = -|P|ga_s^2 + E_k, \quad E_k \simeq \begin{cases} 2\pi^2 g(1-2g) & \text{if } g \ll 1, \\ 8\sqrt{g} & \text{if } g \gg 1. \end{cases} \tag{5.91}$$

As long as $\varepsilon_k > 0$, or $|P| < P_{FM}$, where the critical value P_{FM} is defined by the equation $\varepsilon_k = 0$, so that

$$P_{FM} \simeq \begin{cases} \frac{1}{2}[1 - 2g/(1+3g)] & \text{if } g \ll 1, \\ 2/\pi^2\sqrt{g} & \text{if } g \gg 1, \end{cases} \tag{5.92}$$

the GS configuration will remain trivial. However, beyond the critical value P_{FM}, it holds $\varepsilon_k < 0$, and the creation of kinks will lower the chain's energy. (In this subsection the name kink means a kink configuration for the reference structure with $\theta_0 = 1$. To avoid misunderstanding, we will use the name "superkink" for kink configurations on the background of reference structures with a complex unit cell). A kink-kink repulsion is characterized by the energy

$$v_{\text{int}}(R) \simeq 32\sqrt{g}\exp(-R/d), \quad g \gg 1, \tag{5.93}$$

where $d = a_s\sqrt{g}$, and it sets a limit for the density of kinks. Thus, at $P_{FM} < |P| < \frac{1}{2}$, the chain with a free end will expand or contract in order to lower its potential energy, and the GS configuration will be described by a cnoidal wave, i.e., by an infinite sequence of kinks separated by a distance R_0 which is defined by the equation $v_{\text{int}}(R_0) + \varepsilon_k = 0$, so that

$$R_0 \simeq d \ln\left(\frac{4P_{FM}}{|P| - P_{FM}}\right), \quad g \gg 1. \tag{5.94}$$

The transition from the trivial GS to the "cnoidal wave" GS was firstly considered by Frank and van der Merwe [31]. It is known in literature as the C-IC transition [456, 457]. However, this name may lead to misunderstanding, because at any $g < \infty$ (except the limiting case of $g = \infty$) the transition occurs between two C-phases, the trivial and higher-order commensurate ground states (see next subsection). Therefore, we will call this transition as the "Frank-van-der-Merwe (FvdM) transition". Now we consider it in more details following a quite instructive original work of Frank and van der Merwe [31].

First, let us consider the infinite FK chain and neglect the boundary conditions. Stable configurations of the chain are determined by the equation

$$g(x_{l+1} + x_{l-1} - 2x_l) = \sin x_l, \tag{5.95}$$

which reduces in the continuum limit ($g \gg 1$, $x_l = la_s + u_s$, $l \to x = la_s$, $u_l \to u(x)$) to the pendulum equation

$$\frac{d^2u}{d\widetilde{x}^2} = \sin u, \tag{5.96}$$

where we introduced the dimensionless variable $\widetilde{x} = x/d$, $d = a_s\sqrt{g}$. A solution of Eq. (5.96) describes a configuration with the potential energy

$$E[u] = \sqrt{g}\int d\widetilde{x}\; \varepsilon\left[u(\widetilde{x})\right],$$
$$\varepsilon[u] = \frac{1}{2}\left(\frac{du}{d\widetilde{x}} - Pd\right)^2 + (1-\cos u)\,. \tag{5.97}$$

Equation (5.96) has two important solutions. The first corresponds to the trivial configuration, $u^{(\mathrm{tr})}(\widetilde{x}) = ja_s, j = 0, \pm 1, \ldots$, and has an energy per unit length

$$\varepsilon^{(\mathrm{tr})} \equiv E[u^{(\mathrm{tr})}]/L = P^2d^2/2a_s. \tag{5.98}$$

The second solution describes a cnoidal wave,

$$u^{(\mathrm{cw})}(\widetilde{x}) = \pi + 2\sin^{-1}\mathrm{sn}\left\{-\sigma(\widetilde{x}-\beta)/k, k\right\}, \tag{5.99}$$

where $\mathrm{sn}(z,k)$ is the Jacobian sinus function, k being its modulus ($0 < k \le 1$), β is an arbitrary phase which shows the possibility for free sliding of an infinite chain in the SG limit. The cnoidal wave describes a configuration with a regular sequence of kinks of topological charges $\sigma = -\mathrm{sign}P$, a distance R between kinks being determined by the modulus k,

$$R = 2dk\mathbf{K}(k), \tag{5.100}$$

so that the density of excessive (missing) atoms is

$$\rho(x) = -\frac{u_x}{a_s} = \frac{\sigma}{\pi d}\left\{\frac{(1-k^2)}{k^2} + \frac{1}{2}\left[1-\cos u^{(\mathrm{cw})}(x)\right]\right\}. \tag{5.101}$$

The function $\rho(x)$ is periodic with the period R. The cnoidal wave (5.99) has the energy per one period

$$\varepsilon_{\rm pp}^{\rm (cw)} = \sqrt{g} \int_0^{R/d} d\widetilde{x}\, \varepsilon \left[u^{\rm (cw)}(\widetilde{x}) \right]$$
$$= \sqrt{g} \left\{ P^2 d^2 k \mathbf{K}(k) - 2\pi |P| d - \frac{4}{k}(1-k^2)\mathbf{K}(k) + \frac{8}{k}\mathbf{E}(k) \right\}, \quad (5.102)$$

so that the energy per unit length is equal to

$$\varepsilon^{\rm (cw)} = \varepsilon_{\rm pp}^{\rm (cw)}/R = \varepsilon^{\rm (tr)} - \frac{1}{a_s} \left\{ 2\frac{(1-k^2)}{k^2} + \frac{\pi |P| d}{k\mathbf{K}(k)} - \frac{4\mathbf{E}(k)}{k^2 \mathbf{K}(k)} \right\}. \quad (5.103)$$

Evidently, in the GS configuration the period R must minimize the energy (5.103). Thus, from equation $\partial \varepsilon^{\rm (cw)}/\partial k = 0$ it follows the equation

$$\frac{\mathbf{E}(k)}{k} = \frac{\pi}{4}|P|d \equiv \frac{|P|}{P_{FM}}. \quad (5.104)$$

The solution of Eq. (5.104), $k = k_0$, determines the period R_0 [through Eq. (5.100)] and the energy per unit length,

$$\varepsilon_0^{\rm (cw)} = \varepsilon^{\rm (tr)} - \frac{2}{a_s} \frac{(1-k_0^2)}{k_0^2}, \quad (5.105)$$

of the “cnoidal wave” GS configuration in the continuum limit approximation. Note that Eq. (5.104) has a solution, and therefore the cnoidal wave GS has a lower energy than that of the trivial GS, if and only if $|P| > P_{FM}$ in accordance with previous discussion.

Now let us return to the free-end boundary condition following the elegant approach of Frank and van der Merwe [31]. Consider a stationary configuration of an infinite chain and suppose that the spring which connects the N-th and $(N+1)$-th atoms, is unstressed, $x_{N+1} - x_N = a_{\rm min}$. So, we can break this spring without changing of positions of all the atoms; in a result we obtain two semi-infinite chains in stationary states each having one free end. Thus, we come to the boundary condition $u_{N+1} - u_N = P a_s$, or, in the continuum limit approximation,

$$\left. \frac{du}{d\widetilde{x}} \right|_{\widetilde{x} = \widetilde{x}_{\rm end}} = Pd. \quad (5.106)$$

Substituting the cnoidal wave (5.99) into Eq. (5.106), we obtain the relationship

$$\frac{1}{k} \,\mathrm{dn} \left(\frac{\widetilde{x}_{\rm end} - \beta}{k}, k \right) = \frac{1}{2}|P|d, \quad (5.107)$$

where $\widetilde{x}_{\rm end} = N a_s/d$ is the coordinate of the free end. This boundary condition leads to a number of interesting consequences.

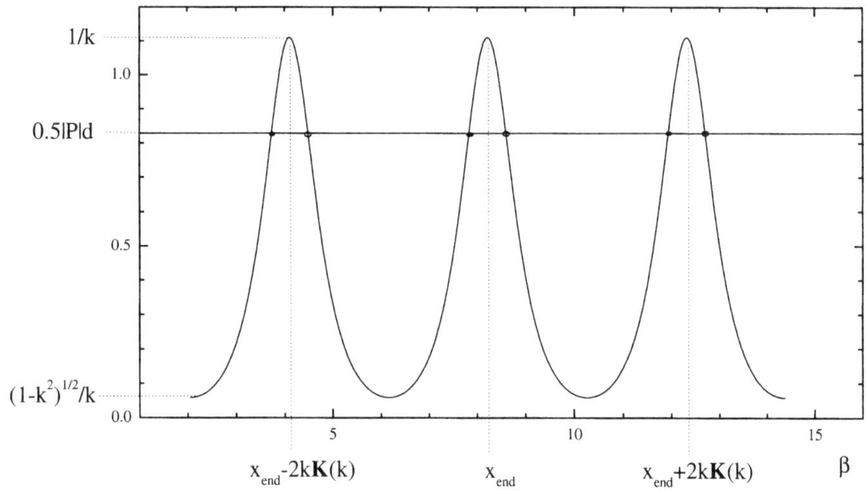

Fig. 5.10. Graphic solution of Eq. (5.107). The filled and open circles correspond to stable and unstable configurations of the chain, correspondingly.

At any $P \neq 0$ the trivial solution, $u^{\rm tr}(\widetilde{x}) = {\rm const}$, does not satisfy Eq. (5.106). This means that closely to the end of the chain (at distances $|x - x_{\rm end}| \leq d$) the trivial GS configuration is always "perturbed", i.e. the atoms near the end are shifted away from the minima of the substrate potential. So, the trivial GS configuration is *always* described by the cnoidal wave (5.99) in the limit $k \to 1$ (or $R_0 \to \infty$), while the energy remains equal to $\varepsilon^{(\rm tr)}$ [Eq. (5.98)].

Equation (5.107) is the equation on the phase β. As shown in Fig. 5.10, this equation has two solutions [both are defined up to an additional constant which is multiplier of $2k\mathbf{K}(k)$], one corresponds to a stable configuration, and another, to unstable one. Therefore, the phase β of the cnoidal wave GS is fixed, this means that *the free end of the FK chain is always pinned* even in the SG limit $g \to \infty$ when PN barriers are absent. Note that this is a consequence of the fact that the SG equation with free boundary conditions is not an exactly integrable equation (contrary to the case of periodic boundary conditions where the SG equation does is exactly integrable). As a result, a sliding mode is absent in the "free-end" FK chain at any model parameters, the chain always requires an activation energy to slide [447, 458]. The potential energy of the chain with free ends is a periodic function (with period a_s) of the location of the end atoms in the chain. The amplitude of oscillations of the potential energy $\varepsilon_{\rm pin}$ is finite in the limit $L \to \infty$. Of course, the energy per unit length is still given by Eq. (5.105) because $\lim_{L\to\infty} \varepsilon_{\rm pin}/L = 0$.

Equation (5.107) has a solution if and only if (see Fig. 5.10)

$$\frac{\sqrt{1-k^2}}{k} < \frac{1}{2}|P|d < \frac{1}{k}. \tag{5.108}$$

In particular, the trivial configuration ($k \to 1$) is permitted as long as

$$|P| < P_{ms} \equiv 2/d. \tag{5.109}$$

Thus, the trivial configuration of the chain with free ends remains stable beyond the misfit P_{FM} up to the second critical value of misfit P_{ms}. Within the interval $P_{FM} < |P| < P_{ms}$ the trivial configuration corresponds to a metastable state of the system, because at $|P| < P_{ms}$ a finite activation energy is to be overcome in order to insert a kink into the chain through its free end. Beyond the metastable limit P_{ms} (i.e., within the interval $P_{ms} < |P| < \frac{1}{2}$) the trivial configuration becomes unstable, because kinks are introduced spontaneously at the free ends of the chain.

5.3.2 Devil's Staircase and Phase Diagram

Now we are able to construct the whole phase diagram for the GS structure of the "free-end" FK chain. As was shown above, in the interval $0 \le |P| < P_{FM}(g)$ the GS is trivial ($\theta = 1$ and $w = 0$), while in the interval $P_{FM}(g) < |P| \le \frac{1}{2}$ the GS corresponds to a cnoidal wave of kinks separated by the distance R_0 so that

$$\theta = \frac{R_0/a_s + \sigma}{R_0/a_s} = 1 + \frac{\sigma a_s}{R_0} \quad \text{and} \quad w = \left\{ \frac{R_0}{R_0 + \sigma a_s} \right\}. \tag{5.110}$$

Thus, at a given value of g, any value of the misfit P (or the pressure Π, or the chemical potential μ) defines a unique value of the concentration θ (or the window number w), and the structure of the GS coincides with that one described in Sect. 5.2.1 for the same value of θ (except the chain's structure at distances $|x - x_{\text{end}}| \le d$ near its free ends where the structure is always perturbed). Clearly, the value $w(g, P)$ can be a rational as well as an irrational number, thus corresponding to C or IC phase. Recalling that kinks in the cnoidal wave can be considered as quasiparticles subjected to a periodic PN potential and interacting via exponential law $v_{\text{int}}(X)$, we again may use the renormalization approach of Joos *et al.* (1983). Now we have to renormalize simultaneously two model parameters, $\left(g^{(i)}, P^{(i)}\right) \to \left(g^{(i+1)}, P^{(i+1)}\right)$, where the elastic constant $g^{(i)}$ is defined by the interaction between quasiparticles, and the misfit parameter $P^{(i)}$ is defined by the distance between quasiparticles instead of $a_{\min}$ in Eq. (5.4). If the renormalization sequence, starting from the point (g, P), is finished at the stable fixed point $\left(g^{(\infty)}, P^{(\infty)}\right)$ with $g^{(\infty)} = \infty$, then the GS of the system corresponds to a IC-phase. Otherwise, the GS is commensurate. Thus, the phase plane (g, P) is separated into regions where the GS has C or IC structure. Clearly, IC phases may exist only in the region $g > g_c \equiv \min_w g_{\text{Aubry}}(w)$.

Using the ideology of the previous subsection, it is easy to guess that as well as the trivial configuration, each commensurate configuration with rational $w = r/s$ will be stable within a finite interval

$$P^-_{ms}(w) < |P| < P^+_{ms}(w), \tag{5.111}$$

and it should correspond to the GS of the system within a more narrow interval

$$P^-_{FM}(w) < |P| < P^+_{FM}(w). \tag{5.112}$$

Indeed, considering the commensurate structure w as a reference one, we can define superkinks (super-antikinks) with topological charges $\sigma = 1$ ($\sigma = -1$). According to Chapter 3, the creation energy of a superkink may be represented as

$$\varepsilon_k^\sigma = -\sigma g a_s^2 \left(|P| - P_0(w)\right) + E_k(w), \tag{5.113}$$

where $P_0(w)$ and $E_k(w)$ are constants which depend on w. The limits $P^\pm_{FM}(w)$ are defined by the equation $\varepsilon_k^\sigma = 0$,

$$P^\pm_{FM}(w) = P_0(w) \pm E_k(w)/g a_s^2. \tag{5.114}$$

As long as P remains between two limits $P^\pm_{FM}(w)$, the GS configuration is independent of P, while beyond the edges of the plateau the creation energy of the corresponding superkink is negative, making the w configuration metastable (see Fig. 5.11). The new GS will correspond to a finite density of superkinks, with the density determined by a competition between the negative creation energy (which favors the creation of as many superkinks as possible), and the positive energy of repulsion of the superkinks (which increases as superkinks come closer together). Thus, the function $w(P)$ consists of a series of "plateaus", or "terraces" on which w is constant as a function of P (see Fig. 5.1). These occur at rational values $w = r/s$, and the lengths of the plateaus tend to be smaller as the period q of the commensurate structure increases due to decreasing of the value $E_k(w)$ in Eq. (5.114).

Aubry and co-workers in a series of works (see Ref. [460] and references therein) have proved that the function $w(P)$ is always continuous, contrary to what one might guess from simply glancing at Fig. 5.1. Therefore, the function $w(P)$ takes irrational numbers when P lies between the neighboring

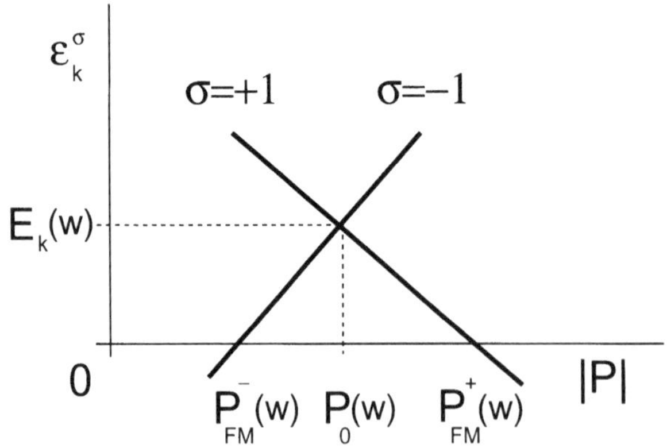

Fig. 5.11. Energy for creation of a superkink ($\sigma = +1$) and a super-antikink ($\sigma = -1$) for $w \neq 0$ reference configuration as functions of the misfit parameter P.

terraces, so that in this case the values w and P are connected in a unique manner.

Analogously, the concentration θ as a function of misfit P locks at all commensurate values $\theta = s/q$, and $\theta(P)$ is a continuous monotonically decreasing curve against P. Following Manderbrot [444], such a curve is known as Devil's staircase. Aubry and co-workers (see Ref. [460], and Refs. [498]–[500]) have proved that at $g < g_c$ the function $\theta(P)$ reaches incommensurate values for a zero measure set of values of P, and the curve $\theta(P)$ is the complete Devil's staircase. [By definition, Devil's staircase is called complete when it is entirely composed of steps, or equivalently when $\theta(P)$ has a zero derivative almost everywhere except an uncountable set of point which forms a Cantor set. Thus, the gaps between the steps of the function $\theta(P)$ is a fractal with zero measure, and magnification of any part of the curve $\theta(P)$ (not within a step) will reproduce the original curve.] When $g > g_c$, Devil's staircase $\theta(P)$ becomes incomplete, that is, irrational values are reached by $\theta(P)$ for a finite measure set of P values (recall that the corresponding map orbits belong to KAM circles). $\theta(P)$ still locks at every rational value but the locking intervals (in terms of P value) become smaller so that the curve $\theta(P)$ seems to be smooth between the largest steps. An important feature is that, in a given interval (P_1, P_2), the sum of the width of all the steps goes to zero when g goes to infinity.

The description given above allows us to draw the zero-temperature phase diagram of the FK system in the parametric plane (g, P) (see Fig. 5.12). Of course, the phase diagram can be constructed only by computer simulations; the most powerful method was proposed by Griffiths and Chou [276]. The region occupied by a GS with a given rational w appears as a "tongue" in

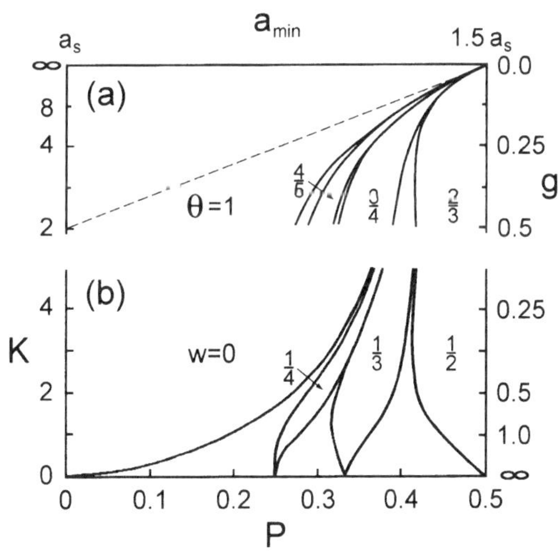

Fig. 5.12. Phase diagram for the standard FK model in (a) the $(a_{\min}, g)$ plane and (b) the (P, K) plane, where $K = g^{-1}$ [276].

the phase plane. Between those tongues which are shown in Fig. 5.12, there are an infinite number of other tongues (which are not shown, most of them being extremely narrow) for every rational value of w, and these fill up, to some extent, the regions between tongues pictured explicitly. In addition, because w is continuous function on P, there must be lines corresponding to irrational values of w lying between the tongues for rational values.

In conclusion, we notice that for the strictly convex $V_{\text{int}}(x)$ all transitions between different structures are continuous ("second order") for the complete as well as for the incomplete Devil's staircase regions because all interactions are repulsive (see Refs. [455]–[457],[460],[501]–[504]).

5.4 Generalizations of the FK Model

In order to be applied to real physical systems, the FK model has to be generalized because usually a substrate potential is not purely sinusoidal as well as interatomic interactions are not purely harmonic. From a general point of view, any deviation away from the exactly integrable SG system leads usually to "increasing of chaos" in the system, i.e., to decreasing of an area occupied by the "KAM-tori" ("KAM-circles" in the two-dimensional map) [see, e.g., the work of Milchev [221] where the harmonic potential is replaced by the Toda potential]. In other words, the regular "KAM-circles" of the symplectic map start to destroy at lower values of the mapping parameter $K \equiv g^{-1}$. For the corresponding model this results in increasing of (a) the PN barriers, (b) the value of $g_{\text{Aubry}}(w)$ and, therefore, (c) the threshold g_c of transition from complete to incomplete Devil's staircase. Besides, the deviation from the standard FK model may lead to (i) appearing of different phases with the same window number, (ii) replacing the continuous transitions between phases by discontinuous ones, and (iii) violation of the kink-antikink symmetry. Below we consider these questions in more details.

5.4.1 On-Site Potential of a General Form

In a general case the periodic substrate potential $V_{\text{sub}}(x)$ can be expressed by a Fourier cosine series

$$V_{\text{sub}}(x) = \sum_{p=1}^{\infty} \beta_p \left[1 - \cos(px)\right]. \tag{5.115}$$

Below we consider only small deviations from the sinusoidal form of the potential so that $\beta_1 \simeq 1$ and $|\beta_p| \ll 1$ for $p \geq 2$. A behavior of the FK model is different for the cases of $\beta_2 < 0$ and $\beta_2 > 0$. When $\beta_2 < 0$, $V_{\text{sub}}(x)$ has a shape of broad wells separated by narrow barriers. In this case the phase diagram in the (g, P) plane is topologically similar, and phase transitions occur on qualitatively the same manner as in the standard FK system [445, 505].

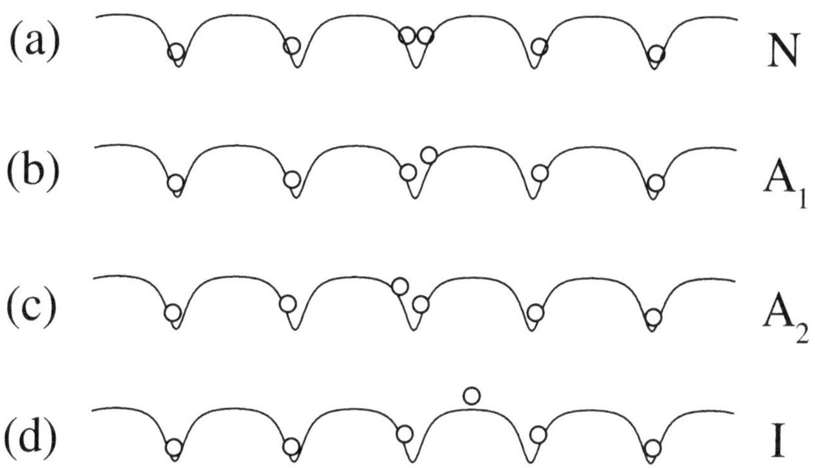

Fig. 5.13. Stable kink configurations for the substrate potential with sharp wells separated by flat barriers. (a) Normal configuration, (b) and (c) Asymmetric configurations, and (d) Inverse configuration.

However, the case of $\beta_2 > 0$ when $V_{\rm sub}(x)$ has a shape of sharp wells separated by flat barriers, is more complicated, because the structure of a kink depends now on the model parameter g [165, 186]. Recall that for the standard FK model a single kink has only two configurations which are stationary. The first (stable) one shown in Fig. 5.13(a) describes a configuration where two atoms occupy the same potential well and locate symmetrically in the bottom of the well. It corresponds to a minimum of the PN potential. In the second (unstable, or saddle) configuration shown in Fig. 5.13(d) an additional atom locates at a top of $V_{\rm sub}(x)$; this configuration corresponds to a maximum of $V_{PN}(X)$. We will call the described situation as the N (i.e., normal) one. Evidently that it always takes place for any $V_{\rm sub}(x)$ at least in the limit $g \to 0$. However, when $V_{\rm sub}(x)$ has sharp wells separated by flat barriers, at certain values of the strength of interatomic potential g, it emerges the I (i.e., inverse) situation where the configuration Fig. 5.13(a) corresponds to a maximum, while the configuration Fig. 5.13(d), to a minimum of the PN potential. As the parameter g increases, the N- and I- cases alternate one another. In addition, between the N- and I- regions there may exist intermediate regions where both configurations of Fig. 5.13(a) and Fig. 5.13(d) correspond to local maxima of function $V_{PN}(X)$, while its minimum occurs at some intermediate asymmetric configuration shown in Fig. 5.13(b) or Fig. 5.13(c) and denoted by A_1 or A_2. Now, if we increase the misfit P beyond the FvdM limit P_{FM} (of course, the value P_{FM} now differs from that one calculated for the standard FK model), the GS of the system will be a cnoidal wave of kinks of N-, I-, or A- types depending on the value of g.

Thus, at $\beta_2 > 0$ a commensurate GS with a window number $w = r/s$ may be one of the following types [466, 478]. The N-type GS is defined as that one for which it exists an integer j such that for any l

$$\begin{aligned} x_{j+l} &= -x_{j-l} \quad \text{if } s \text{ is odd,} \\ x_{j+l} &= -x_{j+1-l} \quad \text{if } s \text{ is even.} \end{aligned} \tag{5.116}$$

For s odd there are atoms at minima of the potential $V_{\rm sub}(x)$, while there are no atoms at minima of $V_{\rm sub}(x)$ for s even. A GS is called I-type if it exists an integer j such that for any l

$$x_{j+l} - \frac{1}{2}a_s = \frac{1}{2}a_s - x_{j-l}. \tag{5.117}$$

In I-type configurations there are atoms at maxima of the potential $V_{\rm sub}$. Both N-type and I-type configurations have reflection symmetry. A GS that does not satisfy any of Eqs. (5.116), (5.117) is called A-type. The symmetry-breaking A-type GS is additionally twice degenerated because it can be considered as constructed by kinks of type A_1 or type A_2, so that the ground states A_1 and A_2 are mirror images of one another. Therefore, the situation is now similar to that one for the double-well substrate potential (see Sect. 3.3.4), where we also had "left" and "right" kinks.

Thus, at $\beta_2 > 0$ the window number w does not uniquely define the configuration. Different types of ground states can be characterized by, e.g., the center-of-mass coordinate of a unit cell defined by Eq. (5.33), (5.34), and by letters N, I, A_1, or A_2. A phase diagram of the "free-end" FK chain is shown in Fig. 5.14. The tongues corresponding to rational values of w are now split by a series of horizontal bars at which there is a phase transition between phases of the same window number but different types. The number of such horizontal bars increases with increasing "order" s of the C-phase. Note that for the double-SG substrate potential (i.e., if $\beta_p = 0$ for $p \geq 3$) the asymmetric (A-type) phases are absent [445].

The transitions between phases of different types but the same w are "phonon-stimulated": with varying the strength g, the minimum phonon frequency $\omega_{\min}$ turns to zero, and the corresponding phase becomes unstable. This transitions may be continuous ("second order") if $\omega_{\min}$ continuously tends to zero, or they may be discontinuous ("first order") when the instability appears abruptly. For example, for the case described in Fig. 5.14, all transitions between N and A phases are discontinuous, whereas those between I and A phases are continuous. If is easy to guess using symmetry reasons that if the A-phases are absent (as in the double-SG model), transitions N$\leftrightarrow$I are always discontinuous [445]. The transitions between phases with different window numbers are continuous because they are "kink-stimulated" with repulsive interaction between kinks. These transitions are analogous to those in the standard FK model. Note only that the limits $P^{\pm}_{FM}(w)$ of existing of an A-phase are determined now by the equation $\varepsilon_k^{(\rm left)} + \varepsilon_k^{(\rm right)} = 0$ because

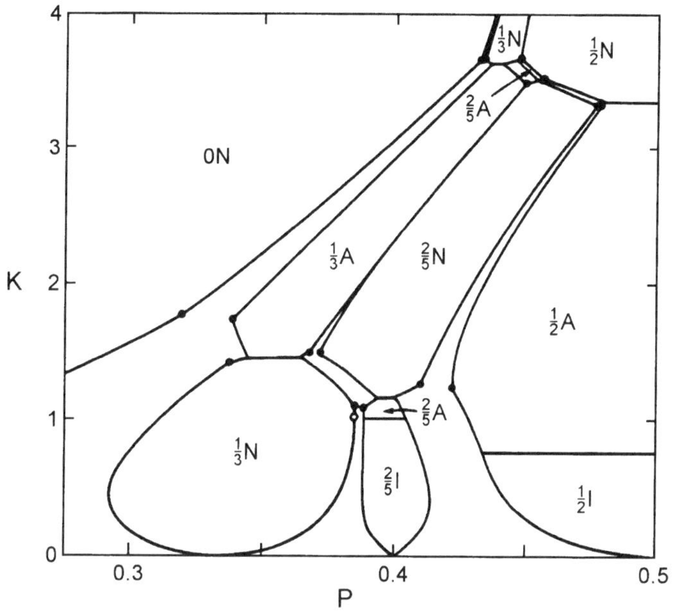

Fig. 5.14. Phase diagram for harmonically interacting atoms with $g = K^{-1}$ and nonsinusoidal substrate potential with $\beta_1 = 1$, $\beta_2 = 1/4$, $\beta_3 = 1/6$, and $\beta_p = 0$ for $p \geq 4$. Numbers correspond to w, letters describe the type of the GS phase [478].

for an asymmetric GS "left" and "right" kinks should follow one another due to topological constraint.

The fixed-density FK chain for the double-SG substrate potential was investigated in details by Black and Satija [488, 506]. They have showed that the phase diagram in the (β_1, β_2)-plane for an irrational window number w is very complex and consists of cascades of pinning-depinning (Aubry) transitions, where the boundary between two phases has a fractal structure (see Fig. 5.15). In addition, in the pinned phase the Aubry transitions to depinning are associated with first-order transitions between two competing (N- and I-) ground states of the system. This implies the existence of a fractal structure in the spectrum for the phonon excitations about the possible ground states. Black and Satija [488] also showed that the fractal phase boundary and the fractal structure can be accounted for by simple additive rules.

5.4.2 Anharmonic Interatomic Potential

In real physical systems an interaction between particles is not purely harmonic. For example, a quantum mechanical ("chemical") bonding of atoms or molecules can be described approximately by an exponential law. The same law describes a repulsion of impurity charges screened by substrate electrons

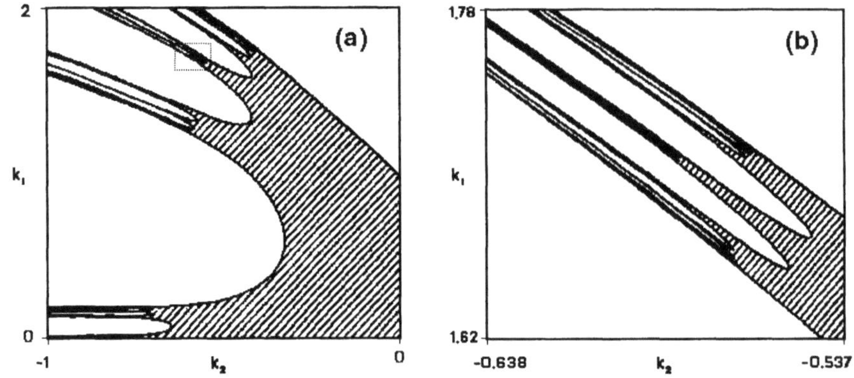

Fig. 5.15. Phase diagram in the (k_2, k_1) plane for the fixed-density FK chain with double-SG substrate potential $V_{\rm sub}(x) = -k_1 \cos x + \frac{1}{4} k_2 \cos 2x$. $g = 1$, $w = (\sqrt{5} - 1)/2$, (b) shows the blow-up of the boxed regime in (a). The dashed area corresponds to unpinned ("truly IC") configurations [488].

(Debye screening). An unscreened repulsion of charged particles can be described by the Coulomb law. Note also a dipole-dipole repulsion of atoms chemically adsorbed on crystal surfaces [36].

We consider a typical example of anharmonic interactions, the exponential (Toda) potential,

$$\widetilde{V}_{\rm int}(x) = V_0 \exp\left[-\beta\left(\frac{x}{a_s} - 1\right)\right], \tag{5.118}$$

where V_0 is the interaction energy of two atoms located at the nearest neighboring minima of the substrate potential and β is the dimensionless anharmonicity. Here we restrict ourselves by the interaction of nearest neighbors only to show qualitatively a role of interaction anharmonicity on the phase diagram. For an anharmonic potential the main parameter of the FK model,

$$g_x = V''_{\rm int}(x), \tag{5.119}$$

is not constant as it was for the standard FK model. This allows us to investigate behavior of the system when the concentration θ is a natural variable of the model, while in the standard FK model the assumption $g = $ Const restricts the consideration to a small interval of the interatomic distances.

To consider the "free-end" FK chain, we have to introduce an external pressure Π, i.e., to add a linear attractive branch to the repulsive potential (5.118),

$$V_{\rm int}(x) = \widetilde{V}_{\rm int}(x) + x\,(\Pi - \Pi_0), \tag{5.120}$$

where the value Π_0 is to be chosen in such a way that for $\Pi = 0$ the potential $V_{\rm int}(x)$ has the minimum at $x = a_s$. Equation (5.120) can be rewritten in the form

$$V_{\rm int}(x) = \widetilde{V}_{\rm int}(x) - x\,\widetilde{V}'_{\rm int}(a_{\rm min}), \qquad (5.121)$$

so that the modified interatomic potential achieves its minimum at $x = a_{\rm min}$. Emphasize that for the "fixed-density" chain this modification is unwarranted.

As for the standard FK model, stationary configurations of the anharmonic model can be obtained as orbits of an area-preserving map [221]

$$\begin{cases} \widetilde{x}_{l+1} = \widetilde{x}_l + \widetilde{p}_l, \\ \exp(-\beta\widetilde{p}_{l+1}/a_s) = \exp(-\beta\widetilde{p}_l/a_s) - (a_s/\beta V_0)\sin\widetilde{x}_l. \end{cases} \qquad (5.122)$$

The presence of anharmonicity leads to a drastic change in the map, making it asymmetric with respect to the sign of $\widetilde{p}$ (compare Figs. 5.2 and 5.16). For $\widetilde{p} < 0$ (i.e., for the configurations corresponding to $a_A < a_s$) the orbits in plane $(\widetilde{x}, \widetilde{p})$ are much less modulated, due to the "hard-core" repulsion of the Toda potential (5.118) at $x \to 0$. Otherwise, for $\widetilde{p} > 0$ ($a_A > a_s$) most orbits reveal discontinuity, that indicates the breakdown of system's integrity (the destruction of KAM-circles), due to decreasing of interaction forces with increasing of interatomic distances. Thus, at constant V_0 and β, the sliding ground states exist only at large enough irrational values of θ, while at low values of θ the IC phases are locked. However, the transition by breaking of analyticity occurs similarly as in the standard FK model [266]. In particular, Lin and Hu [467] have shown that the critical indices in Eq. (5.81) do not depend on the anharmonicity parameter and are equal to those of the standard FK model.

The most important result of anharmonicity is the breaking of kink-antikink symmetry of the standard FK model. This result was firstly obtained in computer simulation by Milchev and Markov [262],[507]–[509]. Indeed, effective interaction forces for a kink (in a region of a local contraction) exceed

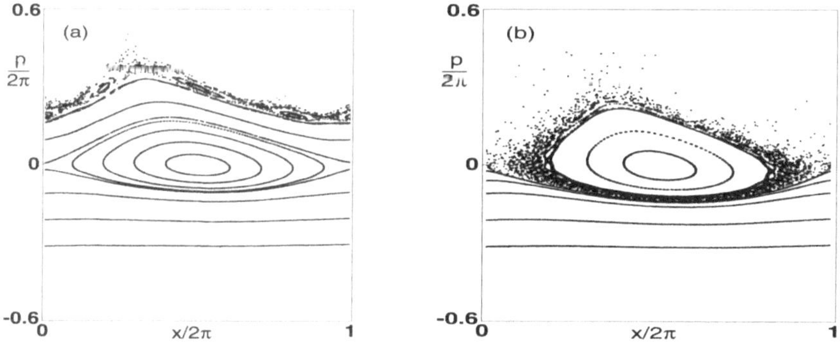

Fig. 5.16. Orbits of the map (5.19) for $\beta = 4\pi$ and (a) $V_0 = 1.260$, (b) $V_0 = 1.031$ [221].

those for an antikink (in the region of a local extension of the atomic chain). As a result, at the same values of parameters V_0 and β, a kink, as compared with an antikink, is characterized by a larger value of the creation energy and by lower values of the effective mass and PN potential height [165]. Analytically this effect can be described in the following way. Let us consider a kink with a topological charge $\sigma = \pm 1$ for the trivial reference GS with $\theta = 1/q$, q being an integer. At the weak-coupling limit, $V_0 \to 0$, using the approach of Sect. 5.2.1, the kink energy can be represented as

$$\varepsilon_k^\sigma \simeq y^2 + 2V_{\rm int}(qa_s - \sigma y) + V_{\rm int}(qa_s - \sigma a_s + 2\sigma y) - 3V_{\rm int}(qa_s), \qquad (5.123)$$

where y is the magnitude of displacements of two atoms located at kink's center (see Fig. 5.17). Expanding $V_{\rm int}$ in small y and then minimizing ε_k^σ with respect to y, we obtain

$$\varepsilon_k^\sigma \simeq [V_{\rm int}(qa_s - \sigma a_s) - V_{\rm int}(qa_s)] - \frac{[V'_{\rm int}(qa_s - \sigma a_s) - V'_{\rm int}(qa_s)]^2}{[1 + V''_{\rm int}(qa_s) + 2V''_{\rm int}(qa_s - \sigma a_s)]}.$$

Similarly we can calculate other Π-independent parameters such as the energy of creation of kink-antikink pair, $\varepsilon_{\rm pair} = \varepsilon_k^{(\sigma=+1)} + \varepsilon_k^{(\sigma=-1)}$, and the amplitude of PN barriers, ε_{PN}^σ.

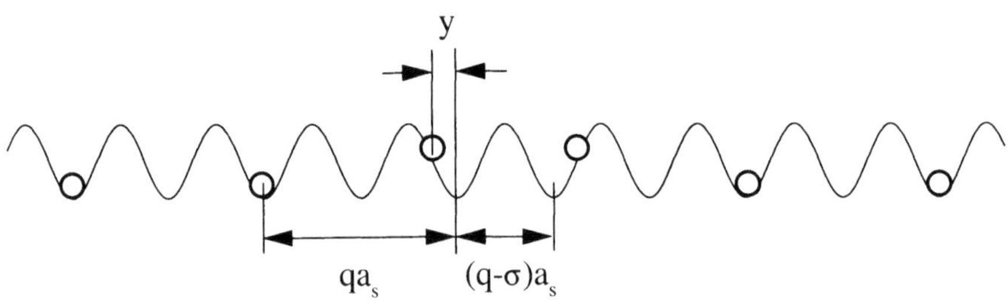

Fig. 5.17. Kink structure for $V_0 \to 0$.

In the opposite case of strong interaction between atoms, $V_0 \to \infty$, we can use the continuum approximation of Sect. 5.2.1. (Recall, however, that if the anharmonicity β exceeds some critical value $\beta_{\rm crit}$, the continuum approximation breaks down even in the limit $V_0 \to \infty$; see Sect. 3.5.1 and Ref. [266]). In this case the motion equation of the system reduces to the form [165, 264]

$$\frac{d^2u}{d\tilde{x}^2}\left(1 - \alpha\frac{du}{d\tilde{x}}\right) = \sin u, \qquad (5.124)$$

where

$$\alpha = -\frac{V'''_{\rm int}(qa_s)}{[V''_{\rm int}(qa_s)]^{3/2}}. \qquad (5.125)$$

From Eq. (5.124) it follows that the kink width depends now on the kink topological charge,

$$d \to d_{eff}^{\sigma} = d\left(1 - \frac{\pi}{3}\sigma\alpha\right), \tag{5.126}$$

and the kink creation energy is equal to

$$\varepsilon_k^{\sigma} = -\sigma a_s V_{\rm int}'(q a_s) + E_k^{\sigma}, \tag{5.127}$$

where the Π-independent contribution E_k^{σ} depends on σ too,

$$E_k^{\sigma} = 4\varepsilon_s \sqrt{g_a}\left(1 + \frac{1}{12}\sigma\alpha\right), \quad g_a = V_{\rm int}''(q a_s). \tag{5.128}$$

Analogously one can calculate the kink effective mass,

$$m^{\sigma} = m\left(1 - \frac{\pi}{6}\sigma\alpha\right), \tag{5.129}$$

and the amplitude of PN potential,

$$\varepsilon_{PN}^{\sigma} \simeq \varepsilon_{PN}(g_a) + \sigma\left(\frac{d\varepsilon_{PN}(g)}{dg}\right)_{g=g_a} \Delta g, \tag{5.130}$$

where the function $\varepsilon_{PN}(g)$ is given by Eq. (5.58) and $\Delta g = (2\pi/3)\alpha g_a$.

The presented above formulas allow us to find kink parameters as functions of the concentration θ in the whole interval $0 \le \theta < \infty$, that is quite important in investigation of mass transport in the "free-density" FK chain (see Sect. 7.4 below). Clearly that such functions $\varphi(\theta)$ (where φ is either m or ε_{PN}) are defined on a countable set of rational numbers θ (because kinks can be introduced for C-phases only) taking at each rational θ two values, the left one, $\varphi(\theta - 0) \equiv \varphi^{(\sigma=-1)}(\theta)$, and the right one, $\varphi(\theta + 0) = \varphi^{(\sigma=+1)}(\theta)$. However, for concentrations in the interval $0 \le \theta < \theta_{\rm Aubry}$, where the irrational value $\theta_{\rm Aubry}$ is introduced as the lowest one which satisfies the inequality $V_{\rm int}''(a_s/\theta_{\rm Aubry}) > g_{\rm Aubry}(\theta_{\rm Aubry})$, the functions $m(\theta)$ and $\varepsilon_{PN}(\theta)$ may be made continuous if we consider the IC ground state as the limit of C-phases with increasing periods. Both functions $m(\theta)$ and $\varepsilon_{PN}(\theta)$ monotonically decrease from $m(0) = m_a$ and $\varepsilon_{PN}(0) = \varepsilon_s$ to zero as θ varies from 0 to ∞ and, as was described above, they undergo jumps of a magnitude

$$\Delta\varphi(\theta) = \varphi(\theta - 0) - \varphi(\theta + 0) \tag{5.131}$$

at every rational θ, so that they look like an inverse Devil's staircase (see Fig. 5.18). The largest jump occurs for the C-structure with a concentration θ close to θ^*, where θ^* is defined by $V_{\rm int}''(a_s/\theta^*) = 1$. Besides, for two reference structures with closely spaced concentrations, a lower jump exhibits the phase with higher-order structure, i.e. the phase with a larger period a. Such a behavior is due to that, the interaction between adjacent unit cells rather

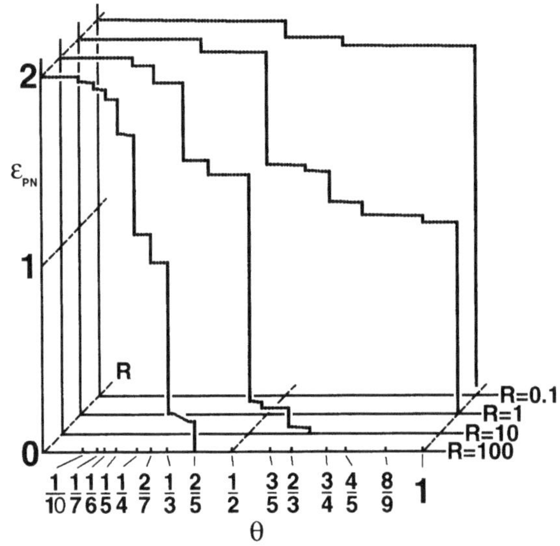

Fig. 5.18. Amplitude of the PN potential ε_{PN} vs. θ for dipole-dipole repulsion $V_{\rm int}(x) = V_0\,(a_s/x)^3$ at $R = 0.1$, 1, 10, and 100, where $R = V_0/\varepsilon_s$ [165].

than between adjacent atoms in one cell is responsible for the jump, because the splitting of antikink and kink parameters is determined by anharmonicity of interaction at the distance $x = a$, so that the anharmonicity obviously decreases with increasing of the order of structure.

As to the function $\varepsilon_{\rm pair}(\theta)$, its behavior is more irregular. Because the energy of creation of a kink-antikink pair is also determined by the interaction of nearest neighboring unit cells and not of atoms, $\varepsilon_{\rm pair}$ takes larger values for simpler structures and lower values for higher-order ones. However, $\varepsilon_{\rm pair}(\theta)$ increases "in average" with increasing of θ due to increasing of the parameter $g_{a_A} = V''_{\rm int}(a_A)$ with decreasing of interatomic distance $a_A = a_s/\theta$.

For the "free-end" FK chain, Eqs. (5.120), (5.121), (5.127), (5.128) help to construct the global phase diagram of the system. Because the Toda potential (5.118) is strictly convex, this diagram is topologically similar to that of the standard FK model (compare Figs. 5.19 and 5.12). Besides, a phase diagram is to be considered for a whole interval $0 < a_{\rm min} < \infty$ due to asymmetry of anharmonic potentials. For example, let us consider the Frank–van-der-Merwe boundaries which separate the trivial GS with $\theta = 1/q$ from the cnoidal wave ground states and are determined by the equation $\varepsilon_k^\sigma = 0$. In the weak-bond limit, $V_0 \to 0$, we obtain the relationship

$$\sigma a_s \widetilde{V}'_{\rm int}(a_{\rm min}) = \Delta\widetilde{V}_{\rm int}(qa_s) + \frac{[\Delta\widetilde{V}'_{\rm int}(qa_s)]^2}{\left[1 + \widetilde{V}''_{\rm int}(qa_s) + 2\widetilde{V}''_{\rm int}(qa_s - \sigma a_s)\right]}, \quad (5.132)$$

where

$$\Delta\widetilde{V}_{\rm int}(qa_s) = \left[\widetilde{V}_{\rm int}(qa_s) - \widetilde{V}_{\rm int}(qa_s - \sigma a_s)\right],$$

Thus, for the Toda potential (5.118) the FvdM curves are described by the equation

$$\frac{a_{\min}}{a_s} \simeq \left(q - \frac{1}{2}\sigma\right) - \psi(\beta) + \sigma\left(\frac{V_0\beta^2}{a_s^2}\right)\left(\frac{e^{\beta}-1}{\beta}\right)\exp\left[\beta\left(\frac{1}{2}+\frac{1}{2}\sigma - q\right)\right], \tag{5.133}$$

where the shift of FvdM boundaries due to anharmonicity is equal to

$$\psi(\beta) = \frac{1}{2} + \frac{1}{\beta}\ln\frac{(1-e^{-\beta})}{\beta} \simeq \begin{cases} \beta/24 & \text{if } \beta \to 0, \\ \frac{1}{2} - \frac{1}{\beta}\ln\beta & \text{if } \beta \to \infty. \end{cases} \tag{5.134}$$

Analogously, in the strong-coupling limit, $V_0 \to \infty$, the equation $\varepsilon_k^{\sigma} = 0$ together with Eqs. (5.121), (5.125), (5.127), (5.128) leads to the relationship

$$a_s \widetilde{V}'_{\text{int}}(a_{\min}) \simeq a_s \widetilde{V}'_{\text{int}}(qa_s) - 8\sqrt{g_a}\,(\sigma + \alpha/12)\,. \tag{5.135}$$

Thus, for the Toda potential (5.118) we obtain an equation

$$a_{\min} \approx qa_s - \frac{8}{\beta}e^{\beta(q-1)/2}\left[\sigma\sqrt{K} - \left(\frac{2}{\pi} - \frac{1}{12}\right)e^{\beta(q-1)/2}K\right], \tag{5.136}$$

where $K = 1/V_0$. The resulting phase diagram is shown in Fig. 5.19.

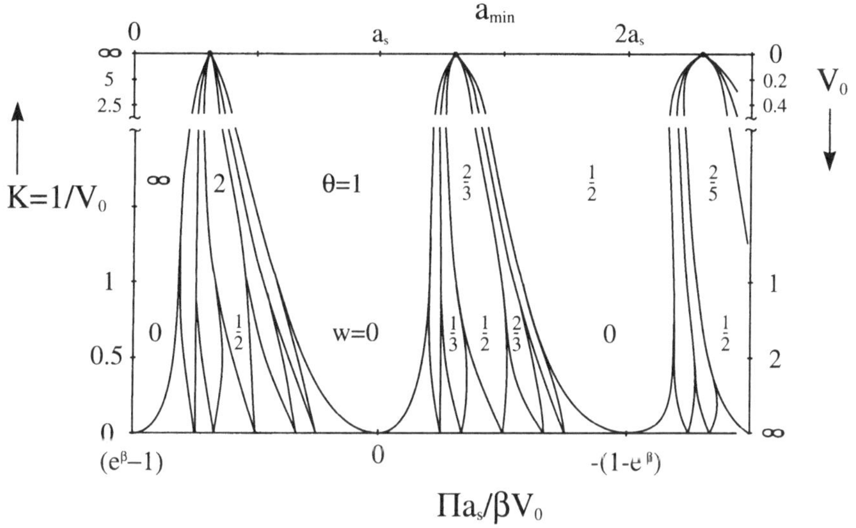

Fig. 5.19. Global phase diagrams for anharmonic FK model with Toda (exponential) interatomic potential (schematically). The commensurate phases are numerated by concentration $\theta = s/q$ and window number $w = \{q/s\}$. The top horizontal axis is linear in $a_{\min}$, for convenience we draw also the "pressure" Π (notice that Π-scale is nonlinear), which is connected with $a_{\min}$ by relationships $\Pi a_s/\beta V_0 = \exp\left[\beta(1 - a_{\min}/a_s)\right] - 1$. The left vertical axis corresponds to $K = V_0^{-1}$, while the right axis shows V_0. Note also that the bottom part of the left axes is linear in K, while the top part of the right axis is linear in V_0.

The FvdM boundaries for the Toda potential were calculated by Milchev and Markov [262],[507]–[509] and by Byrne and Miller [279]. Besides, Lin and Hu [467] have calculated the phase diagram of the Toda–FK system and studied multi-fractal properties of Devil's staircase.

5.4.3 Nonconvex Interaction

If a real physical interaction has an attractive branch, it is not strictly convex as a rule. As typical examples we may mention the double-well (ϕ^4) potential investigated by Aubry *et al.* [510], Marchand *et al.* [286], Byrne and Miller [279], the Lennard-Jones potential [279, 511], the Morse potential [512, 513], and the oscillating potential [282]. All the potentials have one or more inflation points defined by the equation

$$V''_{\text{int}}(a_{\text{inf}}) = 0. \tag{5.137}$$

Thus, now we are meeting with a model having three (or more) competing lengths, a_s, $a_{\min}$ and a_{inf}, and an analytical investigation of the problem becomes very complicated. However, computer simulation makes it possible to understand the main aspects of behavior of the FK model with nonconvex interactions. Namely, it is useful to distinguish between "convex" FK configurations, which use only the convex part of $V_{\text{int}}(x)$, from the "nonconvex" configurations, which use the concave part of $V_{\text{int}}(x)$ at least once in a period of the C-phase. In the "convex" region of the phase plane (i.e., in the region occupied by convex configurations only) a window number uniquely defines the phase, and transitions among these phases are continuous exhibiting Devil's-staircase behavior as was described above.

For nonconvex configurations the parameter $g_l = V''_{\text{int}}(x_{l+1} - x_l)$ becomes negative at least for some pairs of nearest neighboring atoms, and that, clearly, drastically changes the system behavior leading to two new aspects. *First*, now the window number (or the concentration) does not uniquely define the structure due to appearance of modulated (polymerized) structures which, for example, consist of alternating short ("strong") and long ("weak") bonds. The driving force for such a distortion is the decreasing of energy of distorted GS with respect to undistorted one. The transitions among these structures are phonon-stimulated and are usually (but not necessary) continuous. These questions were considered already in Sect. 3.5.2 for the "fixed-density" FK chain with $\theta = 1$. The *second* important aspect is that for a nonconvex reference C-configuration a kink-kink interaction may be (and very often is) attractive at least at a certain separation R^*. Now, if we consider an "excited state" of this configuration as consisting of a regular array of kinks separated by R^*, the excess energy per kink can be written in the form $\Delta\varepsilon = \varepsilon_k + v_{\text{int}}(R^*)$. Then, if the interaction energy between kinks becomes negative for a certain spacial arrangement of the array, $v_{\text{int}}(R^*) < 0$, the condition $\Delta\varepsilon \leq 0$ may be fulfilled even for a positive kink creation energy,

$\varepsilon_k > 0$. Therefore, with changing model parameters such as $a_{\min}$ or Π, the transition between two phases with different window numbers will be discontinuous (first-order), and the new phase will arise abruptly with a finite jump in the kink density. Below we consider in more details two important examples of nonconvex potentials, the Morse and oscillating ones.

The Morse potential shown in Fig. 5.20, is one of the most important ones for real physical systems. It is described by the function

$$V_{\text{int}}(x) = V_{\min}(e^{-2Z} - 2e^{-Z}), \tag{5.138}$$

where

$$Z = \kappa \left(\frac{x}{a_{\min}} - 1 \right).$$

The function (5.138) has a minimum at $x = a_{\min}$, $V_{\text{int}}(a_{\min}) = -V_{\min}$, and tends to zero (as it should be for a realistic interatomic potential) with increasing the distance x. This potential describes, e.g., a "direct" interaction between atoms or molecules chemically adsorbed on crystal surfaces [36]. (Note that physically adsorbed rare gas atoms interact via Lennard-Jones potential, $V_{\text{int}}(x) = V_{\min}\left[(a_{\min}/x)^{12} - 2(a_{\min}/x)^6\right]$, which is qualitatively similar to the Morse potential). The Morse potential has an inflection point at

$$a_{\text{inf}} = a_{\min}\left(1 + \kappa^{-1}\ln 2\right), \tag{5.139}$$

beyond which (at $a_{\text{inf}} < x < \infty$) it is concave, $V''_{\text{int}}(x) < 0$. Near the minimum the Morse potential is harmonic,

$$V_{\text{int}}(x) \simeq -V_{\min} + \frac{g}{2}(x - a_{\min})^2, \quad g = 2V_{\min}(\kappa/a_{\min})^2. \tag{5.140}$$

Expanding the attractive branch of the Morse potential [i.e., the second form in r.h.s. of Eq. (5.138)] in Taylor series up to the linear term, one obtains the Toda potential

$$V_{\text{int}}^{(\text{Toda})}(x) = V_0\left\{\exp\left[-\beta\left(\frac{x}{a_s} - 1\right)\right] + x\frac{\beta}{a_s}\exp(-\beta P_1) + C\right\} \tag{5.141}$$

with the parameters

$$V_0 = V_{\min}\exp(\beta P_1), \tag{5.142}$$

$$\beta = \sqrt{2}\kappa/(1 + P_1), \tag{5.143}$$

and

$$C = -(2 + \beta a_{\min}/a_s)\exp(-\beta P_1), \tag{5.144}$$

where $P_1 = (a_{\min} - a_s)/a_s$ is the natural misfit. Both potentials (5.138) and (5.141) have the same harmonic approximation (5.140) and the same repulsive branches, while their attractive branches (as well as third and higher derivatives) are different. The comparison of behavior of Morse and Toda FK chains helps to distinguish anharmonic effects from the nonconvex ones [512]. Namely, the Morse FK model exhibits two phenomena which were absent in the Toda FK model: the existence of modulated configurations and the possibility of chain's rupture at the antikink core in the case of negative misfits.

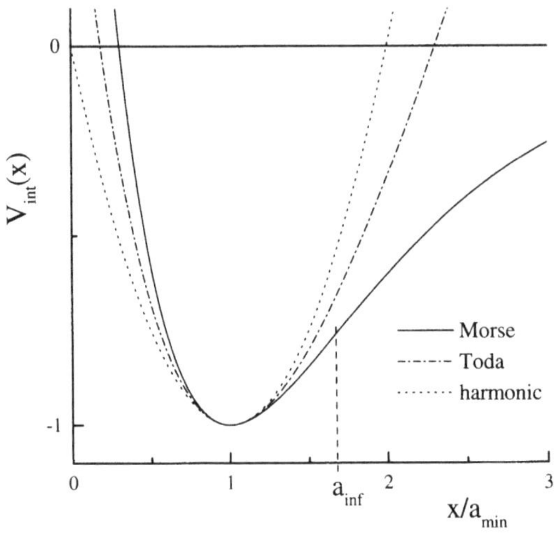

Fig. 5.20. Morse potential (solid curve), its harmonic approximation (dotted curve), and Toda potential (dot-dashed curve).

As above, let us begin with the "fixed-density" FK model. Recall that the harmonic FK chain is characterized by two model parameters, the concentration θ (or the mean interatomic distance $a_A = a_s/\theta$) and the elastic constant g. The anharmonic (e.g., Toda) FK model has three parameters: θ, the strength of interaction V_0 and, additionally, the parameter β which characterizes the anharmonicity and leads to splitting of kink-antikink characteristics. The Morse-FK chain has four model parameters: θ, V_{min} (it plays a role of the strength of interaction), κ (describes the anharmonicity), and a_{min} [or, more rigorously, a_{inf} connected with a_{min} by Eq. (5.139)] which now enters explicitly in the motion equations. Namely the fourth parameter makes the system behavior more rich and complicated.

First, the nonconvex FK configurations may be distorted, or modulated. In particular, the trivial configuration with the concentration $\theta = 1/q$ and mean interatomic distance $a = qa_s$ is nonconvex if $a \leq a_{\text{inf}}$, or if the misfit parameter P_q defined in accordance with Eq. (5.4) as

$$P_q = (a_{\text{min}} - qa_s)/a_s, \tag{5.145}$$

is lower than the critical value $P_{\text{inf}}^{(q)}$,

$$P_{\text{inf}}^{(q)} = -\frac{q}{(1+\kappa/\ln 2)}, \tag{5.146}$$

the last expression being obtained with the help of Eqs. (5.139) and (5.145) where we put $a_{\text{inf}} = qa_s$ [512]. The trivial GS will dimerize (so that short–long bonds alternate) if

$$V''_{\text{int}}(a) \leq -1/4. \tag{5.147}$$

Substitution of Eq. (5.138) into Eq. (5.147) leads to the equation

$$4gz(1-2z) = 1, \quad z \equiv \exp\left[-\kappa\left(\frac{qa_s}{a_{\text{min}}} - 1\right)\right], \tag{5.148}$$

so that a curve which separates the undistorted GS from the dimerized one in the $(a_{\text{min}}, V_{\text{min}})$ plane, is determined by the equation

$$a_{\text{min}}\left[\left(1+\frac{\ln 4}{\kappa}\right) - \frac{1}{\kappa}\ln\left(1+\sqrt{1-a_{\text{min}}^2/\kappa^2 V_{\text{min}}}\right)\right] = qa_s. \tag{5.149}$$

Analogously we can determine the curves separating the trimerized, tetramerized, etc. ground states. They were calculated by Markov and Trayanov [512] and shown in Fig. 5.21. As seen, they start at $P_q = P_{\text{inf}}^{(q)}$. Besides, at $V_{\text{min}} < V_{\text{min}}^{(\text{crit})} \equiv (qa_s/\kappa)^2$ no distortion takes place irrespective of the value of P_q, because a weak interaction favors the undistorted structure.

The second important feature of the "fixed-density" Morse-FK chain is that it cannot exist if the concentration θ is lower than a critical value $\theta^{(\text{crit})}$,

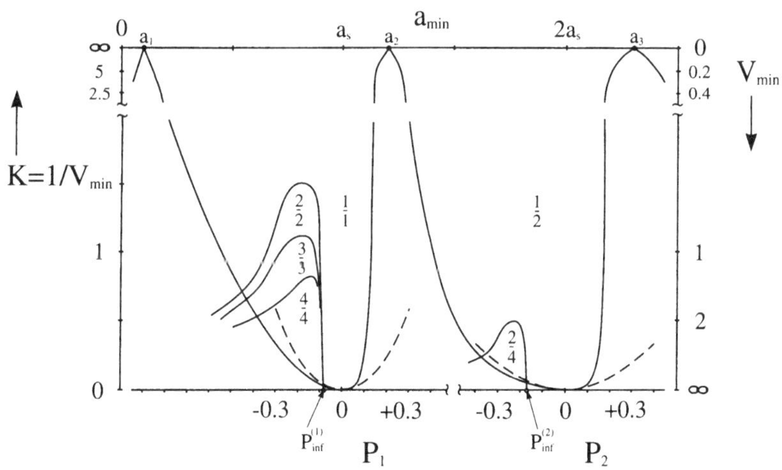

Fig. 5.21. Phase diagram of Morse-FK system for a fixed value of κ (schematically). The commensurate GS phases are indicated by concentration θ. To distinguish the undistorted GS from the dimerized one, we indicated the former by $\theta = \frac{1}{1}$ or $\theta = \frac{1}{2}$, while the dimerized GS is indicated by $\theta = \frac{2}{2}$ or $\theta = \frac{2}{4}$, etc. [512].

or if the mean interatomic distance a_A exceeds the critical value $a_A^{(\mathrm{crit})} = a_s/\theta^{(\mathrm{crit})}$. To show this, let us consider an isolated Morse chain (i.e. without the substrate potential) and apply the external pressure Π, so that $V_{\mathrm{int}}(x)$ is replaced by

$$V_{\mathrm{int}}^{(\Pi)}(x) = V_{\mathrm{int}}(x) + \Pi x. \tag{5.150}$$

This chain can exist in a compressed state ($a_A < a_{\min}$) at $\Pi > 0$, while the expanded GS (with $a_A > a_{\min}$) is forbidden. Indeed, at any negative pressure $\Pi < 0$ we have $V_{\mathrm{int}}^{(\Pi)}(x) \to -\infty$ as $x \to \infty$. In a result, the expanded configuration cannot correspond to the absolute minimum of the potential energy. (However, expanded configurations may exist as metastable states at low enough expansive forces). Thus, at any $\Pi < 0$ the isolated Morse chain breaks up into independent semi-infinite chains with free ends, each of sub-chains taking the GS with $a_A = a_{\min}$.

Returning to the chain placed into the substrate potential, it is easy to guess that now the Morse-FK chain can exist in an expanded state, but only if the expansion does not exceed a critical value, $a_A < a_A^{(\mathrm{crit})}$, while at $a_A > a_A^{(\mathrm{crit})}$ the Morse-FK chain will rupture. To find the value $a_A^{(\mathrm{crit})}$, we first consider the trivial configuration with $\theta = 1$ (or $a_A = a_s$). Clearly, this configuration can exist in a compressed state at any value of $a_{\min}$ which exceeds $a_A \equiv a_s$, because atoms interact via convex repulsive branch of the Morse potential. Now, let us decrease $a_{\min}$ below the value $a_A \equiv a_s$ obtaining an expanded chain. When $a_{\min}$ achieves the value a_{FM}^- which corresponds to the left FvdM boundary,

$$(a_{FM}^- - a_s)/a_s = P_{FM}^- = -|P_{FM}^-|, \tag{5.151}$$

the kink creation energy for the configuration $\theta = 1$ becomes equal to zero. Thus, at $a_{\min} = a_{FM}^- - \delta$ (where $\delta = +0$) an injection of kinks into the chain will decrease the system energy. However, in the "fixed-density" FK chain the creation of kinks of the same topological charge $\sigma = +1$ is forbidden due to topological constrain, kinks can be created by kink-antikink pairs only. Nevertheless, let us create N_p kink-antikink pairs and distribute the kinks uniformly over the chain, while all antikinks (vacancies) let us take together in one place of the chain, thus making one large cluster containing N_p antikinks. The kinks distributed in the chain, will decrease the system energy linearly with N_p due to negative kink creation energy, while the energy of the antikink cluster, E_{cluster}, will increase with N_p. For any convex interatomic potential this process is energetically unfavorable due to $E_{\mathrm{cluster}} \propto N_p^2$ [for example, for the standard FK model $E_{\mathrm{cluster}}(N_p) \approx \frac{1}{2}g(N_p a_s)^2$]. But for the Morse potential the cluster energy $E_{\mathrm{cluster}}(N_p) \approx V_{\mathrm{int}}(N_p a_s) - V_{\mathrm{int}}(a_{\min})$ tends to a finite value ($\approx V_{\min}$) which is independent on N_p as $N_p \to \infty$. Thus, the described process will be energetically favored at $N_p \to \infty$, and an infinite cluster of vacancies will grow in one place of the chain. This means that one of the expanded bonds ruptures, and the chain breaks up into two parts.

So, at $a_{\min} = a_{FM}^{-} - \delta$ the chain with $\theta \leq 1$ cannot exist as a GS because it will always rupture. (Note, however, that the state $\theta = 1$ can still exist as a metastable state with further decreasing of $a_{\min}$ up to the metastable limit a_{ms}^{-}). Further, let us take $a_{\min}$ in the interval $a_{FM}^{-} < a_{\min} < a_{FM}^{+}$ and will try to construct a chain's configuration with $\theta = 1 - \delta$. This state may be considered as a cnoidal wave of antikinks. But now all antikinks again will come together and form one infinite cluster (i.e. rupture), because at $a_{FM}^{-} < a_{\min} < a_{FM}^{+}$ the configuration without antikinks has a lower energy than the cnoidal-wave state, while the energy of rupture is finite. Thus, in a general case the value $a_A^{(\text{crit})}$ coincides with the value $a_A^{(\text{free})}$ which is equal to the mean interatomic distance in the "free-end" FK chain with the same model parameters $V_{\min}$, κ and $a_{\min}$.

We should emphasize that for an infinite FK chain the rupture is possible only for the "fixed-density" chain and only for nonconvex $V_{\text{int}}(x)$ which tends to a finite value at $x \to \infty$. The "free-end" FK chain will never rupture because it can shift its free end in order to introduce kinks and to decrease system energy using the attractive branch of $V_{\text{int}}(x)$. Notice also that the Toda-FK chain, where $E_{\text{cluster}}(N_p) \propto N_p$, may be ruptured but for larger values of its expansion.

Now let us consider kink excitations in the Morse-FK model. If the trivial GS with $\theta = 1/q$ is undistorted, we can use the approximate methods described above. In particular, in the continuum approximation ($V_{\min} \gg 1$) the creation energy of a single kink with a topological charge σ is equal to [see Eqs. (5.125), (5.128), (5.130), (5.138), (5.148)]

$$\varepsilon_k^{\sigma} \simeq \left(\frac{8\kappa}{a_{\min}}\sqrt{2V_{\min}}\right)\left\{1 + \left(\sqrt{z(2z-1)} - 1\right)\right.$$
$$\left. + \sigma\left[\frac{\pi}{2\sqrt{2}}\sqrt{V_{\min}}\,z(z-1) + \frac{1}{4\sqrt{2V_{\min}}}\frac{(4z-1)}{3(2z-1)}\right]\right\}, \tag{5.152}$$

where

$$z = \exp\left[\kappa/(1 + q/P_q)\right], \tag{5.153}$$

and P_q is defined by Eq. (5.145). The dependences of ε_k^{σ} on P_q are shown in Fig. 5.22. It is seen that the energy of antikink creation in the compressed chain ($P_q > 0$) is always smaller than the energy of creation of kink in the expanded chain ($P_q < 0$) at the same absolute value of the misfit. For small values of the misfit if $|t| \ll 1$, where $t \equiv \kappa P_q/q$, Eqs. (5.152) and (5.153) lead to

$$\varepsilon_k^{\sigma} \simeq \left(\frac{8\kappa}{a_{\min}}\sqrt{2V_{\min}}\right)\left\{1 + \sigma\left(\frac{\pi}{2\sqrt{2}}\right)\sqrt{V_{\min}}\left(1 + \frac{3}{2}t\right)t\right.$$
$$\left. + \frac{3}{2}t\left(1 + \frac{5}{12}t\right) + \sigma\frac{1}{4\sqrt{2V_{\min}}}\left(1 - \frac{2}{3}t + t^2\right)\right\}. \tag{5.154}$$

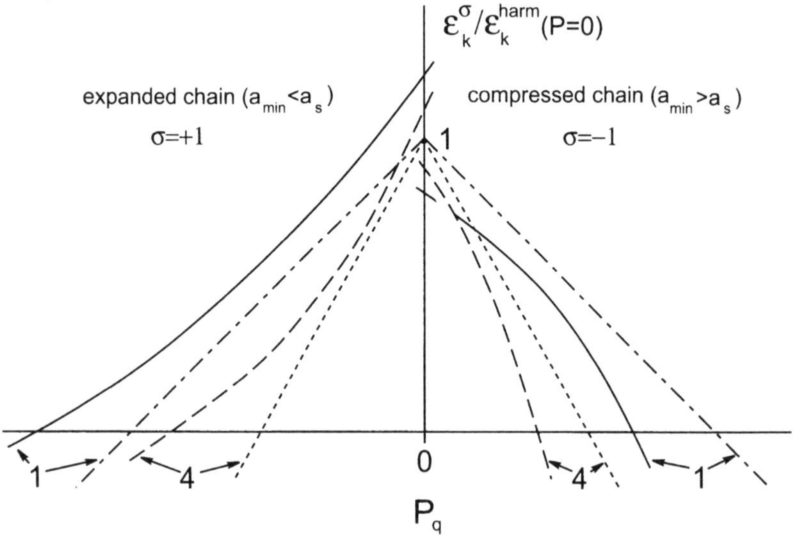

Fig. 5.22. Kink ($\sigma = +1$) and antikink ($\sigma = -1$) creation energy ε_k^σ (normalized by the harmonic-limit value at $P = 0$) versus misfit P at a fixed κ for Morse-FK chain (schematically). The curves cross absissa at the FvdM limits $P_{FM}^{\pm}$. Solid curves correspond to $V_{\min} = 1$, and dashed curves, to $V_{\min} = 4$. Chain and dotted lines illustrate the corresponding harmonic limits [513].

Now we are able to construct the phase diagram for the "free-end" Morse FK chain without an external pressure Π. Note that this case is simpler than the "fixed-density" case considered above, because the "free-end" chain is characterized by three parameters only ($V_{\min}$, κ, and $a_{\min}$). In the weak-bond approximation, $V_{\min} \to 0$, the FvdM limits of existence of the trivial GS with $\theta = 1/q$ are determined by the equation (5.132),

$$V_{\text{int}}(qa_s - \sigma a_s) - V_{\text{int}}(qa_s) = \left[V'_{\text{int}}(qa_s - \sigma a_s) - V'_{\text{int}}(qa_s)\right]^2, \tag{5.155}$$

where we have to take $\sigma = +1$ in order to obtain the left boundary, and $\sigma = -1$ for the right boundary. From Eq. (5.155) it follows that at $V_{\min} \to 0$ the ground states with $\theta = 1/(q-1)$ and $\theta = 1/q$ are separated by the point at which

$$a_{\min} = a_q \equiv -\kappa a_s / \ln z_q, \tag{5.156}$$

where z_q is a solution of the equation

$$z^{q-1}(z+1) = 2e^{-\kappa}. \tag{5.157}$$

Thus, the left boundary of the phase $\theta = 1$ is determined by

$$a_1 = -\frac{\kappa a_s}{\ln(2e^{-\kappa} - 1)} \simeq \begin{cases} \frac{1}{2}a_s(1 - \frac{1}{2}\kappa) & \text{if } \kappa \ll 1, \\ -\kappa a_s / \ln(\kappa_c - \kappa) & \text{if } \kappa_c - \kappa \ll 1, \end{cases} \tag{5.158}$$

where $\kappa_c = \ln 2$, and the right boundary, by

$$a_2 = -\frac{\kappa a_s}{\ln\left[\frac{1}{2}\left(-1+\sqrt{1+8e^{-\kappa}}\right)\right]} \simeq \begin{cases} \frac{3}{2}a_s\left(1-\frac{1}{18}\kappa\right) & \text{if } \kappa \ll 1, \\ a_s(1+\kappa^{-1}\ln 2) & \text{if } \kappa \gg 1. \end{cases} \tag{5.159}$$

From Eq. (5.155) it follows that at small $V_{\min}$ the FvdM boundaries behave linearly with $V_{\min}$ (see Fig. 5.21),

$$a_{\min} \simeq a_q \pm 2V_{\min}\frac{\kappa}{a_s}e^{\kappa}\frac{[\varphi(q-1)-\varphi(q)]^2}{(q-1)\varphi(q-1)-q\varphi(q)}, \tag{5.160}$$

where

$$\varphi(q) = \exp(-\kappa q a_s/a_q)\left[e^{\kappa}\exp(-\kappa q a_s/q_q)-1\right]. \tag{5.161}$$

In the strong-coupling limit, where $K \equiv 1/V_{\min} \to 0$, the FdvM limits are determined by the equation $\varepsilon_k^{\sigma} = 0$, where the kink creation energy is given by Eqs. (5.152) to (5.154). In particular, at $|P_q| \ll 1$ it follows

$$P_{FM}^{\sigma=\pm 1}(q) \simeq \frac{q}{\kappa a_s}\left(-4\sigma\sqrt{2K}-K\right). \tag{5.162}$$

Thus, the FdvM limits corresponded to negative misfit (expansion of the chain) increase in absolute value with increasing of the anharmonicity κ, whereas ones corresponded to positive misfit (compression of the chain), decrease [509, 512].

The phase diagram for the Morse–FK model is shown schematically in Fig. 5.21. In the convex regions of the (P, K) plane, i.e., for small values of K, phase transitions between various phases are continuous and exhibit Devil's staircase structure similarly to the classical FK model. In the nonconvex regions, e.g., at negative P below the value P_{inf}, the transition from the trivial GS to the cnoidal-wave one may be discontinuous [513], so that the kink density changes from zero to a certain value by a jump (see Fig. 5.23). Note that the mean density of kinks is always smaller at negative misfits [509, 512]. As one can see from Fig. 5.21, there exists a value of K above which all transitions are first order. The regions of first-order and second-order transitions are separated by the boundary layer within which the phase diagram is exceedingly complicated with structure appearing on very small length scales.

Markov and Trayanov [512] investigated numerically the metastable limits $P_{ms}^{\pm}$ of existing of the trivial configuration with $\theta = 1$ (see Fig. 5.24). They found that for positive misfits the ratio $\nu = P_{ms}^{+}/P_{FM}^{+}$ as a function of $K \equiv 1/V_{\min}$ steeply decreases starting from the standard FK value $\nu = \pi/2$ at $K \to 0$ and going asymptotically to 1 at $K \to \infty$. This means that at $K \geq 1$ the introduction of antikinks beyond the P_{FM}^{+} boundary takes place practically without overcoming an energy barrier. At negative misfits, Markov and Trayanov [512] brought out an interesting effect that the end parts of the "free-end" Morse-FK chain are always distorted irrespective of the value $V_{\min}$ provided $P < P_{\text{inf}}$, i.e. the ends are modulated even when the middle ("bulk")

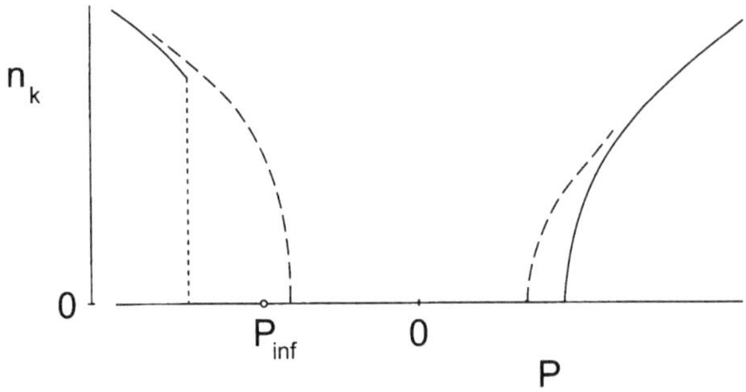

Fig. 5.23. Mean density of kinks n_k versus misfit P at a fixed value of κ for Morse-FK model (schematically). Solid curve: $V_{\min} = 2$, $|P_{FM}^-| > |P_{\inf}|$ (non-convex region), dashed curve: $V_{\min} = 6$, $|P_{FM}^-| < |P_{\inf}|$ (convex region).

part of the chain is undistorted. This shows that the end atom does not climb the slopes of the potential well with increasing misfit (in absolute value), and this excludes the possibility for spontaneous introduction of kinks at the free end. Therefore, the trivial FK configuration exists as a metastable state at any value $P < P_{\inf}$ provided the interatomic potential is weak enough. So, the metastability limit P_{ms}^- below which the trivial configuration cannot exist, disappears at a point where $P_{ms}^- = P_{\inf}$ (see Fig. 5.24).

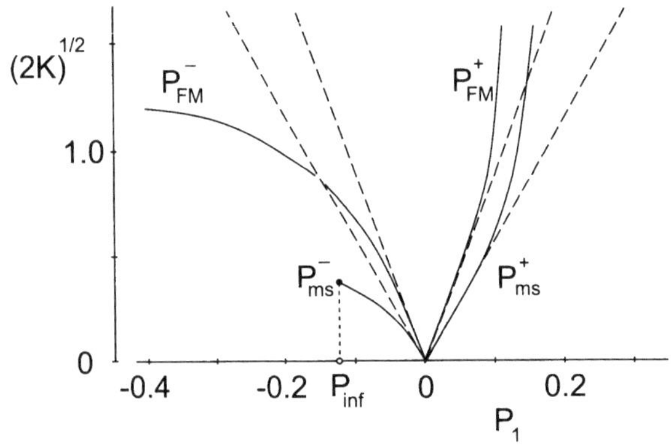

Fig. 5.24. The FvdM misfit limits $P_{FM}^{\pm}$ and the metastable limits $P_{ms}^{\pm}$ of the Morse-FK model for the trivial $\theta = 1$ GS at a fixed value of κ (schematically). $K = 1/V_{\min}$. Dashed lines correspond to harmonic limit [512].

In some physical situations a "free-end" FK chain may undergo an external pressure Π, for example, in the case when the FK system is in contact with a reservoir of atoms (e.g., the vapor phase for adsorption systems) with a fixed chemical potential μ, while the length L of the chain is fixed (e.g., due to a finite size of the substrate). In this case the Morse-FK model has four independent parameters [$V_{\min}$, κ, $a_{\min}$, and Π; note that now $a_{\min}$ is one of the model parameters, while the minimum of interaction is achieved at $x = a_{\min}^{(\Pi)}$ determined by the equation $dV_{\rm int}^{(\Pi)}(x)/dx = 0$. Note also that the variation of Π is restricted by $\Pi \geq 0$]. The phase diagram of this system can be constructed analogously to the previous case with using the potential (5.150) instead of the Morse one.

The theory of the Morse-FK system has an important application in investigation of crystal growth. In particular, in the design of novel devices it is necessary to form an epitaxial interface where misfit dislocations are to be avoided at any cost [514], so that the pseudomorphic state of the film should have the atomic spacing of the overgrowth which is exactly equal to the spacing of substrate atoms. The phase diagram of Fig. 5.21 predicts that in this case we have to choose the combination of substrate and overlayer crystal for which the natural misfit P_1 is negative rather then positive [509, 512].

Another interesting example of the nonconvex interatomic interaction is an oscillating potential

$$V_{\mathrm{int}}(x) = g\,[1 - \cos(x - a_{\mathrm{min}})]. \tag{5.163}$$

The FK model with the potential (5.163) describes a one-dimensional system of classical planar spins with nearest-neighbor chiral interactions in the presence of an external magnetic field. Besides, oscillating potentials arise in the Ruderman-Kittel interaction of magnetic impurities in metals as well as in the "indirect" interaction of atoms adsorbed on a metal surface [36].

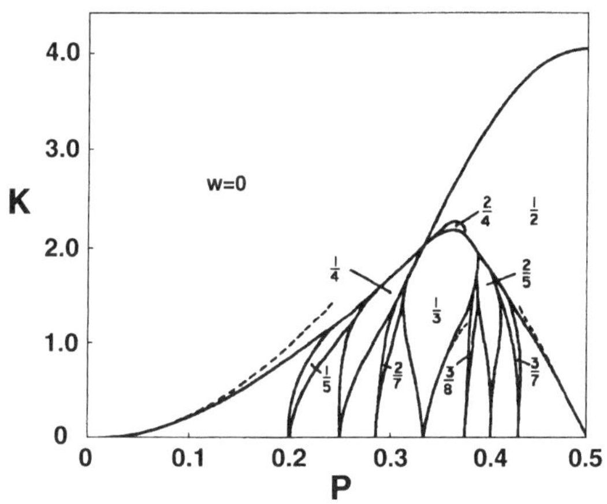

Fig. 5.25. Phase diagram of the FK model with oscillating interatomic potential (5.163) in the (P, K) plane. The phases are numerated by window number w [282].

The global phase diagram of the "free-end" FK model (5.163) was calculated by Yokoi *et al.* [282] using the effective potential method (see Fig. 5.25). In the convex region of the phase diagram, i.e. at $K \to 0$ $(g \to \infty)$ in Fig. 5.25, the model behavior is similar to that of the standard FK system, with continuous transitions between various phases. When the value of g decreases, the lock-in effect on the C-phases becomes stronger and eventually forces the system to make use of the concave part of $V_{\mathrm{int}}(x)$. In the nonconvex region the modulated phases appear (see the trivial GS with $w = \frac{1}{2}$ and the modulated one indicated by $w = \frac{2}{4}$ in Fig. 5.25). The kinks shape in this region may exhibit oscillatory exponential decay, leading to the kink-kink interaction energy which alternates in sign as a function of the distance between them. In a result, the transitions, e.g., $0 \leftrightarrow \frac{1}{4}$ and $0 \leftrightarrow \frac{1}{3}$ are first order because they are driven by the creation of kinks with attractive interaction between them. Note that in this model the phonon-stimulated transition $\frac{1}{2} \leftrightarrow \frac{2}{4}$ is discontinuous too. As g is further decreased, more and more phases disappear through the lock-in process and eventually we are left with only a finite number of phases. Finally, for $g < \frac{1}{4}$ the only trivial GS exists.

An important additional feature of the nonconvex models is the presence of an infinite number of super-degenerate points on the phase diagram, many of which are multi-phase points (see details in Refs. [282, 286]). At these points the GS consists of noninteracting zero-energy kinks with finite kink's density, so that different phases share a common value of the energy. Therefore, at these points the system has residual entropy and violates the third law of thermodynamics.

Thus, the behavior of the one-dimensional FK model with nonsinusoidal substrate potential and nonconvex interatomic interactions may be extremely complicated even at zero temperature. In the next section we will consider the properties of the model if the temperature is nonzero.

6 Statistical Mechanics

In all previous chapters we discussed an isolated FK chain. In the continuum approximation, the motion equations for such a model reduce to the integrable SG equation exhibiting regular dynamics. In constrast, the discrete FK model is not integrable and, in general, its dynamics can be chaotic. When the total energy of the chain is lower than a threshold energy, the system motion is almost regular and the chaotic layers in the phase space are exponentially small. However, for higher energies the system dynamics becomes mostly stochastic, and in the stochastic regime the system rapidly approaches energy equipartition. Then the dynamics of the FK chain should be described by the methods of statistical mechanics, which is the subject of the present chapter.

6.1 Introductory Remarks

The FK model describes a Hamiltonian system and, as usual for such systems, there exists *a threshold energy* $E_{\rm th} \propto a_A^2$ (see, e.g., Ref. [515]) such that when the total energy of the chain is lower than the threshold one, i.e. $E < E_{\rm th}$, the dynamics of the system is almost regular (i.e. the chaotic layers are exponentially small), while at higher energies, when $E > E_{\rm th}$, the system dynamics is mostly stochastic. Goedde *et al.* [515] demonstrated that this transition is rather sharp, and the system rapidly approaches the energy equipartition in the stochastic regime. As a result, this type of the dynamics of the FK chain should be described by statistical mechanics.

Besides, in most of the physical applications of the FK model the atomic chain it describes is not an isolated object but only a part of the whole system. Therefore, there always exists an energy exchange between the FK chain and the substrate, so the chain should be considered as being in contact with a thermal bath, being analyzed by employing the methods of statistical mechanics.

Peierls [516] was the first who had shown that at any temperature $T \neq 0$ a long-range order in one-dimensional systems cannot exist; any ordered state is always destroyed by thermal fluctuations. Therefore, the phase diagram of the FK model at $T \neq 0$ looks simple: there exists only one (disordered) phase and only one phase transition, the continuous *order-disorder transition* at the critical temperature $T_c = 0$. Nevertheless, since the pioneering work of Krumhansl and Schrieffer [517], statistical mechanics of FK-type models is studied very intensively (see, e.g., Refs. [518, 519] to cite a few). This interest is stimulated by the following reasons:

1. A $T \neq 0$ behavior of a condensed-matter system can be understood in terms of its low-energy excitations. In the limit of strong coupling when the SG equation becomes valid, i.e. $g \to \infty$, the system is exactly integrable by the inverse scattering transform, i.e. there exists a canonical transform to the generalized action-angle variables which classifies all types of elementary excitations. Thus, the FK chain in the strong-coupling limit is ideally suited for the phenomenological formulation of classical statistical mechanics of nonlinear systems, because, on the one hand, it is a system of strong-interacting atoms, while on the other hand, it may be rigorously treated as a system of weak-interacting quasi-particles.
2. In the opposite case of weak interatomic interactions, $g \ll 1$, the FK model is far from an integrable system. In this case, the standard map of the model exhibits an infinite (uncountable) number of chaotic orbits. An uncountable fraction of these orbits corresponds to metastable states, which describe the FK configurations with randomly pinned kinks. Thus, excited states of the model are spatially disordered and, therefore, the FK model is a natural physical system to study the glass-like behavior of amorphous solids. Moreover, it is the only model where, to the best of our knowledge, the glass-like behavior emerges intrinsically not being introduced artificially.
3. Due to the one-dimensional nature of the FK model, it has a formally exact solution which is given by the transfer-integral method. Thus, the model is an ideal base to check validity of different approximate methods used in the study of nonlinear systems.
4. Finally, the behavior of the model at $T \neq 0$ is important for its numerous physical applications, as discussed above.

We begin with introducing the general formalism and basic notations which are used throughout this Chapter. Below, we restrict our study by the case of interatomic potential with the equilibrium distance $a_{\min} = a_s$, so that the dimensionless concentration θ is assumed to be close to 1. Usually we consider $\theta \leq 1$, but this assumption is not important due to the kink-antikink symmetry. The physical picture developed below can be applied equally well to the dynamics near of any zero-temperature C-phase with $\theta \neq 1$ taken as a reference structure. The only difference is that in the latter case we have to use the phonon spectrum corresponding to the given C-phase, and the formalism of "superkinks" instead of that of kinks.

6.2 General Formalism

First, we introduce the general formalism and basic notations which are used throughout this Chapter. We notice that when $\theta < 1$ the $T = 0$ the ground state of the model is described by a cnoidal wave of antikinks constructed on the trivial reference phase with $\theta_0 = 1$. We denote the total number of these "residual" ("geometrical") kinks by N_w. At $N_w > 0$ the value

$$\widetilde{w} = N_w/N \tag{6.1}$$

coincides with the window number w defined above in Eq. (5.3). Thus, we have to consider the FK chain of the length

$$L = Ma_s, \tag{6.2}$$

which possesses the total topological charge N_w and has N atoms on M sites of the potential minima, where

$$M = N + N_w. \tag{6.3}$$

When the system temperature is nonzero, the residual kinks are supplemented by thermally exited kink-antikink pairs, the number of which we denote by N_{pair}. So, a system state at $T \neq 0$ is characterized by N_b phonons and/or breathers (the "breather problem" is discussed in the next section), N_k kinks and $N_{\bar{k}}$ antikinks ($N_k = N_{\text{pair}}$ and $N_{\bar{k}} = N_{\text{pair}} + N_w$ at $N_w > 0$, while at $N_w < 0$ we have to put $N_k = N_{\text{pair}} - N_w$ and $N_{\bar{k}} = N_{\text{pair}}$). It is useful to introduce also the total number of kinks,

$$N_{\text{tot}} = N_k + N_{\bar{k}} = 2N_{\text{pair}} + |N_w|, \tag{6.4}$$

and the corresponding concentrations, $n_x = N_x/L$, where the index "x" means either "w", "pair", "k", "$\bar{k}$", "tot", or "b".

For the "fixed-density" FK chain, where the values N and L are fixed by imposing the periodic boundary condition, $x_{N+1} = x_1 + L$, the proper thermodynamic description of the models is in terms of the Helmholtz free energy

$$F(T, L, N) = -k_B T \ln \mathcal{Z}(T, L, N), \tag{6.5}$$

where k_B is Boltzman constant, and $\mathcal{Z}$ describes the canonical ensemble,

$$\mathcal{Z}(T, L, N) = \sum_{\{\xi_i\}} \exp\left[-\beta E(\{\xi_i\}\right], \tag{6.6}$$

$\beta = 1/k_B T$, E is the total energy of the system in the state characterized by independent coordinates ξ_i in the phase space.

For the "free-end" FK chain, when the system is subjected to an external pressure Π, we have to calculate the partition function

$$Y(T, \Pi, N) = \int_0^\infty dL \; \mathcal{Z}(T, L, N) \exp(-\beta \Pi L), \tag{6.7}$$

and the corresponding thermodynamic potential is the Gibbs free energy

$$G(T, \Pi, N) = -k_B T \ln Y(T, \Pi, N) = F + \Pi L. \tag{6.8}$$

The Gibbs energy allows us to find an equilibrium length of the chain generated by the pressure Π,

$$\langle L \rangle = \left(\frac{\partial G}{\partial \Pi} \right)_{T,N}. \tag{6.9}$$

From Eqs. (6.1), (6.2), (6.3), and (6.9) it follows that the mean window number is determined by

$$\langle \widetilde{w} \rangle = -1 + \frac{1}{L_0} \left(\frac{\partial G}{\partial \Pi} \right)_{T,N}, \tag{6.10}$$

where $L_0 = N a_s$ is the chain's length at $\Pi = 0$.

Other thermodynamic potentials can be determined in a standard way: the system entropy is defined by

$$S(E, L, N) = -\left(\frac{\partial F}{\partial T} \right)_{L,N} = -\left(\frac{\partial G}{\partial T} \right)_{\Pi,N}, \tag{6.11}$$

and the total energy is equal to

$$E(S, L, N) = F + TS = \left[\frac{\partial (\beta F)}{\partial \beta} \right]_{L,N}. \tag{6.12}$$

Besides, the chemical potential is introduced as

$$\mu = \left(\frac{\partial F}{\partial N} \right)_{T,L} = \left(\frac{\partial G}{\partial N} \right)_{T,\Pi} = \left(\frac{\partial E}{\partial N} \right)_{S,L}, \tag{6.13}$$

and the specific heat per one atom, as

$$c_N = \frac{1}{N} \frac{dE}{dT} = -\frac{T}{N} \left(\frac{\partial^2 F}{\partial T^2} \right)_{L,N}. \tag{6.14}$$

A major role in static and dynamic behavior of the model plays a dimensionless susceptibility χ which is defined as

$$\chi = \frac{\langle (\Delta \widetilde{N})^2 \rangle}{\langle \widetilde{N} \rangle}, \tag{6.15}$$

where $\widetilde{N}$ is the number of atoms on a fixed length $\widetilde{L}$, $d \ll \widetilde{L} \ll L$, and $\Delta \widetilde{N} = \widetilde{N} - \langle \widetilde{N} \rangle$ is the fluctuation of $\widetilde{N}$. As is well known, for noninteracting atoms (i.e., in limits $g \to 0$ or $T \to \infty$) we have $\chi = 1$, so the deviation of χ from 1 describes the role of nonlinear interactions. In a standard way χ can be expressed through the chemical potential μ,

$$\chi = \left\{ \beta N \left(\frac{\partial \mu}{\partial N} \right)_{T,L} \right\}^{-1}, \tag{6.16}$$

or through the atomic concentration $n = N/L$,

$$\chi = k_B T \left(\frac{\partial \langle n \rangle}{\partial \Pi} \right)_{T,L}. \tag{6.17}$$

For the "free-end" FK chain it is also convenient to define χ through the window number $\widetilde{w}$,

$$\chi = -k_B T a_s n^2 \left(\frac{\partial \langle \widetilde{w} \rangle}{\partial \Pi} \right)_{T,N} = -k_B T \frac{n^2}{N} \left(\frac{\partial^2 G}{\partial \Pi^2} \right)_{T,N}. \tag{6.18}$$

Notice that everywhere in this section the thermodynamic limit $N, L \to \infty$ is implied.

At low temperatures, $k_B T \ll \varepsilon_k$, and zero window number, $N_w = 0$, all the thermodynamic functions are dominated by phonon contributions, since the density of thermally excited kinks is exponentially small. Using the dispersion law for the FK optical phonons, $\omega^2(\kappa) = \omega_0^2 + 2g(1 - \cos\kappa)$, $|\kappa| \leq \pi$, we obtain

$$F_{\text{ph}}^{(0)} = N k_B T \int_{-\pi}^{+\pi} \frac{d\kappa}{2\pi} \ln[\beta \hbar \omega(\kappa)]$$
$$= N k_B T \ln\left(\beta \hbar \sqrt{g/m_a + \omega_0(\omega_0 + \omega_{\max})/2} \right), \tag{6.19}$$

$$s_N \equiv S/N = k_B \left[1 - \ln\left(\beta \hbar \sqrt{g/m_a + \omega_0(\omega_0 + \omega_{\max})/2} \right) \right], \tag{6.20}$$

$$\varepsilon_N \equiv E/N = k_B T, \tag{6.21}$$

$$\mu = k_B T \ln\left(\beta \hbar \sqrt{g/m_a + \omega_0(\omega_0 + \omega_{\max})/2} \right), \tag{6.22}$$

$$c_N = k_B \quad \text{(the Dulong-Petit law)}, \tag{6.23}$$

$$\chi = k_B T \Big/ a_s^2 \left(g + \frac{1}{3} \omega_0^2 N^2 \right) \to 0 \quad \text{for} \quad N \to \infty, \tag{6.24}$$

where $\omega_{\max} = \sqrt{\omega_0^2 + 4g}$ and $\omega_0 = 1$ for the standard FK model. Anharmonic corrections to the phonon contributions can be calculated by a standard perturbation theory if the potential $\delta V(u) = V_{\text{sub}}(u) - (1/2) m_a \omega_0^2 u^2$ is treated as a small perturbation [520].

However, even at low temperatures, kinks are quite important for several features of the FK system. First of all, namely thermally excited kink-antikink pairs destroy the long-range order of a regular FK configuration. As can be seen in Fig. 6.1, where the mean window number $\langle \widetilde{w} \rangle$ is plotted against the pressure Π, at any $T \neq 0$ the Devil's staircase is washed out by thermal fluctuations. However, even at modest temperatures, $k_B T \leq \varepsilon_k$, a few of steps of the Devil's staircase continue to be well defined, i.e., those at $w = 0, \frac{1}{2}, \frac{1}{3}$. As will be shown in Sect. 6.5.4, the "melting" temperature T_m for the structure with a concentration θ depends on the "superkink" creation energy $E_k^\sigma(\theta, \Pi)$, $k_B T_m(\theta) \simeq E_k^\sigma(\theta, \Pi)$. Recall, first, that the value E_k^σ decreases as the "order" of the C-phase (i.e. the number of atom per elementary cell) increases, and second, that the energy to create a kink is largest at the center of a Devil's staircase step and goes to zero as the edges of a step are approached (i.e. at the points where the FvdM transitions occur). Thus, with increasing of system temperature, C-phases characterized by complex unit cells are destroyed first, while the simplest (trivial) C-phase with $w = 0$ ($\theta = 1$) is the last that destroy. Besides, kinks give the main contribution to the susceptibility χ and to "kink-sensitive" static correlation functions as well as they lead to an essential contribution to other thermodynamic properties (for example, kinks are responsible for a Schottky-type peak in the temperature dependence of the specific heat).

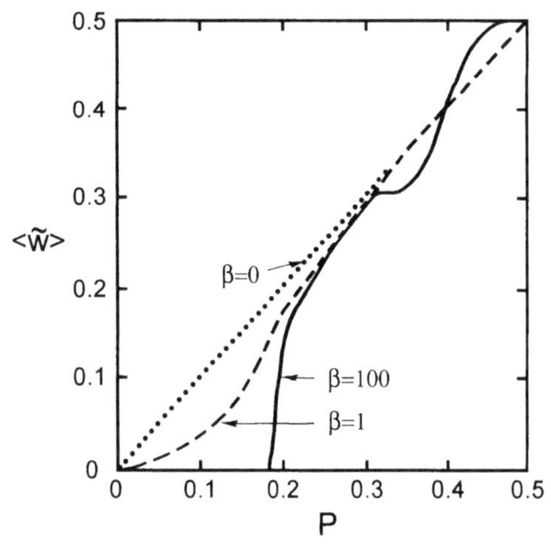

Fig. 6.1. Mean window number $\langle \widetilde{w} \rangle$ as a function of the parameter $P = \Pi / g a_s$ for the standard FK model with $g = 1$ at different temperatures: $\beta = 100$ (solid curve), $\beta = 1$ (dashed curve), and $\beta = 0$ (dotted curve). The dependences were calculated by the transfer-integral technique [521].

In the weak-coupling limit, $g \ll 1$, the low-temperature behavior of the FK model is quite interesting due to strong pinning effects. In this case the kinks can be treated as quasi-particles within the framework of a lattice-gas model with small exponentially decaying interaction of particles. The model exhibits a glass–like behavior with fractal structure of excitation spectrum. As will be shown, the FK model at $g \ll 1$ gives the microscopic justification of the "two-level" system with low gaps, the existence of which was assumed artificially in description of disordered systems [522, 523]. Therefore, the $g \ll 1$ FK model may be used to explain the low-temperature peak in specific heat of amorphous solids.

When the temperature is not small, $k_B T \geq \varepsilon_k$, the number of thermally excited kinks is large, $N_{\text{pair}} \sim N$, and the kink-lattice picture breaks down. In the limit $T \to \infty$ the substrate potential $V_{\text{sub}}(x)$ becomes gradually irrelevant, because it is bounded from below and above. Thus, as a zero approximation we can take the harmonic Hamiltonian ($a_A = L/N$)

$$H_0 = \sum_l \left\{ \frac{1}{2}\dot{x}_l^2 + \frac{g}{2}(x_{l+1} - x_l - a_A)^2 \right\} + E_0, \quad E_0 = \frac{g}{2}N(a_A - a_s)^2, \tag{6.25}$$

which describes acoustic phonons with the dispersion law $\omega^2(k) = 2g\left[1 - \cos(ka_A)\right]$, where $|k| \leq \pi/a_A$ Then, zero-order thermodynamic characteristics are:

$$F = Nk_B T \ln\left(\beta\hbar\sqrt{g/m_a}\right) + E_0, \tag{6.26}$$

$$s_N = k_B\left[1 - \ln\left(\beta\hbar\sqrt{g/m_a}\right)\right], \tag{6.27}$$

$$\varepsilon_N = k_B T + \frac{g}{2}(a_A - a_s)^2, \tag{6.28}$$

$$\mu = k_B T \ln\left(\beta\hbar\sqrt{g/m_a}\right) + \frac{1}{2}g(a_A - a_s)^2, \tag{6.29}$$

$$c_N = k_B, \tag{6.30}$$

$$\chi = k_B T/ga_A^2. \tag{6.31}$$

The single-site potential $V_{\text{sub}}(x)$ can be treated as a small perturbation [520] which leads, for example, to a correction in the specific heat,

$$c_N \approx k_B\left(1 + \frac{1}{2}\beta^2\right). \tag{6.32}$$

Notice that the susceptibility (6.31) has an nonphysical behavior in the limit $T/g \to \infty$. The reason is that mutual atomic displacements are not small, and the harmonic approximation for $V_{\text{int}}(x)$ is not adequate in this case. The account of anharmonicity of interatomic potential restores the correct limit $\chi \to 1$ as $T \to \infty$.

To conclude the introductionary remarks, note that statistical mechanics assumes the applicability of Boltzmann's hypothesis of molecular chaos, or the equal sharing of energy between system degrees of freedom. Of course, the validity of this assumption has not been proved. Moreover, in the limit $g \to \infty$ the SG system itself does not approach the thermal equilibrium since it is the integrable system. We postpone a discussion of this question to the next section because the mechanism of thermalization does not play a role in equilibrium properties of the system (however, it is quite important for dynamical behavior of the model). In this section we simply assume that the FK chain is embedded in a temperature reservoir at a temperature T.

6.3 Weak-Bond Limit: Glass-Like Properties

Due to discreteness effects the FK model always has stable configurations corresponding to minima of the system potential energy. One of these configurations is the ground state of the system, while others describe metastable states. We will call the metastable states by the configurational excitations of the system in order to distinguish them from dynamical excitations such as phonons, breathers, and moving kinks [Vallet *et al.* [528] prefer to use a more rigorous name "discommensurations" for these metastable excitations]. At $g = 0$ the number Ω of configurational excitations of the "fixed-density" FK chain is equal to $\Omega_0 = N!/M!(N-M)! - 1$. With g increasing, Ω decreases up to zero either at $g = \infty$ (for the commensurate structure), or at $g = g_{\text{Aubry}}$ (in the case of incommensurate concentration, see Ref. [524], when $\varepsilon_{PN} = 0$) as shown in Fig. 6.2.

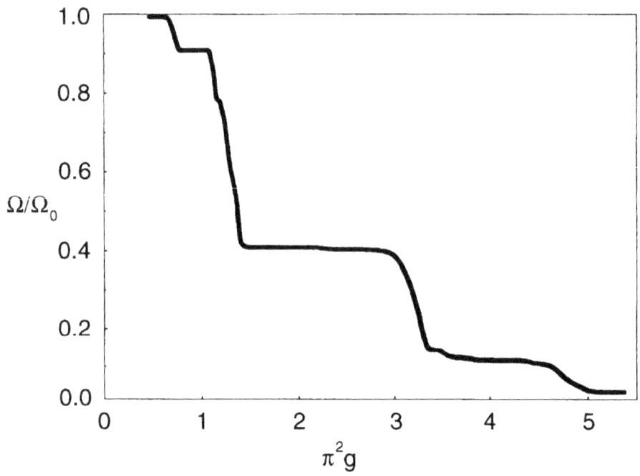

Fig. 6.2. Number of configurational excitations Ω against g for the fixed-density FK chain with $N/M = 13/17$ [525].

Naturally, the following questions emerge: (i) what is a nature of configurational excitations, (ii) what is their energy distribution, and (iii) how large is the contribution of these excitations to thermodynamic characteristics.

These questions can be answered by reducing the FK model to a lattice-gas model, considering strongly pinned kinks as quasi-particles subjected to the periodic PN potential.

6.3.1 Ising-Like Model

Let us numerate the minima of the substrate potential $V_{\text{sub}}(x)$ by an index j, $j = 1, 2, \ldots, M$, and introduce instead of the atomic coordinate x_l, where l is the number of atom, two new variables, the number of the well occupied by the atom l,

$$j_l = \mathrm{int}\left(\frac{x_l}{a_s} + \frac{1}{2}\right), \tag{6.33}$$

and the displacement of this atom from the well bottom,

$$w_l = x_l - j_l a_s. \tag{6.34}$$

From the definition (6.34) it follows that

$$|w_l| < a_s/2. \tag{6.35}$$

For a given consequence $\{j_l\}$ the corresponding set $\{w_l\}$ is uniquely determined by a solution of the stationary equations

$$\frac{\partial U}{\partial w_l} = 0, \tag{6.36}$$

where $U(\{x_l\}) \equiv U(\{j_l\}, \{w_l\})$ is the system potential energy. Therefore, all configurational excitations are uniquely specified by a sequence of increasing integers $\{j_l\}$. Thus, when we neglect by dynamical excitations of the system, the FK model is reduced to a lattice-gas-like model. Further, let us consider the fixed-density FK chain with a concentration $\theta < 1$, and assume that each well can be occupied by not more than one atom. Now, instead of the set $\{j_l\}$, it will be more convenient to use a set $\{s_l\}$, where $s_l = 0, 1, 2, \ldots$ is defined as the number of empty potential wells which lie between the l-th and $(l+1)$-th atoms. From the relationship $N(1 + \langle s_l \rangle) = M$ we obtain

$$\langle s_l \rangle \equiv \frac{1}{N} \sum_{l=1}^{N} s_l = \frac{(1-\theta)}{\theta}. \tag{6.37}$$

So, in this way we come to a spin-like model with constant magnetization [525, 526].

Clearly, the set $\{j_l\}$ corresponded to a configurational excitation, is not an arbitrary sequence of integer numbers. Some of sequences which lead to too large displacements w_l, $|w_l| > \frac{1}{2}a_s$, are to be excluded. Approximately this restriction may be fulfilled if we neglect the configurations with too large deviations of s_l from the average value $\langle s_l \rangle$ given by Eq. (6.37), i.e., if we restrict the available values of s_l by the interval $\langle s_l \rangle - \frac{1}{2}\Delta s \le s_l \le \langle s_l \rangle + \frac{1}{2}\Delta s$. The length Δs of this interval is to be decreased with increasing of the parameter g. In particular, at $g \ll 1$ and $0.7 < \theta < 1$ it seems reasonable to assume that s_l can only be either 0 or 1, so that the FK model reduces to Ising-like model.

To find the parameters of this spin-like model, we have to solve Eqs. (6.36). At $g \ll 1$ the displacements w_l are expected to be small, $|w_l| \ll \frac{1}{2}a_s$, and we may approximate the substrate potential $V_{\mathrm{sub}}(x)$ by an expression

$$V_{\mathrm{sub}}(x_l) \simeq \frac{1}{2}\omega_0^2 w_l^2. \tag{6.38}$$

In a result, the motion equation (6.36) becomes linear in w_l and, therefore, it can be solved by standard methods [525, 527]. Note, however, that the approximation (6.38) does not allow for atoms to move from one potential well to nearest wells and, therefore, this simplified model cannot be used to describe dynamical properties such as mass transport along the FK chain.

Using Eqs. (6.34), (6.38), and the relation $j_{l+1} = j_l + 1 + s_l$, the stationary equation (6.36) takes the form

$$w_{l+1} - 2hw_l + w_{l-1} = \phi_l \tag{6.39}$$

with

$$\phi_l = a_s(s_{l-1} - s_l), \tag{6.40}$$

$$h = 1 + \omega_0^2/2g. \tag{6.41}$$

Introducing the Green function G of the difference operator of the left-hand side of Eq. (6.39), the solution of Eq. (6.39) can be written as

$$w_l = \sum_{l'} G(l - l')\, \phi_{l'}. \tag{6.42}$$

The Fourier transform of Green function can be easily obtained,

$$\widetilde{G}(\kappa) \equiv \sum_l G(l) e^{i\kappa l} = -[2(h - \cos\kappa)]^{-1}, \quad |\kappa| \le \pi, \tag{6.43}$$

from which one gets by contour integration the Green function $G(l)$,

$$G(l) \equiv \frac{1}{2\pi} \int_{-\pi}^{+\pi} d\kappa\, \widetilde{G}(\kappa)\, e^{-i\kappa l} = -\frac{1}{2} \frac{1}{\sqrt{h^2 - 1}} \left(h - \sqrt{h^2 - 1}\right)^{|l|}. \tag{6.44}$$

Then, substituting Eq. (6.42) with Eqs. (6.40) and (6.44) into the expression for the potential energy and omitting constant terms, we finally obtain [525]

$$U = \sum_{l<l'} I(l - l')\, s_l s_{l'}, \tag{6.45}$$

where

$$I(l) = I_0 \exp\left(-|l| a_s/\lambda\right) \tag{6.46}$$

with

$$I_0 = g a_s^2 \omega_0/\omega_{\max} \simeq g a_s^2/(1 + 2g) \tag{6.47}$$

and

$$\lambda = -a_s \Big/ \ln\left(\frac{\omega_{\max} - \omega_0}{\omega_{\max} + \omega_0}\right) \approx -a_s/\ln g, \tag{6.48}$$

$$\omega_{\max} = \sqrt{\omega_0^2 + 4g} \simeq 1 + 2g. \tag{6.49}$$

So, we have obtained the Ising-like model with exponentially decaying coupling constant. Physical interpretation of Eq. (6.45) is trivial if we recall that s_l means the number of antikinks and $I(l)$ describes the antikink-antikink repulsion at $g \ll 1$.

6.3.2 Configurational Excitations

Configurational excitations of the FK model, $\{x_l\}$, may be considered as elements of a set $\widetilde{\mathrm{C}}$. According to Sect. 6.3.1, at $g \ll 1$ the state $\widetilde{\mathrm{C}}$ is isomorphic approximately to the set $\widetilde{\mathrm{M}}$ consisting of configurations $\{s_l\}$ of a spin-like model (6.45). In the set $\widetilde{\mathrm{M}}$ we may define the Bernoulli shift $\hat{S} : \widetilde{\mathrm{M}} \to \widetilde{\mathrm{M}}$ as $(\hat{S}s)_l = s_{l+1}$. The equivalent operator in the set $\widetilde{\mathrm{C}}$ is the mapping operator $\hat{T} : \widetilde{\mathrm{C}} \to \widetilde{\mathrm{C}}$ defined as $(\hat{T}Y)_l = Y_{l+1}$. The model with Bernoulli shift is a classical one exhibiting purely chaotic behavior [474]. Note, however, that due to the restriction (6.35) which forbids some of the configurations, the set $\widetilde{\mathrm{M}}$ forms now a Cantor set [527]–[529].

The set $\widetilde{\mathrm{M}}$ consists of a countable set of regular (crystalline) configurations, an uncountable set of "truly chaotic" (p-normal) configurations and, also, of "mixing" configurations where, for example, a subsequence $\{s_{km}\}$ is regular (for a fixed integer k and all integer m) while the values s_l for $l \neq km$ are random. For the Ising-like model the p-normal configuration may be defined as a random sequence s_l of 1's with probability p and 0's with probability $1-p$, so that each possible finite subsequence of the doubly infinite sequence $\{s_l\}$ appears with probability $p^{n_1}(1-p)^{n_2}$ if n_1 and n_2 denote the number of 1 and 0, respectively, appearing in the subsequence. Clearly, for the fixed-density FK chain the probability p coincides with the average value $\langle s_l \rangle$ given by Eq. (6.37). Note that the p-normality implies that s_l and $s_{l'}$ for $l \neq l'$ are uncorrelated.

Equations (6.34), (6.40), and (6.42) allow us to reconstruct the metastable configurations of the FK model which correspond to the p-normal configurations of the Ising-like model. In particular, Reichert and Schilling [527] have calculated the distribution function $Q_l(x)$ of the l-th nearest-neighbor distances and the pair distribution function $Q(x) = \sum_{l=1}^{\infty} Q_j(x)$. They showed that these functions have a typical glass-like behavior. Namely, $Q(x)$ exhibits a short-range order, i.e., there are more or less pronounced nearest-, next-nearest-, etc. neighbor peaks, and each of these peaks itself can be resolved into finer peaks. However, the long-range order is absent, $\lim_{x\to\infty} Q(x) - n \equiv N/L$.

A detailed numerical study of the low-energy configurational excitations of the classical FK model with the golden-mean window number $w = (\sqrt{5}-1)/2$ below the Aubry transition point ($g < g_{\mathrm{Aubry}}$) has been performed by Zhirov *et al.* [530]. They obtained the following interesting results:

1. The total number of configurational excitations grows exponentially with the length of the FK chain;
2. The energies of these configurations are organized in bands as shown in Fig. 6.3. The low-energy bands have the width much smaller than the distance between bands, while at higher energies the band width increases and eventually nearest bands almost merge into each other. With the increase of $K \equiv g^{-1}$ each band is shifted to lower values in energy, but

the number of states (counted per one atom) in each band is practically independent of K;

3. The number of bands becomes larger with K increasing, and the energies of lowest bands approach exponentially the ground state energy with increasing of the parameter K or the chain length (see Fig. 6.4). If the bands are labelled by the index k in order of increasing energy, then the average energy of k-th low-energy band may be described by an empirical formula

$$\langle \Delta U_k \rangle \approx C \exp\left(-\alpha w^k N \sqrt{\beta k + \lambda^2}\right), \tag{6.50}$$

where $w = (\sqrt{5} - 1)/2$ is the window number, N is the number of atoms in the chain, the numerical values of parameters are $\alpha \approx 0.59$, $C \approx 1$, $\beta \approx 0.12$, and λ is the Lyapunov exponent of the standard map (5.18) computed on the invariant Cantor set of orbits [λ describes also the phonon gap in the spectrum due to which any static displacement perturbation δx_{i0} of atom i_0 decays exponentially along the chain, $\delta x_i \propto \exp(-\lambda |i - i_0|)$];

4. The energies of the configurational excitations form a fractal quasi-degenerate band structure as demonstrated in Fig. 6.5. The fractal structure becomes deeper and deeper with the increase of the chain length.

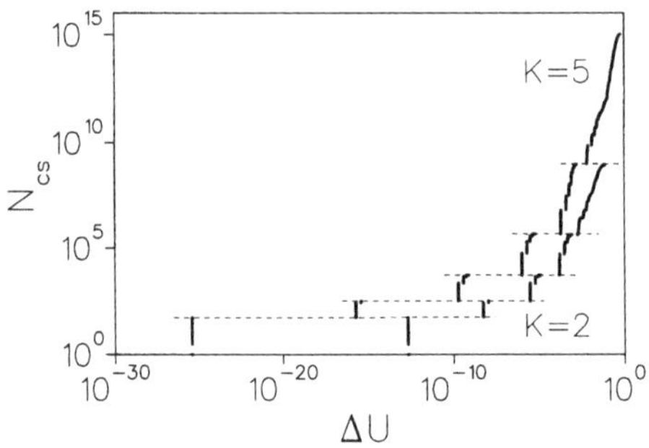

Fig. 6.3. Integrated number of metastable configurational states N_{cs} (per particle) as a function of the energy difference ΔU between E_i and the GS energy $E_{\rm GS}$, for $\theta = 89/55$ and $K = 5$ and $K = 2$, where $K \equiv g^{-1}$. Horizontal dashed lines show the border between the energy bands [530].

To explain the hierarchical structure of configurational excitations, Zhirov *et al.* [530] used the observation that in the ground state a considerable amount of atoms is located very close to the bottoms of the substrate potential. One can see from Fig. 6.6 (left panel) that the values of small deviations are grouped into three well resolved hierarchical levels, and separations along the chain for these atoms are also ordered: the two atoms closest to the bottoms ($|\Delta x| \approx 5\,10^{-25}$) are separated by the distances 55 and 89, eight atoms (including the previous two) with deviations $|\Delta x| \leq 3\,10^{-6}$, by the distances 13 and 21, and 34 atoms with deviations $|\Delta x| \leq 10^{-1}$, by the distances 3 and 5 (notice that all these separations form the Fibonacci sequence s_n, see

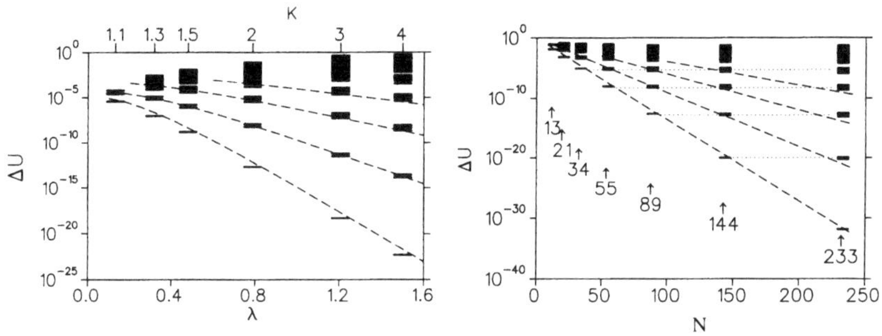

Fig. 6.4. Left panel: band energy spectrum vs. the parameter K (upper scale) and the phonon gap λ (lower scale) for $\theta = 89/55$. Right panel: the band energies vs. the chain length (i.e., for N/M =13/8, 21/13, 34/21, 55/34, 89/55, 144/89, and 233/144). The bands are marked by filled areas. The dashed curves correspond to the semi-empirical expression (6.50) [530].

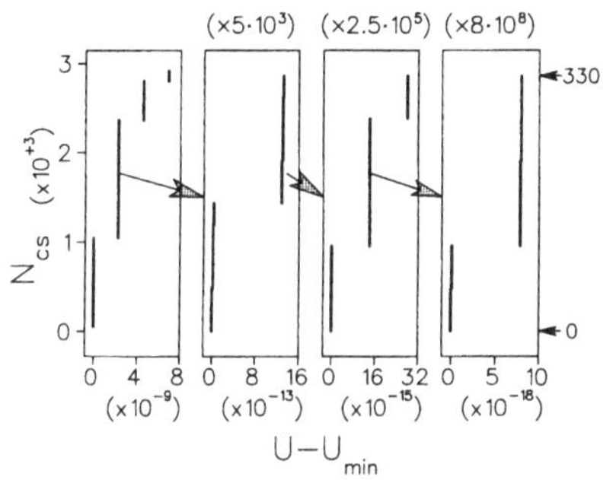

Fig. 6.5. Fractal energy band structure of the third excited band for the chain with $K = 4$ and $\theta = 89/55$. Shown are four hierarchical levels with growing resolution. Here U is the energy per atom and $U_{\min}$ stands for the energy of the most left band in each panel. Vertical scale N_{cs} gives the integrated number of metastable configurations counted from the bottom of the most left band in each panel. The vertical magnification increases in ten times from left to right [530].

Sect. 5.1). In what follows let us call the atoms lying very close to the well bottoms by the "glue" atoms (g-atoms), because the tension forces acting from both sides of any such atom approximately compensate each other. Now, let us cut the chain at glue atoms into fragments, or "bricks". The lowest level of the hierarchy shown in Fig. 6.6 (with deviations $|\Delta x| \leq 10^{-1}$), is built up by bricks of two types consisting of 2 and 4 atoms respectively, so that a 8-atomic brick can be denoted as $8 = (g2g4)$. The next level of the hierarchy (with deviations $|\Delta x| \leq 3\,10^{-6}$) has bricks $12 = (4g2g4)$ and $20 = (4g2g4g2g4)$, the third level of the hierarchy (with deviations $|\Delta x| \leq 10^{-24}$) is composed by the bricks $54 = (20g12g20)$ and $88 = (20g12g20g12g20)$, and so on: if a given level of hierarchy is composed by bricks A and B (with $A < B$), then the next level should be built by bricks $A' = (BgAgB)$ and $B' = (BgAgBgAgB)$.

Now, if we interchange two different bricks, the brick's ends will be only slightly distorted, and the system energy will change on an amount $\Delta E \propto (\Delta x)^2$. An example of the configurational excitation obtained from the GS by a brick's permutation at the third hierarchical level is shown in Fig. 6.6 (right panel) [notice that for the first two levels of hierarchy, the

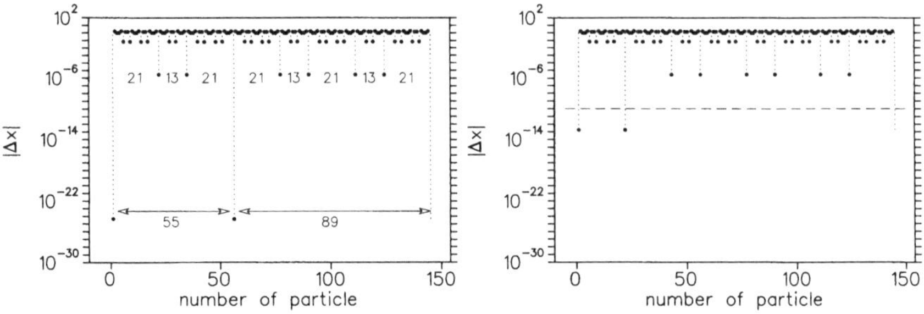

Fig. 6.6. Absolute value of the particle deviations from the nearest bottom of the substrate potential vs. the atomic number for $\theta = 144/89$ and $K = 2$. The left panel is for the ground state, while the right panel corresponds to an excited configuration taken from the first excited energy band with $\Delta U \sim 10^{-20}$, when the deviations of glue atoms from the well bottoms are $|\Delta x| \leq 10^{-10}$ (dashed line) [530].

deviations of glue atoms from the well bottoms are practically the same as in the GS, while at the third hierarchical level (below the dashed line in Fig. 6.6) the deviations of two glue atoms become considerably larger than the ones in the GS]. Since the glue particle deviations are exponentially small and hierarchically ordered, then the corresponding energies of configurations obtained by brick's interchange, are also exponentially small and ordered. The energy change caused by brick's permutation depends on the level of the hierarchy inside which the permutation is done. An analytical approach [530] yields an estimation $\Delta U \propto \exp(-2\lambda n_{\rm min})$, where $n_{\rm min}$ is the number of atoms in the smaller brick at the given level of hierarchy. Moreover, the number of configurations at any level of the hierarchy is combinatorially large, thus the number of configurational excitations grows exponentially with the length of the chain, and most of them are disordered (chaotic).

6.3.3 Two-Level Systems and Specific Heat

To consider energy distribution of configurational excitations, we will follow the works of Pietronero and Strässler [526], Pietronero *et al.* [531], and Reichert and Schilling [527]. Let us define the differences

$$\Delta^{(i,i')} = E_i - E_{i'}, \tag{6.51}$$

where $E_i = E\left(\left\{x_l^{(i)}\right\}\right)$ is the potential energy of the i-th metastable state. The values $\Delta^{(i,i')}$ may be positive or negative on account of the metastability of the configuration i. From the previous discussion it is clear that $\Delta^{(i,i')}$ form a spectrum which is a Cantor set symmetric to zero.

Indeed, the simplest way to obtain a new configuration $\{x_l'\}$ from a given $\{x_l\}$ is just to move one of atoms over a potential barrier of the substrate potential to the nearest neighbor empty potential well with all other atoms fixed and then relaxing the chain. The obtained two sequences $\{j_l\}$ and $\{j_l'\}$ differ only for one number, while the atomic positions w_l and w_l' of course differ for all l but converge exponentially to each other for $l \to \pm\infty$. At $\theta < 1$, when the variables s_l take 0 or 1 only, it is more convenient to denote a configuration by a set of integers m_k, where m_k is the number of consecutive occupied wells following the k-th empty well. For example, the configuration $\ldots m_1; m_2; m_3; \ldots$ means that there is a vacancy, then m_1 occupied wells, a vacancy, m_2 occupied wells, a vacancy, m_3 occupied wells, etc. Clearly that

$$\langle m \rangle = (1 - \theta)^{-1}. \tag{6.52}$$

Using this definition, the simplest transition $\{x_l\} \to \{x_l'\}$ is described as

$$\ldots m_1; m_2; m_3; m_4; \ldots \to \ldots; m_1; m_2 - 1; m_3 + 1; m_4; \ldots . \tag{6.53}$$

According to the Ising Hamiltonian (6.45), (6.46), this transition gives the energy difference

$$\Delta = \Delta_1 + \Delta_2 + \ldots \tag{6.54}$$

with

$$\Delta_1 = I(m_2 - 1) + I(m_3 + 1) - I(m_2) - I(m_3), \tag{6.55}$$

$$\Delta_2 = I(m_2 - 1 + m_1) + I(m_3 + 1 + m_4) - I(m_1 + m_2) - I(m_3 + m_4), \tag{6.56}$$

etc. Because of the exponential decay of the Ising constant $I(m)$, we have $\Delta_1 \gg \Delta_2 \gg \ldots$, and

$$\Delta_n \simeq I_0 (a_s/\lambda)^2 \exp\left(-n\langle m \rangle a_s/\lambda\right). \tag{6.57}$$

As seen, for the configuration with $m_2 = m_3$ we have $\Delta \simeq \Delta_1$. However, if $m_2 = m_3 + 1$, then $\Delta_1 = 0$, i.e. this configuration is degenerated with respect to the nearest-neighbor interaction of holes, and the main contribution to Δ comes from the interaction of second neighbors, so that $\Delta \simeq \Delta_2$. In this way it is clear that the energy spectrum exhibits a hierarchy of "two-level systems" with energy gaps distributed around $\Delta_1, \Delta_2, \ldots$ Analogously, for each configuration i with energy E_i there exists a set of other configurations i' with energy $E_{i'}$ where each $\{j_l^{(i')}\}$ differs from $\{j_l^{(i)}\}$ only locally by one or more numbers, and the transition $i \to i'$ represents the excitation of a two-level system with excitation energy $\Delta^{(i,i')} = E_i - E_i'$. Note that a more detailed investigation [527] shows a scaling behavior of distribution of $\Delta^{(i,i')}$.

Now let us shortly discuss the density of configurational excitations at different concentrations θ. If $\theta = \theta_1 \equiv m_0/(m_0 + 1)$ with $m_0 \geq 2$, the $T = 0$ ground state of the FK chain is described by $m_k = m_0$ for all k. So, at low temperatures, $k_B T \ll \Delta_1$, a majority of excitations has the energy gap $\Delta \simeq \Delta_1$, and only a small fraction of configurational excitations with statistical weight $\sim \exp(-\Delta_1/k_B T)$ leads to the energy gap $\Delta \simeq \Delta_2$. Otherwise, if $\theta = \theta_2 \equiv (2m_0 + 1)/(2m_0 + 3)$ when the $T = 0$ GS has the form $\ldots m_0; m_0 +$

$1; m_0; m_0 + 1; \ldots$, a majority of excitations has $\Delta \simeq \Delta_2$, while only a small fraction yields the gaps $\Delta \simeq \Delta_1$.

The energy of the $T \neq 0$ ground state of the two-level system with gap Δ may be estimated as $E(T) \propto \Delta \exp(-\Delta/k_B T)/[1+\exp(-\Delta/k_B T)]$. That leads to the specific heat of the FK chain

$$c(T) \propto k_B \left(\frac{\Delta}{k_B T}\right)^2 \frac{\exp(\Delta/k_B T)}{[1+\exp(\Delta/k_B T)]^2}, \tag{6.58}$$

which has a maximum at $T \simeq 0.4\,\Delta/k_B$. Because of the fine structure of excitation spectrum of the FK model, the function $c(T)$ should consist of a series of peaks at temperatures corresponding to gaps $\Delta_1, \Delta_2, \ldots$. The intensity of these peaks will, however, strongly depend on the concentration θ. Indeed, with changing of θ the intensity of the peak contributed by the gap Δ_1 has to have maxima at $\theta = \theta_1$, $\theta = \theta_1' \equiv (m_0+1)/(m_0+2)$, …, while its minima are to be achieved at $\theta = \theta_2$, $\theta = \theta_2' \equiv (2m_0+3)/(2m_0+5)$, … On the other hand, the intensity of the peak corresponded to the gap Δ_2, will have maxima at $\theta = \theta_2$, θ_2', …, and minima at $\theta = \theta_1$, θ_1', …

For the analytical calculation of thermodynamic characteristics we need a further simplification of the Ising-like model (6.45). Namely, we have to account only the finite number of interacting holes (for example, the nearest neighbors only) and, besides, to restrict the available values of m_k (for example, at $\theta = \theta_1$ we may assume that m_k takes the values m_0 and $m_0 \pm 1$ only). A lengthy but straightforward calculation [531] confirms the given above intuitive picture for the specific heat $c(T, \theta)$.

The rigorous calculation of the spectrum of low-energy excitations, the specific heat, and the hierarchy of "melting" of the FK chain with incommensurate concentration was preformed by Vallet *et al.* [528, 529]

The described behavior of $c(T, \theta)$ was observed experimentally in studying of ionic conductor hollandite $K_{2\rho}Mg_\rho\ Ti_{\delta-\rho}\ O_{16}$ [532]. Note that the experimental time scale of the measurements has to be large enough in order the transitions over the potential barriers have nonzero probability, and that the temperature behavior of $c(T)$ should be resolved on the background of phonon contributions.

In conclusion, note that the transition (6.53) may be considered as the creation of a (super-) kink-antikink pair which is strongly pinned by the high-amplitude PN potential, so that the value Δ_n corresponds to twice of the energy of n-order superkink. In this way the peaks in $c(T, \theta)$ can be interpreted as contributions owing to creation of kink-antikink pairs. As we will see in Sect. 6.5.2, this approach remains useful in the strong-coupling limit, $g \gg 1$. In the last case, however, the fine structure of $c(T)$ disappears due to vanishing of PN barriers.

6.4 Strong-Bond Limit: Gas of Quasiparticles

In the strong-coupling limit, $g \gg 1$, the FK model reduces to the SG model with small corrections due to discreteness effects. Leaving the discussion of discreteness effects to Sect. 6.4.4, let us start with purely SG model. The SG Hamiltonian is exactly separable into contributions corresponding to elementary excitations (phonons, breathers and kinks) using the inverse scattering theory. Therefore, any excited state can be considered as consisting of $N_{\rm ph}$ linear phonons, $N_{\rm br}$ breathers, N_k kinks and $N_{\bar{k}}$ antikinks (here by kinks we mean the kinks for the reference structure $\theta = 1$ as it is natural for the SG case). Collisions of these quasi-particles are "elastic", i.e., their individuality such as shape, energy and momenta remains unchanged. Thus, it is reasonable to calculate thermodynamic potentials assuming an "ideal gas" of noninteracting quasi-particles which participate in partition function on equal foot.

Introducing formally chemical potentials μ_k, $\mu_{\bar{k}}$ and μ_B (index B corresponds to breathers and/or phonons, see below Sect. 6.4.1), we can calculate the grand canonical function Ξ for the "free-end" FK chain as

$$\widetilde{\Xi}(T, \mu_k, \mu_{\bar{k}}, \mu_B) = \sum_{N_k, N_{\bar{k}}, N_B} \frac{1}{N_k! N_{\bar{k}}! N_B!}$$

$$\exp\left[\beta\left(\mu_k N_k + \mu_{\bar{k}} N_{\bar{k}} + \mu_B N_B\right)\right] \int d\Gamma \exp\left[-\beta E(N_k, N_{\bar{k}}, N_B)\right], \quad (6.59)$$

where the integration $d\Gamma$ is over all independent degrees of freedom of the system in phase space, and $E(N_k, N_{\bar{k}}, N_B)$ is the sum of energies of all constituent modes. The corresponding thermodynamic potential,

$$\widetilde{J}(T, \mu_k, \mu_{\bar{k}}, \mu_B) = -k_B T \ln \widetilde{\Xi}(T, \mu_k, \mu_{\bar{k}}, \mu_B), \quad (6.60)$$

allows to calculate the average numbers of excited quasiparticles,

$$\langle N_b \rangle = -\left(\frac{\partial \widetilde{J}}{\partial \mu_b}\right)_T, \quad b = k, \bar{k}, \text{ or } B. \quad (6.61)$$

The thermal equilibrium in the system is achieved by creation and annihilation of kink-antikink pairs. These processes may be regarded as "chemical reaction" $k + \bar{k} \leftrightarrow B$, and the condition for equilibrium is given by $\mu_k + \mu_{\bar{k}} = \mu_B$ [159]. Since we cannot control the number of breathers and phonons, we have to put $\mu_B = 0$. Thus, in the final step of calculations we must put

$$\mu_k = -\mu_{\bar{k}}, \quad (6.62)$$

and the thermodynamic potential becomes a function of T and $\mu_w \equiv \mu_k$ only,

$$J(T, \mu_w) = \widetilde{J}(T, \mu_w, -\mu_w, 0). \quad (6.63)$$

The value of μ_w corresponds to the total topological charge $N_w = N_{\bar{k}} - N_k$,

$$\langle N_w \rangle = \left(\frac{\partial J}{\partial \mu_w} \right)_T . \tag{6.64}$$

Note that μ_w favors kinks over antikinks if $\mu_w > 0$ and vice versa if $\mu_w < 0$, while at $\mu_w = 0$ it follows $\langle N_w \rangle = 0$.

The phonon contribution to J was calculated above, for phonons $\int d\Gamma_{\text{ph}} = \int_{-\pi}^{+\pi} d\kappa/2\pi$, $E_{\text{ph}} = \hbar\omega(\kappa)$, and $J_B = F_B$ because $\mu_B = 0$, where F_B is given by Eq. (6.19) or (6.26).

Now let us consider kink's contribution to thermodynamic characteristics. Each kink has one degree of freedom in configuration space, the Goldstone mode. Thus, the integration in phase space is to be done over kink's coordinate X_k and its momentum P_k,

$$d\Gamma_k = \frac{dX_k \, dP_k}{2\pi\hbar} . \tag{6.65}$$

The full energy of a kink in nonrelativistic case is equal to

$$E_k = \varepsilon_k + P_k^2/2m, \tag{6.66}$$

where ε_k is kink rest energy and m is kink mass. The integration over X_k in Eq. (6.59) yields the factor L, the chain's total length, while the integration over P_k leads to the factor $z = \left(m k_B T/2\pi\hbar^2\right)^{1/2} \exp(-\beta\varepsilon_k)$. Thus, after straightforward calculations we obtain from Eqs. (6.59) to (6.65) the following kink's contributions:

$$J_s = -k_B T \langle N_{\text{tot}} \rangle, \tag{6.67}$$

$$F_s = -(k_B T \langle N_{\text{tot}} \rangle + \mu_w \langle N_w \rangle). \tag{6.68}$$

The kink concentrations in Eqs. (6.67), (6.68) are determined by relations

$$\langle n_k \rangle = \langle n_{\text{pair}} \rangle \exp(\beta \mu_w), \tag{6.69}$$

$$\langle n_{\bar{k}} \rangle = \langle n_{\text{pair}} \rangle \exp(-\beta \mu_w), \tag{6.70}$$

$$\langle n_{\text{tot}} \rangle = 2 \langle n_{\text{pair}} \rangle \cosh(\beta \mu_w), \tag{6.71}$$

$$\langle n_w \rangle = -2 \langle n_{\text{pair}} \rangle \sinh(\beta \mu_w). \tag{6.72}$$

It is important to notice that while before an averaging we had $N_{\text{tot}} = 2N_{\text{pair}} + |N_w|$, after the averaging we get the relationship [159]

$$\langle n_{\text{tot}} \rangle^2 = \langle 2n_{\text{pair}} \rangle^2 + \langle n_w \rangle^2 . \tag{6.73}$$

Using a coupling of N_w with chain's length L,

$$L = a_s (N + \langle N_w \rangle), \tag{6.74}$$

we may introduce a pressure Π as

$$\Pi = -\left(\frac{\partial F}{\partial L}\right)_{T,N} = \frac{\mu_w}{a_s}. \tag{6.75}$$

It was mentioned above that kinks give the main contribution to the susceptibility χ defined by Eq. (6.18). Indeed, from Eqs. (6.67) to (6.75) it follows that

$$\chi = a_s^2 \langle n \rangle \langle n_{\text{tot}} \rangle = \frac{\langle n_{\text{tot}} \rangle}{\langle n \rangle} \frac{1}{(1+\widetilde{w})^2}. \tag{6.76}$$

The described above naive approach leads to a density of thermally excited kink-antikink pairs, determined by the equation

$$\langle n_{\text{pair}}^{(0)} \rangle \equiv z = \left(\frac{m k_B T}{2\pi\hbar^2}\right)^{1/2} e^{-\beta\varepsilon_k}, \tag{6.77}$$

where we have assumed that kinks and antikinks have identical parameters as it is in the standard FK model with $a_{\min} = a_s$. Note that all the obtained results, Eqs. (6.67) to (6.76), will remain true in the rigorous approach to the problem, and only the last Eq. (6.77) is to be corrected.

For the "fixed-density" FK chain, where N_w is fixed externally by periodic boundary conditions, we should choose μ_w so that $\langle N_w \rangle = N_w$. Clearly that for zero window number, $N_w = 0$, we have to put $\mu_w = 0$. If $N_w \neq 0$, it is convenient to introduce a temperature T^* at which the density of thermally excited kinks is equal to the density of external kinks, $\langle n_{\text{pair}}(T^*) \rangle = |n_w|$. Then, from Eq. (6.72) it follows that at low temperatures, when $\langle n_{\text{pair}} \rangle \ll |n_w|$, $|\mu_w|$ tends to kink creation energy ε_k (see, however, Eq. (6.97) below),

$$\mu_w \simeq -\varepsilon_k \operatorname{sgn}(n_w) \left\{1 + \frac{k_B T}{\varepsilon_k} \ln \frac{|n_w| e^{-\beta\varepsilon_k}}{\langle n_{\text{pair}} \rangle}\right\}, \quad T \ll T^*, \tag{6.78}$$

while at higher temperatures, $T \gg T^*$ (but $k_B T \ll \varepsilon_k$), when $\langle n_{\text{pair}} \rangle \gg |n_w|$, $|\mu_w|$ is close to zero,

$$\mu_w \simeq -k_B T n_w / 2\langle n_{\text{pair}} \rangle. \tag{6.79}$$

Therefore, as was firstly mentioned by Currie *et al.* [159],

$$\lim_{T\to\infty} \lim_{N_w\to 0} \mu_w \neq \lim_{N_w\to 0} \lim_{T\to 0} \mu_w. \tag{6.80}$$

Analogously we may calculate breather's contribution to thermodynamic functions, taking into account that each breather has two degrees of freedom in configuration space, one corresponds to its free motion as a whole, and another describes breather's internal vibrations with a frequency ω_{br}. Recall that a low-frequency breather, $\omega_{\text{br}} \to 0$, may be considered as a coupled (virtual) kink-antikink pair, while a high-frequency breather, $\omega_{\text{br}} \to \omega_0$, may be interpreted as a phonon pair bounded due to unharmonicity effects.

The used above simple approach, first proposed by Krumhansl and Schrieffer [517], silently assumes that the system has

$$N_\Sigma = N_{\rm ph} + 2N_{\rm br} + N_k + N_{\bar{k}} \tag{6.81}$$

degrees of freedom in configuration space, or that a system state is described by a point in phase space of $2N_\Sigma$ dimensions. Evidently, however, that in statistical mechanics a one-dimensional system consisting of N atoms is to be defined in phase space of $2N$ dimensions. So, it must be $N_\Sigma = N$. This problem was firstly pointed out and resolved by Currie *et al.* [159].

6.4.1 Sharing of the Phase Space and Breathers

While normal modes of a linear system can be considered as independent degrees of freedom, the nonlinear modes are not independent. Therefore, the $2N$-dimensional phase space should be shared between different nonlinear modes, and the phase space element $d\Gamma$ in Eq. (6.59) is to be taken in a form

$$d\Gamma = \left(\Pi_k^{N_{\rm tot}} d\Gamma_k\right) \left(\Pi_B^{N_B} d\Gamma_B\right) \Omega(\{\xi_i\}), \tag{6.82}$$

where the function Ω represent the restriction on the phase space available for the quasiparticles due to their "interaction". Although quasi-particle collisions are "elastic" in the SG case, the phases (i.e. coordinates) of quasiparticles are shifted during collisions. Namely this "phase-shift" interaction provides the mechanism for sharing of degrees of freedom among the elementary excitations of nonlinear systems. As an example let us consider following Currie *et al.* [533] and Trullinger [534], the FK chain of length L which is imposed to periodic boundary conditions. In the absence of kinks, the allowed wave vectors k are determined by the condition $Lk_p = 2\pi p$ $(p = 0, \pm 1, \ldots, N/2)$, and the density of phonon states is $\rho_0(k) = L/2\pi$. The presence of a single static kink or antikink distorts the phonon waveform near the kink which results in a "phase shift" of the phonon

$$\delta_0(k) = \pi k/|k| - 2\tan^{-1}(kd). \tag{6.83}$$

For a kink moving with velocity v, the phonon phase shift can be obtained from Eq. (6.83) by a Lorentz boost,

$$\delta(k;v) = \delta_0(k'), \quad k' = \gamma\left[k - (v\omega_0/c^2)\sqrt{1+k^2d^2}\right], \tag{6.84}$$

where $\gamma = 1/\sqrt{1-v^2/c^2}$ and $c = \omega_0 d$. In the presence of kink, the allowed wave vectors k are determined by

$$Lk_p + \delta(k_p;v) = 2\pi p. \tag{6.85}$$

The phonon density of states is then

$$\rho(k)=\frac{dp}{dk}=\rho_0(k)+\Delta\rho(k,v),\quad \Delta\rho(k,v)=\frac{1}{2\pi}\frac{\partial\,\delta(k;v)}{\partial k}. \tag{6.86}$$

Note that $\int_{-\infty}^{+\infty} dk\, \Delta\rho(k,v) = -1$ in the SG case. Thus, a kink "traps" one phonon state due to its very presence, and transforms it into kink's translation (Goldstone) mode.

Analogously, a single breather "traps" two phonon modes. A general expression for the restriction factor Ω through corresponding phase shifts was given by Sasaki [535]. However, this approach meets with significant difficulties known as the breather problem [535]–[541]. Namely, while the number of excited kinks is low at low temperatures and, therefore, the factor Ω is close to 1, the number of excited breathers is of the same order as the number of phonons ($N_{\rm br} \sim \frac{1}{2}N$, $N_{\rm ph} \sim N$) even at $T \to 0$, because both types of excitations are created without energy threshold. Therefore, the sharing of phase space between phonons and breathers is not simple, and it must be taken into account even at zero approximation. In other words, phonons and breathers cannot be considered simultaneously as Maxwell-Boltzmann gas of independent quasiparticles. The physical reason of "large interaction" between breather and phonon is that a "size" of breather tends to infinity as $\omega_{\rm br} \to 0$.

Although the full "phonon-breather-kink" program may be developed in principle, it is more convenient to take as the zero approximation the "phonon-kink" ensemble (and, thus, to ignore breathers at all), or the "breather-kink" ensemble (i.e., ignoring phonons). It is to be emphasized that all the approaches lead to the same final result. The phonon-kink gas approach is useful in classical statistical mechanics. It was used in a majority of works devoted to the problem. In the first approximation phonons are treated as noninteracting quasiparticles (harmonic phonons). The anharmonic corrections are evaluated by the standard perturbation theory of phonons, leading to terms of order T and higher in thermodynamic potentials [538].

Thermodynamics of the Boltzmann gas of breathers and kinks logically follows from the Bethe-ansatz formulation of the quantum SG system. This approach was used by Theodorakopoulos [539, 540] and Sasaki [535]. Note that phonons are incorporated automatically in this approach, because the quantum lowest energy state of breather coincides with renormalized phonon. Below we consider in more details the kink-phonon approach.

6.4.2 Kink-Phonon Interaction

The sharing of phase space between phonons and kinks can be evaluated in two ways. The first method proposed by Tomboulis [542], uses a canonical transformation from the original set of canonical variables to a new set where kink coordinates X_k and their momenta P_k (conjugate to X_k) are introduced as canonical variables. For each kink we have to impose two Dirac's constrains in order to keep the dimensionality of phase space unchanged. This approach was used, e.g., by Miyashita and Maki [543, 544], Fukuma and Takada [545]. We, however, will follow a more simple and "transparent" method of the first works devoted to this problem [533, 534].

As was shown above, the presence of a kink moving with velocity v changes the phonon density of states on the value $\Delta\rho(k;v)$ given by Eq. (6.86). This change results in a contribution to the free energy which is equal to [533, 534]

$$\Delta F(v) = k_B T \int_{-\infty}^{+\infty} dk \; \Delta\rho(k;v) \ln[\beta\hbar\omega(k)] \\ = -k_B T \ln[(1+\gamma)\beta\hbar\omega_0], \tag{6.87}$$

where $\gamma = 1/\sqrt{1-v^2/c^2}$. Each kink and antikink contributes independently to the change in the phonon density of states, since the phase shifts in the SG case are additive. Thus, the total change in the free energy caused by the presence of $N_{\rm tot}$ kinks and antikinks can be expressed as

$$\Delta F_N(v_1, v_2, \ldots, v_{N_{\rm tot}}) = \sum_{j=1}^{N_{\rm tot}} \Delta F(v_j). \tag{6.88}$$

Returning to calculation of the grand canonical partition function (6.59), we note that phonon density of states depends on the number of kinks and their velocities. Therefore, before we can integrate over kink's momenta, we must first integrate over the phonon degrees of freedom. After this integration the function $\widetilde{\Xi}$ takes the form

$$\widetilde{\Xi} = e^{-\beta F_{\rm ph}^{(0)}} \sum_{N_k=0}^{\infty} \sum_{N_{\bar{k}}=0}^{\infty} \frac{e^{\beta(\mu_k N_k + \mu_{\bar{k}} N_{\bar{k}})}}{N_k! N_{\bar{k}}!} \int \Pi_{k=0}^{N_k} \frac{dX_k dP_k}{2\pi\hbar} \Pi_{\bar{k}=0}^{N_{\bar{k}}} \frac{dX_{\bar{k}} dP_{\bar{k}}}{2\pi\hbar} e^{\Theta}$$

where

$$\Theta = \exp\left\{-\beta\left[E_k(P_k)N_k + E_{\bar{k}}(P_{\bar{k}})N_{\bar{k}}\right] - \beta\Delta F_{N_k+N_{\bar{k}}}\right\},$$

where $F_{\rm ph}^{(0)}$ is the unperturbed phonon free energy in the absence of kinks given by Eq. (6.19). Due to additive structure of the function $\Delta F_{N_{\rm tot}}$, Eq. (6.88), we can incorporate the contributions $\Delta F(v_j)$ with the kink energies $E_k(P_j)$, introducing the "renormalized" kink energy

$$E_k^*(P) = E_k(P) + \Delta F(v). \tag{6.89}$$

Recall that for the SG case

$$E_k(P) = E_{\bar{k}}(P) = \varepsilon_k \gamma \tag{6.90}$$

and

$$P = mv\gamma. \tag{6.91}$$

Currie *et al.* [533] proposed to interpret the value $\Delta F(v)$ as a "self-energy" of the kink analogously to self-energy of an electron in crystal which emerges in the polaron theory due to phonon dressing of electrons.

Now, repeating the calculations as had been done above, we come again to Eqs. (6.67) to (6.76) except only that the integration over kink's momenta yield now the factor

$$z \equiv \langle n_{\text{pair}} \rangle = \int_{-\infty}^{+\infty} \frac{dP}{2\pi\hbar} \exp[-\beta E_k^*(P)]. \tag{6.92}$$

This integral may be expressed through modified Bessel functions [534]. In the low-temperature limit, $k_B T \ll \varepsilon_k$, Eq. (6.92) leads to the result

$$\langle n_{\text{pair}} \rangle \simeq \sqrt{\frac{2}{\pi}} \frac{\sqrt{\beta\varepsilon_k}}{d} \left[1 + O\left(\frac{1}{\beta\varepsilon_k}\right)\right] e^{-\beta\varepsilon_k}, \tag{6.93}$$

where $O(\tau) = \frac{5}{8}\tau + \frac{21}{128}\tau^2 + \ldots$ are small corrections [540, 543, 544]. Comparing Eq. (6.93) with Eq. (6.77), we see that phonon-kink sharing of phase space reduces the number of thermally excited kink-antikink pairs by a factor $2\hbar\omega_0/k_B T \ll 1$ (notice that classical mechanics is applicable at $\hbar\omega_0 \ll k_B T$ only).

Thus, finally the thermodynamic functions take the following form:

$$J = F_{\text{ph}}^{(0)} - k_B T \langle N_{\text{tot}} \rangle, \tag{6.94}$$

$$F = F_{\text{ph}}^{(0)} - \left[k_B T \langle N_{\text{tot}} \rangle + \mu_w \langle N_w \rangle\right], \tag{6.95}$$

$$\begin{aligned} E = {} & k_B T (N - \langle N_{\text{tot}} \rangle) + \left(\varepsilon_k + \frac{1}{2} k_B T\right) \langle N_{\text{tot}} \rangle \\ & + \mu_w \left[\beta\mu_w \langle N_{\text{tot}} \rangle + \left(\beta\varepsilon_k - \frac{1}{2}\right) \langle N_w \rangle\right]. \end{aligned} \tag{6.96}$$

Two first terms in Eq. (6.96) have a simple interpretation [159] if we recall that each kink takes away one phonon mode, and that each kink has ε_k rest (potential) energy and $\frac{1}{2} k_B T$ translation (kinetic) energy, while the last term in Eq. (6.96) leads to a small corrections at $\langle n_w \rangle \ll \langle n_{\text{pair}} \rangle$ as well as at $\langle n_{\text{pair}} \rangle \ll \langle n_w \rangle$.

Clearly, this description is restricted to the case when the total number of kinks is low, $\langle N_{\text{tot}} \rangle \ll N$, so that we may ignore the phase space sharing effects between the kinks. To satisfy this condition, it should be

$$\beta\left(\varepsilon_k - |\mu_w|\right) \ll 1, \tag{6.97}$$

i.e., the chemical potential μ_w must have a magnitude less then (and not too close to) the kink creation energy. When $|\mu_w| \geq \varepsilon_k$, it is to be taken into account the kink-kink interaction which is responsible for keeping the kink density finite.

6.4.3 Kink-Kink Interaction

When the total number of kinks increases (with increasing of T or $|\mu_w|$), we have to take into account the sharing of phase space among kinks due to phase shifts at their collisions. Recall (see, e.g., Sect. 1.2) that the position of a kink of velocity v_1 is shifted by an amount

$$\delta(v_1, v_2) = \operatorname{sgn}(v_1 - v_2) \frac{d}{\gamma_1} \ln \frac{\gamma_1 \gamma_2 (1 - v_1 v_2 / c^2) + 1}{\gamma_1 \gamma_2 (1 - v_1 v_2 / c^2) - 1}, \quad \gamma_j = \left(1 - \frac{v_j^2}{c^2}\right)^{-1/2} \tag{6.98}$$

after collision with a kink of v_2 (the effect of interaction is independent of wether the interacting "particles" are kinks or antikinks). In the nonrelativistic limit ($|v_1|, |v_2| \ll c$) we have $\delta(v_1, v_2) = -\delta(v_2, v_1)$; trajectories of kinks in a two-kink collision are sketched in Fig. 6.7. It is difficult to identify kink positions X_j when they come close to each other. Sasaki [546, 547] proposed to assume that kinks move with constant velocities until they "contact" at a time $t = t_c$ (see Fig. 6.7) as if they were hard rods of length

$$\Delta(v_1, v_2) = |\delta(v_1, v_2)| = 2\, d\, \ln(2c/|v_1 - v_2|). \tag{6.99}$$

Regarding $\Delta(v_1, v_2)$ as an excluded volume ("length"), the possible values of X_j are restricted as $|X_1 - X_2| > \Delta(v_1, v_2)$, and we obtain for the factor Ω in Eq. (6.82) the expression

$$\Omega(X_1, X_2, P_1, P_2) = \Theta\left(|X_1 - X_2| - \Delta(v_1, v_2)\right), \tag{6.100}$$

where $\Theta(x)$ is Heaviside step function. Sasaki [546, 547] have shown that the kink-kink interaction reduces the number of thermally excited kinks,

$$\langle n_{\text{pair}} \rangle = \langle n_{\text{pair}} \rangle_0 \left(1 - 4\bar{\Delta} \langle n_{\text{pair}} \rangle_0\right), \tag{6.101}$$

where $\langle n_{\text{pair}} \rangle_0$ is the pair concentration is absence of kink-kink interaction, Eq. (6.93), and

$$\bar{\Delta} = d\, \ln(4\gamma_0 \varepsilon_k / k_B T) + O(k_B T / \varepsilon_k), \tag{6.102}$$

$\gamma_0 = 1.781...$ being the Euler constant. Analogously, the absolute value of kink contribution to the free energy decreases too,

$$F_s = -k_B T \langle N_{\text{tot}} \rangle_0 \left(1 - 2\bar{\Delta} \langle n_{\text{pair}} \rangle_0\right), \tag{6.103}$$

where we have assumed $\mu_w = 0$. Clearly that higher-order collisions will lead to contributions $\propto \exp(-3\beta\varepsilon_k)$ and smaller.

6.4.4 Discreteness Effects

Although the exact separation of energy of an excited state into contributions of "noninteracting" phonons, breathers and kinks is possible only for

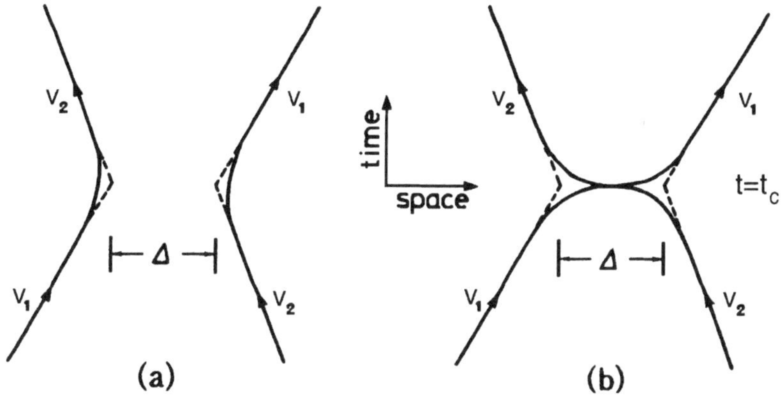

Fig. 6.7. Trajectories of (a) two kinks and (b) kink and antikink in the two-body collision [519].

integrable systems such as the SG one, the physical utility of a configuration-space philosophy remains no less useful and applicable in more general cases. In particular, Trullinger and Sasaki [548], and Willis and Boesch [549] have used the ideal kink-gas approach in investigation of discreteness effects in the FK model.

First, in the discrete FK chain the free motion of the SG kink is replaced by kink motion in the periodic Peierls-Nabarro potential with a height ε_{PN}. Thus, the kink energy E_k depends now not only on its velocity v, but also on the position of its center X,

$$E_k(X, v) \simeq \varepsilon_k^{(FK)} + \frac{1}{2}\varepsilon_{PN}(1 - \cos X) + \frac{1}{2}m^{(FK)}(X)v^2, \tag{6.104}$$

where $\varepsilon_k^{(FK)}$ is the rest energy of FK kink at the bottom of PN well, $m^{(FK)}(X)$ is kink's effective mass, and the nonrelativistic approximation is used ($|v| \ll c$). The deviation of $\varepsilon_k^{(FK)}$ and $m^{(FK)}$ from the SG values $\varepsilon_k^{(SG)}$ and $m^{(SG)}$ were discussed in Sect. 3.1. Note that $m^{(FK)}(X)$ differs very little from $m^{(SG)}$ even at $g \sim 1$ [549].

Second, in consideration of the phonon-kink phase space sharing we have to use the phonon dispersion law for the discrete FK chain, $\omega^2(k) = \omega_0^2 + 2g[1 - \cos(ka_s)]$, for $|k| \leq \pi/a_s$, as well as the kink-phonon phase shift $\delta^{(SG)}(k; v)$ is to be replaced by the phase shift in the discrete FK chain, $\delta^{(FK)}(k; v)$ [166]. Thus, the "self-energy" of a kink is now defined by the integral

$$\Delta F(v) = k_B T \int_{-\pi/a_s}^{+\pi/a_s} \frac{dk}{2\pi} \frac{\partial\, \delta^{(FK)}(k; v)}{\partial k} \ln[\beta\hbar\omega(k)] \tag{6.105}$$

instead of Eq. (6.87). In a result, all previously obtained expressions remain valid except that the factor z, Eq. (6.92), which now should be calculated by the integral

$$z \equiv \langle n_{\text{pair}} \rangle = \frac{1}{a_s} \int_0^{a_s} \frac{dX}{2\pi\hbar}\, m^{(FK)}(X) \int_{-\infty}^{+\infty} dv \exp\left\{-\beta\left[E_k(X,v) + \Delta F(v)\right]\right\}. \tag{6.106}$$

Using Eqs. (6.104) and (6.105), we obtain

$$\langle n_{\text{pair}} \rangle^{(FK)} = \langle n_{\text{pair}} \rangle^{(SG)} I_0\left(\frac{\beta\varepsilon_{PN}}{2}\right) \exp\left[\beta\left(\varepsilon_k^{(SG)} - \varepsilon_k^{(FK)}\right)\right]\left(f^{(FK)}/f^{(SG)}\right),$$

where I_0 is the modified Bessel function, and the last ratio, $f^{(FK)}/f^{(SG)}$, describes the corrections caused by discreteness effects in the kink "self-energy", Eq. (6.105). Computer simulation [549] shows that discreteness effects lead to increasing of $\langle n_{\text{pair}} \rangle$, mainly due to decreasing of the kink creation energy, $\varepsilon_k^{(FK)} < \varepsilon_k^{(SG)}$.

To conclude the discussion of quasiparticle-gas approach, note that the low-temperature results obtained by this method, coincide precisely with the exact ones calculated, for example, by the transfer-integral technique. Moreover, while the transfer-integral method is restricted to equilibrium classical mechanics of one-dimensional systems only, the phenomenological quasiparticle-gas approach admits the consideration of dynamical behavior of the system, two-dimensional versions of the model, quantum effects, etc.

6.5 Statistical Mechanics of the FK Chain

6.5.1 Transfer-Integral Method

The simplest way to obtain formally exact description of an equilibrium state of classical one-dimensional system is to employ the transfer-integral (TI) method originally proposed by Scalapino *et al.* [550]. The idea of the method is to reduce the calculation of partition function to the solution of an eigenvalue integral equation. It was firstly applied to the FK model by Gupta and Sutherland [551] and then further developed in a number of papers (see, e.g., Refs. [159, 496, 520, 552, 553]).

For the standard FK model with Hamiltonian

$$H = \sum_{l=1}^{N} \left[\frac{1}{2}\dot{u}_l^2 + \frac{g}{2}\left(u_{l+1} - u_l\right)^2 + V_{\text{sub}}(u_l)\right], \tag{6.107}$$

where $u_l = x_l - la_s$, the canonical ensemble is defined by the partition function ($m_a = 1$)

$$\mathcal{Z}(T, L, N) = \frac{1}{(2\pi\hbar)^N} \int_{-\infty}^{+\infty} d\dot{u}_1 \int_{-\infty}^{+\infty} d\dot{u}_2 \dots \int_{-\infty}^{+\infty} d\dot{u}_N$$
$$\int_0^{a_s} du_1 \int_{-\infty}^{+\infty} du_2 \dots \int_{-\infty}^{+\infty} du_N e^{-\beta H}. \tag{6.108}$$

Here one of the variables has to be restricted to a finite range, and we confine u_1 to the interval $(0, a_s)$. The "finite-density" FK chain should be subjected to the cyclic boundary conditions,

$$u_{N+l} = u_l + \Delta L, \tag{6.109}$$

where $\Delta L = N_w a_s$. Thus, $\mathcal{Z}$ splits into a configuration part $\mathcal{Z}_x$ and a conjugate momentum part $\mathcal{Z}_{\dot{x}}$,

$$\mathcal{Z} = \mathcal{Z}_{\dot{x}} \mathcal{Z}_x, \quad \mathcal{Z}_{\dot{x}} = \left(\frac{m_a k_B T}{2\pi\hbar^2}\right)^{N/2}. \tag{6.110}$$

Introducing the effective transfer matrix K,

$$K(u, u') = \sqrt{\frac{\beta g}{2\pi}} \exp\left\{-\frac{1}{2}\beta \left[V_{\text{sub}}(u) + V_{\text{sub}}(u') + g(u-u')^2\right]\right\}, \tag{6.111}$$

and incorporating the boundary condition (6.109) with the help of Dirac's δ-function, the configuration partition function can be rewritten in the form

$$\mathcal{Z}_x(T, L, N) = \left(\frac{2\pi}{\beta g}\right)^{N/2} \int_0^{a_s} du_1 \sum_{m=2}^{N+1} \int_{-\infty}^{+\infty} du_m$$
$$\delta(u_{N+1} - u_1 - \Delta L) \sum_{m=1}^{N} K(u_m, u_{m+1}). \tag{6.112}$$

Now let us introduce the "right-hand" eigenfunctions $\psi_\alpha(u)$ and eigenvalues λ_α of K,

$$\int_{-\infty}^{+\infty} K(u, u')\, \psi_\alpha(u')\, du' = \lambda_\alpha \psi_\alpha(u). \tag{6.113}$$

Analogously, we can define the "left-hand" eigenfunctions $\psi_\alpha^*(u)$. Sturm-Liouville theory guaranties that the eigenfunctions $\psi_\alpha(u)$ constitute a complete ortho-normal set in the interval $u \in (-\infty, +\infty)$. Substituting the completeness relation

$$\sum_\alpha \psi_\alpha^*(u) \psi_\alpha(u') = \delta(u - u') \tag{6.114}$$

instead of δ-function in Eq. (6.112) and using repetitively Eq. (6.113), we finally get

$$\mathcal{Z}_x(T, L, N) = \left(\frac{2\pi}{\beta g}\right)^{N/2} \sum_\alpha \lambda_\alpha^N \int_0^{a_s} du_1 \psi_\alpha^*(u_1 + \Delta L) \psi_\alpha(u_1). \tag{6.115}$$

The eigenvalue equation (6.113) is homogeneous Fredholm integral equation of the second kind with symmetric kernel. Owing to periodicity of $V_{\mathrm{sub}}(u)$, the transfer matrix $K(u,u')$ is unchanged if any integer multiple of a_s is added to both u and u'. Therefore, eigenfunctions $\psi_\alpha(u)$ must satisfy Floquet theorem, i.e., they can be written in Bloch form,

$$\psi_\alpha(u) = \eta_\alpha(u)\exp(i\kappa_\alpha u/a_s), \tag{6.116}$$

where $\eta_\alpha(u)$ is periodic in u, and κ_α is a "wave-vector". Analogously to reduced-zone scheme for electrons in crystal, the eigenvalues λ_α occur in bands, which may be labelled by an index p. The eigenvalues within each band are uniquely specified by the wavevector κ, which can be chosen to lie in the first Brillouin zone, $|\kappa| \le \pi$. Further it is convenient to write

$$\lambda_\alpha = \exp[-\beta\,\epsilon_p(\kappa)]. \tag{6.117}$$

Thus, the partition function (6.108) takes the form

$$\mathcal{Z}(T,L,N) = \mathcal{Z}_1(T,N)\,\mathcal{Z}_2(T,L,N) \tag{6.118}$$

with

$$\mathcal{Z}_1(T,N) = \left(\beta\hbar\sqrt{g/m_a}\right)^{-N} \tag{6.119}$$

and

$$\mathcal{Z}_2(T,L,N) = \sum_{p=0}^{\infty}\int_{-\pi}^{+\pi} d\kappa\,\exp\left(-N\beta\epsilon_p(\kappa) - iN_w\kappa\right)C_p(\kappa), \tag{6.120}$$

where

$$C_p(\kappa) = \int_0^{a_s} du\,\eta_{\kappa p}^*(u)\,\eta_{\kappa p}(u). \tag{6.121}$$

The sum over p and integration over κ in Eq. (6.120) may be rewritten as $\sum_p \int d\kappa \ldots = \int d\epsilon\,\rho(\epsilon)\ldots$, where $\rho(\epsilon) = [d\epsilon_p(\kappa)/d\kappa]^{-1}$ is the density of eigenstates. As usual for spectra in one-dimensional systems, $\rho(\epsilon)$ is infinite at edges of each band. In this way it is clear that the eigenstates near the band edges give the main contribution to thermodynamic characteristics. Evidently also that $\mathcal{Z}_2$ is dominated by the contribution coming from the bottom of the lowest band with $p = 0$, and that the irrelevant multiplicative constant $C_p(\kappa)$ may be omitted in the thermodynamic limit $N \to \infty$.

The inconvenient integral over κ can be handled by considering the "temperature-pressure" ensemble $Y(T,\Pi,N)$, which describes the "free-end" FK chain subjected to the external pressure Π. Noting that $L = L_0 + N_w a_s$ and $L_0 = Na_s$, Eq. (6.115) can be rewritten as

$$Y(T,\Pi,N) = \mathcal{Z}_1(T,N)e^{-\beta\Pi L_0}a_s$$
$$\int_{-\pi}^{+\pi} d\kappa\,\exp\left[-N\beta\epsilon_0(\kappa)\right]\int_{-\infty}^{+\infty} dN_w\,\exp\left[-iN_w(\kappa - i\beta\Pi a_s)\right]. \tag{6.122}$$

If we consider $Y(\Pi)$ in the complex Π-plane [551, 552, 554], then for imaginary $\Pi = ib$ (where b is real) the integral over N_w yields δ-function, and we have

$$Y(T,\Pi,N) = \mathcal{Z}_1(T,N)\exp\left[-\beta L_0 \Pi - \beta N \epsilon_0(\kappa^*)\right], \tag{6.123}$$

where $\kappa^* = i\beta a_s \Pi$, and irrelevant multiplicative constants are omitted. Thus, the Gibbs free energy (6.116) takes the form

$$G(T,\Pi,N) = F_1(T,N) + L_0\Pi + N\epsilon_0(\kappa^*), \tag{6.124}$$

where

$$F_1(T,N) = Nk_BT\ln\left(\beta\hbar\sqrt{g/m_a}\right). \tag{6.125}$$

The chain's length L is determined by the pressure Π through the relation

$$L = L_0\left[1 + i\beta\left(\frac{d\epsilon_0(\kappa)}{d\kappa}\right)_{\kappa=\kappa^*}\right], \tag{6.126}$$

and the Helmholtz free energy is equal to

$$F(T,L,N) = G - \Pi L = F_1(T,N) + N\left[\epsilon_0(\kappa) - \kappa\left(\frac{d\epsilon_0(\kappa)}{d\kappa}\right)\right]_{\kappa=\kappa^*},$$

where $\kappa^* = \kappa^*(T,L,N)$ is a solution of Eq. (6.126).

Thus, the free energy of the FK chain without "windows", $N_w = 0$ (or $\Pi = 0$ at $a_{\min} = a_s$), is determined by the energy of bottom of the lowest band, $\epsilon_{\text{bottom}} = \epsilon_0(0)$, while at $N_w \neq 0$ we have to calculate the eigenspectrum dispersion law near the bottom and then to continue it analytically to imaginary values of the wave-vector κ (see Ref. [555]).

It is instructive to investigate two following examples. First, let us consider the harmonic chain of atoms without external potential, $V_{\text{sub}}(x) \equiv 0$. In this case the eigenfunctions of Eq. (6.113) are plane waves, $\eta_\alpha(u) = \text{Const}$, and the eigenvalues are

$$\epsilon_0(\kappa) = \kappa^2/2\beta^2 g a_s^2, \tag{6.127}$$

where we may use now the "extended-zone scheme", $-\infty < \kappa < +\infty$. Substituting Eq. (6.127) into Eq. (6.120) and using $\sum_p \int_{-\pi}^{+\pi} d\kappa \ldots = \int_{-\infty}^{+\infty} d\kappa \ldots$, we get

$$\mathcal{Z}_2(T,L,N) = \left(\frac{2\pi\beta g a_s^2}{N}\right)^{1/2} \exp\left(-\frac{1}{2}\beta g a_s^2 N \widetilde{w}^2\right), \tag{6.128}$$

so that the free energy is equal to

$$F_A(T,L,N) = \frac{1}{2}N g a_s^2 \widetilde{w}^2 + F_1(T,N) + \frac{1}{2}k_BT\ln N + O(N). \tag{6.129}$$

The first term in the right-hand side of Eq. (6.129) describes the energy of compression or expansion of the chain (recall $a_{\min} = a_s$), the second term, F_1, corresponds to acoustic phonon contribution, and the last term ($\propto \ln N$) may be neglected at $N \to \infty$. Equation (6.129) describes the limit $k_B T \to \infty$ of the FK model. At high (but finite) temperatures, $k_B T \gg \varepsilon_s$ (recall $\varepsilon_s = 2$), the substrate potential $V_{\rm sub}(x)$ may be treated as a small perturbation. In particular, Tsuzuki and Sasaki [519] have given the following high-temperature expansion in inverse powers of $\tau = k_B T/\varepsilon_k$:

$$\beta \epsilon_{\rm bottom} \sqrt{g} = \frac{1}{2^3 \tau} - \frac{1}{2^9 \tau^5} - \frac{7}{2^{23} \tau^7} - \frac{29}{2^{32} 3^2 \tau^{11}} + \dots, \tag{6.130}$$

from which one can obtain the free energy for the zero-window case.

Second, eigenvalues $\epsilon_p(\kappa)$ are periodic functions of κ with period 2π and, therefore, they may be expanded in Fourier series. Keeping only the first term in the expansion, we come to the approximation

$$\epsilon_0(\kappa) = \epsilon_c - 2t \cos \kappa, \tag{6.131}$$

where ϵ_c is the center of the lowest band and t is the "overlapping integral of nearest site electron wave-functions" in the tight-binding scheme. Substituting Eq. (6.131) into Eq. (6.120), we obtain

$$\mathcal{Z}_2(T, L, N) = 2\pi \exp(-N\beta\epsilon_c)\, I_{N_w}(2N\beta t), \tag{6.132}$$

where I_N is the modified Bessel function. Currie *et al.* [159] emphasized that it should be clearly distinguished the cases of $N_w = 0$ and $N_w \neq 0$. Namely, at $N_w = 0$ the Bessel function has the asymptotic $I_0(z) \simeq (2\pi z)^{-1/2} e^z$ as $z \to \infty$ [556] leading to the contribution to free energy

$$\begin{aligned} F_2(T, L, N) &= N(\epsilon_c - 2t) + \frac{1}{2} k_B T \ln N + O(N) \\ &= N\epsilon_{\rm bottom} \quad \text{when} \quad N \to \infty. \end{aligned} \tag{6.133}$$

Otherwise, at $N_w \neq 0$, we have to use the asymptotic where both the argument and the order of the Bessel function tend to infinity. However, a simpler way is to use directly Eqs. (6.124) to (6.126) which yield

$$G = F_1 + \varPi L_0 + N\epsilon_c - 2Nt \cosh(\beta a_s \varPi), \tag{6.134}$$

$$N_w = -2N\beta\, t \sinh(\beta a_s \varPi), \tag{6.135}$$

$$F = F_1 + N\epsilon_c - 2Nt \left[\cosh(\beta a_s \widetilde{\varPi}) - (\beta a_s \widetilde{\varPi}) \sinh(\beta a_s \widetilde{\varPi})\right], \tag{6.136}$$

where $\widetilde{\varPi} = \widetilde{\varPi}(T, N_w, N)$ is a solution of Eq. (6.135). Comparing Eq. (6.136) with quasiparticle-gas approximation (6.68) we see that both expressions coincide if we interpret t as

$$t = k_B T (N_{\rm pair}/N), \tag{6.137}$$

and put $\mu_w = a_s \widetilde{\varPi}$ for "window chemical potential", and $F_{\rm ph}^{(0)} = F_1 + N\epsilon_c$ for phonon contribution to the free energy.

In a general case the solution of the transfer-integral equation (6.113) can be found by numerical methods which are described by Schneider and Stoll [520] and Gillan and Holloway [552]. Besides, approximate results can be obtained by reducing the eigenvalue Eq. (6.113) to a pseudo-Schrödinger equation.

6.5.2 The Pseudo-Schrödinger Equation

Using the operator identity [520]

$$\int_{-\infty}^{+\infty} dy\, \exp\left[-b(x-y)^2\right]\, f(y) = \left(\frac{\pi}{b}\right)^{\frac{1}{2}} \exp\left(\frac{1}{4b}\frac{d^2}{dx^2}\right) f(x), \qquad (6.138)$$

the TI equation (6.113) with the kernel (6.111) can be rewritten in the form

$$\exp\left[-\frac{\beta}{2}V_{\text{sub}}(x)\right] \exp\left(\frac{1}{2\beta g}\frac{d^2}{dx^2}\right) \exp\left[-\frac{\beta}{2}V_{\text{sub}}(x)\right] \psi_\alpha(x) = \lambda_\alpha \psi_\alpha(x). \qquad (6.139)$$

Then, combining three exponentials of Eq. (6.139) into a single one, we can write Eq. (6.139) as

$$\exp\left(\frac{1}{2\beta g}\frac{d^2}{dx^2} - \beta V_{\text{sub}}(x) - \beta\widehat{W}\right) \psi_\alpha(x) = \lambda_\alpha \psi_\alpha(x), \qquad (6.140)$$

or, which is equivalent, in the form

$$\left[-\frac{1}{2}\left(\hbar^*\right)^2 \frac{d^2}{dx^2} + V_{\text{sub}}(x) + \widehat{W}\right] \psi_{p\kappa}(x) = \epsilon_p(\kappa)\, \psi_{p\kappa}(x), \qquad (6.141)$$

where we introduced "Planck's constant"

$$\hbar^* = \frac{k_B T}{\sqrt{g/m_a}}. \qquad (6.142)$$

The operator $\widehat{W}$ can be determined by expanding exponents of Eqs. (6.139) and (6.140) into Taylor series, that yields [548, 557]

$$\widehat{W} = \frac{1}{24g}\left\{-\left[V'_{\text{sub}}(x)\right]^2 + \ldots + \frac{2}{\beta^2 g}\left[-\frac{1}{4}V''''_{\text{sub}}(x) + V''_{\text{sub}}(x)\frac{d^2}{dx^2} + \ldots\right]\right\}. \qquad (6.143)$$

Thus, the operator $\widehat{W}$ in Eq. (6.141) may be neglected, if two conditions are satisfied simultaneously: (a) $g \gg 1$, or $d \gg a_s$ (d is the kink width), and (b) $\beta^2 g \geq 1$, or $k_B T < \varepsilon_k$; notice that this leads to $\hbar^* \leq 1$. In this case Eq. (6.141) looks like a one-particle "Schrödinger equation" for an "electron" with a unit mass and "Planck's constant" $\hbar^*$. (The first-order lattice-discreteness corrections to the "Schrödinger equation" were studied by Trullinger and

Sasaki [548]). For the periodic function $V_{\rm sub}(x)$, Eq. (6.141) is known as the Hill equation, and for sinusoidal substrate potential it is called the Mathieu equation which is extensively documented (see, e.g., Ref. [556]).

At very low temperatures, $k_B T \ll \varepsilon_k$, when $\hbar^* \ll 1$, the tight-binding (narrow-band) approximation (6.131) is adequate [533]. The eigenfunctions $\nu_\kappa(x)$ are almost κ independent and strongly localized in potential wells of $V_{\rm sub}(x)$, so that ϵ_c is accurately approximated by the lowest harmonic oscillator level $\epsilon_c = \frac{1}{2}\hbar^*\omega_0$ (recall $\epsilon_c \ll \varepsilon_s = 2$), and the sum $F_1 + N\epsilon_c$ yields the phonon contribution $F_{\rm ph}^{(0)}$ which coincides with Eq. (6.19) in the limit $g \gg 1$. The presence of other potential wells gives rise to exponentially small "tunnelling broadening" of the level into the band with width $4t$. To evaluate the "tunnelling rate" t, we can use the WKB approximation which, however, leads to too crude expression for the pre-exponential factor. The improved WKB approximation [558]–[561] yields the expression

$$t = (\hbar^*\omega_0)(2\beta\varepsilon_k/\pi)^{1/2} e^{-\beta\varepsilon_k}. \tag{6.144}$$

If can be seen that Eqs. (6.136), (6.137), (6.142), (6.144) give the contribution to the free energy which coincides precisely with the kink contribution of Eqs. (6.68), (6.93) obtained by the quasiparticle-gas approach.

For a general case of $V_{\rm sub}(x)$, solutions of the Hill equation (6.141) can be found by standard methods of band theory of solids, for example, by Green function technique [562]–[564]. The most extensively studied is the SG case of zero window number, where the lowest eigenvalue has been obtained in the double-series expansion in $\tau = k_B T/\varepsilon_k$ and $\nu = \exp(-1/\tau)$ by Tsuzuki and Sasaki [519],

$$\frac{\epsilon_{\rm bottom}}{\hbar^*\omega_0} = \frac{1}{2}\left[1 - \frac{1}{2}\tau + O(\tau^2)\right] - \left(\frac{8}{\pi\tau}\right)^{1/2} \nu \left[1 - \frac{7}{8}\tau + O(\tau^2)\right]$$
$$+ \left(\frac{8}{\pi\tau}\right)\nu^2 \left[\ln\frac{4\gamma_0}{\tau} - \frac{5}{4}\tau\left(\ln\frac{4\gamma_0}{\tau} + 1\right) + O(\tau^2)\right] + O(\nu^3), \tag{6.145}$$

where $\ln\gamma_0 = 0.5772$ is the Euler constant. Comparing this expression with the results of quasiparticle-gas approximation, we see that the first term in Eq. (6.145) corresponds to phonon contribution with anharmonic corrections, the second, to kink contribution, and the last term describes kink-kink interactions.

Equation (6.145) allows us to calculate the specific heat of the SG model at low temperatures [519],

$$(c_N - k_B) = (k_B/\sqrt{g})\, h(\tau), \tag{6.146}$$

where the scaling function $h(\tau)$ depends on the dimensionless temperature only,

$$h(\tau) = -\frac{1}{\tau^2}\frac{d^2}{d(\tau^{-1})^2}\left(\frac{\epsilon_{\rm bottom}}{\hbar^*\omega_0}\right). \tag{6.147}$$

The function $h(\tau)$ is shown in Fig. 6.8. As the temperature increases from $T = 0$, the specific heat increases because more and more kinks become excited. This situation holds until the interaction between kinks suppresses the increase of kink density and then the specific heat reaches its maximum. At higher temperatures the kink-gas picture becomes inadequate, and the deviation of $h(\tau)$ from zero may be interpreted as a small contribution to the specific heat of the harmonic chain of atoms, caused by the presence of substrate potential $V_{\rm sub}(x)$. Thus, the maximum in the dependence $c_N(T)$ may be considered as the Schottky-type anomaly due to thermally excited kinks [520, 562].

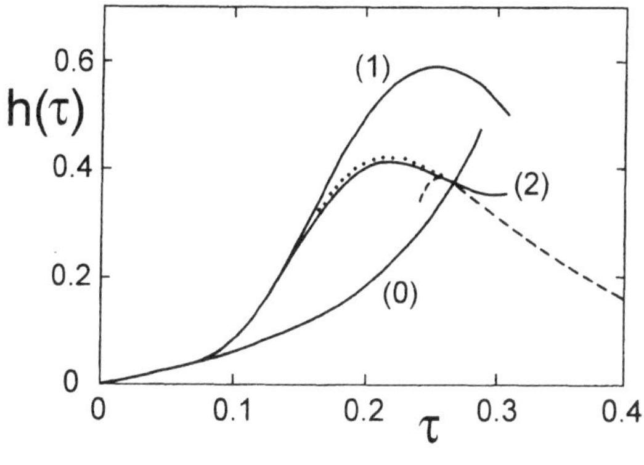

Fig. 6.8. Function $h(\tau)$ for the specific heat of the SG model. Shown are numerical (dotted) and low-temperature approximate (solid) results obtained without tunnelling corrections (0), with the corrections of order ν (1) and ν^2 (2). Dashed curve is the high-temperature approximation [519].

6.5.3 Susceptibility

The susceptibility χ of the FK chain deserves a more detailed discussion due to its important role in mass transport. Besides, χ characterizes an "order" of the system, because $\chi = 0$ for the ordered state (at $T = 0$ for one-dimensional models) and $\chi = 1$ in the totally disordered state (e.g., at $T \to \infty$ or $g \to 0$). According to the definition (6.15), the susceptibility χ describes the mean square fluctuations of the relative displacement $x_{l'+l} - x_{l'}$ from its average value,

$$\Lambda_l \equiv \frac{1}{N} \sum_{l'} \langle [(x_{l'+l} - x_{l'}) - \langle x_{l'+l} - x_{l'} \rangle]^2 \rangle = \chi a_A^2 |l|. \tag{6.148}$$

For the "fixed-density" FK chain of length $L = Na_A$, Eq. (6.148) is to be modified [565]

$$\Lambda_l = \chi a_A^2 l (1 - l/N), \quad l = 1, 2, \ldots, N. \tag{6.149}$$

For the "free-end" FK model with the harmonic $V_{\rm int}(x)$, the window number $\widetilde{w}$ as a function of the pressure Π increases by 1 when Π decreases by $g a_s$ [552],

$$\widetilde{w}(\Pi - kga_s) = \widetilde{w}(\Pi) + k; \quad k = 0, \pm 1, \dots \tag{6.150}$$

Therefore, from Eq. (6.18) it follows that the function $\phi = \chi/n^2$ is the periodic function of Π with period ga_s, and also ϕ is the periodic function of $\widetilde{w}$ with period 1. Due to kink-antikink symmetry the function ϕ is also even on $\widetilde{w}$, $\phi(-\widetilde{w}) = \phi(\widetilde{w})$.

Equations (6.18) and (6.124) connect χ with the lowest-zone eigenvalue of the TI equation by the relationship

$$\chi = \beta(na_s)^2 \frac{d^2\epsilon_0(\kappa)}{d\kappa^2}\bigg|_{\kappa=i\beta\Pi a_s}. \tag{6.151}$$

It is easy to examine that the free-electron approximation (6.127) for $\epsilon_0(\kappa)$ gives the value $\chi = k_BT/ga_A^2$, Eq. (6.31), while the tight-binding approximation (6.131) leads to

$$\chi = \beta(na_s)^2 2t \cosh(\beta a_s \Pi), \tag{6.152}$$

which coincides with Eq. (6.76) if Eq. (6.137) is taken into account. In a general case χ can be calculated in limiting cases only. First, at weak-coupling limit, $g \ll 1$, the FK model reduces approximately to the Ising-like model, Eqs. (6.45) to (6.48). Owing to strong decreasing of the coupling constant $I(l)$ with l, it is reasonable to leave the repulsion of nearest-neighboring vacancies only, i.e. to put $I(l) = 0$ for all $|l| \geq 1$. Thus, we come to the classical Ising model (or Langmuir lattice-gas model) with Hamiltonian

$$H = \sum_{j=1}^{M} (I_0 h_j h_{j+1} - \mu_h h_j), \tag{6.153}$$

where $I_0 \simeq ga_s^2$, h_j means the "number of holes" at the j-th well of the substrate potential ($h_j = 0$ or 1), and the "hole chemical potential" μ_h determines the average value $\langle h_j \rangle \equiv M^{-1} \sum_{j=1}^{M} h_j$, which is coupled with the concentration θ by the relation

$$\langle h_j \rangle = 1 - \theta. \tag{6.154}$$

Assuming periodic boundary conditions, the grand canonical function

$$\Xi(T, L, \mu_h) = \sum_{\{h_j\}} \exp(-\beta H) \tag{6.155}$$

can now be calculated exactly by the TI technique (see, e.g., Ref. [566]). The result is:

$$J(T, L, \mu_h) \equiv -k_BT \ln \Xi = \frac{1}{4} M(I_0 - 2\mu_h) - k_BTM \ln \lambda_+, \tag{6.156}$$

$$\langle N_w \rangle \equiv M\langle h_j \rangle = \frac{1}{2} M[1 + \mathrm{sgn}(\Phi)\cos(2\Phi)], \tag{6.157}$$

$$Q_{ij} \equiv \langle h_i h_j \rangle - \langle h_i \rangle \langle h_j \rangle = \frac{1}{4}\sin^2(2\Phi)\left(\frac{\lambda_-}{\lambda_+}\right)^{|i-j|}, \tag{6.158}$$

where

$$\lambda_\pm = e^{-\beta I_0/4}\cosh\left[\frac{1}{2}\beta(I_0 - \mu_h)\right] \pm \left\{e^{\beta I_0/2} + e^{-\beta I_0/2}\sinh^2\left[\frac{1}{2}\beta(I_0-\mu_h)\right]\right\}^{1/2}, \tag{6.159}$$

$$\tan(2\Phi) = \exp\left(\frac{1}{2}\beta I_0\right) \Big/ \sinh\left[\frac{1}{2}\beta(\mu_h - I_0)\right], \quad |\Phi| < \pi/4. \tag{6.160}$$

Equations (6.157) and (6.160) couple μ_h with the concentration θ through the relation

$$\sin^2 2\Phi = 4\theta(1-\theta). \tag{6.161}$$

Now, using Eqs. (6.158), (6.159), (6.161) and recalling that $N = \sum_{j=1}^{M}(1-h_j)$, we can calculate χ directly from its definition (6.16):

$$\chi = (1-\theta)\left\{1 - 4\theta(1-\theta)\left[1 - \exp(-\beta I_0)\right]\right\}^{1/2}, \quad \theta \le 1. \tag{6.162}$$

Note that the expression (6.162) assumes $k_B T \gg I(1) \simeq g^2 a_s^2$ because we neglected the repulsion of next-nearest neighboring vacancies.

At low temperatures, the "ideal kink-gas" ideology can be used [565]. In particular, we consider the fixed-density FK chain characterized by a C-phase with $\theta = s/q$ at $T = 0$. Recall that on the background of this structure we can define superkinks (shortly s-kinks) and s-antikinks, which are characterized by rest energies ε_{sk} and $\varepsilon_{\bar{sk}}$. At $T \neq 0$ the equilibrium state of the chain contains N_{sk} s-kinks and $N_{\bar{sk}}$ s-antikinks. Because kinks and antikinks can be created by kink-antikink pairs only, we have

$$\langle N_{sk} \rangle = \langle N_{\bar{sk}} \rangle \propto \exp(-\beta E_{sk}), \tag{6.163}$$

where $E_{sk} = \frac{1}{2}(\varepsilon_{sk} + \varepsilon_{\bar{sk}})$. (Unfortunately, the preexponental factor in Eq. (6.163) is known analytically for the trivial GS ($s = 1$) in the case of $g \gg 1$ only.) Now, let us take a section of length $\widetilde{L}$ in the FK chain and assume that $\widetilde{L} \ll L$ but $\widetilde{L}$ is large enough so that it contains $\widetilde{N}$ atoms, $\widetilde{N}_{sk}$ s-kinks and $\widetilde{N}_{\bar{sk}}$ s-antikinks with $\widetilde{N}, \widetilde{N}_{sk}, \widetilde{N}_{\bar{sk}} \gg 1$. Then, suppose that s-kinks and s-antikinks are non-interacting and, therefore, their probability distributions are Poisonian,

$$\langle \widetilde{N}_{sk}^2 \rangle - \langle \widetilde{N}_{sk} \rangle^2 = \langle \widetilde{N}_{sk} \rangle, \tag{6.164}$$

$$\langle \widetilde{N}_{\bar{sk}}^2 \rangle - \langle \widetilde{N}_{\bar{sk}} \rangle^2 = \langle \widetilde{N}_{\bar{sk}} \rangle, \tag{6.165}$$

$$\langle \widetilde{N}_{sk} \widetilde{N}_{\bar{sk}} \rangle = \langle \widetilde{N}_{sk} \rangle \langle \widetilde{N}_{\bar{sk}} \rangle. \tag{6.166}$$

Recalling that each s-kink contains $1/q$ excess atoms, and each s-antikink, the same number of vacancies, we can write the number of atoms on the length $\widetilde{L}$ as

$$\widetilde{N} = \widetilde{N}_0 + \frac{1}{q}(\widetilde{N}_{sk} - \widetilde{N}_{\bar{s}\bar{k}}), \tag{6.167}$$

where $\widetilde{N}_0 = n\widetilde{L}$ is the number of atoms at $T = 0$. Using Eqs. (6.164) to (6.167), from the definition (6.15) we obtain the result

$$\chi = \frac{\langle N_{s,\mathrm{tot}} \rangle}{q^2 \langle N \rangle}, \tag{6.168}$$

where $\langle N_{s,\mathrm{tot}} \rangle = \langle N_{sk} \rangle + \langle N_{\bar{s}\bar{k}} \rangle$, and tilde has been omitted. The obtained expression (6.168) is valid only for small enough temperatures,

$$k_B T \ll E_{sk}, \tag{6.169}$$

when the concentration of thermally excited s-kinks is small so that they can be considered as noninteracting quasiparticles. For the trivial C-phase ($s = 1$) Eq. (6.168) coincides with previously obtained Eq. (6.76) if we mention that $q = 1 + \widetilde{w}$ in this case.

When $k_B T \geq E_{sk}$, the concentration of thermally excited s-kink–s-antikink pairs is so large that they "melt" the reference C-structure. However, if the C-phase was nontrivial, $s \neq 1$, the kink-gas approach remains useful up to temperatures $kT \leq E_k$. For example, let us consider the case of $s = q - 1$ with $q \gg 1$, where the $T = 0$ GS can be treated as a regular structure of antikinks (with the number N_w) defined on the background of the reference structure with $\theta = 1$. At $g \gg 1$ these antikinks are characterized by $d = a_s\sqrt{g}$, $E_k = 8\sqrt{g}$, $\varepsilon_{PN} = (16\pi^4/3)g\exp(-\pi^2\sqrt{g})$, and $v_{\mathrm{int}}(R) = 32\sqrt{g}\exp(-R/d)$, where $R \equiv qa_s \gg d$. Considering the antikinks as quasiparticles, we come to an effective FK model with the elastic constant

$$g^* \equiv a_s^2 v_{\mathrm{int}}''(R)/2\pi^2\varepsilon_{PN} = 3\pi^{-6}g^{-3/2}\exp(\pi^2\sqrt{g} - q/\sqrt{g}). \tag{6.170}$$

Then, we can calculate the s-kink energy E_k^* as $E_k^* = 8\sqrt{g^*}$ if $g^* \gg 1$, or as $E_k^* = 2\pi^2 g^*$ if $g^* \ll 1$. Returning to old units, we have to put $E_{sk} = E_k^*\varepsilon_{PN}/2$, thus obtaining

$$\frac{E_{sk}}{E_k} \simeq \begin{cases} 2g^{-1}\exp(-q/\sqrt{g}) & \text{if } q > \pi^2 g, \\ (8\pi/\sqrt{3})g^{-1/4}\exp(-\pi^2\sqrt{g}/2) & \text{if } \sqrt{g} \ll q < \pi^2 g. \end{cases} \tag{6.171}$$

From Eq. (6.171) we see that if $q \gg \sqrt{g}$, there exists a wide temperature interval,

$$E_{sk} \leq k_B T \ll E_k, \tag{6.172}$$

where the trivial ($\theta = 1$) structure of atoms still exists, while the primary C-phase (when $\theta = (q-1)/q$) is disordered because the lattice of antikinks is already melted and, therefore, the antikinks can be considered as a gas of noninteracting quasiparticles. Now, repeating the calculations as we have done above, we obtain for the interval (6.172) the expression

$$\chi = \frac{\langle N_{\rm tot} \rangle}{\langle N \rangle}, \qquad (6.173)$$

where $\langle N_{\rm tot} \rangle = \left[\langle 2N_{\rm pair} \rangle^2 + \langle N_\omega \rangle^2 \right]^{1/2}$, see Eq. (6.73). From Eq. (6.173) it follows that on the left-hand side of the interval (6.172) the function $\chi(T)$ has a plateau $\chi \simeq \widetilde{w}$, while on the right-hand side, $\chi(T)$ increases exponentially due to thermal excitation of kink-antikink pairs which join those already present from the melted kink lattice (see Fig. 6.9).).

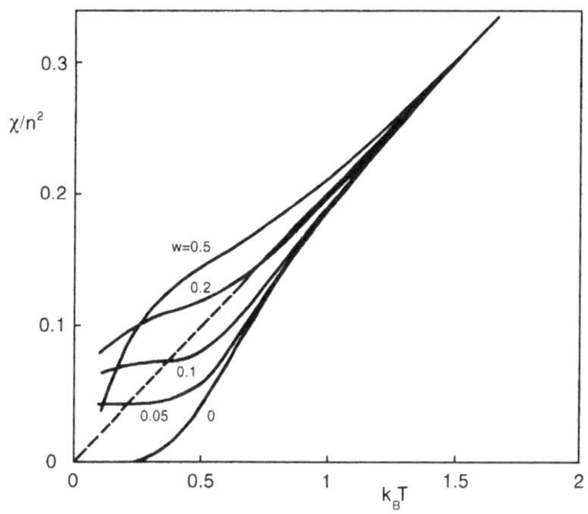

Fig. 6.9. Susceptibility χ as a function of temperature T at different winding numbers $\widetilde{w}$ for the standard FK model with $\Gamma \equiv g a_s^2 = 5$ [552].

With the help of renormalization arguments (Sect. 5.2.1) we may describe similarly the temperature dependence of χ for more complicated structures with $s \neq 1, q-1$. Considering the FK system as a hierarchy of consequently melted superkink lattices, we can divide the temperature interval $0 \leq k_B T < E_k$ into subintervals where the more complex superkink structure is already melted, but the more simple structure still exists and admits definition of corresponding s-kinks and s-antikinks which are approximately noninteracting in this temperature subinterval. So, $\chi(T)$ has to have a series of plateaus at low-temperature sides of these subintervals, and to change exponentially between the plateaus.

With further increase of temperature, $k_B T > E_k$, the function $\chi(T)$ tends to the asymptotic $\chi(T) = k_B T / g a_A^2$. However, in the limit $T/g \to \infty$ this asymptotic has nonphysical behavior $\chi \to \infty$ instead of the correct one $\chi \to 1$. The reason lies in the approximate nature of the standard FK model. Namely, at $k_B T \geq g a_A^2$ the mutual displacements of atoms are not small,

$|x_{l+1} - x_l - a_{\min}|$ could exceed the value a_A, and the harmonic approximation for the interatomic potential $V_{\rm int}(x)$ becomes inadequate. To clarify the high-temperature behavior of $\chi(T)$, let us investigate the exponential interaction

$$V_{\rm int}(x) = gb^2(e^{-r} - 1 + r), \quad r \equiv (x - a_{\min})/b, \tag{6.174}$$

where the inverse length b^{-1} plays the role of anharmonicity parameter,

$$V_{\rm int}(x) \simeq \frac{1}{2} g(x - a_{\min})^2 \left[1 - (x - a_{\min})/3b\right], \quad |x - a_{\min}| \ll b. \tag{6.175}$$

Neglecting the substrate potential at $k_B T \geq E_k \gg \varepsilon_s$ (recall $g \gg 1$), the model with interaction (6.174) corresponds to the Toda system which has the exact solution [270]. The Gibbs free energy of the "free-end" Toda chain has the form

$$G(T, \Pi, N) = -k_B T N \ln\left[(m_a k_B T/2\pi\hbar^2)^{1/2} y \exp(-\beta \Pi a_{\min})\right], \tag{6.176}$$

where

$$y(T, \Pi) = b e^B \Gamma(B + \beta b \Pi)/B^{B+\beta b\Pi}, \tag{6.177}$$

$$B = \beta g b^2, \tag{6.178}$$

and $\Gamma(x)$ is gamma-function. The chain's length is determined by the equations

$$L(T, \Pi, N) = \left(\frac{\partial G}{\partial \Pi}\right)_{T,N} = N(a_{\min} + \delta a), \tag{6.179}$$

$$\delta a(T, \Pi) = b\left[\ln B - \Psi(B + \beta b \Pi)\right], \tag{6.180}$$

where $\Psi(x) = \frac{d}{dx} \ln \Gamma(x)$. Using the definition (6.18), we obtain from Eqs. (6.179) and (6.180) that the susceptibility is equal to

$$\chi = -k_B T \frac{N}{L^2} \frac{\partial L}{\partial \Pi}\bigg|_{\Pi=0} = \frac{\Psi'(B)}{(\ln B - \Psi(B) + a_{\min}/b)^2}. \tag{6.181}$$

From Eq. (6.181) it follows that Eq. (6.31) holds at intermediate temperatures only,

$$E_k \leq k_B T \ll g b^2, \tag{6.182}$$

when $B \gg 1$, $\Psi(B) \simeq \ln B - \frac{1}{2B} - \frac{1}{12B^2}$, and

$$\chi \simeq \frac{k_B T}{g(a_{\min} + \delta a)^2}\left(1 + \frac{1}{2\beta g b^2}\right). \tag{6.183}$$

At high temperatures,

$$k_B T \gg g b^2, \tag{6.184}$$

when $B \ll 1$ and $\Psi(B) \simeq -\gamma_0 - \frac{1}{B}$, γ_0 being the Euler constant, the correct limit $\chi \to 1$ is restored:

$$\chi \simeq \left[1 + (a_{\min}/b)B + B \ln B\right]^{-2} \simeq 1 - 2\beta g b a_{\min}. \tag{6.185}$$

In the limit $b \to 0$ the potential (6.174) reduces to the hard-core potential $V_{\rm int}(x) = V_0 \Theta(a_{\min} - x)$ with $V_0 \to \infty$. In this case Eqs. (6.179), (6.180), (6.181) yield the expression [567]: $\chi = (1 - n a_{\min})^2$, where $n = N/L$.

6.5.4 Hierarchy of Superkink Lattices

According to renormalization arguments of Sect. 5.2.1, the complex commensurate C-structure of the fixed-density FK chain with concentration $\theta = s/q$ (s being the number of atoms per one elementary cell) may be considered as a lattice of kinks constructed on the background of a more simple commensurate structure C_0 called the reference structure. Moreover, the renormalization procedure may be repeated by a required number of times, leading to a set of kink (superkink) lattices. Because of one-dimensionality of the FK model, all these lattices (superstructures) are disordered at any nonzero temperature T. However, at low temperatures this disordering is "small", and the system may be considered as a regular lattice of interacting quasiparticles (atoms, kinks, superkinks, etc.) interpreted as the reference structure characterized by an effective concentration $\widetilde{\theta} = \widetilde{s}/\widetilde{q}$ and an effective external potential with period $\widetilde{a}_s$, with a small number of approximately noninteracting topological excitations (kinks, superkinks, super-superkinks, etc.) which are characterized by a width $\widetilde{d}$ and rest energy $\widetilde{E_k}$. Note that these kinks may be both residual as well as thermally excited. As was shown in Sect. 5.2.1, the more complex (higher-order) is the reference structure, the lower is the energy $\widetilde{E_k}$ of its excitations and, therefore, the larger is to be the number of excitations which are thermally excited at a given temperature T. Intuitively it is clear that when the concentration of excitations becomes too large, they destroy the order of the reference structure, and the given reference structure disappears ("melts"). Here we derive the equation for this "melting" temperature (not be confused with critical temperature T_c which is always zero for one-dimensional system with short-range interactions).

As was shown in Sect. 6.5.3, mutual positions of two quasiparticles separated by a distance x, $x = la_A$, fluctuate with an amplitude

$$\delta(x) = \sqrt{\Lambda_l} = \sqrt{\chi a_A |x|}. \tag{6.186}$$

In order that one can construct topological excitations for the reference structure with $\widetilde{\theta} = \widetilde{s}/\widetilde{q}$, the mutual fluctuations must be small, $\delta(x) \leq \widetilde{a}_s/\widetilde{s}$, at least on a distance of kink width, $\widetilde{x} \geq \widetilde{d}$. In this way we come to the upper limit when the given reference structure may be considered as a regular one; the corresponding equation is

$$\delta(\widetilde{d}) = \widetilde{a}_s/\widetilde{s}, \quad \text{or} \quad \chi(T) a_A \widetilde{s}^2 \widetilde{d} = \widetilde{a}_s^2. \tag{6.187}$$

The solution of Eq. (6.187) defines the temperature $\widetilde{T}_{\text{melt}}$, above which the given reference structure is destroyed due to thermal fluctuations. Taking the expression (6.168) for $\chi(T)$, we see that at $T = \widetilde{T}_{\text{melt}}$ the average distance $\widetilde{R}$ between topological excitations is equal to $\widetilde{d}$, so that at $T > \widetilde{T}_{\text{melt}}$ the total concentration of superkinks becomes so large that they begin to overlap.

For the trivial reference structure ($\widetilde{s} = 1$) the susceptibility $\chi(T)$ is known analytically, and Eq. (6.187) reduces to $(2\beta\widetilde{E}_k/\pi)^{\frac{1}{2}} \exp(-\beta\widetilde{E}_k) = 1$, leading to the result

$$k_B \widetilde{T}_{\text{melt}} \simeq \widetilde{E}_k. \tag{6.188}$$

It is reasonable to expect that the expression (6.188) will remain valid for all superstructures with zero number of residual superkinks (although this statement has not been proved yet for a general case). If the number of residual (nonthermal) superkinks is nonzero, it follows $k_B \widetilde{T}_{\text{melt}} < \widetilde{E}_k$, because residual s-kinks supplement thermally excited s-kink–s-antikink pairs and, therefore, the criterion $\widetilde{R} = \widetilde{d}$ is achieved at a lower temperature.

While in the "fixed-density" FK chain kinks can be created by kink-antikink pairs only, the "free-end" FK chain admits the creation of a single superkink or super-antikink (with the energy ε_{sk} or $\varepsilon_{s\bar{k}}$ respectively) through a free end of the chain. Therefore, Eq. (6.188) should be modified in the "free-end" case to the form

$$k_B \widetilde{T}_{\text{melt}} \simeq \min\{\varepsilon_{sk}, \varepsilon_{s\bar{k}}\}. \tag{6.189}$$

The given reference structure is mostly stable at the center of a step of the $T = 0$ Devil's staircase, while at the edges of steps (i.e., at the points corresponded to the FvdM transitions) the melting temperature reduces to zero.

Thus, the whole temperature interval $0 < k_B T < E_k$ can be divided into m subintervals ($m = 1$ for $s = 1$, $m = 2$ for $s = q - 1$, etc.), where superkink-gas ideology with corresponding reference structures and superkinks, is adequate.

6.5.5 Equal-Time Correlation Functions

Now let us consider spacial correlation functions which provide an information on structure of the FK chain. A two-particle correlation function is defined by

$$Q_{AB}(l - l') = \langle \delta A_l^* \, \delta B_{l'} \rangle, \tag{6.190}$$

where $\delta A_l = A_l - \langle A_l \rangle$, and A_l, B_l are functions of atomic displacement u_l (recall $u_l = x_l - l a_s$) and, may be, its time derivatives $\dot{u}_l, \ddot{u}_l$, etc. Owing to spacial homogeneity of the chain it is convenient to make the Fourier transform

$$\widetilde{A}(\kappa) = \frac{1}{\sqrt{N}} \sum_l e^{-i\kappa l} \delta A_l, \tag{6.191}$$

and to define the functions

$$\widetilde{Q}_{AB}(\kappa) = \langle \widetilde{A}(-\kappa) \, \widetilde{B}(\kappa) \rangle = \sum_l e^{-i\kappa l} Q_{AB}(l). \tag{6.192}$$

The function $\widetilde{Q}_{AA}(\kappa)$ is known as the form-factor of A.

The simplest form has the form-factor of velocity $\dot{u}_l$:

$$Q_{\dot{u}\dot{u}}(l) = (k_B T/m_a)\delta_{l0}; \quad \widetilde{Q}_{\dot{u}\dot{u}}(\kappa) = k_B T/m_a. \tag{6.193}$$

Some of correlation functions are coupled owing to the motion equation

$$\ddot{u}_l + \omega_0^2 s_l - g(u_{l+1} + u_{l-1} - 2u_l) = 0, \tag{6.194}$$

the Fourier transform of which takes the form

$$\ddot{\widetilde{u}}(\kappa) + \omega_0^2 \widetilde{s}(\kappa) + 2g\,(1 - \cos\kappa)\,\widetilde{u}(\kappa) = 0, \tag{6.195}$$

where we have used the notations $s_l \equiv \sin u_l$ and $c_l \equiv \cos x_l$. Indeed, using the equation

$$\langle \dot{u}_l u_{l'} \rangle = \langle \dot{u}_l \rangle \langle u_{l'} \rangle = 0 \tag{6.196}$$

and differentiating it over time, we obtain the relationship

$$\omega_0^2 \widetilde{Q}_{su}(\kappa) + 2g\,(1 - \cos\kappa)\,\widetilde{Q}_{uu}(\kappa) = k_B T. \tag{6.197}$$

Analogously, starting from the identity $\langle \ddot{u}\dot{u} \rangle = 0$ and using Eq. (6.195), we come to the relationships

$$\widetilde{Q}_{\ddot{u}\ddot{u}}(\kappa) = k_B T \left[\omega_0^2 \langle c_l \rangle + 2g(1 - \cos\kappa)\right], \tag{6.198}$$

$$\omega_0^4 \widetilde{Q}_{ss}(\kappa) = 4g^2(1 - \cos\kappa)^2 \widetilde{Q}_{uu}(\kappa) + k_B T \left[\omega_0^2 \langle c_l \rangle - 2g(1 - \cos\kappa)\right]. \tag{6.199}$$

There exist two types of correlation functions depending on whether the A_l and B_l are periodic or nonperiodic functions of u_l. Nonperiodic functions of u_l emerge in analysis of mass (charge, spin, etc.) diffusion and transport described by FK-like models. Below we will consider as an example of such correlation functions the displacement form-factor Q_{uu}. On the other hand, correlation functions of periodic functions $A(u)$ (such as those for $s \equiv \sin u$, $c \equiv \cos u$, $s_{\frac{1}{2}} \equiv \sin\frac{1}{2}u$, $c_{\frac{1}{2}} \equiv \cos\frac{1}{2}u$) are of greatest interest in scattering experiments. It is important that these experiments yield the possibility to isolate kink contributions by choosing scattering vectors (and frequencies in the case of dynamical correlation function) appropriately.

Equal-time correlation functions for the classical FK model can be calculated exactly with the help of TI method. In order to consider the case of $N_\omega \neq 0$, let us take the "temperature-pressure" ensemble in which the correlation function is expressed as

$$Q_{AB}(l_1 - l_2) = Y^{-1} e^{-\beta \Pi L_0} \int \frac{d(\Delta L)}{(2\pi\hbar)^{-N}} \int d\dot{u}_1 \ldots du_N \int du_{N+1}$$
$$\delta A(u_{l_1})\, \delta B(u_{l_2})\, \delta(u_{N+1} - u_1 - \Delta L) \exp[-\beta H(\dot{u}_1 \ldots u_{N+1})] e^{-\beta \Pi \Delta L}.$$

Repeating the calculations just as we did in Sect. 6.5.1, we come to the relation [496, 520]

$$Q_{AB}(l_1 - l_2) = (\mathcal{Z}_1/Y)e^{-\beta\Pi L_0}\sum_{\alpha\beta\gamma}\lambda_\beta^{l_1-1}A_{\beta\gamma}\lambda_\gamma^{l_2-l_1}B_{\gamma\alpha}\lambda_\alpha^{N-l_2+1}$$
$$\int_{-\infty}^{+\infty} d(\Delta L)\, e^{-\beta\Pi\Delta L}C_{\alpha\beta}(\Delta L), \quad (6.200)$$

where

$$A_{\beta\gamma} = \int_{-\infty}^{+\infty} du\, \psi_\beta^*(u)\, \delta A(u)\, \psi_\gamma(u) \equiv \langle\beta|\delta A|\gamma\rangle, \quad (6.201)$$

$$B_{\gamma\alpha} = \int_{-\infty}^{+\infty} du\, \psi_\gamma^*(u)\, \delta B(u)\, \psi_\alpha(u), \quad (6.202)$$

$$C_{\alpha\beta}(\Delta L) = \int_0^{a_s} du\, \psi_\alpha^*(u + \Delta L)\, \psi_\beta(u). \quad (6.203)$$

In the thermodynamic limit, $N \to \infty$, these expressions are simplified to become

$$Q_{AB}(l_1 - l_2) = \sum_{\kappa',p}\langle\kappa^*, 0|\delta A|\kappa', p\rangle\langle\kappa', p|\delta B|\kappa^*, 0\rangle$$
$$\exp\{-\beta|l_1 - l_2|\,[\epsilon_p(\kappa') - \epsilon_0(\kappa^*)]\}, \quad (6.204)$$

where $\kappa^* = i\beta a_s\Pi$. The Fourier transform (6.192) of Eq. (6.204) is:

$$\widetilde{Q}_{AB}(\kappa) = \sum_{\kappa',p}\langle\ldots\rangle\langle\ldots\rangle\frac{\sinh\beta[\epsilon_p(\kappa') - \epsilon_0(\kappa^*)]}{\cosh\beta[\epsilon_p(\kappa') - \epsilon_0(\kappa^*)] - \cos\kappa}. \quad (6.205)$$

Clearly that at large distances the correlation function $Q_{AB}(l)$ is dominated by the TI state with the smallest eigenvalue for which the corresponding matrix elements are nonvanishing.

Putting formally $\delta B \equiv 1$ in Eq. (6.204), we obtain that one-particle average of a function $A(u_l)$ can be expressed as

$$\langle A(u_l)\rangle = \langle\kappa^*, 0|A|\kappa^*, 0\rangle = \int_{-\infty}^{+\infty} du\, A(u)P_0(u), \quad (6.206)$$

where the function

$$P_0(u) = \eta_{\kappa^*,0}^*(u)\eta_{\kappa^*,0}(u) \quad (6.207)$$

serves as one-atomic distribution for the equilibrium state with the given temperature and pressure.

Displacement Form-factor

With the help of band theory of solids it can be shown [520] that at $|\kappa| \ll 1$

$$\sum_{\kappa,p}|\langle\kappa, p|u|\kappa^*, 0\rangle|^2 \propto a_s^2\frac{\partial^2}{\partial\kappa^2}\bigg|_{\kappa=\kappa^*}. \quad (6.208)$$

The substitution (6.208) into (6.205) yields the asymptotic $\widetilde{Q}_{uu}(\kappa) \propto \kappa^{-2}$ at $\kappa \to 0$. A more elegant method to derive the low-momentum asymptotic was proposed by Gillan [565], who used the coupling of $\widetilde{Q}_{uu}(\kappa)$ with the mean square fluctuations of relative displacement Λ_l,

$$\Lambda_l = \frac{2}{N} \sum_{\kappa} [1 - \cos(\kappa l)] \, \widetilde{Q}_{uu}(\kappa). \tag{6.209}$$

In the limit $N \to \infty$ we can put $\sum_{\kappa} \to N \int_{-\pi}^{+\pi} d\kappa/2\pi$. Then, substituting $\Lambda_l = \chi a_A^2 |l|$, we obtain the equation

$$\chi a_A^2 |l| = \frac{2}{\pi} \int_0^{\pi} d\kappa \; [1 - \cos(\kappa l)] \; \widetilde{Q}_{uu}(\kappa). \tag{6.210}$$

It is easy to see that for $l \to \infty$ Eq. (6.210) may be satisfied if and only if

$$\lim_{\kappa \to 0} \widetilde{Q}_{uu}(\kappa) = \chi a_A^2 / \kappa^2. \tag{6.211}$$

Thus, the low-momentum behavior of the displacement form-factor is determined by the susceptibility χ which was calculated above. Analogously it can be shown that $Q_{uu}(l) \simeq -\frac{1}{2}\chi a_A^2 |l|$ for $l \to \pm\infty$.

Form-Factor of Periodic Functions

For periodic functions $A(u)$, the correlation functions may exhibit two different types of behavior, depending on the relative periodicity of $A(u)$ and $V_{\mathrm{sub}}(u)$ [533]. If $A(u)$ has periodicity $a_s = 2\pi$ (or a_s/r, r being an integer), a kink does not change the asymptotic value of A, and the only characteristic length is the kink width d. The TI matrix element $\langle \kappa, p | \delta A | \kappa^*, 0 \rangle$ couples the TI ground state, $|\kappa^*, 0\rangle$, only to the $\kappa = \kappa^*$ states of other bands with $p = 0, \pm 1, \pm 2, \ldots$. The self-coupling of the ground state, $\langle \kappa^*, 0 | \delta A | \kappa^*, 0 \rangle$, simply provides a "Bragg peak". If this matrix element vanishes, the form-factor Q_{AA} is determined by the excited states $p = \pm 1, \pm 2, \ldots$ which are separated from the lowest state by band gaps $\simeq p\hbar^* \omega_0$. According to Eq. (6.204), the contribution from the p-th band decays exponentially, $\propto \exp(-|x|/\xi_p)$, with the correlation length

$$\xi_p = a_A / \beta [\epsilon_p(\kappa^*) - \epsilon_0(\kappa^*)] \simeq a_A / \beta p \hbar^* \omega_0 = d / p\theta. \tag{6.212}$$

Thus, the correlation length remains finite as $T \to 0$. In this case the behavior of Q_{AA} can be explained in terms of phonons only, because kink corrections are exponentially small. In particular, for the SG case (i.e., at $g \gg 1$ and $\widetilde{w} = 0$) at low temperatures, $\tau \equiv k_B T / \varepsilon_k \ll 1$, and at large distances, $|x| \gg d$, correlation functions are found to be [519]

$$Q_{cc}(x) \simeq 8\tau^2 \exp(-2|x|/d), \tag{6.213}$$

$$Q_{ss}(x) \simeq 4\tau \exp(-|x|/d), \tag{6.214}$$

$$Q_{s_{1/2}s_{1/2}}(x) \simeq \tau \exp(-|x|/d). \tag{6.215}$$

The functions Q_{cc} and Q_{ss} were also calculated numerically by TI technique by Schneider and Stoll [520].

On the other hand, if $A(u)$ is periodic with the period ra_s ($r \geq 2$, integer), the correlation function Q_{AA} is kink-sensitive. For this case the TI matrix elements in Eq. (6.204) are nonzero when both κ and κ^* lie in the same lowest band and differ by the wave-vector $2\pi/r$. Clearly that these states will dominate the sum in Eq. (6.204) for large x. Because they are separated by "tunnel-splitting", the correlation length ξ will directly reflect the kink rest energy, and ξ grows exponentially when $T \to 0$. Indeed, using the tight-binding approximation, $\epsilon_0(\kappa) = \epsilon_c - 2t\cos\kappa$, which is appropriate at low temperatures, we find

$$\epsilon_0(\kappa) - \epsilon_0(\kappa^*) = 2t\left\{[1-\cos(2\pi/r)]\cos\kappa^* + \sin(2\pi/r)\sin\kappa^*\right\}. \tag{6.216}$$

Adopting the ideal kink-gas approximation, we can put $t = \langle n_{\text{pair}}\rangle/\beta\langle n\rangle$ and $\kappa^* = \beta a_s \Pi = \beta\mu_w$. Then from (6.204) we obtain for $x = la_a \to \infty$

$$Q_{AA}(x) \simeq |\langle\kappa^* - (2\pi/r)|\delta A|\kappa^*\rangle|^2 \exp(-|x|/\xi)\cos(x/\xi_{\text{osc}}), \tag{6.217}$$

where

$$\xi^{-1} = \langle n_{\text{tot}}\rangle[1-\cos(2\pi/r)], \tag{6.218}$$

$$\xi_{\text{osc}}^{-1} = \langle n_w\rangle \sin(2\pi/r). \tag{6.219}$$

A simple phenomenological interpretation of this behavior is the following (see Refs. [517, 533]). Each kink has a property that its passage changes the variable δA from, e.g., $\pm\delta A_0$ to $\mp\delta A_0$ for $r = 2$ case (for example, for $A(u) = \cos\frac{1}{2}u$, $\delta A_0 = 1$). Assuming a gas of independent kinks with a Poisson separation distribution, and introducing $N_k(x)$ (the number of kinks on the length x), it can be obtained that

$$Q_{AA}(x) \simeq \delta A_0^2\left\langle(-1)^{N_k(x)}\right\rangle \simeq \delta A_0^2 \exp(-2\langle n_{\text{tot}}\rangle|x|), \tag{6.220}$$

that agrees with Eqs. (6.217) – (6.219) for $\widetilde{w} = 0$. Thus, the kink creation energy appears directly in correlation length, $\xi \propto \exp(\beta\varepsilon_k)$. Sasaki [563] have shown that kink-kink interactions do not affect the correlation length ($\xi^{-1} = 4\langle n_{\text{pair}}\rangle$ at $\widetilde{w} = 0$), but do the magnitude of correlations. For example,

$$Q_{c_{1/2}c_{1/2}}(x) \simeq \left[1 + 8(\bar{\Delta}\langle n_{\text{pair}}\rangle)^2\right]\exp(-|x|/\xi), \tag{6.221}$$

where $\bar{\Delta}$ is given by Eq. (6.102).

6.5.6 Generalized FK Models

Finally, we discuss two generalizations of the standard FK model, the first one consists in replacing the pure sinusoidal substrate potential by a periodic potential of a more general shape, and the second generalization deals with nonharmonic interatomic potentials. Due to one-dimensionality of the FK model, any its generalized version has a formally exact TI solution which may be used to calibrate approximate results. For understanding of physical behavior of the model, however, the quasiparticle-gas approach is more appropriate.

Nonsinusoidal Substrate Potential

As was described above, in the quasiparticle-gas approach the problem reduced to calculation of the phonon-kink phase shifts, then to calculation the "renormalized" kink energy (6.89), and, finally, to determination of the concentration $\langle n_{\rm pair} \rangle$ of thermally excited kinks. However, using general scattering theory (namely, properties of Jost functions), De Leonardis and Trullinger [568] have shown that explicit knowledge of kink shape and phase shifts is not needed because all of the quantities entering into thermodynamic functions, can be expressed directly through simple integrals over the function $V_{\rm sub}(x)$. Not going into details, we give here the final results only.

If x_1 and x_2 are the coordinates of two adjacent absolute minima in $V_{\rm sub}(x)$, $V_{\rm sub}(x_1) = V_{\rm sub}(x_2) = 0$, and $V_{\rm sub}(x)$ achieves its maximum at $x = x_m$, so that the barrier is symmetric about this point, then at $g \gg 1$ and low temperatures ($k_B T \ll \varepsilon_k$)

$$\langle n_{\rm pair} \rangle = S \left(\frac{2}{\pi} \right)^{1/2} \frac{1}{d} (8\beta\sqrt{g})^{1/2} \exp(-\beta\varepsilon_k), \tag{6.222}$$

where $d = a_s\sqrt{g}$, and the temperature-independent "shape" factor S is given by

$$S = [(x_2 - x_1)/a_s]\, \omega_0 \phi, \tag{6.223}$$

where

$$\phi = \frac{1}{2} \exp\left\{ \int_{x_m}^{x_2} dx \left(\frac{1}{\sqrt{2V_{\rm sub}(x)}} - \frac{1}{x_2 - x} \right) \right\}, \tag{6.224}$$

and $\omega_0^2 = V''_{\rm sub}(x_2)/m_a$. If the shape of kink $u_{\rm kink}(x)$ approaches x_2 according to the law $|u_{\rm kink}(x) - x_2| \simeq u_0 \exp(-x/d)$ as $x \to \infty$, then the factor ϕ is equal to $\phi = u_0/(x_2 - x_1)$. Entered into Eq. (6.222) the kink rest energy ε_k for a general shape of substrate potential is determined by the equation (see Sect. 3.3)

$$\varepsilon_k = \sqrt{g} \int_{x_1}^{x_2} dx\, \sqrt{2V_{\rm sub}(x)}. \tag{6.225}$$

For the SG case, $S = 1$, and we recover to Eq. (6.93).

Internal (shape) Modes. First let us consider the case when the potential $V_{\rm sub}(x)$ has no additional absolute minima between x_1 and $x_2 = x_1 + a_s$ (but may have relative minima). In this case the topological structure of kinks is the same as those for the standard FK model. However, in some cases a kink may have additionally one more internal degrees of freedom, the so-called "shape modes" (see Sect. 3.3.2). If this is the case, the total change of phonon states due to kink very presence, $\Delta\rho(k)$, given by Eq. (6.86), satisfies now Levinson's theorem (or the Friedel sum rule),

$$\int_{-\infty}^{+\infty} dk\ \Delta\rho(k) = -(1+s), \tag{6.226}$$

where s is the number of shape models. The Helmholtz free energy is still given by Eq. (6.68), $F = F_{\rm ph}^{(0)} - k_B T\langle N_{\rm tot}\rangle$ for the zero window-number case, but the full energy E can be rewritten now in the form

$$E = k_B T\left[N - (s+1)\langle N_{\rm tot}\rangle\right] + \left(\varepsilon_k + \frac{1}{2}k_B T + s k_B T\right)\langle N_{\rm tot}\rangle. \tag{6.227}$$

This means that now each kink takes away $(s+1)$ phonon modes from the phonon spectra; one mode is transferred into the Goldstone mode (or the PN mode in the discrete FK chain), and other s modes are modified into the shape modes.

One of interesting examples which exhibits the existence of exactly one shape mode, is the double-barrier substrate potential where the function $V_{\rm sub}(x)$ has one local minimum at the middle distance between x_1 and $x_2 = x_1 + a_s$. In this case the usual FK kink (2π-kink) can be represented as a coupled pair of two subkinks (π-kinks), and the shape mode corresponds to relative oscillations of positions of these kinks. When the depth of local minimum is close to zero (i.e., to the absolute minimum of $V_{\rm sub}(x)$), the mean distance R between the subkinks is large, and an effective potential which couples π-kinks into 2π -kink, is small. In a result, the amplitude of mutual vibrations of π -kinks may be large enough even at low temperatures, and, therefore, we have to take into account an anharmonicity of the shape mode [207, 569, 570]. As may be understood from a general point of view, the anharmonic corrections will increase the number of thermally excited kinks (comparing with the "harmonic" approximation) with increasing of R or T.

Polikink Systems. Another situation emerges when the substrate potential has one or more additional absolute minima between x_1 and $x_2 = x_1 + a_s$, say, at the point $x = x_0$. In this case the usual 2π-kink of the FK model splits ("dissociates") into two (or more) independent subkinks, one "links" the minima at x_1 and x_0, and the second, the minima at x_0 and x_2 (see Sect. 3.3.4). Depending on the particular shape of $V_{\rm sub}(x)$, the subkinks may be called in one case as "large" and "small" kinks, or in another case as "left" and "right" kinks, etc. Because different types of kinks may overcome distinct barriers, they may be characterized by different rest energies, e.g., by $\varepsilon_{\rm SK}$ and $\varepsilon_{\rm LK}$ with $\varepsilon_{\rm SK} < \varepsilon_{\rm LK}$.

The new feature that enters the kink-gas phenomenology of polikink system is a topological constraint on the manner in which the various types of kinks can be sequentially arranged in the system. For example, the small kink can be followed by a small antikink or a large kink and not by the second small kink or large antikink. Taking this fact into account, it can be shown [185, 571, 572] that the free energy is again determined by Eq. (6.68) where we have to substitute

$$\langle N_{\rm tot} \rangle = \sum_j \langle N_{\rm tot}^j \rangle, \tag{6.228}$$

the index j distinguishing different types of kinks. The expression for the full energy E, Eq. (6.227), naturally generalizes to the form

$$E = k_B T \left[N - \sum_j (s_j + 1) \langle N_{\rm tot}^j \rangle \right] + \sum_j \left(\varepsilon_k^j + \frac{1}{2} k_B T + s_j k_B T \right) \langle N_{\rm tot}^j \rangle. \tag{6.229}$$

Note, however, that if the curvatures of the substrate potential at different absolute minima (e.g., at $x = x_1$ and $x = x_0$) are different, there exist two (or more) different phonon branches which yield a "weighted" contribution to the phonon free energy $F_{\rm ph}^{(0)}$ [572].

The kink concentrations can be calculated by formulas (6.222) to (6.225) with appropriate limits of integration. Because the equilibrium concentrations are determined by their Boltzmann factors, $\langle n_{\rm pair}^j \rangle \propto \exp(-\beta \varepsilon_k^j)$, it is clear that thermodynamic characteristics (free energy, entropy, etc.) are dominated by the "small" kinks (i.e., by the kinks with lower rest energy).

On the other hand, when atomic displacements are involved, they are accomplished by the creation of equal number of small and large kinks. For example, the displacement from 0 to a_s requires one small and one large kink. Thus evidently that the number of large kinks will be a restricted factor in, e.g., the displacement form-factor or mass transport along the chain. For example, in expression for the susceptibility χ we have to replace the value $\langle N_{\rm tot} \rangle$ by the geometrical average [185],

$$\frac{1}{\langle N_{\rm tot}^{\rm av} \rangle} = \frac{1}{\langle N_{\rm SK,tot} \rangle} + \frac{1}{\langle N_{\rm LK,tot} \rangle}. \tag{6.230}$$

The correlation functions were studied in Ref. [571]. It was shown that, if we introduce the "kink-detecting" function $A_j(u_l)$ in such a way that the product of its values at two lattice sites, namely $A_j(u_l) A_j(u_{l'})$, is sensitive to j-th kinks between the l-th and l'-th sites, but totally insensitive to any other types of kinks in the same region of the chain, then $Q_{A_j A_j}(x) \propto \exp(-|x|/\xi_j)$, where the correlation length ξ_j is given by $\xi_j^{-1} = 2 \langle n_{\rm tot}^j \rangle$ in the zero window-number case.

Anharmonic Interatomic Interaction

When the interaction between atoms is anharmonic but remains short-ranged and strictly convex, we have no new phenomena in system behavior. Statistical mechanics of the chain can be described in the same manner as for harmonic interactions, and the only difference is that the kink rest energy ε_k has to be replaced by the arithmetic average of kink and antikink rest energies which are different for anharmonic potentials.

An interesting behavior exhibits the FK chain with long-range interaction when, for example, all atoms in the chain interact according to Kac-Baker law

$$V_{\text{int}}(x_l - x_{l'}) = g\frac{(1-r)}{2r} r^{|l-l'|}(u_l - u_{l'})^2, \tag{6.231}$$

where $r = \exp(-\alpha)$, and α^{-1} essentially defines the interaction range. The magnitude of the potential (6.231) decreases exponentially with increasing of the interatomic distance $x_l - x_{l'} \approx (l-l')a_s$. The SG model and the double SG model with the interaction (6.231) have been studied by Croitoru [290] (see also Ref. [573]). Using the technique of Sarker and Krumhansl [288], the model can be reduced to an effective model which contains only nearest-neighbor interactions (see Sect. 3.5.3). The kink rest energy ε_k for the model (6.231) tends to the SG value if $r \to 0$, while at $r \to 1$ we have $\varepsilon_k \to \infty$. Therefore, at any $r < 1$ the model behavior is similar to that of the standard FK model. However, in the Van-der-Waals limit, $r \to 1$, the kink creation energy becomes infinite and their concentration tends to zero. So, there is no excitations which may destroy the order in the chain at $r \to 1$. The investigation [290] shows that, analogously to one-dimensional Ising model and ϕ^4 model with long-range interactions, the $r = 1$ FK model also has the continuous order-disorder phase transition which occurs at a *finite* temperature T_c ($k_B T_c = 2\pi^2 g$ if $g \to 0$ and $k_B T_c = 4g$ when $g \to \infty$).

7 Thermalized Dynamics

In the previous chapters we discussed the properties of an isolated FK chain as a Hamiltonian system. However, for modelling realistic physical objects we should recall that the effective FK model describes only a part of the whole physical system while the remainder, with its own degrees of freedom and other types of excitations, can be taken into account indirectly through an effective thermostat field. In this chapter we discuss different concepts and methods for studying the thermalized dynamics of the FK chain.

7.1 Basic Concepts and Formalism

In order to understand the requirements for the study of the thermalized dynamics, we consider an example of adsorption systems where an external potential $V_{\mathrm{sub}}(x)$ is created by the atoms which form a crystal surface; similarly, in the case of a crowdion, the periodic potential is produced by the nearest rows of metal atoms, etc. Because the positions of the substrate atoms are not fixed, in a general case their motion has to lead to the following effects:

(1) The substrate atoms should feel and response to the configurations of atoms in the FK chain. Such a "polaronic effect" will lead to increasing of the amplitude of the external potential $V_{\mathrm{sub}}(x)$;

(2) At a nonzero temperature the substrate atoms are vibrating with some amplitudes. This leads to a dependence of the amplitude of the substrate potential ε_s on the temperature (the Debye-Waller effect);

(3) There always exists an energy exchange between the FK chain and the substrate. This exchange leads to an additional damping, i.e. to the external friction η_{ext}. For example, in the case of adatomic chains the external friction is the main damping mechanism, $\eta_{\mathrm{ext}} \gg \eta_{\mathrm{int}}$.

Thus, the substrate in fact plays a twofold role: first, it produces an effective external potential on the atoms of the primary chain modelled by the FK model, and second, it creates a mechanism for the energy exchange between the atoms of the FK chain and the substrate degrees of freedom. In other words, the FK chain is usually a nonconservative system, and the substrate plays a role of a thermostat at a certain temperature T.

To describe the energy exchange mechanisms mentioned above, we have to introduce an *effective friction force* acting on the atoms of the FK chain.

In a general case, the friction force $F_l^{(\mathrm{fr})}$ can be taken in the following form,

$$F_l^{(\mathrm{fr})}(t) = -m_a \sum_{l'} \int_0^\infty d\tau \mathcal{N}_{ll'}(\tau)\, \dot{x}_{l'}(t-\tau), \tag{7.1}$$

where the dot stands for a derivative in time, and the response function

$$\mathcal{N}_{ll'}(\tau) \equiv \mathcal{N}(\dot{x}_l, \ddot{x}_l, \ldots, \dot{x}_{l'}, \ddot{x}_{l'}, \ldots; \tau) \tag{7.2}$$

is, in a general case, nonlocal, nonlinear, and non-Markovian. The calculation of this function as well as the solution of the corresponding motion equations is an extremely complicated problem. To simplify it, one usually assumes that the operator $\mathcal{N}$ is Markovian and local, and, moreover, it does not depend on the position of a given atom relative to the substrate potential, i.e., $\mathcal{N}_{ll'}(\tau) = 2\eta\delta_{ll'}\delta(\tau)$, so that the force (7.1) reduces to the standard viscous friction,

$$F_l^{(\mathrm{fr})}(t) = -m_a \eta \dot{x}_l(t). \tag{7.3}$$

However, we should realize that the parameter η in Eq. (7.3) is in fact an "effective" friction coefficient which should be calculated by means of a suitable averaging of the response function over all atomic trajectories. Thus, in real systems the value of η should be estimated rather than exactly calculated.

Besides the energy flux from the FK chain to the substrate caused by the damping force (7.3), the backward flux of the energy to the chain, which is usually treated as $T \neq 0$ also exists. This effect may be modelled by introducing a random force (noise) $\delta F_l(t)$ with zero mean value, $\langle \delta F_l(t) \rangle = 0$, acting on the l-th atom of the FK chain from the substrate. Amplitude of the external noise is determined by the fluctuation-dissipation theorem which guaranties the existence of thermal equilibrium, i.e. the equal shearing of the kinetic energy between all degrees of freedom, $\frac{1}{2} m_a \langle \dot{x}_l^2 \rangle = \frac{1}{2} k_B T$, at the equilibrium state. Depending on the physical model under consideration, the external noise may be additive and/or multiplicative. Thus, the problem reduces to the analysis of dynamics of the FK chain governed by a system of Langevin-type equations.

Below, we consider the dynamics of the FK chain when it is disturbed out of the $T \neq 0$ equilibrium state. Dynamic properties of physical systems are widely studied experimentally with spectroscopic techniques, where the dynamic correlation function $Q(k, \omega)$ is measured directly. The function $Q(k, \omega)$ contains much more detailed information than the equal-time correlation function, because the dynamic correlation function has peaks which correspond to phonon, breather, and kink excitations. For example, kinks' contribution into $Q(k, \omega)$ leads to the so-called "central peak", the shape of which yields the information on the mechanism of the kinks' motion. Besides, a response of the system on a small external perturbation describes transport properties of the FK chain, which is quite important in practical applications of the model. Recall that the deviation of the FK system from the exactly integrable SG model can often be treated as a small perturbation in studying equilibrium properties of the model. For dynamic characteristics, however, this deviation plays the main role leading to energy exchange between different degrees of freedom and, therefore, determining the rate of dynamical processes.

The dynamic properties of the FK model are very complicated and may be described usually in the framework of some approximate method. Note, however, that often the only way to check the results of different approximate approaches is to compare their predictions with molecular dynamics results.

7.1.1 Basic Formulas

Let us outline first the basic technique used in calculation of dynamic correlation functions. For a set of dynamical variables

$$\{A_\alpha\}, \quad A_\alpha(t) \equiv A_\alpha(\{z(t)\}), \tag{7.4}$$

where $\{z\} \equiv (z_1, z_2, \ldots, z_N)$ denotes a $2N$-dimensional vector in the phase space of all atoms, $z_l \equiv (u_l, p_l)$, $p_l = m_a \dot{u}_l$, $l = 1, \ldots, N$, the correlation functions are defined as

$$Q_{\alpha\beta}(t) = \langle \delta A_\alpha^*(t)\, \delta A_\beta(0) \rangle, \quad \delta A = A - \langle A \rangle. \tag{7.5}$$

Here $\langle \ldots \rangle$ stands for the averaging over the equilibrium state,

$$\langle A(z) \rangle = \int d\Gamma\; A(z)\, W_{\text{eq}}(z), \tag{7.6}$$

where $d\Gamma = L^{-N} dz_1 \ldots dz_N$, $dz_l = du_l dp_l$, and W_{eq} is the equilibrium Maxwell-Boltzmann distribution function,

$$W_{\text{eq}}(z) = R_{\text{eq}}(u_1, \ldots, u_N) \prod_{l=1}^{N} W_M(p_l), \tag{7.7}$$

$$R_{\text{eq}}(u_1, \ldots, u_N) = Q_N^{-1} \exp[-\beta U(u_1, \ldots, u_N)], \tag{7.8}$$

$$Q_N = L^{-N} \int du_1 \ldots du_N\; e^{-\beta U}, \tag{7.9}$$

$$W_M(p) = (2\pi m_a k_B T)^{-1/2} \exp(-\beta p^2/2m_a), \tag{7.10}$$

$\beta \equiv (k_B T)^{-1}$, and $U = U_{\text{sub}} + U_{\text{int}}$ is the total potential energy of the system.

Assuming that the system is in contact with a thermostat, the time evolution of the system is described by a set of Langevin equations,

$$m_a \ddot{u}_l + m_a \eta \dot{u}_l - F_l = \delta F_l(t), \tag{7.11}$$

where η describes the rate of energy exchange between the system and the thermal bath, $F_l = -\partial U/\partial u_l = F_l^{(\text{sub})} + F_l^{(\text{int})}$, and the fluctuation force δF_l satisfies the equations $\langle \delta F_l(t) \rangle = 0$ and $\langle \delta F_l(t)\ \delta F_{l'}(t') \rangle = 2\eta m_a k_B T \delta_{ll'} \delta(t - t')$. Equation (7.11) is equivalent to the Fokker-Planck-Kramers (FPK) equation for the distribution function $W(z, t|z', 0)$ which describes the probability of the transition from a configuration z' at time $t = 0$ to a configuration z at time t (see, e.g., Ref. [574]),

$$\dot{W}(z;t|z';0) = \mathcal{L}(z)\ W(z;t|z',0), \tag{7.12}$$

where the FPK evolution operator is defined by

$$\mathcal{L}(z) \equiv \sum_l \left[-\frac{p_l}{m_a}\frac{\partial}{\partial u_l} - F_l \frac{\partial}{\partial p_l} + \eta \frac{\partial}{\partial p_l}\left(p_l + m_a k_B T \frac{\partial}{\partial p_l} \right) \right]. \tag{7.13}$$

In the overdamped case, $\eta \gg \omega_{\max}$, where $\omega_{\max}$ is the maximum frequency of vibrations in the system, we may neglect the inertia term in Eq. (7.11), and the FPK equation reduces to the Smoluchowsky equation [575]–[577]

$$\dot{R}(\{u\};t|\{u'\};0) = \mathcal{L}_{\text{sm}}(\{u\})\ R(\{u\};t|\{u'\};0), \tag{7.14}$$

$$\mathcal{L}_{\text{sm}}(\{u\}) \equiv D_f \sum_l \left(-\beta \frac{\partial}{\partial u_l} F_l + \frac{\partial^2}{\partial u_l^2} \right) = D_f \sum_l \frac{\partial}{\partial x_l} e^{-\beta U} \frac{\partial}{\partial x_l} e^{+\beta U}, \tag{7.15}$$

where $D_f \equiv k_B T / m_a \eta$. The functions W and R are coupled by the relationship

$$W(z;t|\ldots) = \left[\prod_{l=1}^{N} W_M(p_l) \right] \left[1 + \sum_l \frac{p_l}{m_a \eta} \left(\beta F_l - \frac{\partial}{\partial x_l} \right) \right] R(\{u\};t|\ldots). \tag{7.16}$$

Defining the Green function W_G as a solution of the FPK equation with the initial condition $W_G(z;t|z';0)|_{t=0} = \delta(z - z')$, the correlation function (7.5) can be represented as

$$Q_{\alpha\beta}(t) = \int d\Gamma\ d\Gamma'\ \delta A_\alpha^*(z)\ W_G(z;t|z';0)\ \delta A_\beta(z')\ W_{\text{eq}}(z'). \tag{7.17}$$

It is convenient to introduce also the reduced distributions $W^{(s)}$ according to the equation

$$W^{(s)}(z_1, \ldots, z_s; t|\ldots) = L^{-(N-s)} \int dz_{s+1} \ldots dz_N\ W(z_1, \ldots, z_N; t|\ldots), \tag{7.18}$$

where $1 \le s \le N$ and $W^{(N)} \equiv W$. The reduced distributions are normalized by

$$L^{-s} \int dz_1 \ldots dz_s\ W^{(s)} = 1. \tag{7.19}$$

Define also the correlation distributions $\Phi^{(s)}$ as

$$W^{(2)}(z_1, z_2; t) = W^{(1)}(z_1; t)\, W^{(1)}(z_2; t) + \Phi^{(2)}(z_1, z_2; t), \tag{7.20}$$

$$W^{(3)}(123) = W^{(1)}(1)\, W^{(1)}(2)\, W^{(1)}(3) + W^{(1)}(1)\, \Phi^{(2)}(23) \\ + W^{(1)}(2)\, \Phi^{(2)}(31) + W^{(1)}(3)\, \Phi^{(2)}(12) + \Phi^{(3)}(123), \tag{7.21}$$

etc. The motion equations for the functions $W^{(s)}$ form a set of N coupled equations known as the BBGKY hierarchy, an approximate solution of which is usually obtained by truncation of the chain at some step r by putting $\Phi^{(s)} = 0$ for all $s \geq r$. When the dynamical variable A depends on the coordinate of a single atom only, $A = A(z_l)$, the correlation function is expressed through the first two reduced distributions,

$$Q_{AA}(l, l'; t) = \frac{\delta_{ll'}}{L^2} \int dz^2\, \delta A^*(z_l)\, W_G^{(1)}(z_l; t|z_l'; 0)\, \delta A(z_l')\, W_{\text{eq}}^{(1)}(z_l')$$
$$+ \frac{(1 - \delta_{ll'})}{L^4} \int dz^4\, \delta A^*(z_l)\, \Phi_G^{(2)}(z_l, z_{l'}; t|z_l', z_{l'}'; 0)\, \delta A(z_{l'}')\, W_{\text{eq}}^{(2)}(z_l', z_{l'}'),$$

where $dz^2 = dz_l dz_l'$ and $dz^4 = dz_l dz_{l'} dz_l' dz_{l'}'$.

7.1.2 Mori Technique

In investigation of transport properties it is necessary to calculate the function

$$\bar{Q}(\bar{\omega}) = \int_{-\infty}^{+\infty} \frac{d\omega}{2\pi i} \frac{\widehat{Q}(\omega)}{\omega - \bar{\omega}}, \quad \text{Im}\, \bar{\omega} \neq 0, \tag{7.22}$$

where $\widehat{Q}(\omega)$ is the Fourier transform of the correlation function over time,

$$\widehat{Q}(\omega) = \int_{-\infty}^{+\infty} dt\, e^{i\omega t} Q(t). \tag{7.23}$$

The function $\bar{Q}(\bar{\omega})$ is analytical on the whole complex plane $\bar{\omega}$ with a cut along the real axis; in the upper half-plane $\bar{Q}(\bar{\omega})$ coincides with the usual Laplace transform of $Q(t)$,

$$\bar{Q}(\bar{\omega}) = \int_0^{+\infty} dt\, e^{i\bar{\omega} t} Q(t), \quad \text{Im}\, \bar{\omega} > 0. \tag{7.24}$$

The Green function W_G may formally be introduced by the relation

$$W_G(z; t|z'; 0) = \exp\left[\mathcal{L}(z)\, t\right] \delta(z - z'). \tag{7.25}$$

Using Eqs. (7.17) and (7.25), the correlation function (7.5) may be rewritten as

$$Q_{\alpha\beta}(t) = \langle \delta A_\alpha^*(0)\, \delta A_\beta(-t) \rangle$$
$$= \int d\Gamma d\Gamma'\, \delta A_\alpha^*(z) \left[e^{-\mathcal{L}(z')t}\, \delta(z - z') \right] \delta A_\beta(z')\, W_{\text{eq}}(z). \qquad (7.26)$$

Taking the integral over $d\Gamma'$ by parts (see, e.g., Ref. [574]), we obtain

$$Q_{\alpha\beta}(t) = \int d\Gamma d\Gamma'\, \delta A_\alpha^*(z)\, \delta(z - z') \left[e^{-\mathcal{L}^+(z')t} \delta A_\beta(z') \right] W_{\text{eq}}(z), \qquad (7.27)$$

where the adjoint operator $\mathcal{L}^+$ corresponds to the backward evolution of W according to the Kolmogorov equation

$$\dot{W}(z;t|z';0) = \mathcal{L}^+(z')\, W(z;t|z';0). \qquad (7.28)$$

For the FPK equation, $\mathcal{L}^+$ has the form

$$\mathcal{L}^+(z) \equiv \sum_l \left[\frac{p_l}{m_a} \frac{\partial}{\partial u_l} + F_l \frac{\partial}{\partial p_l} + \eta \left(-p_l \frac{\partial}{\partial p_l} + m_a k_B T \frac{\partial^2}{\partial p_l^2} \right) \right], \qquad (7.29)$$

and in the overdamped (Smoluchowsky) case, it takes the form

$$\mathcal{L}^+_{\text{sm}}(\{u\}) \equiv D_f \sum_l \left(\beta F_l \frac{\partial}{\partial u_l} + \frac{\partial^2}{\partial u_l^2} \right) = D_f \sum_l e^{+\beta U} \frac{\partial}{\partial x_l} e^{-\beta U} \frac{\partial}{\partial x_l}. \qquad (7.30)$$

Now, defining a scalar product of two functions $A(z)$ and $B(z)$ as

$$\langle A|B \rangle \equiv \langle A^* B \rangle \equiv \int d\Gamma\, A^*(z)\, B(z)\, W_{\text{eq}}(z), \qquad (7.31)$$

the correlation function (7.27) may be rewritten as

$$Q_{\alpha\beta}(t) = \langle \delta A_\alpha | e^{-\mathcal{L}^+ t} \delta A_\beta \rangle. \qquad (7.32)$$

The Laplace transform of Eq. (7.32) is

$$\bar{Q}_{\alpha\beta}(\bar{\omega}) = i \langle \delta A_\alpha | (\bar{\omega} + i\mathcal{L}^+)^{-1} | \delta A_\beta \rangle. \qquad (7.33)$$

The function (7.33) can now be calculated with the help of the Zwanzig-Mori technique [578]–[581] (see also Ref. [582], part V). Namely, introducing the projection operator

$$\mathcal{P} = \sum_{\alpha,\beta} |\delta A_\alpha\rangle [Q^{-1}(0)]_{\alpha\beta} \langle \delta A_\beta|, \qquad (7.34)$$

where

$$Q_{\alpha\beta}(0) \equiv \langle \delta A_\alpha | \delta A_\beta \rangle = Q_{\alpha\beta}(t)|_{t=0}, \qquad (7.35)$$

the matrix (7.33) may be presented in the following form,

$$\bar{Q}(\bar{\omega}) = i \{ \bar{\omega} - [\Omega + M(\bar{\omega})] \}^{-1} Q(0). \qquad (7.36)$$

Here $\Omega = -i \langle \mathcal{L}^+ \delta A | \delta A \rangle Q^{-1}(0)$, and M is the so-called memory function,

$$M(\bar{\omega}) = \langle \delta A | (i\mathcal{L}^M)(\bar{\omega} + i\mathcal{L}^M)^{-1} (i\mathcal{L}^M) | \delta A \rangle Q^{-1}(0), \qquad (7.37)$$

where $\mathcal{L}^M = (1 - \mathcal{P}) \mathcal{L}^+ (1 - \mathcal{P})$.

7.1.3 Diffusion Coefficients

Diffusion properties of the system are determined by the velocity correlation function

$$Q_{\dot{u}\dot{u}}(l-l';t) = \langle \dot{u}_l(t)\, \dot{u}_{l'}(0)\rangle. \tag{7.38}$$

As usual, we make the Fourier transforms over spacial coordinates,

$$\widetilde{u}(\kappa,t) = \frac{1}{\sqrt{N}}\sum_{l=1}^{N} e^{-i\kappa l} u_l(t) = \widetilde{u}^*(-\kappa,t), \quad |\kappa| \le \pi, \tag{7.39}$$

so that

$$\widetilde{Q}_{\dot{u}\dot{u}}(\kappa;t) = \langle \dot{\widetilde{u}}(-\kappa;t)\, \dot{\widetilde{u}}(\kappa;0)\rangle = \sum_l e^{i\kappa l} Q_{\dot{u}\dot{u}}(l;t). \tag{7.40}$$

Note that the Laplace transforms of the velocity and displacement correlation functions are coupled by the relationship

$$\bar{Q}_{\dot{u}\dot{u}}(l;\bar{\omega}) = \bar{\omega}^2 \bar{Q}_{uu}(l;\widetilde{\omega}), \tag{7.41}$$

which follows from the equation

$$Q_{\dot{u}\dot{u}}(l;t-t') = \frac{\partial^2}{\partial t\, \partial t'} Q_{uu}(l;t-t').$$

Now we may introduce the following three diffusion coefficients.
(a) *The self-diffusion coefficient* D_s is defined as

$$D_s = \lim_{\bar{\omega}\to 0+i0} \bar{\mathcal{D}}_s(\bar{\omega}), \tag{7.42}$$

where $\bar{\mathcal{D}}_s(\bar{\omega})$ is the Laplace transform of the function

$$\mathcal{D}_s(t) = Q_{\dot{u}\dot{u}}(0;t) = \frac{1}{N}\sum_{\kappa}^{N} \widetilde{Q}_{\dot{u}\dot{u}}(\kappa;t). \tag{7.43}$$

When $D_s \neq 0$, the self-diffusion coefficient determines a mean-square displacement of a given target atom on a long-time scale,

$$\langle [u_l(t) - u_l(0)]^2 \rangle \simeq 2D_s t, \quad t\to\infty. \tag{7.44}$$

The function $\bar{\mathcal{D}}_s(\bar{\omega})$ describes incoherent scattering experiments.
(b) *The collective diffusion coefficient* D_μ is introduced as

$$D_\mu = \lim_{\bar{\omega}\to 0+i0} \mathcal{D}_\mu(\bar{\omega}), \tag{7.45}$$

$\bar{\mathcal{D}}_\mu(\bar{\omega})$ being the Laplace transform of the function

$$\mathcal{D}_\mu(t) = \sum_{l=1}^{N} Q_{\dot{u}\dot{u}}(l;t) = \lim_{\kappa \to 0} \widetilde{Q}_{\dot{u}\dot{u}}(\kappa;t). \tag{7.46}$$

Because $\mathcal{D}_\mu(t)$ may be represented as

$$\mathcal{D}_\mu(t) = Q_{\dot{Y}\dot{Y}}(t), \tag{7.47}$$

where

$$Y(t) = \frac{1}{\sqrt{N}} \sum_l u_l(t), \tag{7.48}$$

the coefficient D_μ describes a long-time dynamics of the center of mass of the chain,

$$\langle [Y(t) - Y(0)]^2 \rangle \simeq 2D_\mu t, \quad t \to \infty. \tag{7.49}$$

From the definitions (7.43) and (7.46) it follows that

$$\mathcal{D}_\mu(t) = \mathcal{D}_s(t) + \frac{1}{N} \sum_{l \neq l'} \langle \dot{u}_l(t)\, \dot{u}_{l'}(0) \rangle. \tag{7.50}$$

Thus, the function $\mathcal{D}_\mu(t)$ includes the correlated motion of interacting atoms; as a result $D_\mu > D_s$ usually. Because both $\mathcal{D}_s(t)$ and $\mathcal{D}_\mu(t)$ are real and even functions, we have

$$\widehat{\mathcal{D}}_\nu(\omega) = \widehat{\mathcal{D}}^*_\nu(-\omega) = \pm 2 \,\mathrm{Re}\, \bar{\mathcal{D}}_\nu(\omega \pm i0), \quad \nu = s, \mu, \tag{7.51}$$

and

$$D_\nu = \frac{1}{2} \lim_{\omega \to 0} \widehat{\mathcal{D}}_\nu(\omega), \quad \nu = s, \mu. \tag{7.52}$$

The coefficient D_μ is coupled with the mobility coefficient B by the Einstein relationship. Namely, the function

$$B(t) = \begin{cases} \beta \mathcal{D}_\mu(t), & \text{if } t \geq 0, \\ 0, & \text{if } t < 0 \end{cases} \tag{7.53}$$

describes the linear response of the chain on a small external force $F(t)$ acting on each atom,

$$\langle j(t) \rangle = \int_0^{+\infty} d\tau\, B(\tau)\, F(t - \tau), \quad F \to 0, \tag{7.54}$$

where $j(t) = N^{-1} \sum_{l=1}^{N} \dot{u}_l(t)$ is the atomic velocity averaged over the system. From Eqs. (7.53), (7.54) we see that $B(t)$ is a generalized susceptibility and therefore it may be calculated by the Kubo technique [583, 584]. Defining the atomic flux by the equation

$$J(x,t) = \sum_l \dot{u}_l(t)\, \delta(x - x_l(t)), \tag{7.55}$$

we obtain that the force $F(t)$ leads to the average flux $\langle J(t)\rangle = n\langle j(t)\rangle$. Thus, when the particles consisting the FK chain have an electric charge e, the frequency-dependent conductivity of the chain is equal to

$$\sigma(\omega) = ne^2\widehat{B}(\omega), \quad \widehat{B}(\omega) = \beta\bar{\mathcal{D}}_\mu(\omega + i0). \tag{7.56}$$

(c) *The chemical diffusion coefficient* D_c is defined by the relationship

$$D_c = D_\mu/\chi, \tag{7.57}$$

where χ is the static susceptibility studied above in Sect. 6.5.3. D_c determines the atomic flux in a nonequilibrium state, when the concentration of atoms deviates from the equilibrium value n. Namely, when the atomic density

$$\rho(x,t) = \sum_l \delta(x - x_l(t)) \tag{7.58}$$

slowly deviates from the equilibrium density $\rho_{\rm eq}(x)$, the flux of atoms is equal to

$$\langle\langle J(x,t)\rangle\rangle \simeq -D_c\frac{\partial}{\partial x}\langle\langle\rho(x,t)\rangle\rangle. \tag{7.59}$$

Here the symbol $\langle\langle\ldots\rangle\rangle$ stands for the average over a microscopic distance much larger than a. Eq. (7.59) is known as the first Fick law. If we perform the Fourier and Laplace transforms over x and t, i.e.

$$\bar{\widetilde{\rho}}(k,\bar{\omega}) = \int_0^{+\infty} dt\, e^{i\bar{\omega}t} \int_{-\infty}^{+\infty} dx\, e^{-ikx}\rho(x,t), \tag{7.60}$$

Eq. (7.59) takes the form $\bar{\widetilde{J}}(k\bar{\omega}) \simeq ikD_c\bar{\widetilde{\rho}}(k,\bar{\omega})$. This equation becomes exact in the limit $k \to 0$, $\bar{\omega} \to 0 + i0$, which is called the hydrodynamic regime.

7.1.4 Noninteracting Atoms

For the system of noninteracting atoms, $V_{\rm int} \equiv 0$, all the diffusion coefficients coincide,

$$\mathcal{D}_s(t) = \mathcal{D}_\mu(t) = \mathcal{D}(t), \quad D_s = D_\mu = D_c = D. \tag{7.61}$$

In particular, for the trivial case of a free gas of Brownian particles, when $V_{\rm sub}(x) \equiv 0$, we have

$$\bar{\mathcal{D}}(\bar{\omega}) = \bar{\mathcal{D}}_f(\bar{\omega}) \equiv D_f\,(1 - i\bar{\omega}/\eta)^{-1}, \quad D = D_f \equiv k_BT/m_a\eta. \tag{7.62}$$

However, when $V_{\rm sub}(x) \neq 0$, the exact solution is known for two limiting cases only. First, in the overdamped case, $\eta \to \infty$, the diffusion coefficient is equal to [585]–[589] (when $\eta \to \infty$)

$$D = D_f\left\{\left(\int_0^{a_s}\frac{dx}{a_s}\,e^{-\beta V_{\rm sub}(x)}\right)\left(\int_0^{a_s}\frac{dx}{a_s}\,e^{+\beta V_{\rm sub}(x)}\right)\right\}^{-1}. \tag{7.63}$$

For the sinusoidal substrate potential Eq. (7.63) yields

$$D = \frac{D_f}{I_0^2(\beta\varepsilon_s/2)} \simeq \begin{cases} D_f(1-\frac{1}{8}\beta^2\varepsilon_s^2) & \text{for } k_BT \gg \varepsilon_s, \\ (\pi\varepsilon_s/m_a\eta)\exp(-\beta\varepsilon_s) & \text{for } k_BT \ll \varepsilon_s. \end{cases} \tag{7.64}$$

Second, in the underdamped case, $\eta \to 0$, D is determined by the relation [590, 591]

$$D = D_f \left(\frac{2}{\pi m_a k_B T}\right)^{1/2} \left(\int_{\varepsilon_s}^{\infty} \frac{dE}{\langle v(E)\rangle} e^{-\beta E}\right) \left(\int_0^{a_s} \frac{dx}{a_s} e^{-\beta V_{\text{sub}}(x)}\right)^{-1}, \tag{7.65}$$

for $\eta \to 0$, where

$$\langle v(E)\rangle = \int_0^{a_s} \frac{dx}{a_s} \left| \left(\frac{2}{m_a}[E - V_{\text{sub}}(x)]\right)^{1/2} \right|. \tag{7.66}$$

For the sinusoidal $V_{\text{sub}}(x)$, Eq. (7.65) yields the result [591]

$$D = D_f \frac{\sqrt{\pi\beta\varepsilon_s}\exp(-\beta\varepsilon_s/2)}{2I_0(\beta\varepsilon_s/2)} \int_0^{+\infty} dt \frac{e^{-\beta\varepsilon_s t}}{\sqrt{1+t}\,\mathbf{E}(1/\sqrt{1+t})} \tag{7.67}$$

$$\simeq (\pi/2) D_f \exp(-\beta\varepsilon_s) \quad \text{for} \quad k_BT \ll \varepsilon_s. \tag{7.68}$$

In the high-friction case the value D may be represented as a power series in η^{-1} [592]–[594], while in the low-friction case the power series in $\sqrt{\eta}$ may be constructed [591, 595, 596]. For intermediate values of η the diffusion coefficient is to be calculated numerically [595], [597]–[602].

For a low temperature, $k_BT \ll \varepsilon_s$, the value D may be represented in the Arrhenius form,

$$D = \mathcal{R}\lambda^2, \quad \mathcal{R} = \mathcal{R}_0 \exp(-\beta\varepsilon_s), \tag{7.69}$$

where $\mathcal{R}$ is the escape rate of an atom from the bottom of the potential well, and λ is the mean distance of atomic jumps. The pre-exponential factor $\mathcal{R}_0$ in Eq. (7.69) may be calculated by the Kramers theory [603, 604]

$$\mathcal{R}_0 \simeq \begin{cases} \beta\varepsilon_s\eta & \text{if } \eta < \eta_l \equiv \omega_0/2\pi\beta\varepsilon_s, \\ \omega_0/2\pi & \text{if } \eta_l < \eta < \omega_*, \\ \omega_0\omega_*/2\pi\eta & \text{if } \eta > \omega_*, \end{cases} \tag{7.70}$$

where $\omega_0^2 = V''_{\text{sub}}(0)/m_a$ and $\omega_*^2 = -V''_{\text{sub}}(\frac{1}{2}a_s)/m_a$. The value of λ may be estimated as [605]

$$\lambda \simeq \begin{cases} a_s\eta_l/\eta & \text{if } \eta < \eta_l, \\ a_s & \text{if } \eta > \eta_l. \end{cases} \tag{7.71}$$

We notice that for $\eta_l < \eta < \omega_*$, $D \simeq a_s^2(\omega_0/2\pi)e^{-\beta\varepsilon_s}$, so that the diffusion coefficient does not depend on the friction η, and this defines the range of validity of the transition state theory.

7.1.5 Interacting Atoms

When the atoms interact between themselves, all diffusion coefficients are different, and all of them depend on the concentration of atoms n. The role of interaction reduces to the following factors:

(i) Interaction produces an order in atomic arrangement (in one-dimensional systems this order is short-ranged). This fact results in $\chi \neq 1$ and, therefore, $D_c(n) \neq D_\mu(n)$ if $n \neq 0$;
(ii) Every atom feels a potential produced by other atoms and, therefore, the effective potential for a given atom depends on positions of the neighboring atoms. In a result, both D_s and D_μ are functions of n. This effect is of the first-order in $V_{\rm int}$ and may be approximately described by mean-field-like theories;
(iii) The motion of a given atom gives rise to motions of the neighboring atoms. Such a collective motion (sometimes called also by concerned, or consistent, or relay-raced motion) is an analog of the "polaronic effect" in the solid-state theory. It results in that $D_s(n) \neq D_\mu(n)$ if $n \neq 0$ (usually $D_s < D_\mu$). At low temperatures such a collective motion can be described as the motion of a kink as an entity. So, when the kink consists of r atoms ($r \approx d/a_s = \sqrt{g}$), the BBGKY hierarchy of motion equations has to be truncated at the r-th step as a minimal approximation. In this case, therefore, the phenomenological approach is more suitable;
(iv) Due to nonintegrability of the FK model, atomic motion is always accompanied by energy exchange between different modes leading to intrinsic chaotization of its dynamics. The accounting of this effect is a very complicated problem. Approximately it may be taken into account if we assume that the friction coefficient η includes the intrinsic friction $\eta_{\rm int}$, i.e., $\eta = \eta_{\rm ext} + \eta_{\rm int}$.

The factors (i) and (ii) have a static nature, while the last two factors (iii) and (iv) are purely dynamic.

The correct accounting of $V_{\rm int}$ in consideration of system dynamics is the very complicated problem even for the $V_{\rm sub} \equiv 0$ case. At high temperatures the perturbation theory based on the Mori technique is used typically. Below we outline in brief the main approaches to the problem.

By applying the Mori procedure repeatedly, a correlation function may be represented in the continued-fraction form [606, 607]. In particular, for the correlation function $Q_{\dot{Y}\dot{Y}}$ this procedure gives [608]

$$\bar{\mathcal{D}}_\mu(\bar{\omega}) = D_f \frac{\eta}{-i\bar{\omega} + b_1 + \frac{c_1}{-i\bar{\omega}+b_2+\dots\frac{c_p}{-i\bar{\omega}+\Delta_{p+1}(\bar{\omega})}}}, \tag{7.72}$$

where

$$\begin{aligned}
b_1 &= \eta, \\
b_2 &= 0, \\
&\dots, \\
c_1 &= (\beta/m_a N) \sum_{ll'} \langle F_l^{(\mathrm{sub})} F_{l'}^{(\mathrm{sub})} \rangle, \\
c_2 &= (c_1 m_a^2 N)^{-1} \sum_l \left[\left\langle \left(F_l'^{(\mathrm{sub})}\right)^2 \right\rangle - \left\langle F_l'^{(\mathrm{sub})} \right\rangle^2 \right],
\end{aligned} \tag{7.73}$$

etc., where

$$F_l'^{(\mathrm{sub})} = -\frac{\partial^2 U_{\mathrm{sub}}}{\partial x_l^2}.$$

Notice that the interatomic potential V_{int} does not occur explicitly in Eqs. (7.73), it only appears in the averages over the stationary distribution W_{eq}. The averages in $c_1, c_2, \dots$ may be calculated, e.g., by the transfer-integral technique. In numerical calculations the continued fraction (7.72) has to be truncated. Geisel [608] used the truncation at $p = 2$ putting $\Delta_3 = \eta$ which gives accurate results for a high enough temperature ($\beta\varepsilon_s \ll 2$) in the large-friction limit. For the standard FK model the coefficients c_1 and c_2 in the second-order perturbation theory in $\beta\varepsilon_s$ are equal to [608]

$$c_1 = \frac{\beta\varepsilon_s^2}{8m_a} \left(\frac{2\pi}{a_s}\right)^2 \frac{\sinh(2\pi^2/\beta g a_s^2)}{\cosh(2\pi^2/\beta g a_s^2) - \cos(2\pi a_A/a_s)}, \tag{7.74}$$

$$c_2 = \frac{1}{8c_1} \left(\frac{\varepsilon_s}{m_a}\right)^2 \left(\frac{2\pi}{a_s}\right)^4. \tag{7.75}$$

Then, we introduce the correlation functions

$$G_\nu^W(x, p; x', p'; t) = \langle \delta f_\nu(x, p, t)\, \delta f_\nu(x', p', 0) \rangle, \quad \nu = s, \mu \tag{7.76}$$

for the set of dynamical variables

$$f_s(x, p, t) = f(x, p; z_l(t)) \equiv \delta(x - x_l(t))\, \delta(p - p_l(t)) \tag{7.77}$$

and

$$f_\mu(x, p, t) = N^{-1/2} \sum_{l=1}^N f(x, p; z_l(t)). \tag{7.78}$$

Then the diffusion coefficients may be calculated as

$$\bar{\mathcal{D}}_\nu(\bar{\omega}) = m_a^{-2} \lim_{k\to 0} \int dp\, dp'\, pp'\, \bar{\widetilde{G}}_\nu^W (k, p; k, p'; \bar{\omega}), \quad \nu = s, \mu, \tag{7.79}$$

where $\widetilde{G}$ is the spacial Fourier transform of G,

$$\widetilde{G}(k, p; k', p'; t) = \int dx dx' e^{-ikx + ik'x'} G(x, p; x', p'; t). \tag{7.80}$$

Note that the variables k and k' in $\widetilde{G}$ are varying within the interval $(-\infty, +\infty)$; they should not be mixed with the wave vector κ in the function $\widetilde{Q}$ which is restricted by the first Brillouin zone, $|\kappa| \le \pi$.

Using the Mori technique, Munakata and Tsurui [609] obtained the following approximate equations for the functions G_ν^W:

$$\dot{G}_\nu^W(x,p;x',p';t) = \mathcal{L}_\nu(x,p)\, G_\nu^W(x,p;x',p';t), \quad \nu = s, \mu, \tag{7.81}$$

where

$$\mathcal{L}_s(x,p) = -\frac{p}{m_a}\frac{\partial}{\partial x} - F_{\text{eff}}^{(\text{sub})}(x)\frac{\partial}{\partial p} + \eta\frac{\partial}{\partial p}\left(p + m_a k_B T\frac{\partial}{\partial p}\right), \tag{7.82}$$

$$\begin{aligned} \mathcal{L}_\mu(x,p)\, G_\mu^W(x,p;x',p';t) &= \mathcal{L}_s(x,p)\, G_\mu^W(x,p;x',p';t) \\ -\frac{\partial}{\partial p}\rho_{\text{eq}}(x)\, W_M(p) \int dx''dp'' F_{\text{eff}}^{(\text{int})}&(x,x'')\, G_\mu^W(x'',p'';x',p';t). \end{aligned} \tag{7.83}$$

Here $\rho_{\text{eq}}^{(x)}$ is the equilibrium density of atoms,

$$\rho_{\text{eq}}(x) = \left\langle \sum_{l=1}^N \delta(x - x_l) \right\rangle, \tag{7.84}$$

$F_{\text{eff}}^{(\text{sub})}(x) = -\frac{\partial}{\partial x}V_{\text{sub}}^{\text{eff}}(x)$, the effective substrate potential is defined by the equation

$$\rho_{\text{eq}}(x) = n\exp\left[-\beta V_{\text{sub}}^{\text{eff}}(x)\right]\left\{\int_0^{a_s}\frac{dx}{a_s}\exp\left[-\beta V_{\text{sub}}^{\text{eff}}(x)\right]\right\}^{-1}, \tag{7.85}$$

and the effective interatomic interaction in Eq. (7.83) is introduced as

$$\beta F_{\text{eff}}^{(\text{int})}(x,x') = \frac{\partial}{\partial x}\phi(x,x'), \tag{7.86}$$

where $\phi(x,x')$ is the direct correlation distribution for the inhomogeneous system. The function $\phi(x,x')$ satisfies the integral equation

$$\phi(x,x') = \psi(x,x') - \int dx''\, \phi(x,x'')\, \rho_{\text{eq}}(x'')\, \psi(x'',x'), \tag{7.87}$$

where $\psi(x,x')$ is determined by the pairwise correlation distribution introduced in Eq. (7.21),

$$\psi(x,x') = \int dp\, dp'\, \Phi_{\text{eq}}^{(2)}(x,p;x',p')\left[R_{\text{eq}}^{(1)}(x)\, R_{\text{eq}}^{(1)}(x')\right]^{-1}. \tag{7.88}$$

Thus, the interatomic interaction renormalizes the external substrate potential, $V_{\text{sub}} \to V_{\text{sub}}^{\text{eff}}$, and also leads to modification of pairwise interactions due to emerging of the mean field term of the Vlasov type in Eq. (7.83) for G_μ.

According to Eq. (7.41), the diffusion coefficients may be expressed through the displacement correlation function. Using the following simple relationships,

$$\delta u_l = \delta x_l = \delta \left\{ \int dx\, x\, \delta(x - x_l) \right\} = \delta \left\{ i \lim_{k \to 0} \frac{\partial}{\partial k} \int dx\, e^{-ikx} \delta(x - x_l) \right\}, \tag{7.89}$$

Eq. (7.79) may be rewritten in the form

$$\bar{\mathcal{D}}_\nu(\bar{\omega}) = \bar{\omega}^2 \lim_{k,k' \to 0} \frac{\partial^2}{\partial k\, \partial k'} \,\bar{\widetilde{G}}{}_\nu^R(k, k'; \bar{\omega}), \quad \nu = s, \mu, \tag{7.90}$$

where the reduced correlation function

$$G_\nu^R(x, x'; t) = \langle \delta\varphi_\nu(x, t)\, \delta\varphi_\nu(x', 0) \rangle, \quad \nu = s, \mu \tag{7.91}$$

is defined for the variables

$$\varphi_\nu(x, t) = \int dp\, f_\nu(x, p, t), \quad \nu = s, \mu, \tag{7.92}$$

so that $\varphi_s(x, t) = \delta(x - x_l(t))$ and $\varphi_\mu(x, t) = N^{-1/2} \sum_l \delta(x - x_l(t))$. Note also that $\langle \varphi_\mu(x, t) \rangle = N^{-1/2} \rho_{\text{eq}}(x)$, and that the function $F(k, t) \equiv \widetilde{G}_\mu^R(k, k; t)$ is called the intermediate scattering function. Because

$$\widetilde{G}^R(k, k'; t) \propto \text{Const}(t)\, kk' + O(k^3),$$

Eq. (7.90) may be rewritten in a form

$$\bar{\mathcal{D}}_\nu(\bar{\omega}) = \frac{1}{2} \bar{\omega}^2 \lim_{k \to 0} \frac{\partial^2}{\partial k^2} \,\bar{\widetilde{G}}{}_\nu^R(k, k; \bar{\omega}), \quad \nu = s, \mu. \tag{7.93}$$

Using the Mori representation (7.36) for G^R and taking into account that Ω and M vanish not slowly than k^2 as $k \to 0$, we obtain

$$D_\nu = \frac{i}{2} \lim_{\bar{\omega} \to 0 + i0} \lim_{k \to 0} \frac{\partial^2}{\partial k^2} \left[\Omega_\nu^R(k, k) + M_\nu^R(k, k; \bar{\omega}) \right] \widetilde{G}_\nu^R(k, k; 0). \tag{7.94}$$

When the memory function vanishes faster than k^2 for $k \to 0$, Eq. (7.94) may be rewritten in a simpler form,

$$D_\nu = \lim_{q \to 0} \left[k^{-2} \langle \mathcal{L}^+ \delta\widetilde{\varphi}_\nu(k) | \delta\widetilde{\varphi}_\nu(k) \rangle \right], \quad \nu = s, \mu. \tag{7.95}$$

The describes approach is very useful in the overdamped case, when evolution of the system is described by the Smoluchowsky operator $\mathcal{L}_{\text{sm}}$. Note also that $\widetilde{G}_s^R(k, k; 0) = \langle e^{-ikx_l} e^{+ikx_l} \rangle = 1$, while the function $\widetilde{G}_\mu^R$ at $t = 0$ defines the static structure factor

$$\widetilde{\chi}(k) \equiv \widetilde{G}_\mu^R(k, k; 0), \tag{7.96}$$

so that

$$\chi = \lim_{k \to 0} \widetilde{\chi}(k) \tag{7.97}$$

due to $\chi = \langle (\delta N)^2 \rangle / N = \int \int dx dx'\, G_\mu^R(x, x'; 0)$.

7.2 Diffusion of a Single Kink

First let us discuss the system dynamics when the chain contains a single kink only. This situation is described by the periodic boundary conditions, $u_{N+1} = u_1 \pm a_s$, when the temperature is assumed to be low enough, $k_B T \ll \varepsilon_k$, so that the probability of the thermal creation of a kink-antikink pair is negligible. So, the problem is to derive and solve a stochastic motion equation for the kink collective coordinate $X(t)$. The kink diffusion coefficient can then be obtained as

$$D_k = \lim_{t \to \infty} \frac{1}{2t} \langle [X(t) - X(0)]^2 \rangle, \tag{7.98}$$

or, more generally, as

$$D_k(\bar{\omega}) = \int_0^{\infty} dt e^{i\bar{\omega} t} \langle \dot{X}(t) \dot{X}(0) \rangle, \quad \mathrm{Im}\, \bar{\omega} > 0. \tag{7.99}$$

The physical reason of a diffusional motion for the kink is the effective coupling of the FK chain with the thermostat, i.e. the existence of the nonzero "external" friction coefficient η_{ext}. However, the long-time-scale dynamics of the kinks might be also diffusional even for an isolated chain. Indeed, as we have shown above, any deviation of the model from the integrable case of pure SG system such as nonsinusoidal substrate potential, anharmonic interatomic interactions and discreteness of the atomic chain will destroy the exact integrability of the system. Therefore, besides the external chaos induced by the substrate, the dynamics of the FK chain has to exhibit its own "intrinsic chaotization". This effect may be described approximately by introducing an "intrinsic" friction coefficient η_{int}. It is clear that η_{int} cannot be easily calculated, but it can be estimated by a perturbation technique.

It is interesting that the kink dynamics exhibits two types of diffusion, namely, the *conventional* or *viscous* diffusion and *anomalous* diffusion. To explain the latter mechanism of the kink diffusional motion, we recall that any collision of a kink with other excitations such as phonons or breathers causes a phase shift of the kink, i.e. the displacement of the kink's coordinate. If such collisions occur randomly in time, the kink will undergo a Brownian random walk, however, keeping its averaged velocity unchanged because such collisions are almost elastic (or even completely elastic for the SG limit). It is important that this diffusion mechanism exists even in the integrable SG model where the viscous diffusion is absent (if, of course, we suppose that the mechanism which makes the collisions can be modelled as a random process, e.g., due to coupling with a thermal bath of phonons). A physical reason for the anomalous diffusion is based on the fact that a kink is an extended object with its own width, whereas a usual particle cannot exhibit this type of motion being a point-like object which does not suffer a shift of its location after a collision. The coefficient of the anomalous diffusion can be calculated in the random phase approximation as it is usually assumed in the case of the friction for a particle linearly coupled with a thermostat (see, e.g., Ref. [610]).

In real physical systems the kink diffusion coefficient is determined by both mechanisms mentioned above, and one can expect that D_k is determined by the anomalous diffusion coefficient D_a for short-time scales, $t \ll \eta^{-1}$, and by the viscous diffusion coefficient D_η for $t \gg \eta^{-1}$. Besides, in a strongly discrete chain, when the amplitude of the PN potential exceeds the energy of kink's thermalized motion, $E_{PN} > k_B T$, the kink diffusion becomes thermally activated according to the Arrhenius law, $D_k \propto \exp\left(-E_{PN}/k_B T\right)$.

7.2.1 Langevin Equation

In the presence of the viscous friction (7.3) and the additive stochastic force, the motion equation for the classical FK chain is changed to be

$$m_a \ddot{x}_l + m_a \eta \dot{x}_l - g(x_{l+1} + x_{l-1} - 2x_l) + V'_{\rm sub}(x_l) = \delta F_l(t), \qquad (7.100)$$

where $\dot{x}_l \equiv dx_l/dt$. The fluctuation-dissipation theorem says that the self-correlation function of the fluctuation force δF_l should satisfy the relation

$$\langle \delta F_l(t)\, \delta F_l(t') \rangle = 2\eta m_a k_B T\, \delta(t-t'), \qquad (7.101)$$

while the cross-correlation function, $\langle \delta F_l\, \delta F_{l'} \rangle$ for $l \neq l'$, can be defined in an arbitrary manner. In particular, it is natural to suppose that the spacial correlations decay exponentially with a correlation length λ_F,

$$\langle \delta F_l(t)\, \delta F_{l'}(t') \rangle = 2\eta m_a k_B T \exp\left(-|l-l'|a_s/\lambda_F\right)\, \delta(t-t'). \qquad (7.102)$$

Below we consider the SG limit, i.e. $g \gg 1$ and $V'_{\rm int}(x) = \sin x$, when we should take $x_l = la_s + u_l$, $la_s \to x$, $u_l(t) \to u(x,t)$, $\delta F_l(t) \to \delta F(x,t)$, so that the Langevin equation (7.100) becomes (recall $m_a = 1$)

$$\frac{\partial^2 u}{\partial t^2} - d^2 \frac{\partial^2 u}{\partial x^2} + \eta \frac{\partial u}{\partial t} + \sin u = \delta F(x,t). \qquad (7.103)$$

In order to write Eq. (7.102) in the continuum limit, we have to use additionally the rules $\sum_l \to \int dx/a_s$ and $\delta_{ll'} \to a_s \delta(x-x')$, thus obtaining

$$\langle \delta F(x,t)\, \delta F(x',t') \rangle = 2\eta k_B T \frac{\exp(-|x-x'|/\lambda_F)}{2\lambda_F[1-\exp(-a_s/2\lambda_F)]} \delta(t-t'). \qquad (7.104)$$

For a spatially uncorrelated random force, $\lambda_F \to 0$, Eq. (7.104) reduces to

$$\langle \delta F(x,t)\, \delta F(x',t') \rangle = 2\eta k_B T a_s \delta(x-x')\, \delta(t-t'), \qquad (7.105)$$

while for the coherent external noise, $\lambda_F \to \infty$, this leads to the relation

$$\langle \delta F(x,t)\, \delta F(x',t') \rangle = 2\eta k_B T \delta(t-t'). \qquad (7.106)$$

In dimensionless units, when $\widetilde{x} = x/d$ and $\widetilde{t} = \omega_0 t$, the Langevin equation (7.103) takes the form (we omit all the tildes in what follows),

$$\frac{\partial^2 u}{\partial t^2} - \frac{\partial^2 u}{\partial x^2} + \sin u = f(x,t;u,u_t) \equiv \delta F(x,t) - \eta \frac{\partial u}{\partial t}. \tag{7.107}$$

If the perturbation f is small, the solution of Eq. (7.107) can be obtained by the perturbation technique [27, 86]. Namely, looking for a solution in the form of a nonrelativistic kink,

$$u(x,t) = 4\tan^{-1}\exp\{-\sigma[x - X(t)]\}, \tag{7.108}$$

we obtain the following equation for the kink's coordinate $X(t)$,

$$\frac{d^2X}{dt^2} = -\frac{\sigma}{4}\int_{-\infty}^{\infty} dx\, \frac{f(x,t;u,u_t)}{\cosh[x - X(t)]}. \tag{7.109}$$

Thus, the effect of perturbations reduces to modulations of the kink's coordinate and velocity while the kink's shape is assumed to be unchanged (the adiabatic approximation). Equations (7.107) and (7.109) lead to the Langevin equation for $X(t)$ (see, e.g., Refs. [244, 611, 612])

$$m\frac{d^2X}{dt^2} + m\eta\frac{dX}{dt} = \delta F_k(t) \tag{7.110}$$

with the kink fluctuation force satisfying the relation

$$\langle \delta F_k(t)\, \delta F_k(t')\rangle = 2\eta m^* k_B T\, \delta(t - t'). \tag{7.111}$$

Here m is the kink's mass, and the effective mass m^* is defined as

$$\begin{aligned} m^* &= \int_{-\infty}^{\infty} dx \int_{-\infty}^{\infty} dx'\; u(x)\, u(x')\, \langle \delta F(x)\, \delta F(x')\rangle \\ &= \frac{m a_s}{2\lambda_F[1 - \exp(-a_s/2\lambda_F)]}\, \xi\left(2, \frac{1 + \lambda_F/a_s}{2\lambda_F/a_s}\right), \end{aligned} \tag{7.112}$$

where $\xi(s,v) = \sum_{n=0}^{\infty}(n+v)^{-s}$ is the generalized Riemann zeta-function. For a large-time scale, $t \gg \eta^{-1}$, the Langevin equation describes the Brownian kink motion,

$$\langle X^2(t)\rangle = 2D_\eta t, \quad D_\eta = \left(\frac{m^*}{m}\right)\frac{k_B T}{m\eta}, \tag{7.113}$$

where D_η is the diffusion coefficient.

Thus, if the fluctuation force is spatially uncorrelated, i.e. $\lambda_F \ll d$, from Eq. (7.112) we have $m = m^*$ and the mean kinetic energy of a nonrelativistic kink, $\frac{1}{2}m\langle \dot{X}^2(t)\rangle$, is equal to the thermal energy $\frac{1}{2}k_B T$ in the equilibrium state. In this case the kink diffusion coefficient is equal to [146, 612, 613]

$$D_\eta = \frac{k_B T}{m\eta}. \tag{7.114}$$

In the opposite case of spatially correlated fluctuation force, i.e. for $\lambda_F \gg d$, we find $m^* = (\pi^2/2)m$, and the thermal energy of a single kink is modified to be

$$\frac{1}{2}m\langle \dot{X}^2(t)\rangle = \frac{\pi^2}{4}k_B T. \tag{7.115}$$

The reason for such a renormalization is that a kink is an extended object and, therefore, its coupling to an external source of noise varies to be determined by a ratio of the kink's width d and the noise correlation length λ_F. In the case of a coherent external noise the kink diffusion is determined by the expression [146, 241, 612, 614]

$$D_\eta = \frac{\pi^2}{16}\frac{k_B T}{m\eta}. \tag{7.116}$$

The result (7.116) has been verified with the help of molecular dynamics simulations [241]–[243] which showed that the assumption of the preserved kink's shape works with a good accuracy.

The kink diffusion for the case of uncorrelated noise was studied in detail by Quintero *et al.* [615] in the continuum limit approximation, when the motion equation reduces to the SG form (7.103) with the correlator (7.105). The solution was looked for in the form

$$u(x,t) = u_{SG}[x - X(t)] + \int_{-\infty}^{+\infty} dk\, A_k(t)\Psi_k[x - X(t)], \tag{7.117}$$

where $u_{SG}(x)$ is the shape of (nonrelativistic) SG kink, $X(t)$ is its position, and $\{\Psi_k(x)\}$ is the complete orthonormal set of eigenfunctions of the linear Schrödinger-type equation (3.51) of Sect. 3.3.2. Recall that $\Psi_0(x)$ corresponds to the zero-frequency Goldstone mode, while others describe phonon modes disturbed by the kink. Substitution of the ansatz (7.117) into Eq. (7.103) results in a system of stochastic second-order differential equations for the new variables $X(t)$ and $A_k(t)$. Then, defining the parameter ϵ as $\epsilon = (2\eta k_B T)^{1/2}$ and assuming that ϵ is small at low temperatures, the functions $X(t)$ and $A_k(t)$ are expanded in powers of ϵ, i.e. $X(t) = \sum_{n=1}^{\infty} \epsilon^n X_n(t)$ and $A_k(t) = \sum_{n=1}^{\infty} \epsilon^n A_k^n(t)$. By substituting these expansions into the system of equations mentioned above and grouping together the terms of the same order of ϵ, one obtains the hierarchy of equations, the lowest orders of which can then be solved. After quite lengthy but straightforward calculation the authors found the second-order in T correction to the kink diffusion coefficient (7.114),

$$D_\eta = \frac{k_B T}{m\eta}\left(1 + C\frac{k_B T}{\varepsilon_k}\right), \tag{7.118}$$

where $\varepsilon_k = 8\sqrt{g}$ is the kink rest energy, and the numerical factor C is equal to $C \approx 1.733$ (for the overdamped case, $\eta \to \infty$, Quintero *et al.* [616] had found with the same approach somewhat a smaller value for the numerical constant C, $C \approx 1.244$). The dependence (7.118) quite well agrees with the results of simulation for temperatures $k_B T < 0.1\,\varepsilon_k$.

The result (7.118) is somewhat surprising. Indeed, the damping and stochastic force in Eq. (7.103) destroy the exact integrability of the SG equation, which usually results in the kink-phonon interaction. Such an interaction typically leads to kink "dressing" by a phonon cloud (polaronic effect), so that the kink effective mass and viscosity should increase, $m \to m_{\text{eff}} > m$ and $\eta \to \eta_{\text{eff}} > \eta$. Indeed, Quintero *et al.* [615] obtained that the average kink's kinetic energy is larger than the one corresponding to the equipartition principle,

$$\frac{1}{2} m \langle \dot{X}^2(t) \rangle = \frac{1}{2} k_B T \left(1 + C' \frac{k_B T}{\varepsilon_k} \right) \tag{7.119}$$

(here $C' \approx 2.744$), which may be explained by an increase of the kink effective mass. One could expect that the kink diffusivity $\sim k_B T / m_{\text{eff}} \eta_{\text{eff}}$ will decrease comparing with the value given by Eq. (7.114). The reason why the effect is just opposite, is in that the kink is not a point-like object, but an extended one.

In the discrete FK system, we have to take into account oscillations of kink's shape when it moves in the PN relief. These oscillations result in radiation of phonons, so that the kink-phonon interaction leads to an intrinsic damping of kink motion. Approximately this effect can be accounted by using in Eqs. (7.114) or (7.116) a temperature-dependent friction coefficient,

$$\eta(T) \approx \eta(0) + \alpha T, \tag{7.120}$$

where α is some numerical coefficient [617].

In some physical problems the fluctuation force acting on the FK chain from the substrate is modelled by a multiplicative external noise. In this case the corresponding Langevin equation takes the form

$$\frac{\partial^2 u}{\partial t^2} - d^2 \frac{\partial^2 u}{\partial x^2} + \eta \frac{\partial u}{\partial t} + \sin u = \delta V(x,t)\; \sin u, \tag{7.121}$$

where $\delta V(x,t)$ is usually assumed to be Gaussian with zero mean value and the correlation function

$$\langle \delta V(x,t)\, \delta V(x',t') \rangle = \mu \delta(t-t') \delta(x-x'), \tag{7.122}$$

μ being a measure of the noise intensity, $\mu \propto T$. The perturbation theory applied to this kind of problems shows that the multiplicative noise leads to a similar Brownian motion of the kink with a diffusion coefficient different from that calculated for an additive noise [241, 618].

7.2.2 Intrinsic Viscosity

In the previous section we have assumed that the viscous friction η has its origin in the energy exchange between the atoms of the chain and the substrate, i.e. it corresponds to the external friction η_{ext}. It is clear, however, that a Brownian kink diffusion should exist also in an isolated FK chain, and the main reason for that is nonintegrability of the primary model. If the corresponding response function (7.1), (7.2) is approximated by a local function, and the fluctuation force, by an additive uncorrelated noise, the kink diffusion coefficient is calculated to be

$$D_\eta = \frac{k_B T}{m\eta_{\rm int}}, \tag{7.123}$$

so that the problem itself reduces to the calculation of the intrinsic friction coefficient $\eta_{\rm int}$.

Although the value $\eta_{\rm int}$ cannot be calculated exactly, it can be estimated considering the momentum exchange at a kink collision with other excitations such as phonons and/or breathers. These calculations are usually based on perturbation techniques such as the Mori technique and memory-function approach, the inverse scattering transform, the technique in which the kink's coordinate $X(t)$ is treated as a canonical variable, etc [619]–[625].

As an example, let us below describe briefly the approach used by Bar'yakhtar *et al.* [625], where the continuum limit of the FK model with nonsinusoidal substrate potential was investigated. As usual, the field variable $u(x,t)$ is presented in the form

$$u(x,t) = u_k(x - vt) + \phi(x,t), \tag{7.124}$$

where u_k corresponds to a slowly moving unperturbed kink, and $\phi(x,t)$ describes the phonon field accompanying the kink motion. Substituting Eq. (7.124) into the system Hamiltonian and expanding $V_{\rm sub}(u_k + \phi)$ into the Taylor series in ϕ, the Hamiltonian can be presented in the form $H = H_k + H_{\rm ph} + H_{\rm int}$, where H_k corresponds to an isolated kink, $H_{\rm ph}$ describes the phonon subsystem, and the third term,

$$H_{\rm int} = \sum_{n=3}^{\infty} H_n; \quad H_n \propto \frac{1}{n!} \int dx \left(\frac{\delta^n V_{\rm sub}(u)}{\delta u^n} \right) \Bigg|_{u=u_k} \phi^n, \tag{7.125}$$

is responsible for inelastic scattering of phonons on the moving kink. Then, let us introduce the complete set of functions $\Psi_\alpha(x)$ found as eigenfunctions of the pseudo-Schrödinger equation (3.52); we denote the corresponding eigenvalues as ω_α. The set $\{\Psi_\alpha(x)\}$ consists of the Goldstone mode, the shape modes (if any), and the continuum spectrum modes. Using this basis, we introduce new canonical variables ξ_α by the expansion

$$\phi(x,t) = \sum_\alpha \xi_\alpha \Psi_\alpha(x)\, e^{i\omega_\alpha t}, \tag{7.126}$$

and rewrite $H_{\rm int}$ in the terms of ξ_α. Then one can calculate the probability of the n-phonon inelastic scattering process and obtain the corresponding contribution to the rate of the energy exchange $\eta_{\rm int}$. For the pure SG model such a procedure gives the trivial result $\eta_{\rm int} = 0$ because the contributions of all the orders compensate each other, as it should be for an exactly integrable model where the kink motion is not accompanied by radiation [619, 625]. Otherwise, for a nonintegrable case, the lowest-order contribution to the inelastic scattering comes from the three-phonon scattering and it leads to a viscous friction coefficient

$$\eta_{\rm int} \approx C_1 \omega_0 \left(\frac{k_B T}{\varepsilon_k} \right)^2. \tag{7.127}$$

Substituting Eq. (7.127) into Eq. (7.123), we obtain the diffusion coefficient for a slowly moving kink in the isolated FK chain,

$$D_\eta = C d^2 \omega_0 \left(\frac{\varepsilon_k}{k_B T} \right). \tag{7.128}$$

The numerical factors C_1 and C in Eqs. (7.127) and (7.128) depend on the particular model under consideration and, for example, in the case of the ϕ^4 model $C \approx 20$ [620].

Equations (7.127) and (7.128) describe the classical behavior of the chain. At extremely low temperatures, i.e. for $k_B T \ll \hbar\omega_0$, we should employ the quantum statistics for phonons which leads to the result [625] $\eta_{\rm int} \propto \exp(-C_2 \hbar\omega_0 / k_B T)$, and $D_k \propto T^2 \exp(C_2 \hbar\omega_0 / k_B T)$, where C_2 is another numerical constant. The quantum effects in kink diffusion were also considered by Alamoudi *et al.* [626].

7.2.3 Anomalous Diffusion

When a kink collides with other excitations, it suffers a phase shift, or a spatial displacement $\delta(k)$ without a change of its momentum. As a result, the kink's coordinate which has an initial value X_0, evolves according to the equation

$$X(t) = X_0 + V_0 t + \delta X(t), \quad \delta X(t) = \int dk\, \delta(k)\, \nu_t(k), \tag{7.129}$$

where V_0 is the initial kink velocity, and $\nu_t(k)\, dk$ stands for the number of kink's collisions with excitations having the wavenumbers between k and $k + dk$. The function $\nu_t(k)$ is defined as $\nu_t(k) = \rho(k) n_t(k)$, where $\rho(k)$ is the density of the phonon states with the wavenumber k, and $n_t(k)$ is the number of collisions with the mode having the wavenumber k. Note that such an approach can be based on a phonon picture of the low-energy excitations [620], [627]–[631], as well as on an alternative description in terms of breathers [624, 632, 633] (both the approaches lead to identical results). Here we follow the work of Theodorakopoulos and Weller [624].

Let us assume that the low-energy excitations constitute a heat bath, i.e. that the kink-phonon collisions occur in a random manner and, moreover, that for the time interval between the collisions the heat bath "regenerate" its equilibrium state. In this case, due to a series of spatial shifts $\{\delta(k)\}$ of the kink, the kink dynamics will be diffusional to be considered as a random walk with the mean velocity V_0,

$$\langle (X(t) - X_0 - V_0 t)^2 \rangle = 2D_a t. \tag{7.130}$$

Since the shift $\delta(k)$ is proportional to the squared amplitude of the scattered phonon (or breather), the anomalous diffusion constant D_a is proportional to T^2. Indeed, the fluctuations of the kink's position with respect to the thermal average of $X(t)$ is equal to

$$\delta X(t) = \int dk \; \delta(k) \; \delta\nu_t(k), \tag{7.131}$$

and thus

$$\langle [\delta X(t)]^2 \rangle = \int dk \, dk' \; \delta(k) \; \delta(k') \; \langle \delta\nu_t(k) \; \delta\nu_t(k') \rangle. \tag{7.132}$$

Taking the phonons as a thermal bath, we obtain

$$\langle \delta\nu_t(k) \; \delta\nu_t(k') \rangle = L^{-1} |v(k)| \; t\rho_0 \; \delta(k-k') \langle [\delta n(k)]^2 \rangle, \tag{7.133}$$

where $v(k)$ is the group velocity of phonons with the wavenumber k, $v(k) = d\omega_{\mathrm{ph}}(k)/dk$, $\omega_{\mathrm{ph}}(k) = \sqrt{\omega_0^2 + c^2 k^2}$, and $\rho_0(k) = \rho_0 \equiv L/2\pi$, L being the chain length. Using the classical limit of the Bose-Einstein statistics,

$$\langle [\delta n(k)]^2 \rangle = [k_B T / \hbar \omega_{\mathrm{ph}}(k)]^2,$$

we get from Eqs. (7.132) and (7.133) the result

$$\langle [\delta X(t)]^2] \rangle = 2D_a t, \quad D_a = \widetilde{C} d^2 \omega_0 \left(\frac{k_B T}{\varepsilon_k} \right)^2. \tag{7.134}$$

Here the numerical factor $\widetilde{C}$ depends on the model under consideration, for example, $\widetilde{C} = 2/3\pi$ for the SG system and $\widetilde{C} = 0.916$ for the ϕ^4 model [620, 634].

Anomalous (nondissipative) kink diffusion was first investigated by Wada and Schrieffer [627] for the ϕ^4 model. It should be emphasized that this diffusion mechanism assumes the existence of an "external" thermalization which produces the low-energy heat bath. The only mechanism of such a thermalization is the energy exchange between different degrees of freedom of the system, which emerges due to nonintegrability of the isolated FK chain and/or due to a coupling of the chain with the substrate. As a result, the anomalous diffusion defined by (7.130) exists only on short-time scales, $t \ll \eta^{-1}$, where $\eta = \eta_{\mathrm{ext}} + \eta_{\mathrm{int}}$, while for $t \gg \eta^{-1}$ the kink dynamics should be viscous leading to the standard expression $\langle [X(t) - X_0]^2 \rangle = 2D_\eta t$ with $D_\eta = k_B T / m\eta$ [622], [635]–[637].

As was verified by Theodorakopoulos and Weller [624], the dominant contribution to D_a comes from a relatively narrow band of the phonon wavenumbers, $k \sim 0.05\,\pi/a_s$. Therefore, the continuum limit approximation yields a correct value of D_a for the discrete FK model as well. Besides, a small number of phonons in the vicinity of the Brillouin-zone edges gives rise to inelastic kink-phonon scattering [629] leading to a small contribution to the coefficient η. The anomalous diffusion was observed for the SG model in the molecular dynamics simulations by Theodorakopoulos and Weller [624]. Notice, however, that a discretization procedure applied to the continuum equation always destroys integrability of the model and could be a factor for the subsequent viscous diffusion of kinks.

7.2.4 Kink Diffusion Coefficient

In a general case, the kink diffusion coefficient $D_k(\omega)$ is determined by Eq. (7.99). Using the Kubo-Mori technique [580, 584], the coefficient $D_k(\omega)$ can be expressed in the form [637]

$$D_k(\omega) = \frac{k_B T}{m[\eta(\omega) - i\omega]}, \tag{7.135}$$

where $\eta(\omega)$ is the total generalized friction coefficient. Expanding $\eta(\omega)$ in ω,

$$\eta(\omega) = \eta_0 - i\eta_1 \left(\frac{\omega}{\omega_0}\right) + \eta_2 \left(\frac{\omega}{\omega_0}\right)^2 + \dots, \tag{7.136}$$

the real part of the diffusion coefficient can be presented as

$$\mathrm{Re}\, D_k(\omega) \approx \frac{k_B T}{m} \frac{\eta_0 + \eta_2(\omega/\omega_0)^2}{\eta_0^2 + (\omega/\omega_0)^2[(\omega_0 + \eta_1)^2 + 2\eta_0\eta_2]}. \tag{7.137}$$

Thus, the viscous diffusion is characterized by the coefficient

$$D_k \approx \frac{k_B T}{m\eta_0}, \quad \eta_0 = \lim_{\omega \to 0} \eta(\omega). \tag{7.138}$$

It dominates for low frequencies, $\omega \ll \omega^*$, or long-time scales, $t \gg t^* = 2\pi/\omega^*$, where $\omega^* = \omega_0\sqrt{\eta_0/\eta_2}$. For the SG model we have $\eta_0 = 0$ and $\omega^* = 0$, so that the viscous diffusion is naturally absent. Otherwise, at high frequencies when $\omega \gg \omega^*$ (or short time scales) the anomalous diffusion dominates. Comparing $D_k(\omega)$ for $\omega \gg \omega^*$ with the value D_a given by Eq. (7.134), we can estimate the coefficient η_2 as $\eta_2 \sim \omega_0(k_B T/\varepsilon_k)$ and then the crossover frequency ω^* is found to be

$$\omega^* \approx \left(\frac{\omega_0\eta_0\varepsilon_k}{k_B T}\right)^{1/2}. \tag{7.139}$$

For the isolated FK chain, when, according to Eq. (7.127), $\eta_0 \sim \omega_0(k_BT/\varepsilon_k)^2$, we obtain $\omega^* \sim \omega_0(k_BT/\varepsilon_k)^{1/2}$. Thus, if $\omega \neq 0$ and the temperature T increases, a crossover from the anomalous diffusion $D_k \approx D_a$ to the standard (viscous) diffusion $D_k \approx D_\eta$ should take place. This effect was investigated by Ogata and Wada [620] for the ϕ^4 model (see Fig.7.1). However, if the FK chain is coupled with a thermostat and $\eta_0 \neq 0$ as $T \to 0$, the inverse consequence should take place.

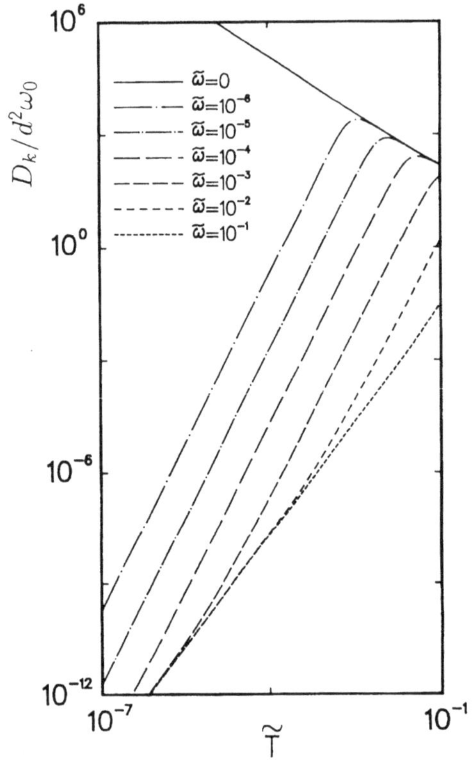

Fig. 7.1. Log–log plot of the real part of $D_k(\omega)/d^2\omega_0$ for the ϕ^4 model, as a function of the dimensionless temperature $\widetilde{T} = \frac{2}{3}k_BT/\varepsilon_k$ for several values of the dimensionless frequency $\widetilde{\omega} = \omega/\omega_0$. In the region $\widetilde{T} < \widetilde{\omega}^2$ the diffusion coefficient is approximately $D_a \propto T^2$, while in the region $\widetilde{T} > \widetilde{\omega}^2$, $D_k(\omega)$ approaches to the curve $D_\eta \propto T^{-1}$ [620].

Up to now we have neglected by the existence of the PN relief in the FK chain. In a strongly discrete FK chain, when $g \leq 1$, the amplitude of the PN potential may exceed the kink thermal energy, and the kink diffusion will become activated. Phenomenologically, in such a case the Langevin equation (7.110) for the kink's coordinate $X(t)$ has to be replaced by the equation [174, 613, 638, 639]

$$m\ddot{X} + m\eta\dot{X} + V'_{PN}(X) = \delta F_k(t), \tag{7.140}$$

where $V_{PN}(X) \approx \frac{1}{2}\varepsilon_{PN}(1 - \cos X)$, $\delta F_k(t)$ is the Gaussian force, $\langle \delta F_k(t)\rangle = 0$ and $\langle \delta F_k(t)\, \delta F_k(t')\rangle = 2\eta m k_B T \delta(t - t')$, and m is the kink mass defined by Eq. (3.10). The damping coefficient η in Eq. (7.140) coincides with the external friction coefficient for an isolated atom, provided the intrinsic kink's friction may be neglected. Indeed, this can be easy checked by performing the operation $m\sum_{l=-\infty}^{+\infty}\ldots$ for both sides of the motion equation (7.11) and taking into account the definition of the kink coordinate X. The applicability of the Langevin equation (7.140) has been carefully tested by Cattuto *et al.* [640] in the framework of the overdamped ϕ^4-model. It was shown that at very high discreteness, $g \ll 1$, one should take also into account a weak dependence of the kink mass and the kink's damping coefficient on the kink coordinate X, because both $m(X)$ and $\eta(X)$ are actually the periodic functions with minima at X corresponding to bottoms of the PN potential.

Equation (7.140) leads to the Arrhenius form for the diffusion coefficient at $t \to \infty$ (or $\omega \to 0$),

$$D_k = D_0 \exp(-\varepsilon_{PN}/k_B T), \tag{7.141}$$

where the pre-exponential factor D_0 can be calculated approximately with the help of the Kramers theory,

$$D_0 \approx \begin{cases} a_s^2 \omega_{PN}/2\pi & \text{if } \eta^* \ll \eta < \omega_{PN}^*, \\ a_s^2 \omega_{PN}\omega_{PN}^*/2\pi\eta & \text{if } \eta > \omega_{PN}^*, \end{cases} \tag{7.142}$$

where

$$\omega_{PN} = \left(\frac{V_{PN}''(0)}{m}\right)^{1/2}, \quad \omega_{PN}^* = \left(-\frac{V_{PN}''(a_s/2)}{m}\right)^{1/2},$$

and

$$\eta^* = \frac{\omega_{PN} k_B T}{2\pi\varepsilon_{PN}}.$$

The case of a low friction, $\eta < \eta^*$, is rather complicated to be investigated analytically. Numerically, an activated kink diffusion (7.141) was observed in the molecular-dynamics simulations for the ϕ^4 model by Combs and Yip [641], and for the undamped standard FK model by Holloway and Gillan [642], Gillan and Holloway [552].

In application of the theory discussed above to realistic physical objects the following remarks should be taken into account:

(i) The total "generalized" friction coefficient $\eta(\omega)$ consists of the intrinsic contribution η_{int} and the external (substrate) contribution η_{ext}, both η_{int} and η_{ext} depend, in a general case, on the frequency ω and the temperature T. In some cases, for example, for the chain of atoms chemically adsorbed on a crystal surface, the intrinsic friction is negligible, $\eta_{\text{int}} \ll \eta_{\text{ext}}$ [184];

(ii) The intrinsic friction η_{int} has an origin in radiation-induced effects due to nonintegrability of the FK chain unlike the completely integrable SG equation. This nonintegrability is caused by all the factors such as discreteness effects, a nonsinusoidal shape of the substrate potential, anharmonicity of the interatomic interaction, the possible presence of impurities, etc.

7.3 Dynamic Correlation Functions

Dynamic correlation functions describe experimentally measured spectra. Namely, for the system being in the equilibrium state, the differential cross-section of inelastic scattering of a particle (e.g., a photon, electron, neutron, ion, etc.) with changing of its energy ($\varepsilon_{\text{fin}} = \varepsilon_{\text{ini}} - \hbar\omega$) and momentum ($k_{\text{fin}} = k_{\text{ini}} - k$) is expressed through the dynamic correlation function,

$$\frac{d^2\sigma}{d\Omega\, d\varepsilon} \propto \left(\frac{k_{\text{fin}}}{k_{\text{ini}}}\right) \widehat{\widetilde{Q}}_{\text{AA}}(k;\omega). \tag{7.143}$$

Here Ω is the cone angle, and $\widehat{\widetilde{Q}}_{\text{AA}}(k;\omega)$ is the Fourier transform of the structure factor $Q_{\text{AA}}(x,t)$,

$$\widehat{\widetilde{Q}}_{\text{AA}}(k;\omega) = \int_{-\infty}^{+\infty} dt\, e^{i\omega t} \widetilde{Q}_{\text{AA}}(k;t), \tag{7.144}$$

$$\widetilde{Q}_{\text{AA}}(k;t) = \langle \delta\widetilde{A}(-k;t)\, \delta\widetilde{A}(k;0)\rangle, \tag{7.145}$$

$$\delta\widetilde{A}(k;t) = \frac{1}{\sqrt{N}} \sum_l \exp(-ikla_s)[A_l(t) - \langle A_l(t)\rangle]. \tag{7.146}$$

In the continuum limit we have to put

$$A(x;t) = a_s \sum_l \delta(x - la_s)\, A_l(t) \tag{7.147}$$

and

$$\delta\widetilde{A}(k;t) = \frac{1}{\sqrt{La_s}} \int dx\, e^{-ikx}[A(x;t) - \langle A(x;t)\rangle]. \tag{7.148}$$

The dynamic correlation function is connected with the static one studied above in Sect. 6.5.5 by the relationship

$$\widetilde{Q}(k) = \widetilde{Q}(k;t=0) = \int_{-\infty}^{+\infty} \frac{d\omega}{2\pi}\, \widehat{\widetilde{Q}}(k;\omega). \tag{7.149}$$

The variable A in the correlation function is related to the basic variables entering the system Hamiltonian. For example, light scattering arises from fluctuations of the atomic density, so we have to put

$$A(x;t) = \rho(x;t) \equiv \sum_l \delta(x - x_l(t)). \tag{7.150}$$

When the FK model describes a ferromagnetic chain of spins, then

$$A_l = s_l \equiv \sin u_l \quad \text{or} \quad A_l = c_l \equiv \cos u_l, \tag{7.151}$$

while for the antiferromagnetic chain, suitable variables are [630, 632, 633]

$$A_l = s_{\frac{1}{2},l} \equiv \sin\left(\frac{u_l}{2}\right) \quad \text{or} \quad A_l = c_{\frac{1}{2},l} \equiv \cos\left(\frac{u_l}{2}\right). \tag{7.152}$$

It is natural to expect that all elementary excitations of the FK chain such as phonons, breathers and kinks will participate in scattering processes leading to different peaks (resonances) at different k and ω. Unfortunately, there is no a firmly based analytical technique to calculate dynamic correlation functions for a system of interacting particles evolving according to Newton's or Langevin equations, therefore computer simulations remain to be the most powerful method for their calculation. However, at low- or high-temperature limits, the dynamic correlation functions admit approximate analytical treatment.

A more simple is the high-temperature case, $k_B T \gg \varepsilon_k$, where the substrate potential is irrelevant, kinks do not play a role, and the FK chain behaves as a weakly interacting phonon gas, associated with nearly harmonic large amplitude oscillations with the acoustic dispersion law, $\omega_T^2(k) = 2g\,[1 - \cos(ka_A)]$. In this case the correlation functions are expected to have peaks corresponded to resonances close to $\omega = \pm\omega_T(k)$. This prediction has been confirmed by molecular-dynamics simulation [643].

In the low-temperature limit, $k_B T \ll \varepsilon_k$, we have to expect a weakly-damped resonances at $\omega = \pm\omega_A(k)$, where $\omega_A(k)$ corresponds to the optic-phonon branch with the self-consistent frequency ($\theta \simeq 1$)

$$\omega_A^2(k) = \omega_0^2\,\langle\cos u_l\rangle + 2g\,[1 - \cos(ka_s)].$$

Thus, the high-frequency peaks in dynamic correlation functions are expected to be predominantly due to a one-phonon response, with possible anharmonic broadening and contributions from higher-order multiphonon processes [643, 644].

If A is odd function of u such as $A = u^2$ or $A = c \equiv \cos u$, the one-phonon peak is absent, and the first nonvanishing contribution to Q_{AA} comes from two-phonon processes. For example, the anharmonic perturbation theory yields the following expression for the function Q_{cc} [643]:

$$\hat{\tilde{Q}}_{cc}(k;\omega) \propto \left(\frac{k_B T}{\varepsilon_k}\right)^2 \sum_{k'} \frac{\delta(\omega \pm \Delta\omega_+) + \delta(\omega \pm \Delta\omega_-)}{\omega_A^2(k')\,\omega_A^2(k-k')}, \tag{7.153}$$

where $\Delta\omega_\pm = [\omega_A(k') \pm \omega_A(k-k')]$. The first term in the right-hand-side of Eq. (7.153) gives a high-frequency resonance at $\omega \approx 2\omega_A(k)$, while the second term predicts a low-frequency peak at $\omega \approx 0$. So, in this case the two-phonon difference processes yield a contribution to the "central peak". This contribution, however, could be significant at large k only, $kd \geq 1$ [643, 644], because at small k the dominant contribution to the central peak comes from kinks (see below).

Another nontopological excitation, the breather, can participate in dynamic correlation functions as well. The internal breather vibrations with a frequency $\omega_{\rm br}$ $(0 < \omega_{\rm br} < \omega_0)$ may lead to a high-frequency peak, while the Goldstone mode of the breather may give a contribution to the central peak. However, the breather's contributions are not separated clearly from the phonon and kink contributions (see discussion in Refs. [518, 643, 644]).

Now we discuss kink's contributions to the dynamic correlation functions. Clearly that they may be essential at low temperatures only. As an example, we consider light scattering experiments, where we are dealing with the density structure factor $Q_{\rho\rho}$. In the continuum approximation the fluctuations of the atomic density are described by the function

$$\rho(x;t) \simeq -\frac{1}{a_s}\frac{\partial u}{\partial x}. \tag{7.154}$$

To study kink's contribution to $Q_{\rho\rho}$, let us neglect phonon fluctuations and, moreover, ignore the intrinsic kink structure, putting, for simplicity,

$$u(x;t) \simeq \sum_j a_s \sigma_j \Theta(x - X_j(t)), \tag{7.155}$$

where $\Theta(x)$ is the Hevisade (step) function and the sum stands to a summation over all kinks. The approximation (7.155) is valid on the spacial scale $x \geq d$, i.e. for small scattering wave-vectors, $k \leq d^{-1}$. The function $\rho(x;t)$ is now given by

$$\rho(x;t) \simeq \sum_j \sigma_j\, \delta(x - X_j(t)). \tag{7.156}$$

Substituting Eq. (7.156) into Eqs. (7.145) and (7.148), we obtain

$$\begin{aligned}\widetilde{Q}_{\rho\rho}(k;t) &\simeq L^{-1} \sum_j \langle \exp\left(ik[X_j(t) - X_j(0)]\right)\rangle \\ &= L^{-1} \sum_j \exp\left(-\frac{1}{2}k^2 \langle [X_j(t) - X_j(0)]^2\rangle\right),\end{aligned} \tag{7.157}$$

where we have neglected the kink-kink correlations according to the basic assumption of the kink-gas phenomenological approach.

For the free (ballistic) motion of kinks we have

$$\langle [X_j(t) - X_j(0)]^2\rangle = \langle [\dot{X}_j(0)\, t]^2\rangle = \frac{k_B T}{m} t^2. \tag{7.158}$$

The Fourier transform of Eq. (7.157) with Eq. (7.158) yields the dynamic correlation function

$$\widehat{\widetilde{Q}}_{\rho\rho}(k;\omega) \propto \exp\left(-\frac{m\omega^2}{2k_B T k^2}\right). \tag{7.159}$$

More accurate calculations (see Refs. [73, 643], [645]–[654]), which take into account the effects of the kink shape, lead to the following expression,

$$\hat{\hat{Q}}_{AA}(k;\omega) = (2\pi k)^{-1} n_{tot} \bar{\gamma}^{-2} P(\bar{v}) |B(k/\bar{\gamma})|^2, \tag{7.160}$$

where n_{tot} is the total concentration of kinks, $\bar{v} = \omega/k$,

$$\bar{\gamma} = \gamma(\bar{v}) \equiv (1 - \bar{v}^2/c^2)^{-1/2}, \tag{7.161}$$

the function

$$P(v) = \gamma^3 \exp(-\gamma\varepsilon_k/k_B T) \left[\int_{-c}^{+c} dv\, \gamma^3 \exp(-\gamma\varepsilon_k/k_B T)\right]^{-1} \tag{7.162}$$

describes the velocity distribution of an ideal relativistic gas of kinks, and the kink form-factor B associated with the variable A (recall that for the light scattering $A(x;t) = \rho(x;t)$) is defined as

$$B(q) = \int_{-\infty}^{+\infty} dx\, e^{-qx} A[u_k(x)], \tag{7.163}$$

$u_k(x)$ being the kink's shape, so that $B(q)$ decays on a scale proportional to the inverse kink width. In the nonrelativistic limit, $k_B T \ll \varepsilon_k$, Eq. (7.160) predicts the Gaussian peak (7.159), which is restricted to small k and ω values. Note that at higher temperatures this peak may split due to relativistic effects in $P(v)$.

Otherwise, if kink's dynamics is Brownian, $\langle [X_j(t) - X_j(0)]^2 \rangle \approx 2D_k t$, Eq. (7.157) yields the Lorentzian central peak [630], [632]–[634],

$$\hat{\hat{Q}}_{AA}(k;\omega) \propto \frac{D_k(\omega)\, k^2}{\omega^2 + [D_k(\omega)\, k^2]^2}. \tag{7.164}$$

Thus, kink's contribution yields the central peaks around $k = 0$ and $\omega = 0$ in dynamic correlation functions. Their intensities are proportional to the kink density n_{tot}. The width of the peak in the k-direction is roughly the inverse of the kink width d. For the ballistic motion of kinks the peak width in the ω-direction is proportional to the wavevector k. The diffusive motion of kinks broadens the central peak, the broadening being larger for larger k.

When phonon contributions are taken into account, the intensity of the central peak decreases. On the other hand, the phonon peaks at $\omega = \pm\omega_A(k)$ are broadened owing to presence of kinks because the kinks destroy the coherence of the phonon wave due to phase shifts at their collisions [653, 654].

The behavior of other dynamic correlation functions such as Q_{ss}, Q_{cc}, $Q_{s/2,s/2}$ is similar to that described above (see Ref. [519] and references therein).

Molecular-dynamics simulations [643, 644] verified that namely kinks are responsible for the central peak at $k \leq d^{-1}$. Therefore, measuring the shape of the central peak (Gaussian or Lorentzian), one can make a conclusion on character of kink's motion (ballistic or diffusion). Moreover, in the latter case we may find the diffusion coefficient $D_k(\omega)$ and observe the crossover from the viscous (dissipative) diffusion to the anomalous (nondissipative) diffusion with changing k and ω values [635]. Note that the neutron spin-echo experiments with the one-dimensional antiferromagnet $(CD_3)_4NMnCl_3$ (TMMC) yield the Lorentzian form of the central peak [655, 656].

Besides the central peak, in strongly discrete FK chain, where kink's diffusion is thermally activated due to existence of large PN barriers, $\varepsilon_{PN} \geq k_B T$, kinks yield additional peaks at $\omega = \pm\omega_{\rm PN}$ [641].

The kink-sensitive correlation functions such as $Q_{c/2,c/2}$, were studied in Refs. [517, 537, 657, 658]. In the quasiparticle-gas approximation it can be estimated as

$$Q(x;t) \propto \left\langle (-1)^{N(x,t)} \right\rangle \simeq \exp\left(-\langle N(x;t)\rangle\right), \tag{7.165}$$

where $N(x,t)$ is the total number of kinks and antikinks whose trajectories in x–t plane cross the line segment $(0,0)$–(x,t). For the ballistic kink's motion $\langle N(x,t)\rangle$ is defined by the expression

$$\langle N(x;t)\rangle = n_{\rm tot} \int dv\; P(v)\; |x - vt|, \tag{7.166}$$

and can be expressed in terms of the error function. When the kink motion is diffusive, Eq. (7.166) should be slightly modified [632].

7.4 Mass Transport Problem

Investigation of mass or charge transport is very important for practical applications of the FK model. A flux of atoms may be caused by a gradient of atomic concentration or by applying of an external driving force. The former is described by the chemical diffusion coefficient D_c, and the latter, by the mobility B, or the collective diffusion coefficient D_μ. Besides, motion of a given (target) atom is described by the self-diffusion coefficient D_s. At high temperatures all diffusion coefficients can be calculated with the help of a perturbation theory. At low temperatures, however, we have to use either molecular dynamics simulations, or the phenomenological approach based on the quasiparticle-gas ideology. Below we consider the mass transport coefficients only, although the FK model may be used as well to find other kinetic coefficients, for example, the thermal conductivity of the FK chain [553].

In order to show general approaches to the problem, we begin this section with a more simple case of a free atomic chain, when the substrate potential is absent at all. Then we describe the perturbative approaches which work at high temperatures, and the phenomenological approach which can be used at low T. Finally, we discuss the temperature and concentration dependencies of diffusion coefficients.

7.4.1 Diffusion in a Homogeneous Gas

Because of complexity of the transport problem, let us discuss first the $V_{\rm sub}(x) \equiv 0$ case. In this case from Eqs. (7.11) and (7.48) it follows that due to pairwise character of interatomic interactions the motion equation for the coordinate $Y(t)$ does not include the interaction $V_{\rm int}(x)$. Therefore, from Eq. (7.47) we have

$$\bar{\mathcal{D}}_{\mu}(\bar{\omega}) = \bar{\mathcal{D}}_{f}(\bar{\omega}). \tag{7.167}$$

Eq. (7.167) can be obtained also from the continued fraction expansion (7.72) to (7.75). Thus, the interaction does not influence the collective diffusion and conductivity of the homogeneous gas. However, the chemical diffusion and self diffusion are essentially modified by the interactions.

Chemical Diffusivity

The chemical diffusion coefficient (7.57) for the homogeneous gas is equal to

$$D_c = D_f/\chi_0. \tag{7.168}$$

For the harmonic, Toda, and hard-core interatomic potentials the expressions for the susceptibility χ_0 have been given above in Sect. 6.5.3. Besides, for the homogeneous gas the static structure factor may be calculated by carrying out an expansion in small $\beta V_{\rm int}(x)$. To the second order in $\beta V_{\rm int}(x)$, $\widetilde{\chi}_0(k)$ is equal to [659]

$$\widetilde{\chi}_0(k) \simeq \left\{ 1 + n\beta \widetilde{V}_{\rm int}(k) - \frac{n}{2}\beta^2 \int_{-\infty}^{+\infty} \frac{dk'}{2\pi}\, \widetilde{V}^{*}_{\rm int}(k'+k/2)\, \widetilde{V}_{\rm int}(k'-k/2) \right\}^{-1}, \tag{7.169}$$

where

$$\widetilde{V}_{\rm int}(k) = \int_{-\infty}^{+\infty} dx\, e^{-ikx} V_{\rm int}(x). \tag{7.170}$$

Expression (7.169) is valid provided the interaction $V_{\rm int}(x)$ is sufficiently small compared with $k_B T$ for all x. For arbitrary $V_{\rm int}(x)$ but low atomic density n the value $\widetilde{\chi}_0(k)$ may be found by means of a virial expansion [659],

$$\widetilde{\chi}_0(k) = 1 + n \int_{-\infty}^{+\infty} dx\, e^{-ikx} \left[e^{-\beta V_{\rm int}(x)} - 1 \right] + O(n^2). \tag{7.171}$$

Note also that for the harmonic and hard-core potentials the static structure factor is known exactly. For the harmonic potential, $V_{\rm int}(x) = \frac{1}{2}g(x-a_A)^2$, the substitution $x_l = la_A + w_l$ yields

$$\begin{aligned}\widetilde{\chi}_0(k) &= \sum_l \langle e^{ik(x_l - x_0)} \rangle = \sum_l e^{ikla_A} \langle e^{ik(w_l - w_0)} \rangle \\ &= \sum_l e^{ikla_A} e^{-\frac{1}{2}k^2 \langle (w_l - w_0)^2 \rangle} = \sum_l e^{ikla_A} \exp[-(k^2/2\beta q)|l|] = \\ &= \sinh(k^2/2\beta g)/[\cosh(k^2/2\beta g) - \cos(ka_A)]. \end{aligned} \tag{7.172}$$

For the hard-core potential,

$$V_{\rm int}(x) = \begin{cases} \infty \text{ if } |x| < a_{\rm hc}, \\ 0 \;\text{ if } |x| > a_{\rm hc}, \end{cases} \tag{7.173}$$

where $\frac{1}{2}a_{\rm hc}$ is the hard-core radius, $\widetilde{\chi}_0(k)$ is equal to (e.g., see Ref. [660])

$$\widetilde{\chi}_0(k) = 1 + 2\,\mathrm{Re}\left\{e^{-ika_{\rm hc}}[1 + ik(a_{\rm hc} - a_A)] - 1\right\}^{-1}. \tag{7.174}$$

Thus, the chemical diffusion coefficient D_c depends on the sign of interaction: for repulsion interaction ($V_{\rm int} > 0$, $\chi < 1$) D_c is larger, while for attractive interaction ($V_{\rm int} < 0$, $\chi > 1$), D_c is lower than D_f.

Self-Diffusion Coefficient

Calculation of the self-diffusion coefficient D_s is a delicately problem even in the $V_{\rm sub} = 0$ case. In the overdamped case, $\eta \to \infty$, it is convenient to use the Mori technique and Eq. (7.94). When the interaction is weak, the expansion up to the terms of the second order in $\beta V_{\rm int}(x)$ yields [659, 661]

$$\Omega_s(k,k) = -iD_f k^2,$$

$$M_s(k,k;\bar{\omega}) \simeq in(\beta D_f)^2 \int_{-\infty}^{+\infty} \frac{dk'}{2\pi} \frac{(kk')^2 |\widetilde{V}_{\rm int}(k')|^2}{-i\bar{\omega} + D_f[k'^2 + (k-k')^2]}, \tag{7.175}$$

so that the memory function lowers the self-diffusion coefficient as follows,

$$D_s \simeq D_f \left\{1 - \frac{1}{6} n\beta^2 \int_{-\infty}^{+\infty} \frac{dk}{2\pi}\, |\widetilde{V}_{\rm int}(k)|^2 \right\}. \tag{7.176}$$

Thus, collective nature of the atoms motion, i.e. the factor (iii) mentioned above in Sect. 7.1.5, hinders the motion of a given atom for either sign of the interaction, contrary to the behavior of the chemical diffusion coefficient D_c. Besides, Dieterich and Peschel [659] calculated the memory function in the limit of low atomic density, but arbitrary $V_{\rm int}(x)$.

When the friction η is not large, the calculation of D_s becomes too complicated because of inertia effects [661],[662]–[667]. However, for the harmonic interatomic potential the value D_s may be calculated exactly.

Harmonic Chain

For the harmonic potential, $V_{\rm int}(x) = \frac{1}{2}g(x - a_A^2)$, the motion equations for the Fourier variables (7.39) decouple,

$$\ddot{\widetilde{u}}(\kappa,t) + \eta\dot{\widetilde{u}}(\kappa,t) + \omega^2(\kappa)\widetilde{u}(\kappa,t) = \delta\widetilde{F}(\kappa,t)/m_a, \tag{7.177}$$

$$\langle \delta\widetilde{F}(\kappa,t)\,\delta\widetilde{F}(\kappa',t')\rangle = 2\eta m_a k_B T \delta_{\kappa,-\kappa'}\delta(t-t'), \tag{7.178}$$

$$\omega^2(\kappa) = \frac{1}{2}\omega_{\max}^2(1-\cos\kappa), \quad \omega_{\max}^2 = 4g/m_a, \tag{7.179}$$

and the correlation function $Q_{\dot{u}\dot{u}}$ can be calculated exactly,

$$\bar{\tilde{Q}}_{\dot{u}\dot{u}}(\kappa,\bar{\omega}) = D_f \frac{-i\bar{\omega}\eta}{\omega^2(\kappa) - i\bar{\omega}\eta - \bar{\omega}^2}. \tag{7.180}$$

From Eq. (7.43) we obtain that

$$\bar{\mathcal{D}}_s(\bar{\omega}) = \int_{-\pi}^{+\pi} \frac{d\kappa}{2\pi}\,\bar{\tilde{Q}}_{\dot{u}\dot{u}}(\kappa;\bar{\omega}) = \bar{\mathcal{D}}_f(\bar{\omega})\Psi(\bar{\omega}), \tag{7.181}$$

where

$$\Psi(\bar{\omega}) = \left\{ \frac{-i\bar{\omega}(\eta - i\bar{\omega})}{\omega_{\max}^2 - i\bar{\omega}(\eta - i\bar{\omega})} \right\}^{1/2}. \tag{7.182}$$

Notice that

$$\lim_{\bar{\omega}\to 0+i0}\lim_{g\to 0}\bar{\mathcal{D}}_s(\bar{\omega}) = D_f \neq \lim_{g\to 0}\lim_{\bar{\omega}\to 0+i0}\bar{\mathcal{D}}_s(\bar{\omega}) = 0. \tag{7.183}$$

Thus, for the harmonic chain $D_s = 0$ exactly. Dynamics of a target atom is described by the following equations [668, 669]

$$\langle (x_l(t) - x_l(0))^2 \rangle \simeq \begin{cases} (k_B T/m_a)t^2 & \text{for } t \ll t_1, \\ 2(k_B T/m_a\omega_1)t & \text{for } t_1 \ll t \ll t_2, \\ 4D_f(\eta t/\pi\omega_{\max}^2)^{1/2} & \text{for } t \gg t_2, \end{cases} \tag{7.184}$$

where $t_1 = \omega_1^{-1}$, $\omega_1 = \max(\eta, \omega_{\max})$, and $t_2 = \omega_2^{-1}$, $\omega_2 = \min(\eta, \omega_{\max}^2/\eta)$.

The sub-diffusional asymptotic behavior $\langle (x_l(t) - x_l(0))^2\rangle \propto \sqrt{t}$ at $t \to \infty$ is the result of one-dimensionality of the system. Because the interatomic potential is unbounded, $V_{\rm int}(x) \to \infty$ for $x \to \pm\infty$, the atoms cannot pass one another in the one-dimensional system. The Lennard–Jones, Toda, and Morse potentials satisfy this condition as well and, therefore, they will lead to the $\sqrt{t}$ long-time dynamics too. In real physical systems, however, the interatomic potential depends on the absolute value of the distance, $V_{\rm int}(x) = V_{\rm int}(|x|)$, and also it should vanish at infinity, $V_{\rm int}(x) \to 0$ when $x \to \infty$. Thus, if $V_{\rm int}(0) \neq \infty$, the atoms may pass one another, and the self-diffusion coefficient is to be nonzero,

$$D_s \sim D_f \exp\left\{-(na_{\min})[\beta V_{\rm int}(0)]^2\right\}. \tag{7.185}$$

7.4.2 Approximate Methods

Now we return to an inhomogeneous system. At high temperatures its diffusional properties can be studied by two variants of the perturbation theory. First, we may start from the $V_{\rm sub}(x) = 0$ limit (see Sect. 7.4.1) and then treat $\beta V_{\rm sub}(x)$ perturbatively. Second, we may begin with the $V_{\rm int}(x) = 0$ case (see Sect. 7.1.4) and consider $\beta V_{\rm int}(x)$ as a small perturbation. At a low temperature, however, when the main contribution to diffusional properties is expected to come from kinks, both perturbative approaches fail because kinks can not be obtained adequately by perturbation techniques, so that we have to use a phenomenological approach.

Perturbative Expansions

When the intermediate scattering function for a homogeneous system, $F_0(k;t) \equiv \widetilde{G}^R_\mu(k,k;t)$, is known, the substrate potential can be accounted by the perturbation technique. Applying the Mori projection formalism to $Q_{\dot{Y}\dot{Y}}$, Dieterich and Peschel [659] have obtained the following expression to second order in $\beta V_{\rm sub}(x)$:

$$\bar{\mathcal{D}}_\mu(\bar{\omega}) \simeq D_f \left[1 - i\bar{\omega}/\eta + D_f \sum_j \left(\beta k_j |V(k_j)|\right)^2 \bar{F}_0(k_j;\bar{\omega}) \right]^{-1} . \tag{7.186}$$

Here $k_j = j(2\pi/a_s)$ is the reciprocal lattice vector, $j = 0, \pm 1, \ldots$, and the substrate potential is expanded into the Fourier series, $V_{\rm sub}(x) = \sum_j V(k_j) e^{ik_j x}$, so that for the sinusoidal $V_{\rm sub}(x)$ it follows $V(k_j) = \frac{1}{4}\varepsilon_s(\delta_{j,1} + \delta_{j,-1})$. The Mori representation for F_0 is the following [659]

$$\bar{F}_0(k;\bar{\omega}) = \frac{i\widetilde{\chi}_0(k)}{[\bar{\omega} - \Omega_0(k) - M_0(k;\bar{\omega})]}, \tag{7.187}$$

where $\Omega_0(k) = -iD_f k^2/\widetilde{\chi}_0(k)$. Note that in the overdamped case ($\eta \to \infty$) the memory function M_0 to second order in $\beta V_{\rm int}$ is given by the expression [659]

$$M_0(k;\bar{\omega}) \simeq \frac{i}{2} n (\beta D_f)^2 \int \frac{dk'}{2\pi} \frac{f^2(k,k')}{[-i\bar{\omega} + D_f(k^2/2 + 2k'^2)]}, \tag{7.188}$$

where

$$f(k,k') = k \left[\left(\frac{1}{2}k + k'\right) \widetilde{V}_{\rm int}\left(\frac{1}{2}k + k'\right) + \left(\frac{1}{2}k - k'\right) \widetilde{V}^*_{\rm int}\left(\frac{1}{2}k - k'\right) \right].$$

If we neglect by the memory function M_0 (i.e., if we ignore the correlated motion of atoms), D_μ takes the form [670]

$$D_\mu \simeq D_f \left[1 + \beta^2 \sum_j |V(k_j)|^2 \, \widetilde{\chi}_0^2(k_j) \right]^{-1} . \tag{7.189}$$

For the standard FK model Eqs. (7.189) and (7.172) coincide with the continued fraction result given by Eqs. (7.72) to (7.75).

Thus, the change in chain's conductivity to second order in $\beta V_{\rm sub}(x)$ is determined by the static structure factor $\widetilde{\chi}_0(k)$ taken at the reciprocal substrate vectors k_j. The function $\widetilde{\chi}_0(k)$ has maxima at the wavevectors which are multiplier of $k_A \equiv 2\pi/a_A$ reflecting the short-range order owing to the interatomic potential $V_{\rm int}(x)$ [see, e.g., Eq. (7.172) for the harmonic $V_{\rm int}$]. From Eq. (7.189) it follows that the collective diffusion coefficient D_μ is suppressed if the dominant k_j coincides with k_A, i.e. when the pair interaction favors a short-range order which is commensurate with the substrate periodicity. Otherwise, D_μ is enhanced if $\widetilde{\chi}_0(k_j) < 1$, i.e. when the interatomic and substrate periodicity is incommensurate.

Clearly that $\beta V_{\rm sub}(x)$ can be treated perturbatically provided $k_B T \gg \varepsilon_s$ only. Besides, at $k_B T \leq E_k$, E_k being the kink creation energy, a collective motion of atoms corresponding to motion of a kink as an entity, should play the main role in mass transport. However, this effect is completely ignored in Eq. (7.189), and is not accounted adequately in Eq. (7.186) too, because the $\beta V_{\rm sub}$ perturbation expansion cannot lead to a correct description of kinks which are nonlinear excitations. Thus, a range of validity of the described perturbation approach is determined by the inequality

$$k_B T \gg \max\,(\varepsilon_s, E_k). \tag{7.190}$$

Mean–Field Approximation

If the interaction potential $V_{\rm int}(x)$ can be treated as a small perturbation, we may truncate the BBGKY hierarchy of equations for the distribution functions at some step r. The simplest mean-field (MF) theory supposes that

$$W^{(N)}(\{z_l\};t) = \prod_{l=1}^{N} W^{(1)}(z_l;t),$$

where $W^{(1)}$ should satisfy the one-particle FPK equation with a self-consistent one-particle potential $V_{\rm sub}^{\rm MF}(x)$ defined by the equation

$$V_{\rm sub}^{\rm MF}(x) = V_{\rm sub}(x) + \int_{-\infty}^{+\infty} dx' \; V_{\rm int}(x-x')\, \rho(x',t). \tag{7.191}$$

This approximation is adequate as far as $\beta V_{\rm int}(x) \ll 1$ for all x.

When the interatomic potential is unbounded as, e.g., the harmonic one, the MF approximation may be modified in the way proposed by Trullinger *et al.* [671] and Guyer and Miller [496]. Namely, the distribution is looked for in a form

$$W(z;t) = W_{\rm eq}(z) \prod_{l=1}^{N} h(z_l;t). \qquad (7.192)$$

The ansatz of Eq. (7.192) retains exactly the equilibrium correlations between atoms, while the motion of different atoms is again assumed to be independent. Therefore, this approximation works at $k_B T \gg E_k$ only. In solid-state theory, such type of approximation is known as the dynamic MF, or random-phase approximation (RPA). The RPA approach leads to motion equations which are similar to Eqs. (7.81)–(7.85) obtained by Munakata and Tsurui [609].

The RPA approximation was firstly applied to the standard FK model with $\theta = 1$ in the overdamped limit ($\eta \to \infty$) by Trullinger *et al.* [671] and by Guyer and Miller [496]. They have obtained the collective diffusion coefficient D_μ with the help of response theory, assuming that a small driving force F is applied to each atom and then calculating the resulting flux of atoms in the steady state. It follows that in the limit $F \to 0$ the value D_μ is determined by Eq. (7.63) where, however, we have to substitute an effective one-atomic potential $V_{\rm sub}^{\rm eff}(x)$ instead of the bare potential $V_{\rm sub}(x)$. The potential $V_{\rm sub}^{\rm eff}(x)$ is given by Eq. (7.85), $V_{\rm sub}^{\rm eff}(x) \propto -k_B T \ln \rho_{\rm eq}(x)$. Thus, the computation of D_μ reduces to finding an equilibrium atomic density $\rho_{\rm eq}(x)$. The function $\rho_{\rm eq}(x)$ can be calculated exactly with the aid of transfer-integral technique [496, 671]. Besides, at high temperatures, $k_B T \gg \varepsilon_s$, the perturbation expansion in $\beta V_{\rm sub}(x)$ can be employed; the result coincides with that given by Eq. (7.189). Finally, if the interatomic potential is short-ranged, e.g., exponential,

$$V_{\rm int}(x) = V_0 e^{-|x|/a_{\rm int}}, \qquad (7.193)$$

then at low and intermediate atomic concentrations, when the range of pair forces is comparable with or smaller than the mean atomic separation, $a_{\rm int} \leq a_A$, a virial-type approach can be used [670]. Namely, truncating the BBGKY hierarchy for distribution functions $R_{\rm eq}^{(s)}(u_1, \ldots, u_s)$ by putting

$$R_{\rm eq}^{(2)}(x,x') \simeq R_{\rm eq}^{(1)}(x)\, R_{\rm eq}^{(1)}(x')\, \exp[-\beta V_{\rm int}(x-x')], \qquad (7.194)$$

we obtain an integral equation for $V_{\rm sub}^{\rm eff}(x)$,

$$V_{\rm sub}^{\rm eff}(x) = V_{\rm sub}(x) + k_B T \int_{-\infty}^{+\infty} dx' \left\{1 - e^{-\beta V_{\rm int}(x-x')}\right\} \rho_{\rm eq}(x'), \qquad (7.195)$$

which is to be solved self-consistently by iteration. Clearly that Eq. (7.195) reduces to Eq. (7.191) if $\beta V_{\rm int}(x) \ll 1$ for all x.

When the friction η is not very large, we may solve the FPK equation following the Munakata-Tsurui approach (7.81) to (7.88). Finite-damping corrections to the Smoluchowsky equation correct up to the order η^{-3} have been calculated by Schneider *et al.* [673] and by Lee and Trullinger [674].

The main effect of pair interaction consists in renormalization of the effective one-particle potential which each atom sees. As follows from Eqs. (7.191) and (7.195), the interatomic interaction lowers the barrier height, $\varepsilon_s \to \varepsilon_s^{\text{eff}} < \varepsilon_s$, independently on the character of interaction (repulsion or attraction) provided the function $V_{\text{int}}(x)$ is strictly convex $[V_{\text{int}}''(x) > 0]$ and slowly varies on distances $x \sim a_A$. Compared with the noninteracting case, the value D_μ increases with increasing of the strength of interaction $g_A = V_{\text{int}}''(a_A)$ and decreasing of the temperature T. These conclusions were confirmed by Munakata and Tsurui [609] for the exponential repulsion potential (7.193) with $V_0 > 0$ by numerical solution of the FPK equation (7.83) using an expansion of G_μ^W in Fourier series over coordinates and Hermite polynomials over momenta (see Fig. 7.2).

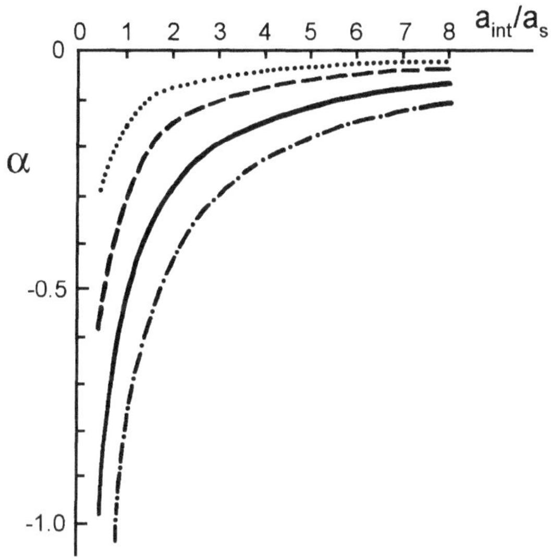

Fig. 7.2. Amplitude of the effective one-particle potential $\alpha = (\varepsilon_s^{\text{eff}} - \varepsilon_s)/\varepsilon_s$ for the exponential interaction (7.193) vs. a_{int} at different temperatures $k_BT/\varepsilon_s = 2.0$, 1.0, 0.5, and 0.25 (dotted, dashed, solid, and dot-dashed curves, respectively), and $V_0 = 2\pi\varepsilon_s$, $\eta = \omega_0/2\pi$ [609].

For the hard-core interaction (7.173) the situation is more complicated [659, 670, 675]: the conductivity can be enhanced or suppressed compared to the independent particle case, depending on the ratio of the hard-core diameter a_{hc} to the substrate constant a_s (see Fig. 7.3). Notice that the results are periodic in a_{hc} with the period a_s. At high temperatures this behavior follows from Eqs. (7.189) and (7.174). Otherwise, in the low temperature limit the density $\rho_{\text{eq}}(x)$ may be taken to be a sum of δ-functions located at the potential minima $x_m = ma_s$. Then, for the $a_{\text{hc}} < a_s$ case the effective potential can be found from Eq. (7.195) which results in the following expression for the collective diffusion coefficient [670],

$$D_\mu \simeq D_{\mu 0} \exp\left[\theta \operatorname{sgn}\left(\frac{1}{2}a_s - a_{\rm hc}\right)\right], \quad a_{\rm hc} < a_s, \qquad (7.196)$$

where $\theta = na_s$ and D_{μ_0} stands for the diffusion coefficient at $n = 0$. It is interesting to compare the dependence (7.196) with predictions of the Langmuir lattice-gas model, which forbids the double occupancy of wells. The latter gives $D_\mu = D_{\mu_0}(1 - \theta)$, that is similar to the result of Eq. (7.196) for the case when $\frac{1}{2}a_s < a_{\rm hc} < a_s$, i.e. when the jump of an atom from a site to the nearest neighboring occupied site is forbidden. On the other hand, when $0 < a_{\rm hc} < \frac{1}{2}a_s$ so that two atoms can sit in the same well, the escape rate will be enhanced by their mutual repulsion. According to Eq. (7.196), the hard-core interaction does not influence the activation energy for jumps (because of $\varepsilon_s^{\rm eff} - \varepsilon_s \sim k_B T$), but rather enters the pre-exponential factor.

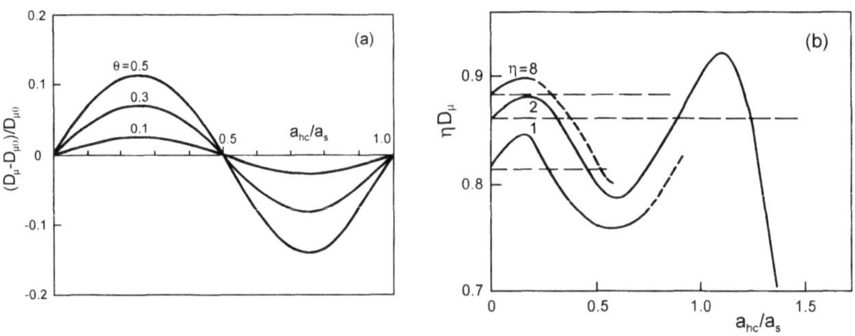

Fig. 7.3. Collective diffusion coefficient D_μ vs. the hard-core diameter $a_{\rm hc}$: (a) Results for different atomic concentrations, $\theta = 0.1$, 0.3, and 0.5, obtained for the overdamped case at $k_BT/\varepsilon_s = 0.5$ [670]; (b) Dependencies ηD_μ vs. $a_{\rm hc}/a_s$ for $\eta = 1$, 2, and 8. The curves are obtained by numerical solution of the FPK equation for two atoms subjected into the sinusoidal potential which is closed to form a ring after four periods, so that $\theta = 1/2$ [675].

For more realistic interatomic potentials, for example, if a Coulomb repulsion potential is added to the hard-core potential, the oscillatory behavior of D_μ on $a_{\rm hc}$ is smeared out and the maximum in $D_\mu(a_{\rm hc})$ is suppressed because the Coulomb repulsion prevents the atoms from reaching the hard-core distance [675].

Above we have considered the collective diffusion coefficient D_μ only. The chemical diffusion coefficient $D_c = D_\mu/\chi$ may be easily obtained by using the results of Sect. 6.5.3 for the susceptibility χ. The calculation of the self-diffusion coefficient D_s, however, is a more complicated problem owing to significant role of the memory function (i.e., the collective motion of atoms). This question will be studied below in Sect. 7.4.4.

7.4.3 Phenomenological Approach

At low temperatures, $k_B T \leq E_k$, when both of the described above perturbation approaches fail, we may use a quasiparticle-gas approximation. Let us suppose that the FK chain contains $N_{\rm tot} = N_k + N_{\bar{k}}$ kinks and antikinks with coordinates $X_j(t)$ and topological charges σ_j. The atomic displacements can be written in the following form,

$$u_l(t) = u_l^{\rm ph}(t) + \sum_{j=1}^{N_{\rm tot}} u_l^{\sigma_j}[X_j(t)], \tag{7.197}$$

where u^σ describes the shape of a slowly moving kink, and $u^{\rm ph}$ corresponds to localized phonons. For $\theta_0 = 1$ and $g \gg 1$ the kink shape is

$$u_l^\sigma(X) \simeq 4\tan^{-1}\exp[-\sigma(la_s - X)/d], \tag{7.198}$$

where $d = a_s\sqrt{g}$ and $a_s = 2\pi$. From Eqs. (7.197), (7.198) we obtain the expression for the atomic velocities

$$\dot{u}_l(t) = \dot{u}_l^{\rm ph}(t) + \sum_{j=1}^{N_{\rm tot}} \dot{X}_j(t)\frac{2\sigma_j}{d}\,\mathrm{sech}\,\{\sigma_j[X_j(t) - la_s]/d\}\,. \tag{7.199}$$

Substituting Eq. (7.199) into Eqs. (7.45)–7.48) for $\mathcal{D}_\mu(\bar{\omega})$, we obtain three types of terms. First, there is a contribution from the phonon correlation function $\langle \dot{u}^{\rm ph}(t)\, \dot{u}^{\rm ph}(0)\rangle$ of the form

$$\mathcal{D}_\mu^{\rm ph}(\bar{\omega}) = D_f \sum_\kappa \frac{i\bar{\omega}\eta}{i\bar{\omega}\eta + \bar{\omega}^2 - \omega_{\rm ph}^2(\kappa)}, \tag{7.200}$$

where κ numerates the phonon modes with frequencies $\omega_{\rm ph}(\kappa)$. Because the phonon spectrum of the FK chain is optical, $\omega_{\rm ph} > 0$, the contribution (7.200) tends to zero at the limit $\bar{\omega} \to 0$. Second, the mixed correlation functions $\langle \dot{u}^{\rm ph}(t)\, \dot{X}_j(0)\rangle$ describe the kink-phonon interactions. The rigorous calculation of these functions is too complicated. In the phenomenological approach, however, we may suppose that this interaction is taken into account indirectly, if the kink concentration $n_{\rm tot}$ and the friction coefficient for the moving kink are calculated in a way which includes the kink-phonon interaction. Finally, to calculate the third contribution, we assume that the kink concentration is small, $n_{\rm tot} \ll n$, so that kinks can be treated as independent quasiparticles, and that

$$\int_0^\infty dt\, \langle \dot{X}_j(t)\, \dot{X}_{j'}(0)\rangle = \delta_{jj'} D_k, \tag{7.201}$$

where D_k is the diffusion coefficient for a single kink. Then, substituting Eq. (7.199) into Eqs. (7.45), (7.46), (7.38) and using the continuum limit, we obtain approximately that

$$D_\mu \simeq \frac{1}{N}\left(\langle N_k\rangle D_k + \langle N_{\bar{k}}\rangle D_{\bar{k}}\right). \tag{7.202}$$

Thus, the collective diffusion of the FK chain is directly proportional to the total kink concentration, $D_\mu \propto \langle n_{\text{tot}}\rangle$ [113, 613, 638]. For the standard FK model with the $\theta_0 = 1$ reference structure in the SG limit ($g \gg 1$) we can take $D_k = D_{\bar{k}} = k_B T/m\eta$ (m being the kink mass) and $\langle n_{\text{tot}}\rangle$ from Eqs. (6.73), (6.93), thus obtaining

$$D_\mu = D_f \pi^{3/2}\sqrt{2}(\beta E_k)^{1/2}\exp(-\beta E_k). \tag{7.203}$$

This result was first obtained by Büttiker and Landauer [676]. It is interesting that the formal application of the RPA approximation of Sect. 7.4.2 for the $\beta E_k \ll 1$ and $g \gg 1$ case, where the transfer-integral technique leads to the Schrödinger equation which can be evaluated analytically, leads to the result similar to Eq. (7.203) except an incorrect factor $(\beta E_k)^{3/2}$ instead of the correct one $(\beta E_k)^{1/2}$ [496, 671].

The low-temperature conductivity of the FK system can be calculated more rigorously with the aid of generalized rate theory [247, 525], [676]–[678]. Namely, dynamics of the N-atomic chain can be described as motion of a point in the N-dimensional configuration space. If we define the hyper-surface $E = U(x_1, \ldots, x_N)$ in the $(N+1)$ -dimensional space $(x_1, \ldots, x_N, E)$, where U is the total potential energy of the system, the point will move on this hyper-surface. The motion of the point is described by the Langevin equation which includes damping and stochastic forces acting on mobile atoms from the thermal bath. The hyper-surface has an infinite number of minima (absolute and relative), maxima and saddle points. Then, let us introduce the adiabatic trajectory as a curve which connects the nearest-neighboring absolute minima and passes through the saddle point characterized by the minimum height of E. If we assume that the system moves strictly along the adiabatic trajectory, its low-temperature dynamics can be described with the aid of the one-dimensional Kramers theory [603]. Small fluctuations of the system trajectory from the adiabatic one can be accounted in the framework of the generalized rate theory [575], [679]–[683] (see also Ref. [604]).

At low temperatures the rate of dynamical processes will be described by the Arrhenius law, $\mathcal{R} = \mathcal{R}_0 \exp(-\beta\varepsilon_a)$, and the activation energy ε_a is given by the difference between the total potential energy in the unstable saddle-point configuration which has to be crossed, and the configuration corresponding to the absolute minimum of U. The pre-exponential factor $\mathcal{R}_0$ is determined by the curvature of the hyper-surface at the minimum and saddle points and by the external friction η. Usually the damping is large enough so that when the system moves from one configuration to another, it completely relaxes (thermalizes) in the new configuration before a next jump takes place. However, if the FK chain is isolated ($\eta_{\text{ext}} = 0$) and its parameters are close to the completely integrable SG case (i.e. $g \gg 1$), so that the intrinsic friction is very small, $\eta_{\text{int}} \ll \omega_0$, the chain dynamics will be more complicated due to existing of "long jumps" of the point through several minima due to inertia effects.

In applying of this ideology to calculation of conductivity of the FK chain one may suppose that a constant driven force F is applied to each atom and then use the response theory. The force will drive kinks to the right and antikinks to the left up to their eventual recombination by collision with other kinks or annihilation at chain's ends. In the steady (time-independent) state the constant flux of atoms is supported by continuous thermal creation of new kink-antikink pairs. Thus, we have to calculate an adiabatic trajectory which describes the creation of a k-$\bar{k}$ pair and the subsequent motion of the kink and antikink in the PN potential in the opposite directions.

For the FK chain with a weak interatomic interactions ($g \ll 1$) we have $E_k \ll \varepsilon_{PN}$, so that the activation energy ε_a is determined by the height of the PN barriers $\varepsilon_{PN} \leq \varepsilon_s$, $D_\mu \propto \exp(-\beta\varepsilon_{PN})$ [525]. Otherwise, when the coupling of atoms is strong ($g \geq 1$), we have $E_k > \varepsilon_{PN}$, and the atomic flux is restricted by the nucleation rate of k-$\bar{k}$ pairs, $D_\mu \propto \exp(-\beta E_k)$ [676, 677].

For the standard FK model which has the kink-antikink symmetry, $D_k = D_{\bar{k}}$, Eq. (7.202) can be rewritten in the form

$$D_\mu \simeq \chi D_k, \tag{7.204}$$

where $\chi = \langle n_{\rm tot}\rangle / \langle n \rangle$. The same result may be obtained in a general case if the concentration of residual ("geometrical") kinks excess the concentration of thermally excited kinks, i.e. when $\langle n_w \rangle \gg \langle n_{\rm pair} \rangle$. Equations (7.204) and (7.57) lead to the relationship

$$D_c \simeq D_k \ \ (\text{or} \ \ D_{\bar{k}}). \tag{7.205}$$

The result (7.205) has a simple physical interpretation. Indeed, at $k_B T \leq E_k$ the mass transport along the chain is carried out by kinks. Because the concentration of the kinks is linearly coupled with the concentration of atoms, the ratio of the flux of kinks to the gradient of kink's concentration is exactly equal to the same ratio of the atomic flux to the atomic gradient. Moreover, this statement [and, therefore, Eq. (7.205)] remains true for any structure of kinks, i.e., for kinks constructed on any reference commensurate structure θ_0 provided the temperature is lower than the melting temperature of a given reference structure (see Sect. 6.5.4). Thus, the calculation of D_c reduces to the calculation of the kink diffusion coefficient D_k for a given θ_0, i.e., to the calculation of kink parameters m and $V_{\rm PN}(x)$ and then to solution of the Langevin equation

$$m\ddot{X} + m\eta\dot{X} + V'_{PN}(X) = \delta F(t) \tag{7.206}$$

with $\langle \delta F(t) \rangle = 0$ and $\langle \delta F(t)\, \delta F(t') \rangle = 2\eta m k_B T \delta(t-t')$. When the intrinsic friction may be neglected, the coefficient η in Eq. (7.206) coincides with the external friction coefficient for an isolated atom, that can be easy obtained by performing the operation $mq\sum_{l=-\infty}^{+\infty}\dots$ for both sides of the motion equation (7.11) and taking into account the definition of the kink coordinate X. In a general case we have to put $\eta = \eta_{\text{ext}} + \eta_{\text{int}}$. Clearly, a more complicated is the elementary cell of the reference structure θ_0, the higher is the value η_{int} for kinks defined on this structure.

Equation (7.205) is valid provided the kink concentration is so small that we can neglect the kink-kink interaction. With increasing of n_{tot} (due to increasing of T or $|\theta - \theta_0|$) the k-k interactions should be taken into account. When the energy of interaction $\varepsilon_{\text{int}} = v_{\text{int}}(R)$, R being the mean distance between the kinks, is small, $\beta\varepsilon_{\text{int}} \ll 1$, this interaction may be accounted by the perturbation theory of Sect. 7.4.2 treating kinks as quasiparticles. Otherwise, when $\beta\varepsilon_{\text{int}} \geq 1$, the k-k interaction will lead to formation of a kink lattice (kink superstructure), and the coefficient D_k in Eq. (7.205) is to be taken as the diffusion coefficient of the superkink defined on this superstructure. Note that $D_c(\theta) < D_c(0)$ owing to inequality $\varepsilon_{PN} < \varepsilon_s$.

When the coefficient D_c is known, the collective diffusion coefficient D_μ can be obtained as $D_\mu = \chi D_c$ for any coverage θ. Note that $D_\mu(\theta)$ may be higher or lower than $D_\mu(0)$ depending on the value of the susceptibility χ at the given θ. Note also that Eq. (7.204) can be obtained directly from Eqs. (7.47), (7.48) if we rewrite $Y(t)$ in a form

$$Y(t) \simeq \mathrm{C} + \frac{1}{q\sqrt{N}}\sum_{j=1}^{N_{\text{tot}}} X_j(t) = \mathrm{C} + \left(\frac{N_{\text{tot}}}{q\sqrt{N}}\right)\frac{1}{N_{\text{tot}}}\sum_{j=1}^{N_{\text{tot}}} X_j(t) \tag{7.207}$$

(where C is a constant) for any structure with a period $a = qa_s$, and then recall that $\chi \simeq \left(\langle N_{\text{sk}}\rangle + \langle N_{\bar{\text{sk}}}\rangle\right)/Nq^2$ at $k_BT \leq E_{\text{sk}}$, E_{sk} being the superkink creation energy.

7.4.4 Self-Diffusion Coefficient

In Sect. 7.4.1 we have shown that for the $V_{\text{sub}} \equiv 0$ case the self-diffusion coefficient is zero provided the interatomic potential is unbounded such as the harmonic, Toda, Morse, etc., potentials. Clearly this result will not be changed if we include the substrate potential by the $\beta V_{\text{sub}}(x)$ perturbation theory of Sect. 7.4.2. However, the $\beta V_{\text{int}}(x)$ perturbation theory, which does not include the correlated motion of atoms, leads to a nonzero value of D_s (for example, the MF approximation results in $D_s = D_\mu$). Besides, the low-temperature phenomenological approach leads to a nonzero value of D_s too. Namely, substituting expression (7.197) for the case $g \gg 1$ into Eqs. (7.42), (7.43) and neglecting by kink-kink correlations, we obtain

$$D_s \simeq (a/d)D_\mu, \quad d \gg a. \tag{7.208}$$

Recall, however, that kinks of the same topological charge repel each other according to the exponential law at large distances (see Sect. 3.1), and collide as elastic balls at short distances (see Fig. 6.7). Considering the subsystem of kinks only (or antikinks only) we see that the same reasons that led to zero self-diffusion coefficient in the free harmonic chain, works now as well. Therefore, the expression (7.208) should correspond to nonzero frequencies $\omega > \omega_2$, or short times $t < t_2$, where the value of t_2 may be estimated in the same way as in Eq. (7.184), if we will consider the kinks as quasiparticles and apply the reasons of Sect. 7.4.1.

Thus, the long-time dynamics of a given atom in the FK chain with unbounded interatomic potential is sub-diffusional,

$$\langle (u_l(t) - u_l(0))^2 \rangle = \alpha\sqrt{t}, \quad t \to \infty. \tag{7.209}$$

The coefficient α in Eq. (7.209) has been calculated by Gunther and Imry [684] and by Gillan [565]. Here we follow the latter work. The Fourier transform of atomic flux (7.55) can be presented approximately as

$$\begin{aligned} \widetilde{J}(k,t) &= \int dx\, e^{-ikx} J(x,t) = \sum_l e^{-ikx_l(t)}\, \dot{u}_l(t) \\ &\simeq \sum_l e^{-ikla_A}\, \dot{u}_l(t) = N^{1/2}\, \dot{\widetilde{u}}(ka_A, t). \end{aligned} \tag{7.210}$$

Using Eq. (7.210) and the phenomenological law (7.59), the velocity correlation function (7.40) can be rewritten as

$$\widetilde{Q}_{\dot{u}\dot{u}}(\kappa,t) \simeq \left(\frac{\kappa}{a_A}\right)^2 D_c^2 \frac{1}{N} \langle \delta\widetilde{\rho}(-\kappa/a_A, t)\, \delta\rho(\kappa/a_A, 0)\rangle. \tag{7.211}$$

According to Eq. (7.91), the density-density correlation function is equal to NG_μ^R. However, from the general theory of correlation functions (e.g., Ref. [582]) we know that the exact conservation law

$$\frac{\partial \rho(x,t)}{\partial t} + \mathrm{div} J(x,t) = 0 \tag{7.212}$$

together with the phenomenological law (7.59) lead to the following expression for G_μ^R,

$$\bar{\widetilde{G}}{}_\mu^R(k,k;\bar{\omega}) \simeq \frac{i\chi}{(\bar{\omega} + ik^2 D_c)}, \quad \bar{\omega} \to 0 + i0, \quad k \to 0. \tag{7.213}$$

So, in the hydrodynamic regime ($\bar{\omega} \to 0 + i0$, $ka_A \ll 1$) the function $Q_{\dot{u}\dot{u}}$ takes the form [565]

$$\bar{\widetilde{Q}}_{\dot{u}\dot{u}}(\kappa,\bar{\omega}) = -D_\mu \frac{-i\bar{\omega}}{-i\bar{\omega} + (\kappa/a_A)^2 D_c}. \tag{7.214}$$

Then, performing the summation over κ in Eq. (7.43) by putting $\kappa = \nu\,\Delta\kappa$, $\Delta\kappa = 2\pi/N$, $\nu = -\frac{1}{2}N + 1, \ldots, +\frac{1}{2}N$, and assuming that $-i\bar{\omega}$ is small but finite, so that we can extend the sum over ν to $\pm\infty$, we obtain

$$\bar{\mathcal{D}}_s(\bar{\omega}) = -D_\mu \frac{1}{N} \sum_{-\infty}^{+\infty} \frac{A}{A+\nu^2} = -D_\mu \frac{\pi\sqrt{A}}{N\tanh(\pi\sqrt{A})}, \tag{7.215}$$

where $A = N^2(-i\bar{\omega}/D_c)(a_A/2\pi)^2$. In the limit $N \to \infty$, Eq. (7.215) yields

$$\bar{\mathcal{D}}_s(\bar{\omega}) = -\frac{1}{2}\chi a_A(-i\bar{\omega}D_c)^{1/2}, \quad \bar{\omega} \to 0 + i0. \tag{7.216}$$

Thus, the $t \to \infty$ dynamics of a target atom is determined by Eq. (7.209) with the coefficient α given by

$$\alpha = 2\chi a_A \sqrt{D_c/\pi}. \tag{7.217}$$

The result (7.209), (7.217) was confirmed by Gillan and Holloway [552] in molecular dynamics simulation for the standard FK model. Note, however, that when the interatomic potential is bounded ($V_{\rm int}(x) < \infty$ for all x), then D_s is nonzero, and its value may be obtained, e.g., from Eqs. (7.79)–(7.82).

7.4.5 Properties of the Diffusion Coefficients

Now we can predict the behavior of the diffusion coefficients D_μ and D_c as functions of the system temperature T. For definiteness, consider the practically important case of $\theta \le 1$, for example, when $\theta = (q-1)/q$ with $q \gg 1$, and suppose that the system is close to the standard FK model so that an anharmonicity of the interatomic interactions is weak, i.e.,

$$\alpha = -(\varepsilon_s/2)^{1/2} V_{\rm int}'''(a_A)/[V_{\rm int}''(a_A)]^{3/2} \ll 1.$$

At $T = 0$ the ground state of the system corresponds to the commensurate structure with the period $a = qa_s$. Topologically stable excitations of the $T = 0$ GS are superkinks (super-antikinks) with the creation energy E_{sk} and the effective mass m_{sk}. The adiabatic motion of the superkinks is carried out in the PN potential $V_{sPN}(X) \simeq \frac{1}{2}\varepsilon_{sPN}[1-\cos(2\pi X/a)]$ with the height ε_{sPN}, and small vibrations of superkinks near the PN potential minima are characterized by the frequency $\omega_{sPN} = (\varepsilon_{SPN}/2q^2 m_{sk})^{1/2}$. Let η_{sk} corresponds to an effective friction coefficient for a moving superkink.

According to renormalization arguments of Sect. 5.2.1, the described GS may be treated as a regular lattice of the $\theta = 1$ antikinks. When the temperature increases above the melting temperature T_{sk} ($k_B T_{sk} \simeq E_{sk}$, see Sect. 6.5.4), this kink lattice is disordered due to thermal creation of a large number of sk-$\overline{sk}$ pairs. According to the kink-lattice hierarchy ideology of Sect. 6.5.4, within the temperature interval $E_{sk} < k_B T < E_k$ the equilibrium state of the system can be considered as the commensurate structure with the period a_s in which, however, there exist N_w residual kinks and $N_{\rm pair}$ thermally created k-$\bar{k}$ pairs. Let E_k denotes the creation energy of the $\theta = 1$ kink, m_k denotes its effective mass, $V_{PN}(X) \simeq \frac{1}{2}\varepsilon_{PN}[1-\cos(2\pi X/a_s)]$ denotes the

PN potential for moving $\theta = 1$ kink, $\omega_{PN} = (\varepsilon_{PN}/2m_k)^{\frac{1}{2}}$ denotes the PN frequency, and η_k denotes the corresponding friction. The superkink and kink parameters satisfy two inequalities, $\varepsilon_{sPN} < \varepsilon_{PN} < \varepsilon_s$ and $E_{sk} < E_k$. For definiteness, assume also that E_{sk} is the lowest energy parameter of the system (that is always true at least for $q \gg 1$), and that the anharmonicity of $V_{\text{int}}(x)$ is so weak that the energy $\varepsilon_{\text{anh}} = \varepsilon_s/2\alpha^2$ is the largest of the parameters. Now we can describe the functions $D_\mu(T)$ and $D_c(T)$ for various temperature intervals.

At a very low temperature,

$$0 < k_B T < E_{sk}, \tag{7.218}$$

the mass transport along the chain is carried out by superkinks. So, according to the phenomenological approach of Sect. 7.4.3, the chemical diffusion coefficient is equal to

$$D_c \simeq \mathcal{R}_{sk} a^2 \exp(-\beta \varepsilon_{sPN}), \tag{7.219}$$

where

$$\mathcal{R}_{sk} \simeq \begin{cases} \omega_{sPN}/2\pi & \text{if } \eta_{sk} < \omega_{sPN}, \\ \omega_{sPN}^2/2\pi\eta_{sk} & \text{if } \eta_{sk} > \omega_{sPN}, \end{cases} \tag{7.220}$$

while the collective diffusion coefficient D_μ is determined by an equation

$$D_\mu \simeq D_c \frac{\langle N_{sk} \rangle + \langle N_{\overline{sk}} \rangle}{q^2 \langle N \rangle} \propto \exp[-\beta(\varepsilon_{sPN} + E_{sk})]. \tag{7.221}$$

When the temperature increases, $k_B T \sim E_{sk}$, the number of thermally excited sk-$\overline{sk}$ pairs increases too, and their mutual attraction will decrease the real value of D_c compared with that given by Eq. (7.219).

At low temperatures,

$$E_{sk} < k_B T < \min\,(\varepsilon_{PN}, E_k), \tag{7.222}$$

superkinks are destroyed by thermal fluctuations, but the $\theta = 1$ kinks still exist, and they response now for the mass transport. Neglecting kink's interactions, we have

$$D_c \simeq \mathcal{R}_k a_s^2 \exp(-\beta \varepsilon_{PN}), \tag{7.223}$$

$$\mathcal{R}_k \simeq \begin{cases} \omega_{PN}/2\pi & \text{if } \eta_k < \omega_{PN}, \\ \omega_{PN}^2/2\pi\eta_k & \text{if } \eta_k > \omega_{PN}, \end{cases} \tag{7.224}$$

$$D_\mu \simeq D_c \frac{\langle n_{\text{tot}} \rangle}{\langle n \rangle} \propto \exp[-\beta(\varepsilon_{PN} + E_k)]. \tag{7.225}$$

The mutual repulsion of the residual kinks will slightly increase D_c compared with Eq. (7.223). Notice that both D_c and D_μ have the Arrhenius form but with different activation energies.

Analogously we can describe a more complicated situation of $\theta = s/q$ with $2 \leq s \leq q-2$, when the kink-lattice hierarchy consists of more than two temperature intervals.

The limits of an intermediate temperature interval depend on the model parameter g_A. Namely, when $g_A \geq 1$, there is the temperature interval

$$\varepsilon_{PN} < k_B T < E_k, \tag{7.226}$$

within which the kinks still exist, but their motion is not activated. The phenomenological approach predicts for this case a behavior

$$D_c \simeq k_B T / m_k \eta_k \tag{7.227}$$

and

$$D_\mu \simeq D_c \frac{\langle n_{\text{tot}} \rangle}{\langle n \rangle} \propto \exp(-\beta E_k). \tag{7.228}$$

Analytical predictions of Eq. (7.223) and (7.227) are in agreement with the molecular dynamics results of Holloway and Gillan [642], Gillan and Holloway [552]. These simulations have been carried out for the undamped FK chain with $g = 0.127$ so that $\varepsilon_{PN} \simeq 1.10$, $\varepsilon_k \simeq 2.5$ and $m_k \simeq 0.7$. The results can be explained by Eq. (7.227) within the interval (7.226) if we put $\eta_k \simeq 0.028\,\omega_0$ (for the undamped system η_k corresponds to the intrinsic friction only), and by Eq. (7.223) for the temperature interval (7.172) where, however, the simulation gives the pre-exponential factor $\simeq 0.4$, while Eq. (7.224) leads to the value $\simeq 0.1$. This disagreement may arise due to using the one-dimensional theory in Eqs. (7.220) and (7.224) instead of the more rigorous generalized rate theory.

Otherwise, if $g_A \ll 1$, the intermediate temperature interval is

$$E_k < k_B T < \varepsilon_{PN}, \tag{7.229}$$

and in this case the diffusion coefficients can be obtained numerically only with the help, e.g., of the RPA approach of Sect. 7.4.2. It may be expected that D_c and D_μ will have the Arrhenius behavior with an activation energy $\varepsilon_s^{\text{eff}}$, where $\varepsilon_{\text{PN}} < \varepsilon_s^{\text{eff}} < \varepsilon_s$.

At high temperatures we may use the perturbation theory approach of Sect. 7.4.2. In particular, for a temperature interval

$$\max(\varepsilon_s, E_k) < k_B T < \varepsilon_{\text{anh}} \tag{7.230}$$

the perturbation expansion in $\beta V_{\text{sub}}(x)$ gives

$$D_\mu \simeq D_f \left\{ 1 + \frac{1}{8} \left[\frac{\beta \varepsilon_s \sinh(1/2\beta g)}{\cosh(1/2\beta g) - \cos(2\pi a_A / a_s)} \right]^2 \right\}^{-1}, \tag{7.231}$$

and an analogous expression for D_c with the factor $g a_A^2 / m_a \eta$ instead of D_f. Finally, at very high temperatures, $k_B T > \varepsilon_{\text{anh}}$, both $V_{\text{sub}}(x)$ and $V_{\text{int}}(x)$

become irrelevant, and the diffusion coefficients D_μ and D_c are to be close to the value $D_f \equiv k_B T/m_a \eta$.

Now let us consider the behavior of diffusion coefficients as functions of the coverage $\theta \equiv a_s/a_A$ for the standard FK model. At high temperatures we can use the perturbation expansion in $\beta V_{\rm sub}(x)$. From Eq. (7.231) we see that the collective diffusion coefficient $D_\mu(\theta)$ shows oscillations as a function of θ, achieving minima for the trivial ground states with $\theta = 1/q$ $(q = 1, 2, \ldots)$, where the atoms are distributed over the bottoms of the substrate potential wells. On the other hand, maxima of $D_\mu(\theta)$ occur at $\theta = 2/(2q-1)$, when the atoms in the ground state are displaced from the well bottoms bringing them closer to the potential tops. In a result, the absolute maximum of the dc conductivity σ is expected to be for a concentration $\theta \simeq 0.75 - 0.80$ [608]. Maxima and minima of $D_\mu(\theta)$ become more pronounced with increasing of the interatomic interaction and decreasing of the temperature.

At low temperatures, $k_B T < E_k$, the phenomenological ideology leads to a similar behavior [685]. Indeed, let us consider the FK system with a coverage θ which is close to the value $\theta_0 = 1$, $|\theta - \theta_0| \ll 1$, so that the interaction between the residual kinks is so small that they do not form a kink lattice at the given temperature. In this case the chemical diffusion coefficient D_c is equal to D_k and it is approximately independent on θ. [Note that this statement is in agreement with numerical simulation of Gillan and Holloway [552]. However, $D_c(\theta)$ may slowly increase with increasing of $|\theta - \theta_0|$ due to k-k repulsion of residual kinks]. Because the number of thermally excited kinks, $2N_{\rm pair}$, is approximately independent on θ, while the number of residual kinks $|N_w|$ linearly increases with $|\theta - \theta_0|$, the susceptibility $\chi \simeq \langle N_{\rm tot}\rangle/N$ as a function of θ should have a local minimum at $\theta = \theta_0$. Therefore, $D_\mu(\theta)$ will have a minimum at $\theta = \theta_0$ too. Analogously, $D_\mu(\theta)$ will have local minima at those commensurate coverages θ_0 whose melting temperature is larger than T. Clearly that between the local minima the function $D_\mu(\theta)$ will have local maxima. At high temperatures this criterion is fulfilled for the trivial coverages $\theta_0 = 1/q$ only. But with decreasing of the temperature T, new and new local minima of the function $D_\mu(\theta)$ will emerge corresponding to higher-order commensurate structures, and in the limit $T \to 0$ the function $D_\mu(\theta)$ should have minima at every rational coverage θ (of course, all the diffusion coefficients tend to zero according to Arrhenius law when $T \to 0$).

However, the exploiting the standard FK model for the whole interval of coverages from 0 to 1 (or even to ∞) is objectionable because a real interatomic potential $V_{\rm int}(x)$ cannot be approximated by one harmonic function when the mean distance between atoms varies from ∞ to a_s (or even to 0). Now let us consider a more realistic situation, when $V_{\rm int}(x)$ is anharmonic (e.g., the exponential or power function). When the system temperature T is lower than the "melting" temperature $T_{\rm melt}(\theta_0)$ for a given structure θ_0, then the mass transport along the chain is carried out by kinks (local contractions of the chain) if $\theta = \theta_0 + \delta$, where $\delta \to +0$, or by antikinks (local extensions) if $\theta = \theta_0 - \delta$. Thus, $D_c(\theta_0 + \delta) = D_k$ and $D_c(\theta_0 - \delta) = D_{\bar{k}}$. But for an anharmonic $V_{\rm int}(x)$, kink and antikink are characterized by different parameters (see Sect. 3.5.1). For example, the PN barriers for kinks are lower than those for antikinks. So, when the coverage θ increases passing though θ_0, the activation energy for chemical diffusion decreases jumplike. From the dependence $\varepsilon_{PN}(\theta)$, see Fig. 5.18, it follows that the dependence $D_c(\theta)$ should have jumps similar to Devil's staircase: the value of D_c should rise sharply each time whenever the coverage θ exceeds the θ_0 value that characterizes a structure commensurate with the substrate, having at a given T a "melting" temperature that exceeds T [165]. Note that usually in Eqs. (7.219), (7.220) both the diffusion activation energy and, owing to a decrease of the free path length, the pre-exponential factor decrease simultaneously with increasing concentration (the so-called compensation effect). Clearly that the jump in $D_c(\theta)$ at a given $\theta = \theta_0$ exists provided $T < T_{\rm melt}(\theta_0)$; when the temperature increases above $T_{\rm melt}(\theta_0)$, the jump will disappear. Thus, the Devil's staircase will smoothen with increasing temperature since only jumps at the coverages corresponding to simple commensurate structures ($\theta_0 = 1, \frac{1}{2}$, etc.) will "survive".

All the approaches described above, can be directly applied to the FK model with nonsinusoidal substrate potential. In particular, in the phenomenological approach we have to use kink parameters E_k, ε_{PN}, etc., which were calculated for the given shape of $V_{\rm sub}(x)$. For example, Woafo *et al.* [686] studied the kink diffusion for the deformable substrate potential. However, some novel features emerge in polikink systems, where the model admits the existence of different types of kinks. In particular, the activation energy for the low-temperature chemical diffusion coefficient is determined mainly by those kinks which undergo the larger PN potential.

8 Driven Dynamics

In this chapter we consider the FK chain driven by an external force. Two important features of the model which make these problems nontrivial are, first of all, the interaction between atoms in the chain and, second, the presence of an on-site periodic potential. One of the important characteristics of the driven dynamics is the mobility as a function of the applied force, which is of great interest for applications.

8.1 Introductory Remarks

There are many systems where a set of particles is driven by a constant external force (also called the 'direct current' force, or the dc force). The best known example is the electronic or ionic conductivity of solids, where the force acting on electrons or ions appear due to an applied electric field. Other examples are CDW systems and Josephson-junction arrays. One more important example corresponds to tribology problems, where a thin layer of atoms (of a few Angström width) is confined between two solid substrates. When the substrates are in a relative motion, the lubricant film is driven due to the frictional contacts with the surfaces. Finally, a more specific example from a physical point of view, is the problem of traffic flow of cars on roads. The two important ingredients of the driven models are, first, that the atoms interact between themselves, and second, that the atoms are subjected to an external potential. If one of these ingredients is absent, the problem becomes trivial: the motion of noninteracting atoms (which is equivalent to the motion of a single atom) in an external potential is already studied in details as will be described in Sect. 8.2, while the model without an external potential reduces to the nondriven system in the moving frame.

In the present Chapter we consider the FK model when the atoms are driven by the dc force (ac–driven systems are briefly considered in Sect. 8.10; see also the review paper by Floría and Mazo [487] and references therein). Above in Sect. 7.4 we already discussed the coefficient $B_0 = \beta D_\mu$ which defines the linear response of the FK chain, i.e., the steady-state flux of atoms in response to a constant driven force F in the limit $F \to 0$. Now we will consider the case of large forces, when the system response is strongly *nonlinear*. In this case the mobility B is a function of the applied force F, because the total one-atomic potential

$$V_{\mathrm{tot}}(x) = V_{\mathrm{sub}}(x) - F\,x \tag{8.1}$$

depends on F. For instance, at very high fields, $F \gg F_s$, where

$$F_s \equiv \pi\varepsilon_s/a_s, \tag{8.2}$$

for the sinusoidal substrate potential, the barriers of $V_{\mathrm{tot}}(x)$ become completely degraded and the chain behaves as a homogeneous one, so that $B \to B_f$, where $B_f = \beta D_f = (m_a\eta)^{-1}$. To calculate the dependence $B(F)$ in a general case, one has to find a steady-state solution W_F of the FPK (or Smoluchowsky) equation with periodic boundary condition, $W_F^{(1)}(x+a_s,p) = W_F^{(1)}(x,p)$, for the total potential energy $U = \sum_l V_{\mathrm{tot}}(x_l) + U_{\mathrm{int}}$. Then $B(F) = j_F/F$, where

$$j_F = \left\{ \int_0^{a_s} \frac{dx}{a_s} \int_{-\infty}^{+\infty} dp\, \frac{p}{m_a} W_F^{(1)}(x,p) \right\} \left\{ \int_0^{a_s} \frac{dx}{a_s} \int_{-\infty}^{+\infty} dp\, W_F^{(1)}(x,p) \right\}^{-1}.$$

The main questions of interest for the driven systems are the $B(F)$ dependence for the steady state, i.e. for the case of the adiabatically slow change of the force F, and the dynamical transitions between different steady states, first of all at the onset of sliding, i.e. the transition from the locked state to the sliding state. The dependence $j_F = B(F)\,F$ corresponds to the voltage-current characteristic in the ionic conductors and Josephson-junction arrays, while in the CDW systems this dependence is connected with the current-voltage characteristic. In the context of tribology, the driving force F may be interpreted as the frictional force for a given sliding velocity.

8.2 Nonlinear Response of Noninteracting Atoms

Let us consider first a more simple case of noninteracting atoms. The Brownian motion of the noninteracting atoms in a periodic potential driven by an external constant force F has been studied in a number of works (see, e.g., Refs. [601, 602, 673, 687], and also Refs. [595, 596] and references therein). It was shown that the crossover from the low-F mobility $B_0 = \beta D_\mu$ to the high-F mobility B_f strongly depends on the friction coefficient η and the system temperature T. Indeed, the total potential (8.1) experienced by the particle is the sum of the periodic potential and the potential $-Fx$ due to the driving force, i.e. it corresponds to a corrugated surface, whose average slope is determined by F as shown in Fig. 8.1. At small forces the potential has local minima, therefore the particle is static and its mobility is zero at temperature $T = 0$. On the contrary for large forces, $F \geq F_s$, there are no stable positions, and the particle slides over the corrugated potential, approaching its maximum mobility $B_f = (m_a\eta)^{-1}$.

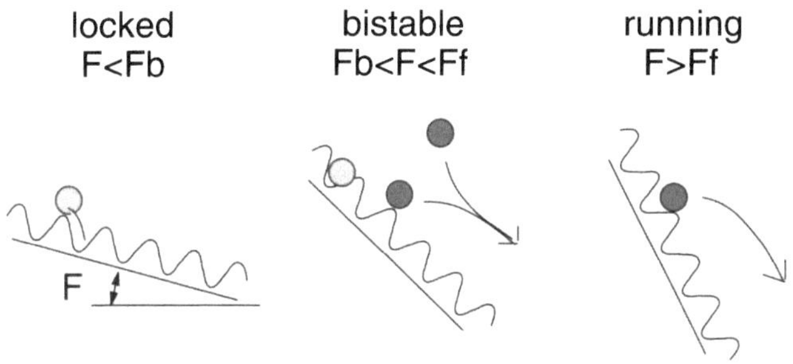

Fig. 8.1. Bistability of an atom driven in the inclined periodic potential.

8.2.1 Overdamped Case

The locked-to-sliding transition at $F = F_s$ is an example of the *saddle-node bifurcation*, a basic concept in the theory of dynamical systems with a few degree of freedom. The depinning force F_s is simply given by the maximum slope of the substrate potential $V_{\rm sub}(x)$, $F_s = \max_x V'_{\rm sub}(x) = V'_{\rm sub}(u_0)$. Below the depinning threshold, $0 < F < F_s$, there are two static solutions ($\dot{x} = 0$) per period: one is stable ($x_s = u_F + ja_s$, where $u_F = \sin^{-1} F$ and j is an integer) and the other (x_u) is unstable. In the *overdamped* limit, $\eta \to \infty$, all system trajectories converge to one of the equivalent attractors as $t \to \infty$, $x(t) - x_s \propto \exp(-t/\tau')$, where the relaxation time is $\tau' = [m_a/V''_{\rm sub}(u_F)]^{1/2}$. As F increases, the stable and unstable solutions collide into a marginally stable static solution and the saddle-node bifurcation takes place at $F = F_s$, above which no static solution exists anymore. Close to the bifurcation the stable static solution behaves as $u_F(F) - u_0 \propto (F_s - F)^{1/2}$, and the relaxation time τ' diverges at the depinning transition,

$$\tau' \propto (F_s - F)^{-1/2}. \tag{8.3}$$

Above the depinning transition but close to it, the trajectories of the overdamped system spend most of their time in the neighborhoods of u_0, where the atomic velocity is close to zero. The time τ spent near the "sticking point" u_0 diverges as

$$\tau \sim \int_{u_0-\delta}^{u_0+\delta} \frac{dx}{\dot{x}} \propto (F - F_s)^{-1/2}, \tag{8.4}$$

where $0 < \delta < a_s$. Thus, the depinning transition of the single-particle system is characterized by two characteristic times, τ' (below the transition) and τ (above the transition), both diverging with the exponent $\psi = 1/2$. When the force decreases above the transition, the average velocity $\bar{v}$ tends to zero as

$$\bar{v} \propto \tau^{-1} \propto (F - F_s)^{1/2}. \tag{8.5}$$

The Smoluchowsky equation for one atom in the inclined sinusoidal potential (8.1) has an exact steady-state solution [595] which yields

$$\frac{B}{B_f} = \frac{k_B T}{F a_s} \frac{A''}{A_+ A_- - A' A''}, \tag{8.6}$$

where

$$A_\pm = \int_0^{a_s} \frac{dx}{a_s} \exp\left[\pm \frac{V_{\text{tot}}(x)}{k_B T}\right], \tag{8.7}$$

$$A' = \int_0^{a_s} \frac{dx}{a_s} \int_0^{x} \frac{dx'}{a_s} \exp\left[\frac{-V_{\text{tot}}(x) + V_{\text{tot}}(x')}{k_B T}\right], \tag{8.8}$$

and

$$A'' = 1 - \exp\left(-\frac{F a_s}{k_B T}\right). \tag{8.9}$$

In the linear response limit, $F \to 0$, Eq. (8.6) reduces to the expression $B/B_f = I_0^{-2}(\varepsilon_s/2k_BT)$, where I_0 is the modified Bessel function [see Eq. (7.64) above in Sect. 7.1.4], while in the $T \to 0$ limit it gives

$$\frac{B}{B_f} = \begin{cases} 0 & \text{for } F < F_s, \\ \left[1 - (F_s/F)^2\right]^{1/2} & \text{for } F > F_s. \end{cases} \tag{8.10}$$

It is interesting to note that at the threshold force, $F \to F_s$, the effective diffusion coefficient, defined as $D = \lim_{t\to\infty} \left[\langle x^2(t)\rangle - \langle x(t)\rangle^2\right]/2t$, is greatly enhanced as compared to free thermal diffusion [688].

8.2.2 Underdamped Case

In the *underdamped* case, $\eta < \eta_c \approx 1.193\,\omega_0$, where $\omega_0 = \left(2\pi^2\varepsilon_s/m_a a_s^2\right)^{1/2}$ is the frequency of vibration at the minimum of the substrate potential ($\omega_0 = 1$ in our system of units), the system may have a running solution even if the minima of the potential $V_{\text{tot}}(x)$ exist. Indeed, because of its momentum the particle may overcome the next hill, which is lower than the one from which it was falling due to the $-Fx$ contribution to the potential, if the gain in potential energy is greater than the energy dissipated during this motion (see Fig. 8.1). The critical force for the backward *sliding-to-locked* transition may occur at a lower force, $F_b < F_s$ (see the parameter regions of η and F in Fig. 8.2). The backward threshold force F_b can be found by balancing the gain in energy due to the driving force and the energy loss due to dissipation. When the particle moves the distance a_s (one period of the external potential), it gains the energy $E_{\text{gain}} = F a_s$ and losses some energy E_{loss},

$$E_{\text{loss}} = \int_0^{\tau_w} dt\, F_{\text{fric}}(t)\, v(t) = \int_0^{\tau_w} dt\, m_a \eta\, v^2(t) = m_a \eta \int_0^{a_s} dx\, v(x), \tag{8.11}$$

where τ_w is the period of the washboard potential (i.e. the time of motion for the distance a_s) and $F_{\text{fric}}(t) = m_a \eta\, v(t)$ is the external frictional force that

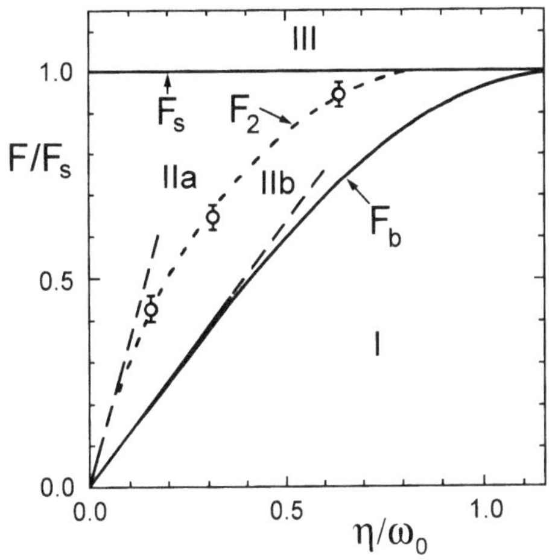

Fig. 8.2. Phase diagram of the different solutions of the one-particle FPK equation in an inclined sinusoidal potential without noise ($T = 0$). Shown are: locked solution (I), running solution (III), and coexistence of locked and running solutions (II). If an infinitesimally small noise is added, the bistability region (II) splits into running (IIa) and locked (IIb) ones. The dashed lines correspond to Eqs. (8.12) and (8.13) [689].

causes the losses. In the regime of the steady motion these energies must be equal to each other, $E_{\text{gain}} = E_{\text{loss}}$. Thus, the backward threshold force for the transition from the sliding (running) motion to the locked (pinned) state is determined by the condition $F_b = \min(E_{\text{loss}})/a_s$. The minimal losses are achieved when the particle has zero velocity on top of the total external potential (8.1). In the limit $\eta \to 0$ when $F \to 0$, from the energy conservation law, $(1/2)m_a v^2(t) + (1/2)\,\varepsilon_s\{1 - \cos[2\pi x(t)/a_s]\} = \varepsilon_s$, we can find the particle velocity $v(x)$ and then substitute it into Eq. (8.11). This yields

$$F_b \approx \frac{m_a \eta}{a_s}\left(\frac{\varepsilon_s}{m_a}\right)^{1/2} \int_0^{a_s} dx \left[1 - \cos\left(\frac{2\pi x}{a_s}\right)\right]^{1/2} = C'\eta\,(\varepsilon_s m_a)^{1/2}\,,$$

and then

$$F_b \approx C\left(\frac{\eta}{\omega_0}\right) F_s \quad \text{for} \quad \eta \ll \omega_0, \tag{8.12}$$

where $C' \equiv (2\pi)^{-1}\int_0^{2\pi} dy\,(1-\cos y)^{1/2} = 2\sqrt{2}/\pi \approx 0.9$ and $C = C'\sqrt{2} = 4/\pi$ are numerical constants which depend on the shape of the external potential in a general case. Equation (8.12) may be rewritten as $F_b = m_a\eta\bar{v} = (2/\pi)\,m_a\eta v_m$, where $\bar{v} = a_s^{-1}\int_0^{a_s} dx\,v(x)$, and $v_m = (2\varepsilon_s/m_a)^{1/2} = \pi\bar{v}/2$ is the maximum velocity achieved by the particle when it moves near the bottom of the external potential $V_{\text{tot}}(x)$. The average particle velocity, $\langle v\rangle = \tau_w^{-1}\int_0^{\tau_w} dt\,v(t) = a_s/\tau_w$, tends to zero when $F \to F_b$, because $\tau_w \to \infty$ in this limit, $\tau_w \propto (F - F_b)^{-1/2}$ and $\langle v\rangle \propto (F - F_b)^{1/2}$, similarly to Eqs. (8.5) and (8.4) of the overdamped limit. Although the transition from the sliding motion to the locked state is continuous, the velocity drops quite sharp at the threshold force $F = F_b$.

As the particle is either locked or running, depending on its initial velocity, the system exhibits bistability and the transition between these two states shows hysteresis due to inertia mechanism [690]. However, the motion of a single particle driven by an external force shows hysteresis only for zero temperature. The bistability disappears in the presence of an external noise, $T > 0$, because the fluctuations may kick the particle out of the locked or running state [689]–[692].

For a low temperature, $T \to 0$ but $T > 0$, the bistable region II (see Fig. 8.2) is split into the running and locked subregions by a curve $F_2(\eta)$ shown by dashed curve in Fig. 8.2 (see also Refs. [595], [693]–[695]). For low damping the threshold force F_2 is given by the expression

$$F_2(\eta) \approx C''(\eta/\omega_0)F_s \quad \text{for} \quad \eta \ll \omega_0, \tag{8.13}$$

where $C'' \approx 3.3576$ is a numerical constant [595, 693] (we notice that the constant C'' can be estimated analytically as $C'' \approx 2 + \sqrt{2} \approx 3.4$, see [696]). When $F < F_2$, the system is mainly in the locked state and undergoes a creep (thermally activated) motion with $B(F) \approx B_0$, while for $F > F_2$ the system is mainly in the running (sliding) state and $B(F) \approx B_f$. Thus, when the force F increases crossing the value F_2, the mobility $B(F)$ changes sharply from B_0 to B_f as shown in Fig. 8.3. Emphasize that at any small but nonzero temperature, the forward locked-to-running transition occurs at the threshold force $F_2 < F_s$, and only at $T = 0$ the threshold force changes to F_s. When the temperature T is increased, this transition is smeared out. Thus, there are no hysteresis at any $T > 0$. However, if F changes with a *finite* rate, then the hysteresis exists as will be explained below. Moreover, even a very small additional ac driving restores the hysteresis, independent on the forcing frequency and system temperature. The hysteresis mechanism in the latter case is related to the occurrence of long (multiple) atomic jumps, whose

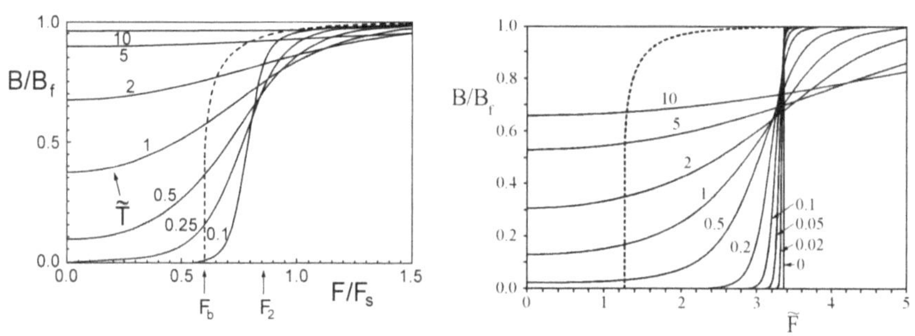

Fig. 8.3. Stationary mobility of a single atom in a sinusoidal potential vs. the external force F, for different dimensionless temperatures $\widetilde{T} = 2k_BT/\varepsilon_s$ ($\widetilde{T} = 0^+ - 10$). Dashed curve corresponds to the stationary mobility of the noiseless running atoms ($T = 0$) given by Eq. (8.24). Left panel: B/B_f vs. F/F_s for $\eta = 0.5$ calculated by the matrix continued fraction method (Vollmer and Risken, 1983). Right panel: B/B_f vs. $\widetilde{F} \equiv F\omega_0/F_s\eta$ calculated analytically in the low-friction limit ($\eta \to 0$) with the help of Eq. (8.25), for $\widetilde{T} < 0.1$, and Eq. (8.17), for $\widetilde{T} > 0.1$ [595].

length and time duration distribution decays according to a universal power law (contrary to the exponential decay in the purely dc–driven system).

In the underdamped case, analytical results are known for the $\eta \to 0$ limit only. Because they will be used for the FK model as well, below we briefly present these results following the monograph by Risken [595]. In the limit $\eta \to 0$ (simultaneously one has to put $F \to 0$ keeping the ratio F/η at a finite value) it is useful to rewrite the FPK equation in the energy-coordinate variables, where the energy variable is defined by $E = (1/2)m_a v^2 + (1/2)\,\varepsilon_s[1 - \cos(2\pi x/a_s)]$, and then to look for the distribution function $W_F(\sigma, E, x)$ for the steady-state solution, which depends additionally on the sign of the velocity $\sigma = \mathrm{sgn}(v/F)$, so that $\sigma = +1$ for atoms moving in the force direction ($v > 0$) and $\sigma = -1$ for atoms moving against the force ($v < 0$). Averaging over the "fast" variable x, one obtains the reduced distribution $W_F(\sigma, E)$, which may be presented at low temperatures $k_B T \ll \varepsilon_s$ in the form

$$W_F(\sigma, E) \propto \exp\left[-V_{\mathrm{eff}}(\sigma, E)/k_B T\right], \tag{8.14}$$

where

$$V_{\mathrm{eff}}(\sigma, E) = \begin{cases} E & \text{for } 0 \le E \le \varepsilon_s, \\ E - \sigma\,(F/\eta)\, g(E) & \text{for } \quad E > \varepsilon_s, \end{cases} \tag{8.15}$$

and the function $g(E)$ is given by the integral

$$g(E) = \frac{1}{m_a}\int_{\varepsilon_s}^{E} \frac{dE'}{\bar{v}(E')}, \quad \bar{v}(E) = \frac{2\sqrt{2}}{\pi}\left(\frac{E}{m_a}\right)^{1/2} \mathbf{E}\left[\left(\frac{\varepsilon_s}{E}\right)^{1/2}\right] \tag{8.16}$$

for the sinusoidal external potential.

The function $V_{\mathrm{eff}}(\sigma, E)$ is shown in Fig. 8.4. It has a minimum $V_L = 0$ at $E = 0$ which describes the locked state, and additionally, if the force

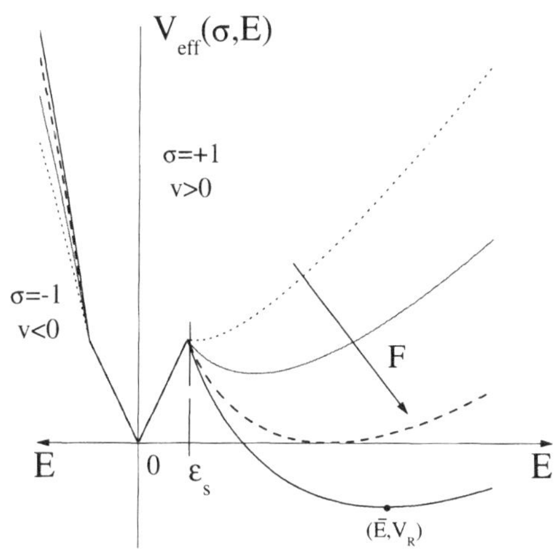

Fig. 8.4. Family of the effective potentials $V_{\mathrm{eff}}(\sigma, E)$ for different values of the force F. Dotted curve is for the force $F = F_b$ corresponding to the spinodal, when nontrivial minimum of the potential $V_{\mathrm{eff}}(+1, E)$ disappears. Dashed curve corresponds to the binodal, when the energies of locked and running states coincide ($F = F_2$).

is not too small, $F > F_b$, a minimum $V_R = V_{\text{eff}}(+1, \bar{E})$ at $E = \bar{E}$ which corresponds to the running state. The minima are separated by a maximum $V_m = V_{\text{eff}}(+1, \varepsilon_s) = \varepsilon_s$ at $E = \varepsilon_s$. The absolute minimum corresponds to the ground state, while the local minimum, to the metastable state. The threshold force F_2 is just determined by the condition that the energies of both states are equal one another, $V_L = V_R$. Thus, for forces $F < F_2$ the system is mainly in the locked state, while for $F > F_2$ the atom is mainly in the running state. At low temperatures, $k_B T \ll \varepsilon_s$, the transition from the metastable state to the ground state is an activated process, and a rate of the transition is given by Kramers expressions (7.69, 7.70) of Sect. 7.1.4. Thus, it is exponentially small, $\mathcal{R} \sim (\eta \,\Delta V / k_B T) \exp(-\Delta V / k_B T)$, where $\Delta V = V_m - V_{L,R}$. Therefore, for any small but nonzero rate R of changing of the driving force, the system has to exhibit hysteresis, the width of which grows with an increase of R and a decrease of T.

At not too low temperatures, $\widetilde{T} \equiv 2k_B T/\varepsilon_s > 0.1$, this approach leads to the steady-state mobility

$$\frac{B}{B_f} = C + D \left(\frac{\eta}{\omega_0} \frac{1}{\widetilde{\widetilde{T}}} \right)^{1/2}, \tag{8.17}$$

where the coefficients C and D depend on F and T,

$$C = \frac{A_3}{A_0 + A_1}, \quad D = \kappa \frac{2\sqrt{2}}{\sqrt{\pi}} \left[\frac{A_4}{A_0 + A_1} - \frac{F}{m_a \eta \,\bar{v}_s} \frac{A_2 A_3}{(A_0 + A_1)^2} \right]. \tag{8.18}$$

Here $\kappa = 0.855(4)$ is a numerical constant, $\bar{v}_s = \bar{v}(\varepsilon_s) = (4/\pi)(a_s/2\pi)\,\omega_0$, $A_0 = (\pi k_B T / 2m_a)^{1/2} \, I_0 \left(\widetilde{T}^{-1} \right)$, and the coefficients A_1 to A_4 are defined by the integrals

$$A_1 = \int_{\varepsilon_s}^{\infty} dE \; \bar{v}'(E) \; \exp(-E/k_B T) \; [\cosh h(E) - 1], \tag{8.19}$$

$$A_2 = \int_{\varepsilon_s}^{\infty} dE \; \bar{v}'(E) \; \exp(-E/k_B T) \sinh h(E), \tag{8.20}$$

$$A_3 = m_a^{-1} \int_{\varepsilon_s}^{\infty} dE \; \bar{v}^{-1}(E) \; \exp(-E/k_B T) \cosh h(E), \tag{8.21}$$

and

$$A_4 = (m_a \bar{v}_s)^{-1} \int_{\varepsilon_s}^{\infty} dE \; \exp(-E/k_B T) \cosh h(E), \tag{8.22}$$

where $h(E) = F g(E) / \eta \, k_B T$ and

$$\bar{v}'(E) = \frac{d\,\bar{v}(E)}{dE} = \frac{\sqrt{2}}{\pi} \left(\frac{1}{m_a E} \right)^{1/2} \mathbf{K} \left[\left(\frac{\varepsilon_s}{E} \right)^{1/2} \right]. \tag{8.23}$$

The case of zero temperature shown by broken curve in Fig. 8.3, corresponds to the solution of the deterministic motion equation

$$F = m_a \eta \, \bar{v}(\bar{E}), \tag{8.24}$$

where the energy $\bar{E}$ corresponds to the (local) minimum of the function $V_{\text{eff}}(E)$ in the region $E > \varepsilon_s$, i.e. to the energy of the atomic motion in the running state. Equation (8.24) just describes the balancing of the energy gain due to the driving force and energy loss due to the friction, as discussed above at the beginning of this subsection [see Eqs. (8.11) and (8.12)]. Because the washboard period is equal to $\tau_w = \int_0^{a_s} dx/v(x,E) = m_a a_s \bar{v}'(\bar{E})$ according to Eq. (8.23), the $T = 0$ mobility B_R is $B_R/B_f = \eta/F\bar{v}'(\bar{E})$. In particular, for the force $F = F_2$, when the locked and running states are characterized by the same energy V_{eff}, it was found that $\bar{E}/\varepsilon_s \approx 3.32955$ and $B_R/B_f \approx 0.9960$. Finally, at very low temperatures, $0 < \widetilde{T} < 0.1$, the mobility is given by an approximate expression [595]

$$\frac{B}{B_f} \approx \frac{B_R}{B_f} \left\{ 1 + \left(\frac{\widetilde{T} B_R}{2\pi B_f} \right)^{1/2} \exp\left[\frac{V_R}{k_B T} \right] \right\}^{-1} . \tag{8.25}$$

The dependencies $B(F)$ calculated with the help of Eqs. (8.17) to (8.25) for different temperatures are presented in Fig. 8.3 (right panel), while the dependencies of the mobility B on temperature for different forces are shown in Fig. 8.5.

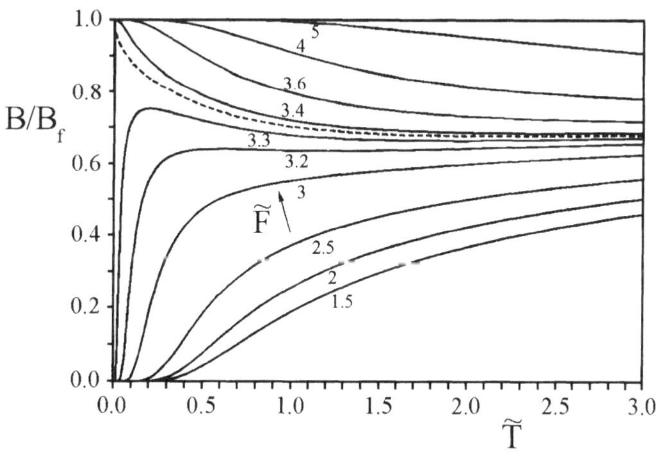

Fig. 8.5. Mobility B/B_f vs. the dimensionless temperature $\widetilde{T}$, for for a single atom in a sinusoidal potential and different values of the dimensionless force $\widetilde{F} = F\omega_0/F_s\eta$ [595]. Note that at $T \to 0$ the mobility B is exponentially small at $F < F_2$ and is close to B_f at $F > F_2$, while at $F = F_2$ we have $B/B_f \approx 0.9960$.

8.3 Overdamped FK Model

In the case of large or intermediate damping, the chain of interacting atoms behaves similarly to the considered above case of noninteracting atoms. Moreover, in the absence of the external noise ($T = 0$) several exact results are available, such as the uniqueness of the solution, the existence of the dynamical hull function, and the universal scaling law for the locked-to-sliding transition.

At high damping one can neglect the inertial term, and the motion equation reduces to

$$\dot{x}_l = g\,(x_{l+1} + x_{l-1} - 2x_l) - \sin x_l + F, \tag{8.26}$$

where the dimensionless time is used ($t \to \eta t$).

Most of analytical results in the overdamped limit have been obtained from the "no-passing" rule formulated by Middleton [697]–[700]. The rule states that the dissipative dynamics preserves the chain's order: if $\{x_l(t_0)\} < \{y_l(t_0)\}$ at an initial time moment t_0, then $\{x_l(t)\} < \{y_l(t)\}$ for all later times $t > t_0$, provided the interatomic interaction is convex [e.g., harmonic as in the classical FK model (8.26)]. Indeed, suppose that the rule does not hold, and denote by t' the earliest time moment at which some atom (with an index l') of the "lesser" configuration $\{x_l(t)\}$ reaches the l'-th atom of the "greater" configuration $\{y_l(t)\}$, $x_{l'}(t') = y_{l'}(t')$. Then from Eq. (8.26) one obtains for the relative velocity $\dot{x}_{l'}(t') - \dot{y}_{l'}(t') = g\,[x_{l'+1}(t') - y_{l'+1}(t') + x_{l'-1}(t') - y_{l'-1}(t')] < 0$, so that the crossing of the configurations is not possible. Moreover, even if two configurations "touch" each other at more than one point [$x_l(t_0) \le y_l(t_0)$ for all l] but both configurations do not coincide asymptotically either as $l \to \infty$ or as $l \to -\infty$, they will separate at later times $t > t_0$, $\{x_l(t)\} < \{y_l(t)\}$.

Let us define the *width* $W(t)$ of a configuration $\{x_l(t)\}$ as the maximum deviation of the length of a finite segment of the configuration from the average length, $W(t) = \max_{l,j} |x_{l+j}(t) - x_l(t) - jwa_s|$. The value of W provides a measure of the degree of spacial regularity of the configuration. In what follows we will assume that all configurations under consideration have a bounded width. An immediate consequence of the no-passing rule is that a rotationally ordered configuration (see Sect. 5.1) remains rotationally ordered at any later time (note also that for these configurations $W(t)/a_s < 1$). The next important consequence is the uniqueness of the average velocity $\bar{v} = \langle\langle \dot{x}_l \rangle_x\rangle_t$ [here $\bar{v} = \langle v(t)\rangle_t = \lim_{\tau\to\infty} \tau^{-1} \int_{t_0}^{t_0+\tau} dt\, v(t)$ and $v(t) = \langle \dot{x}_l(t)\rangle_x = \lim_{N\to\infty} N^{-1} \sum_{l=l_0}^{l_0+N-1} \dot{x}_l(t)$]: the velocity $\bar{v}$ does not depend on the initial configuration (but it depends, of course, on the window number w, the elastic constant g, and the applied force F). Indeed, let us take two initial configurations $\{x_l(t_0)\} < \{y_l(t_0)\}$, both having the same window number. Then, one can always find a pair of integers i and j such that $\{x'_l(t_0)\} = \sigma_{i,j}\{x_l(t_0)\} > \{y_l(t_0)\}$, where the transformation $\sigma_{i,j}$ was defined above in Sect. 5.1 by Eq. (5.5), and from the no-passing rule we obtain $\{x_l(t)\} < \{y_l(t)\} < \{x'_l(t)\}$ for all times $t > t_0$, thus the long-term average velocity of both trajectories must be the same.

Because of uniqueness of the average velocity $\bar{v}$, for a given set of parameters w and g any *steady-state* solution of the overdamped FK model is either locked (pinned) with $\bar{v} = 0$, or it is sliding (running) with $\bar{v} > 0$, independently on the initial state. The state is pinned at $F < F_{PN}$ and it is sliding for $F > F_{PN}$. *The sliding steady state is asymptotically unique* up to time translations. The rigorous proof of this statement has been given by Middleton [697, 699] (see also a review paper by Floría and Mazo [487]). The idea is the following: one has to construct a solution for which, at any time, the velocities of all atoms are positive; then this solution is used to bound any arbitrary solution from above and below, and finally, one shows that these bounds approach each other as $t \to \infty$ due to the no-passing rule. Moreover, it follows that *the sliding steady state is rotationally ordered and has positive velocities everywhere at all times.*

Let $\{x_l(t)\}$ be any steady-state solution and i and j, two arbitrary integers. As $\{x_{l+i}(t)\}$ is also a steady-state solution, there is some time τ such that $\{x_{l+i}(t)\} = \{x_l(t+\tau)\}$ for all t. Thus, the average velocity $v(t)$ is periodic with the period τ (or less), and from the relationships

$$\bar{v}\tau = \langle x_l(t+\tau) - x_l(t)\rangle = \langle x_{l+i}(t) - x_l(t)\rangle = wia_s$$

it follows that

$$\left\{x_{l+i}\left(t - \frac{iwa_s}{\bar{v}}\right)\right\} = \{x_l(t)\} \tag{8.27}$$

for all t. In a similar manner one can derive that

$$\left\{x_l\left(t - \frac{ja_s}{\bar{v}}\right) + ja_s\right\} = \{x_l(t)\} \tag{8.28}$$

for all t. Both results (8.27) and (8.28) may be combined in the statement that any steady-state solution with the window number w and the average velocity $\bar{v}$ is invariant under the transformation

$$\sigma_{i,j,(iw+j)a_s/\bar{v}}\{x_l(t)\} = \{x_l(t)\} \tag{8.29}$$

for all t, where $\sigma_{i,j,\tau}$ is defined by $\sigma_{i,j,\tau}\{x_l(t)\} = \{x_{l+i}(t-\tau) + ja_s\}$.

All the above described properties yield the result that any steady-state solution in the sliding regime can be presented in the form [701, 702]

$$x_l(t) = h(lwa_s + \bar{v}t + \beta) \tag{8.30}$$

for all l and t, where $h(x)$ is a uniquely defined real function called the *dynamical hull function*, and the phase β is any real number. The function

$h(x)$ is an analytical and monotonic strictly increasing function, and the function $\phi(x) = h(x) - x$ is periodic with the period a_s. The set of sliding states in the sliding regime forms a continuum owing to the phase β in Eq. (8.30). As a result, *the sliding structures must be undefectible*, they do not admit defects. For a commensurate structure with $w = r/s$ the velocity $v(t)$ is a periodic function with the period $a_s/s\bar{v}$, while for an incommensurate structure the velocity is constant, $v(t) = \bar{v}$. Finally, in the overdamped system the average velocity $\bar{v}$ is always a monotonically increasing function of F.

The depinning transition in the overdamped system proceeds essentially in the same manner as for the single-particle system, through the saddle-node bifurcation at $F = F_{PN}$. For a *commensurate* structure, $w = r/s$, the problem reduces to that of a system with a finite number s of degrees of freedom. At the depinning transition, the characteristic times τ' and τ both diverge with the same exponent $\psi = 1/2$ similarly to Eqs. (8.3), (8.4). In the sliding regime, $F > F_{PN}$, the existence of the dynamical hull function $h(x)$ permits the analysis in terms of a single degree of freedom,

$$\dot{x} = g(x), \tag{8.31}$$

where $g(x) = \bar{v}h'(z)$ and $z = h^{-1}(x)$. The function $g(x)$ is a periodic function, $g(x + a_s) = g(x)$, and, close to the transition, possesses s local minima per period (see Fig. 8.6, left panel) associated with the atomic positions of the threshold static configuration at $F = F_{PN}$. Close to the transition the atomic motion slows down in the neighborhood of those positions, and the average velocity tends to zero as $\bar{v} \propto (F - F_{PN})^{1/2}$ analogously to the single-particle case.

In the case of an *incommensurate* structure but above the Aubry transition, $g > g_{\text{Aubry}}$, we have $F_{PN} = 0$, and the system slides with the constant

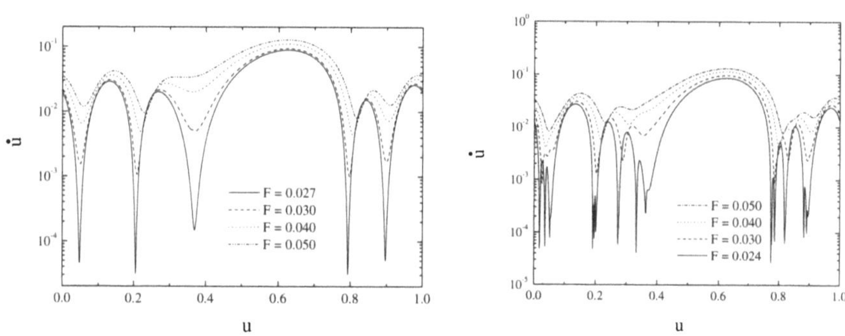

Fig. 8.6. Atomic velocities versus their positions in the sliding steady state [Eq. (8.31), here we plot $u = x/a_s$] for different values of the driving force F close above the depinning PN force for the overdamped FK model. Left panel: commensurate structure with $w = 3/5$ for $g = 0.5$ ($F_{PN} \approx 0.026908$); Right panel: "incommensurate" structure with $w = 144/233$ for $g = 0.5$ ($F_{PN} \approx 0.023620$) [487].

velocity $v(t) = F$ under any nonzero applied force (the intrinsic phonon damping is irrelevant in the overdamped system). On the other hand, below the Aubry transition, $g < g_{\rm Aubry}$, we have $F_{PN} > 0$, and the chain is pinned at forces $F < F_{PN}$. The problem reduces in this case to one with infinitely many degrees of freedom. According to numerical results by Coppersmith [524] and Coppersmith and Fisher [702], the characteristic time τ' (which is the inverse of the square root of the lowest eigenvalue of the corresponding elastic matrix) at $F \to F_{PN}$ still scales according to Eq. (8.3) as in the single-particle system, but now there is a distribution of eigenvectors with eigenvalues approaching zero (soft modes). These soft modes are highly localized and corresponds to atoms which are close to hopping over the substrate potential maxima.

Above the depinning transition, $F > F_{PN}$, the analysis again reduces to that of the one-variable equation (8.31). As in the commensurate case, the function $g(x)$ (see Fig. 8.6, right panel) exhibits local minima with values approaching zero linearly in $F - F_{PN}$. Thus, there is a diverging sticking time $\propto (F - F_{PN})^{-1/2}$ associated with each local minimum of $g(x)$. The important difference with respect to the commensurate case is now that *the number of local minima diverges* in the limit $F \to F_{PN}$ as

$$\propto (F - F_{PN})^{-\psi_\delta} \tag{8.32}$$

with an exponent $0 < \psi_\delta < 1/2$. Therefore, the behavior of the average velocity $\bar{v}$ of an incommensurate structure at the depinning transition is characterized by a critical exponent ψ_v,

$$\bar{v} \propto (F - F_{PN})^{\psi_v}, \tag{8.33}$$

where $\psi_v = 1/2 + \psi_\delta$ so that $1/2 < \psi_v < 1$. Intuitively speaking, the divergent number of local minima of $g(x)$ corresponds to the atomic positions of the threshold static incommensurate configuration at $F = F_{PN}$. The exponent ψ_δ depends both on the window number w and the elastic constant g. One could expect that ψ_δ is highest for the golden-mean window number, and that it increases with g approaching the value $\frac{1}{2}$ at $g = g_{\rm Aubry}$, consistently with the linear behavior ($\bar{v} = F$) above the Aubry transition.

A simple case of the locked-to-sliding transition in the system with one kink in the highly-discrete limit ($g \ll 1$) was considered in detail by Carpio and Bonilla [703], where scaling laws and asymptotic of kink's tails at the threshold $F \sim F_{\rm PN}$ were found.

Finally, we would like to mention the work by Zheng *et al.* [704], where a specific example of the "gradient-driving" chain was considered. In this model the motion equation has the form

$$\dot{x}_l = g\,(x_{l+1} + x_{l-1} - 2x_l) - \sin x_l + F_l[x], \tag{8.34}$$

where the driving force F_l depends on the atomic coordinates of the NN atoms.

$$F_l[x] = r\,(x_{l+1} - x_{l-1} - 2a_{\min}) \tag{8.35}$$

(here r is a coefficient and $a_{\min}$ is the equilibrium spring's length). For the trivial commensurate concentration, when $x_l = la_s$ and $a_{\min} = a_s$, the driving force vanishes. But if the chain's state contains geometrical kinks, they will drive themselves with a constant velocity in the steady state (note that this model has no true ground state). The properties of this model are much similar to those of the dc-driving case described above.

At low temperatures ($k_BT < E_k$) one may use the quasiparticle-gas approach. Recall that in this approach the chain's conductivity is supported due to creation of kink-antikink pairs with their consequent motion in the opposite directions. For the highly discrete FK chain, $g \ll 1$, the atomic flux is restricted by an overcoming of kinks over the PN barriers. With increasing of the driven force F the PN barriers are lowered [together with the original barriers of $V_{\rm tot}(x)$], resulting in increasing of the single-kink mobility. So, the crossover from $B_0 = \beta D_\mu$ to B_f is expected to occur at forces lower or equal to $F \sim \pi \varepsilon_{\rm PN}/a_s$ similarly to Eq. (8.2).

Otherwise, in the SG limit, $g \gg 1$, the restricting factor is the rate of creation of new kink-antikink pairs. When the force F increases, the energy threshold for the creation will decrease from $2E_k$ to $\sim 2E_k - Fx^*$, where x^* is the saddle point determined by a solution of an equation (see Fig. 8.7)

$$\frac{d}{dx}\left[v_{\rm int}(x) - F\,x\right] = 0, \tag{8.36}$$

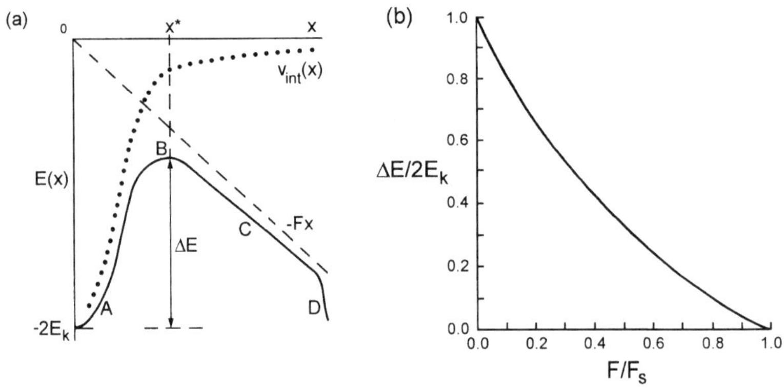

Fig. 8.7. (a) Lowering of the activation energy ΔE for the creation of kink-antikink pair in presence of a driven force F. In the range (A) an energy must be supplied to pull the kink and antikink apart until a critical separation x^* is reached given by the width of the critical nucleus (B). In the range (C) the energy change is dominated by the applied force, rather than by the kink-antikink interaction. At (D) recombination with another kink terminates the motion of one of the original partners. (b) Activation energy ΔE as a function of F for the SG model [676].

$v_{\rm int}(x)$ being the energy of kink-antikink attraction. Thus, the external force lowers the saddle-point energy increasing the rate of creation of kink-antikink pairs. This approach was firstly used by Kazantsev and Pokrovsky [247] for the overdamped SG model, and then it was further developed by Büttiker and Landauer [676], Munakata [705], Marchesoni [706], and Büttiker and Christen [678]. The atomic flux was calculated within the generalized rate theory which takes into account the multi-dimensional nature of the saddle point. The rate theory is adequate provided $F < F_s$ only, i.e., as far as the system dynamics has activated nature. The results of calculation are presented in Fig. 8.8.

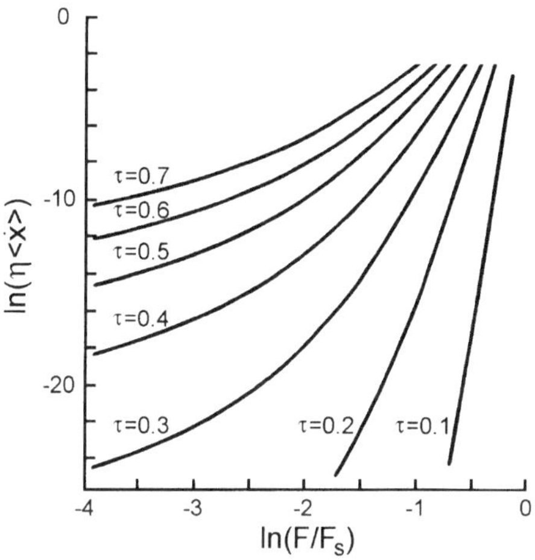

Fig. 8.8. Average velocity as a function of the applied force for different values of low temperatures. The curves are labelled by the values of $\tau = 8k_BT/E_k$ [676].

At high temperatures, $k_BT > E_k$, the conductivity of the FK chain may be calculated with the aid of the RPA approximation (see Sect. 7.4.2) using the ansatz (7.192). Trullinger *et al.* [671] and Guyer and Miller [496] have shown that the time-independent N-particle Smoluchowsky equation can be reduced to one-particle equation with an effective one-atomic potential $V_{\rm eff}(x)$ which is determined by the atomic density $\rho(x)$. Because $\rho(x)$ itself depends on $V_{\rm eff}$ (through the solution of the Smoluchowsky equation) and both $V_{\rm eff}(x)$ and $\rho(x)$ depend on the external force F, the problem should be solved in a self-consistent fashion. Numerical results obtained by the transfer-integral method, shows that at low external field and at high field the atomic flux is proportional to the field (Ohm's law), but the high-F conductivity may be many orders of magnitude greater than the low-F one. The low-F and high-F regimes are separated by a region of very nonlinear conductivity, and the crossover takes place at forces $F \leq F_s$ similarly to the case of noninteracting atoms.

8.4 Driven Kink

In the continuum limit approximation the system dynamics in the presence of an external force is described by the perturbed SG-type equation

$$\frac{\partial^2 u}{\partial t^2} - \frac{\partial^2 u}{\partial x^2} + \eta \frac{\partial u}{\partial t} + V'_{\text{sub}}(u) = \epsilon f(x,t), \tag{8.37}$$

where $V'_{\text{sub}}(u) = dV_{\text{sub}}(u)/du = \sin u$ for the sinusoidal substrate potential. We already discussed the equation of this type in Sect. 3.4.1, where the interaction of kinks with impurities was considered.

Using a simple version of the collective-coordinate approach, we can look for a solution of Eq. (8.37) in the form of a perturbed SG kink,

$$u_{\text{k}}(x,t) = 4 \arctan \exp \left\{ -\frac{\sigma \left[x - X(t)\right]}{d\sqrt{1 - \dot{X}^2(t)/c^2}} \right\}, \tag{8.38}$$

where $X(t)$ is the kink's coordinate and d is its width. Following the method by McLaughlin and Scott [86], first we have to compute the variation of the energy and momentum of the unpertubed SG system due to the perturbation $\epsilon f(x,t) - \eta\, u_t$, then to calculate those variations in terms of the unknown function $X(t)$ using the ansatz (8.38), and finally to make these two ways of calculation consistent. This approach finally results in the equation

$$\left[1 - \dot{X}^2(t)/c^2\right]^{-1} M\!\left[\dot{X}(t)\right] \ddot{X}(t) + \eta\, M\!\left[\dot{X}(t)\right] \dot{X}(t) = F(t), \tag{8.39}$$

where

$$F(t) = 2\sigma\epsilon \int_{-\infty}^{+\infty} dx\, f(x,t) \text{sech}^2 \left[x - X(t)\right], \tag{8.40}$$

$M(v) = m/\sqrt{1 - v^2/c^2}$ and $m = 2/(\pi^2 \sqrt{g})$ is the kink rest mass. In particular, for the damped motion of the dc driven kink, when $\epsilon f(x,t) = F$, we obtain for its steady state the velocity

$$v_{\text{k}} = F/\left[M(v_{\text{k}})\, \eta\right]. \tag{8.41}$$

This approach, however, completely neglects the phonon radiation which should emerge due to destroying of the exact integrability of the SG equation, first of all because of discreteness of the FK model. In the next Sect. 8.5 we will show that the discreteness results in the instability of the fast kink. A nonsinusoidal shape of the substrate potential or the nonharmonic character of the interatomic interaction could also change the simple relation (8.41). The latter effect will be described below in Sect. 8.6, where we show that the anharmonicity of the interaction results in a drastic change of system behavior, allowing the existence of supersonic and multiple kinks. Below we briefly discuss the role of the former effect, i.e. the kink dynamics in the case of the nonsinusoidal substrate potential.

If the substrate potential is nonsinusoidal but symmetric as those discussed in detail in Sect. 3.3, then the zero-temperature dynamics of the dc-driven kink can still be described by Eq. (8.41), provided the corresponding kink's rest mass is used (in the case of nonsymmetric substrate potential a specific "ratchet" dynamics typically emerges as will be discussed in the next Chapter 9). However, a new interesting effect appears at *non-zero temperature* in the case of multiple-well substrate potential, namely for the double-well potential (3.67) (see Fig. 3.13 in Sect. 3.3.4), when the potential has two distinct minima over the period $a_s = 2\pi$, one at $x_{01} = 0$ and another at $x_{02} = \pi$, both with the same energy but different curvatures. Recall that in this system the kink's shape is asymmetric, because its tails lie in the valleys of different curvatures. The model admits two types of kinks, called the "left kink" (LK), which has the long-range left-hand tail (with the atoms locating in the "wider", or "shallower" potential minimum) and the sharp right-hand tail (with the atoms in the "narrower" minimum), and the "right kink" (RK) which is just reverse. Although the energies of both minima are equal one another, their *entropies* at nonzero temperature are different, the GS with the stronger curvature is characterized by a lower entropy. Thus, at $T > 0$ the GS with the larger curvature has a higher *free energy* and, e.g., the LK will be pushed to the right by the "entropic force" $F_e \propto T$, so that in the final configuration all atoms will mostly occupy the shallower valley. This effect was first described by Savin *et al.* [707]. In the framework of the ϕ^4-type model, the entropic force was calculated by Costantini and Marchesoni [709]. The authors solved the pseudo-Schrödinger equation of the transfer-integral method (see Sect. 6.5.2) and obtained that

$$F_e(T) \propto -\left(\frac{k_B T}{d}\right) \frac{[\omega_{\min}^{(L)} - \omega_{\min}^{(R)}]}{[\omega_{\min}^{(L)} + \omega_{\min}^{(R)}]}, \tag{8.42}$$

where $\omega_{\min}^{(L,R)}$ are the minimum phonon frequencies in the left- and right-hand-side GS's correspondingly. Now, if one applies a dc force F, the total force $F + F_e(T)$ may be either positive or negative depending on temperature, and thus it will drive the kink either to the right or to the left.

In the FK-type model with the double-well potential (3.67), the effective substrate potential takes the shape of the double-barrier potential (see Sect. 3.3.5) at nonzero temperatures. Costantini and Marchesoni [709] discussed an interesting case of the double-barrier substrate potential with narrow global minima and shallow local (metastable) minima,

$$V_{\text{sub}}(u) \propto \frac{1 - B\cos(2u)}{(1 + A\cos u)}, \tag{8.43}$$

as shown in Fig. 8.9 (A and B are some parameters). Now the $T = 0$ GS corresponds to the configuration with all atoms located at the deepest min-

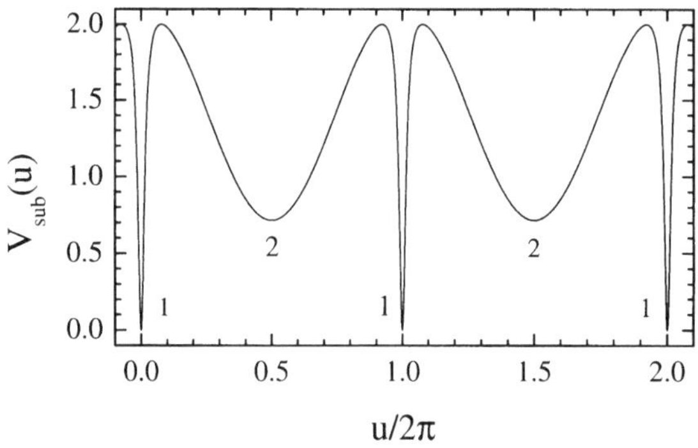

Fig. 8.9. The shape of the substrate potential (8.43) at $A = 0.995$ and $B = 1.01$.

imum of $V_{\text{sub}}(u)$ (labelled as 1 in Fig. 8.9), and the elementary topological excitation corresponds to the 2-π double kink with the tails located in the adjacent global minima. But if the temperature increases, the free energies of the states with atoms in the minima 1 and 2, $F_i(T) = E_i - TS_i$, will change with different rates, and at some critical temperature $T = T_c$ these free energies become equal one another. Therefore, the distance between the subkinks of the double kink will increase when $T \to T_c$, and at $T = T_c$ it will dissociate into two separate π-kinks. Moreover, with further increase of temperature these π-kinks should again be coupled into 2π double kinks, but now of the different shape, with the tails located at the local minima 2, because thermal fluctuations stabilize the valleys 2 over the valleys 1 at $T > T_c$.

8.5 Instability of Fast Kinks

Below in this and next Sections we consider an interesting case of low damping, and begin with a more simple case of a single driven kink in the classical FK model, when the interatomic interaction is harmonic. At zero temperature the kink starts to move at the force $F = F_{PN}$. Considering the kink as a rigid quasiparticle, one could find its velocity as $v_{\text{k}} = F/m_{\text{k}}\eta$. However, the following three new issues emerge. First, the kink is a deformable quasiparticle, so that its mass depends on the kink velocity. Second, the kink moves on the background of the harmonic chain and, thus, it may excite phonons in the chain; this should lead to an increase of the damping coefficient. Third, the kink may have its own degrees of freedom which, being excited, may destroy the steady kink motion at high velocity. The first two issues were already discussed in previous Chapters, so here we will discuss mainly the third effect.

The motion of a single kink which starts with a high speed in the highly discrete undamped FK model was studied by Peyrard and Kruskal [171]. They found two important phenomena. First, during its motion the kink experiences a strong interaction with phonons; therefore, at resonance conditions (i.e., with kink velocities in certain intervals) the kink motion is strongly damped and its velocity decreases quickly, while outside these resonance intervals the damping is very low, and the kink moves with such a velocity for a long time. The velocity-locking effect due to resonances of the washboard frequency of the driven kink with phonons in a finite (short) FK chain was studied also by Ustinov *et al.* [710] and by Watanabe *et al.* [102]. Second, Peyrard and Kruskal observed that the very fast kink is unstable, while a pair of two coupled kinks can be stable and move as a whole with very high velocity and practically without damping. Let us consider the instability of the driven kink in more details following the work by Braun *et al.* [711].

For the classical driven FK model, the motion equation is

$$\ddot{x}_l + \eta \dot{x}_l - g\left(x_{l+1} + x_{l-1} - 2x_l\right) + \sin x_l = F. \tag{8.44}$$

For a large value of the elastic constant, $g \gg 1$, we can use the continuum limit approximation, so that Eq. (8.44) reduces to the form

$$\frac{\partial^2 u}{\partial t^2} - d^2 \frac{\partial^2 u}{\partial x^2} + \eta \frac{\partial u}{\partial t} + \sin u = F, \tag{8.45}$$

where $d = a_s\sqrt{g}$ is the width of the static kink (in our system of units $d = c$). For the case of $F = 0$ and $\eta = 0$, this equation reduces to the SG equation. The kink (excessive atom) solution has the form $u(x,t) = 4\tan^{-1}\exp[-(x-vt)/d_{\mathrm{k}}(v)]$. The SG kink can move with any velocity $|v| < c$ and is characterized by the width $d_{\mathrm{k}}(v) = d\left(1 - v^2/c^2\right)^{1/2}$ and the effective mass $m_{\mathrm{k}}(v) = (4/\pi d)\left(1 - v^2/c^2\right)^{-1/2}$. Considering the kink as a stable quasi-particle and assuming that its parameters at nonzero F and η are the same as those of the SG kink, the steady-state kink velocity can be found approximately from the equation $v_{\mathrm{k}} = F/\eta m_{\mathrm{k}}(v_{\mathrm{k}})$, that leads to the dependence

$$v_{\mathrm{k}}^{(\mathrm{SG})}(F) = c\frac{\pi F}{\sqrt{(\pi F)^2 + (4\eta)^2}}. \tag{8.46}$$

The dependence (8.46) is shown in Fig. 8.10 by the dotted curve.

Looking for a travelling–wave solution of Eq. (8.45), $u(x,t) = u_{\mathrm{k}}(x-vt) \equiv u_{\mathrm{k}}(z)$, we obtain the equation

$$\left(c^2 - v^2\right)\frac{d^2 u_{\mathrm{k}}}{dz^2} + \eta v \frac{du_{\mathrm{k}}}{dz} - \sin u_{\mathrm{k}} + F = 0. \tag{8.47}$$

A kink solution corresponds to a separatrix of Eq. (8.47). Using the dimensionless variables $\widetilde{z} = z/d$, $\widetilde{v} = v/c$, and introducing $\xi = u_{\mathrm{k}}(\widetilde{z})$ and $w(\xi) = u_{\mathrm{k}}'(\widetilde{z})$, Eq. (8.47) can be rewritten as

$$\left[\left(1 - \widetilde{v}^2\right)\frac{dw}{d\xi} + \eta\widetilde{v}\right] w(\xi) - \sin\xi + F = 0. \tag{8.48}$$

A kink solution has to satisfy the boundary conditions $u_{\mathrm{k}}(-\infty) = u_F + 2\pi$, $u_{\mathrm{k}}(+\infty) = u_F$, $u_{\mathrm{k}}'(\pm\infty) = 0$, or $w(u_F + 2\pi) = w(u_F) = 0$. The kink shape can

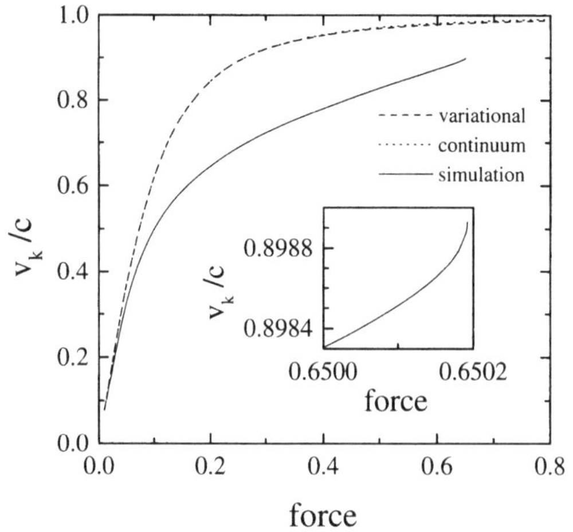

Fig. 8.10. The kink velocity v_k vs. the driving force F for $g = 1$, $\eta = 0.1$, and $N = 128$. The dashed curve corresponds to the SG dependence, and the dotted curve, to the kink velocity in the continuum limit approximation.

be restored from the function $w(\xi)$ by the integration $z = \int_{u_F+\pi}^{u_k(z)} d\xi\, w^{-1}(\xi)$. From Eq. (8.48) it follows that the function $w(\xi)$ has to satisfy the equations

$$w'(u_F + 2\pi) = (-\eta v + D)/2(c^2 - v^2),$$
$$w'(u_F) = (-\eta v - D)/2(c^2 - v^2), \tag{8.49}$$

where $w' \equiv dw/d\xi$, $D = \left[(\eta v)^2 + 4(c^2 - v^2)\sqrt{1 - F^2}\right]^{1/2}$. From Eq. (8.49) it directly follows that the shape of driven kink is asymmetric at $F \neq 0$, the kink "head" is more sharp, while its "tail" is more extended than those of the SG kink. However, in the continuum limit approximation the kink shape is monotonic for all values of the driving force.

Numerical simulation shows that the kink velocity in the discrete case is lower than that in the continuum limit approximation because of additional (intrinsic) damping of kink motion due to discreteness effects (see Fig. 8.10) [notice that in a finite (short) FK chain the resonances of the washboard frequency of the driven kink with phonons result in steps in the $v_k(F)$ dependence due to velocity-locking effect, see Refs. [710, 102]. The main new issue is that, while in the continuum limit approximation the steady-state solution of the moving kink exists for any force $F < F_s$ ($F_s = 1$ in our system of units), in the discrete chain such a solution exists only for $F < F_{\rm crit} < F_s$. At larger forces, $F > F_{\rm crit}$, the sharp transition to the running state takes place. Figure 8.11 demonstrates the force $F_{\rm crit}$ and the critical kink velocity $v_{\rm crit}$ as functions of the external damping η for a fixed value of the elastic constant g. The instability exists only for the underdamped case, $\eta < \eta_{\rm max}$, when the kink can reach the critical velocity, which is close to the sound speed c, at some force $F_{\rm crit} < F_s$. At large frictions the maximum kink velocity is lower than c even in the $F \to 1$ limit, and the transition to the running state takes place at $F = F_s$ only, simply because the minima of the external potential disappear and all atoms go to the running state simultaneously.

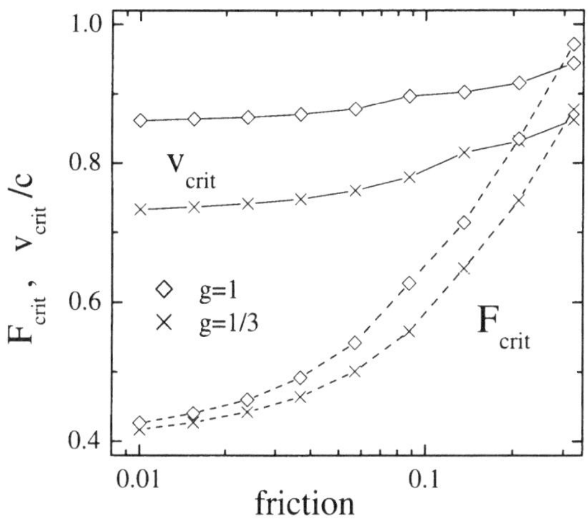

Fig. 8.11. The critical velocity $v_{\rm crit}/c$ (solid) and the critical force $F_{\rm crit}$ (dashed) vs. the damping coefficient η for two values of the elastic constant: $g = 1$ (diamonds) and $g = 1/3$ (crosses), and $N = 256$ [711].

A detailed numerical study of the discrete driven FK model has been done by Braun *et al.* [711]. First, it was found that the steady-state kink motion is always "automodel", i.e. it satisfies the condition

$$u_{l+1}(t) = u_l(t-\tau) + a_s, \tag{8.50}$$

so that each atom repeats the trajectory of the previous atom with the time delay $\tau = a_s/v_{\rm k}$, i.e. the system dynamics can be described by a dynamical hull function. Second, due to discreteness effects the kink tail has a complicated intrinsic structure, it shows spacial oscillations with the wavevector k defined by the resonance of the washboard frequency $\omega = v_{\rm k}$ with phonons. The shape of running kink at different force values is shown in Fig. 8.12. These dependencies can be presented as functions of one independent variable $z(F) = l - t/\tau(F)$, where l is the atomic index. Due to the automodel condition (8.50), the dependence $u_l(t)$ calculated for a time interval $0 < t < N\tau$ for a given atom only, allows us to restore the evolution of all atoms in the chain.

The periodicity of kink's tail may be explained similarly to the work of Peyrard and Kruskal [171]. Recall that without the external periodic potential the only excitations of the model are acoustic phonons with the spectrum $\omega_{\rm ph}(k) = ck$, where $c = 2\pi\sqrt{g}$ is the sound velocity. However, with presence of the external potential the spectrum becomes optical,

$$\omega_{\rm ph}^2(k) = \omega_{\rm min}^2 + 2g(1 - \cos 2\pi k), \tag{8.51}$$

where $\omega_{\rm min}^2 = \left(1 - F^2\right)^{1/2}$ and the wavevector k must belong to the first Brillouin zone, $|k| < 1/2$, so the phonon frequencies lie within the interval $\omega_{\rm min} \le \omega_{\rm ph}(k) \le \omega_{\rm max}$, where $\omega_{\rm max}^2 = \omega_{\rm min}^2 + 4g$. In the frame co-moving with the kink, the phonon spectrum is modified due to Doppler effect, $\Omega_{\rm ph}(k) =$

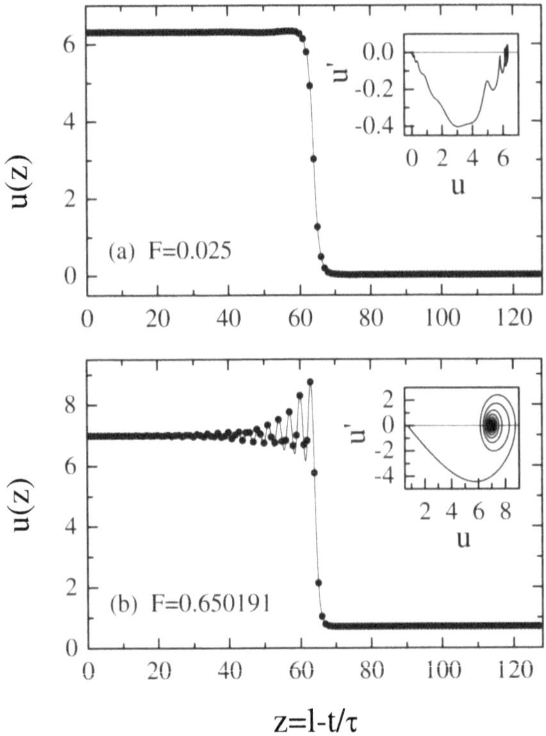

Fig. 8.12. Kink's shape in the discrete chain as a function of $z = l - t/\tau(F)$ for (a) a small force $F = 0.025$ and (b) the critical force $F = 0.650191$. $g = 1$, $\eta = 0.1$, $N = 128$. Black circles show instantaneous atomic positions and the solid curve corresponds to the "automodel" curve. Inset: the corresponding separatrices [711].

$\omega_{\rm ph}(k) - kv_{\rm k}$. The kink may be followed by a standing wave (the wave co-moving with the kink with the same phase velocity) if

$$\Omega_{\rm ph}(k) = 0. \qquad (8.52)$$

This equation always has one or more solutions as shown in Fig. 8.13 in the extended Brillouin zone scheme, $|k| < \infty$. At large kink velocities, $v_{\rm k} > 2\,\omega_{\max}$, this solution corresponds to the wavevector within the first Brillouin zone [see the curve (c) in Fig. 8.13]. At lower kink velocities the solution belongs to the second Brillouin zone [the curve (b) in Fig. 8.13], then to the third Brillouin zone, *etc.* In the restricted Brillouin zone scheme, where $|k| < 1/2$, we have to look for solutions of the equation $\Omega_{\rm ph}(k) = nv_{\rm k}$, where $n = 0, \pm 1, \ldots$ corresponds to the Brillouin zone number. The solution with $n = 0$ corresponds to the resonance of the washboard frequency $\omega_{\rm wash} = v_{\rm k}$ with phonons, so the solutions with $n \neq 1$ may be called as "super-resonances" [712]. The dependence $k_{\rm res}(v_{\rm k})$ obtained by numerical solution of Eq. (8.52) is shown in Fig. 8.15. Note that at small kink velocities Eq. (8.52) has more than one solution. For example, for the case shown by the curve (a) in Fig. 8.13, the lowest root corresponds to the oscillation behind the kink (the group velocity of phonons, $v_{\rm gr} = d\omega_{\rm ph}(k)/dk$, is negative), and the second one, to the oscillation ahead of the kink ($v_{\rm gr} > 0$).

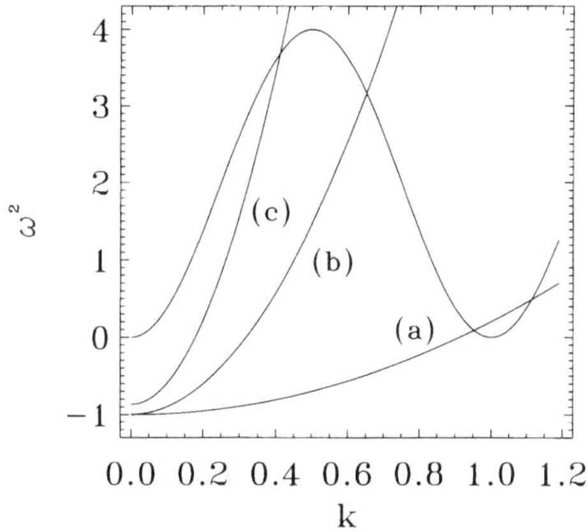

Fig. 8.13. Graphical solution of Eq. (8.52) in the extended Brillouin zone scheme for (a) $F = 0.025$, (b) $F = 0.1$, and (c) $F = 0.5$.

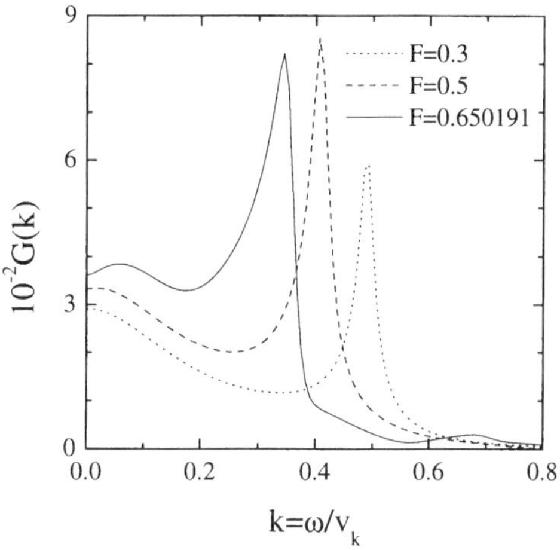

Fig. 8.14. Fourier transform $G(k)$ of the kink shape for $g = 1$, $\eta = 0.1$, and different forces: (a) $F = 0.3$ ($v_{\rm k} = 4.543$), the dotted curve; (b) $F = 0.5$ ($v_{\rm k} = 5.204$), the dashed curve; and (c) $F = 0.650191$ ($v_{\rm k} = 5.648$), the solid curve.

To show that the described resonances do are responsible for tail's oscillation, one may calculate the spacial Fourier transform of kink shape,

$$G(k) = \tau^{-1} \int_0^\tau dt \left| \sum_{l=1}^{N} \dot{u}_l(t) e^{i2\pi lk} \right|. \tag{8.53}$$

The function $G(k)$ is shown in Fig. 8.14, and the positions of the maxima of $G(k)$ for different kink velocities are plotted by symbols in Fig. 8.15. One can see that the short-wave component of kink tail oscillation is characterized by the wavevector which coincides with that obtained from the solution of Eq. (8.52) (see triangles in Fig. 8.15).

To study the resonances in more details, one may also calculate the response function $F(k, \omega)$ which describes a response of the steady-state solution to a small noise. This calculation also shows that namely the excitation

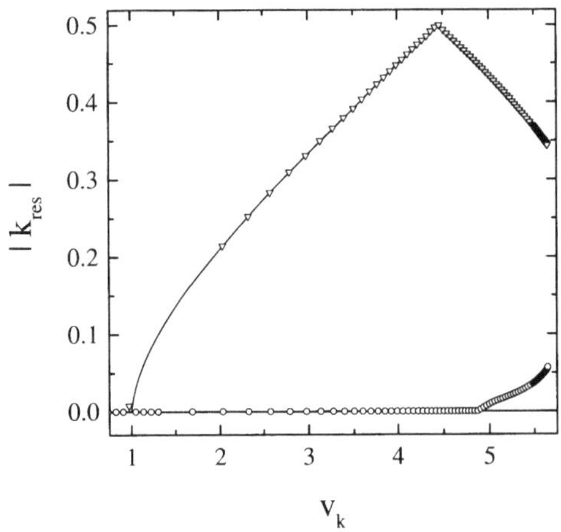

Fig. 8.15. The dependence $k_{\rm res}(v_{\rm k})$, where $k_{\rm res}$ is the solution of Eq. (8.52). Triangles and circles show the positions of maxima of the Fourier transform $G(k)$ of kink's shape. $g = 1$, $\eta = 0.1$.

of phonons is responsible for the oscillating kink tail. However, it shows one more very important result: close to the critical velocity the resonance is spreaded. This indicates that just before the instability, the excitation becomes spatially localized. Another indication that the kink shape has a complicated structure just close to the critical velocity, follows from the Fourier transform (8.53) of $\dot{u}_l(t)$. As seen from Fig. 8.14, close to $v_{\rm crit}$ the function $G(k)$ has an additional peak corresponded to spacial oscillation of the tail with a small wavevector $k < 0.1$ and frequency $\omega \ll \omega_{\rm min}$. This new branch is also shown in Fig. 8.15 by circles. These effects can be interpreted as an indication of appearance of a shape mode (discrete breather) of the moving discrete kink just before its destroying. The same conclusion follows also from the Floquet analysis described in detail by Braun *et al.* [711].

In this context we recall that due to nonlinearity of the FK model, it supports, additionally to phonons and kinks, the existence of localized nonlinear excitations, the discrete breathers (DB's), which do not carry mass along the chain, but at a high amplitude of their vibration they may decay into a kink-antikink pair. These newly created kinks and antikinks then move independently and thus increase the system mobility. Besides, one more issue of the FK model is the existence of linear excitations of the kink itself, the shape (internal) kink modes. The shape mode can be treated as a discrete breather captured by the kink as was described by Sepulchre and MacKay [713]. The essential difference of the captured discrete breather from the free one is that due to the kink, the captured breather is a linearly stable excitation, while the free discrete breather needs a nonzero threshold energy to be excited.

Thus, at a large driving force, a localized shape mode (discrete breather) is excited in the kink tail. With increasing the kink velocity, the amplitude of the DB increases too, and at some critical velocity the DB decays into the kink-antikink pair. *The fast kink becomes unstable and emits antikink(s).*

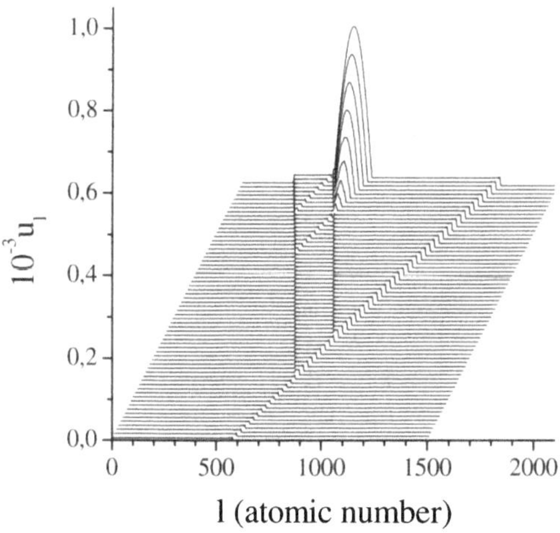

Fig. 8.16. Evolution of the system at $F = 0.651$ starting from the initial state corresponded to the steady state for $F = 0.65$, when $v_k \approx v_{\text{crit}}$. The system parameters are $g = 1$, $\eta = 0.1$, and $N = 1500$.

The emission of kink-antikink pairs in the tail of the fast kink leads to the sharp transition to the running state. Details of this transition are shown in Fig. 8.16, where we plot $u_l(t)$ versus the index l for different time moments, each next curve being slightly shifted upstairs and to the right. One can see that the first event is the creation of a new kink-antikink pair in the tail of primary kink. The newly created antikink moves to the left, while the primary kink and the newly created kink produce the "double" kink which moves as a whole. Then one more kink-antikink pair is created in the tail of the double kink; again the second antikink moves to the left, while the new kink plus the double kink produce the "triple" kink. Then, the first *antikink* creates behind itself one more kink-antikink pair. The kink from this lastly created pair moves to the right and finally meets with the second antikink. After their collision an avalanche starts to grow. Figure 8.17 shows also the evolution of the total system velocity $v_{\text{tot}}(t) = \sum_l \dot{u}_l(t)$ during this process. When the first kink-antikink pair is created, v_{tot} increases in two times (see details in inset of Fig. 8.17), then it again increases at next creation events, and finally, when the avalanche starts to grow, v_{tot} begins to increase linearly with time with the velocity $2c$, so that both fronts of the running domain move with the sound speed.

Simulation shows that in the running domain, the atomic velocities are almost constant and are approximately equal to the maximum atomic velocity F/η, and that ahead of the fronts of the running domain, there are the running triple kink and the double antikink which move with almost sound speed. The shape of growing domain of running atoms, $\rho_l(t) = -[u_{l+1}(t) - u_l(t)]/(2\pi)^2$, which describes the density of excessive atoms in comparison with the commensurate structure, has a cosine profile, $\bar{\rho}(x) \propto \sin k_r(x - x_c)$, where x_c is the center of the running domain and $k_r^2 = -\langle \cos u(x,t) \rangle \ll 1$.

Finally, notice that at a very low damping, $\eta \ll \omega_0$, and/or in a very discrete system, $g \ll 1$ so that F_{PN} is close to F_s, the kink may reach

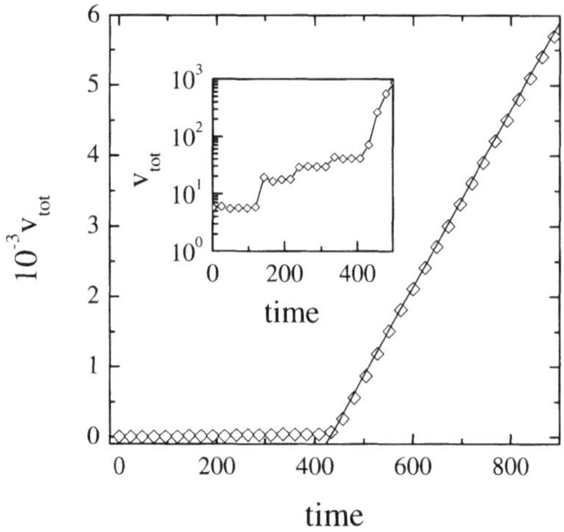

Fig. 8.17. Dependence $v_{\rm tot}(t)$ for the system evolution at $F = 0.651$. Solid line is the fit $v_{\rm tot}(t) = 2c\,(t - t_0)$. Inset: the same in the logarithmic scale.

the critical velocity just when it starts to move, and the locked-to-running transition will occur at $F = F_{PN}$.

Above we discussed the case of zero temperature. At low but nonzero T the kink undergoes a creep (thermally activated) motion at low forces, where $B \propto \exp(-\varepsilon_{PN}/k_BT)$, and begins to slide at higher forces. The mobility B in this case may be calculated approximately with the help of Eqs. (8.17) to (8.25) of the single-particle case described above in Sect. 8.2.2, if one considers the kink as a rigid quasiparticle of mass $m_{\rm k}$ moving in the periodic PN potential. In particular, the transition between the regimes of creep motion and kink-sliding motion takes place either at the threshold force F_{PN}, or at the force $F_{2PN} \approx C''(\eta/\omega_{PN})F_{PN}$ similarly to Eq. (8.13) of the single-atom case, whichever is smaller [714]. Besides, if the system size is not too small, there are thermally excited kink-antikink pairs, therefore the domain of running atoms starts to grow immediately when the fast kink becomes unstable.

8.6 Supersonic and Multiple Kinks

In a realistic system the interatomic interaction is anharmonic usually. As was shown above, the anharmonicity destroys the kink-antikink symmetry of the classical FK model. This may change, for a high enough degree of the anharmonicity, the scenario of the transition to the running state. For example, if the critical force for the antikink is smaller than that for the kink, the avalanche of running atoms will start to grow immediately after creation of a new kink-antikink pair in the tail of the running primary kink. Besides, for the driven FK model, the anharmonicity leads to two new interesting effects. Recall that in the classical FK model the topological excitations are always subsonic, the kink cannot propagate with a velocity $v_{\rm k}$ larger than the sound speed c because of Lorentz contraction of kink's width. Moreover, in the classical FK model, the kinks of the same topological charge repel from one another and, therefore, they cannot carry a multiple topological charge. As we show in the present Section, in the *anharmonic* model both supersonic kinks and multiple kinks may exist.

The motion of topological solitons with *supersonic* velocities was firstly predicted analytically by Kosevich and Kovalev [273] in the FK model with some specific interatomic interaction in the continuum limit approximation (see Sect. 3.5.1). Later, supersonic topological solitons were observed by Bishop *et al.* [715] in molecular dynamics study of polyacetylene. Then the supersonic kinks were studied numerically in the discrete ϕ^4 model with anharmonic interaction by Savin [271]. It was shown that for certain supersonic kink velocities, when its width matches with that of solitons of Boussinesq equation, the kink propagates almost without energy losses. Later similar results were obtained by Zolotaryuk *et al.* [272] for the exponentially interacting atoms subjected to the sinusoidal substrate potential (see details in Sect. 3.5.1).

Multiple fast (but subsonic) kinks were firstly observed numerically by Peyrard and Kruskal [171] in the classical undamped highly discrete FK model (see Sect. 3.2.4). The authors explained the stability of the double kink as the result of "compensation" of the waves emitted by two single kinks when these waves happen to be out-of-phase. The waves suppress each other, so the composite double kink propagates almost without radiation. This problem was studied later by Savin *et al.* [179], where a hierarchy of the double kink states characterized by different distances between the two single kinks, was found numerically. The authors observed these states to be dynamically stable for certain (preferred) values of the velocity. The same results were obtained earlier by Ustinov *et al.* [716] for the *driven* underdamped FK chain with harmonic interatomic interaction (see Fig. 8.18). The authors observed numerically several "bunched" states of two 2π-kinks, which differ from each other by the number of the oscillations trapped between the two bunched kinks. Such "resonant" bunched states exist for certain intervals of the driving force which may overlap. The velocity of the double kink is always higher than that of the single kink due to reducing the radiation losses. Unfortunately, the stability of these bunched states was not studied in detail, although the authors tried to explain the formation of the two-kink bound state through the interaction mediated by kink's oscillating tails. Then, Alfimov *et al.* [302] have shown that multiple kinks exist also in continuum systems with *nonlocal* interaction. The bounded states of kinks (subsonic as well as supersonic) in the case of anharmonic interaction were also studied numerically by Zolotaryuk *et al.* [272]. It was found that these multiple kinks are dynamically stable but asymptotically unstable (see Sect. 3.5.1). The dynamics of the generalized FK chain *driven* by a dc external force was studied numerically by Braun *et al.* [190, 717], where the existence of supersonic

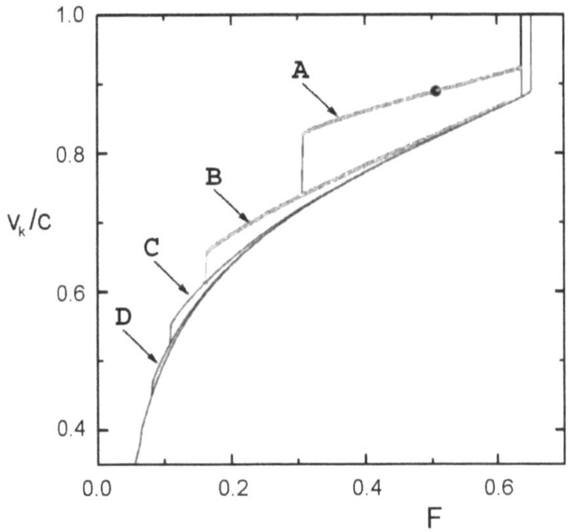

Fig. 8.18. The velocity v_k of the 4π-kink vs. the force F for the damped driven FK model with the harmonic interaction between the atoms ($g = 1$, $\eta = 0.1$, and $N = 100$). The arrows correspond to different bunched states: (A) the pure 4π-kink (no separation between the constituent 2π-kinks); (B) the first bunched state (one oscillation trapped between two 2π-kinks); (C) the second bunched state (two oscillations trapped between the 2π-kinks); and (D) the third bunched state (three oscillations trapped between the 2π-kinks) [716].

kinks and multiple (double and triple at least) kinks was observed. Finally, the existence of multiple kinks for certain kink velocities in the *discrete* FK-type model was proven rigorously by Champneys and Kivshar [182].

Below we will show, following the work by Braun [718], that the FK model admits both *supersonic kinks* and *multiple kinks*, if the following three conditions are satisfied: first, the model must be *discrete*; second, because kink's motion in a discrete chain is damped due to radiation of phonons, one must apply *a driving force* to support the motion; and third, the interatomic interaction must be *anharmonic*. As an example, we consider the FK model

$$\ddot{x}_l(t) + \eta \dot{x}_l(t) + \frac{\partial}{\partial x_l} \left\{ V\left[x_{l+1}(t) - x_l(t)\right] + V\left[x_l(t) - x_{l-1}(t)\right] \right\} + \sin x_l(t) = F, \tag{8.54}$$

with exponential interatomic interaction

$$V(x) = V_0 e^{-\beta x}. \tag{8.55}$$

In the continuum-limit approximation ([7]; see Sect. 1.1), the motion equation reduces to the form

$$\ddot{u} + \eta\dot{u} - d^2u''\left(1 - \alpha d u'\right) + \sin u - F - h^2\left[\ddot{u}'' + \left(u'\right)^2 \sin u - u''\cos u\right] = 0, \tag{8.56}$$

where $g = V''(a_s) = V_0\beta^2\exp(-\beta a_s)$ is the elastic constant,

$$\alpha = -(a_s/d)\left[V'''(a_s)/V''(a_s)\right] = \beta/\sqrt{g}$$

is the anharmonicity parameter, and $h^2 = a_s^2/12 = \pi^2/3$ is the discreteness parameter. A travelling–wave solution, $u(x,t) = u_{\rm k}(x - vt) \equiv u_{\rm k}(z)$, should satisfy the equation

$$\begin{aligned} h^2v^2u_{\rm k}'''' + \left(c^2 - v^2 - h^2\cos u_{\rm k}\right)u_{\rm k}'' + h^2\left(u_{\rm k}'\right)^2\sin u_{\rm k} \\ -\alpha d^3u_{\rm k}''u_{\rm k}' + \eta v u_{\rm k}' - \sin u_{\rm k} + F = 0, \end{aligned} \tag{8.57}$$

where $c = 2\pi\sqrt{g}$ is the sound speed (recall $c = d$). A kink solution corresponds to a separatrix of Eq. (8.57). Using the dimensionless variables $\widetilde{z} = z/d$, $\widetilde{v} = v/c$, $\widetilde{h} = h/d = (12g)^{-1/2}$, and introducing new variables $\xi = u_{\rm k}(\widetilde{z})$ and $w(\xi) = u_{\rm k}'(\widetilde{z})$, Eq. (8.57) can be rewritten as

$$\begin{aligned} &\left\{\widetilde{h}^2\widetilde{v}^2\left[w'''(\xi)w^2(\xi) + 4w''(\xi)w'(\xi)w(\xi) + \left[w'(\xi)\right]^3\right]\right. \\ &- \alpha w'(\xi)w(\xi) + \left(1 - \widetilde{v}^2 - \widetilde{h}^2\cos\xi\right)w'(\xi) \\ &+ \left.\left(\widetilde{h}^2\sin\xi\right)w(\xi) + \eta\widetilde{v}\right\}w(\xi) - \sin\xi + F = 0. \end{aligned} \tag{8.58}$$

A (multiple) kink solution has to satisfy the boundary conditions $u_{\rm k}(-\infty) = u_F + 2\pi p$, $u_{\rm k}(+\infty) = u_F$, $u_{\rm k}'(\pm\infty) = 0$, or $w(u_F + 2\pi p) = w(u_F) = 0$, where p is the topological charge of the kink ($p = 1$ for the single kink, $p = 2$ for the double kink, etc).

Near kink's tails, $z \to \pm\infty$, e.g. for $\xi = u_F + 2\pi n + \epsilon$, where $|\epsilon| \ll 1$, we can use the expansion

$$w(\xi) = a_1\epsilon + \frac{1}{2}a_2\epsilon^2 + \frac{1}{6}a_3\epsilon^3 + \ldots, \tag{8.59}$$

where $a_1 = w'(u_F)$, $a_2 = w''(u_F)$, *etc.* Substituting this expansion into Eq. (8.58) and grouping together the terms of the same power of ϵ, we obtain the following equation for a_1,

$$\widetilde{h}^2\widetilde{v}^2a_1^4 + (1 - \widetilde{v}^2 - \widetilde{h}^2\cos u_F)a_1^2 + \eta\widetilde{v}a_1 - \cos u_F = 0, \tag{8.60}$$

which always has two roots, one positive and one negative, for *any* kink velocity v. Then, equating the terms for higher powers of ϵ, we obtain the relations that uniquely determine the coefficients a_2, a_3, *etc.* Thus, in the continuum limit approximation the separatrix solution of Eq. (8.58) is *unique*, i.e., for a given set of system parameters *the model has either the single-kink solution or the double-kink one*, but never both solutions simultaneously.

It is easy to show that in the case of $F = \eta = h = 0$, the continuum-limit equation (8.57) corresponds to an extremum of the following energy functional [264],

$$E[u(z)] = \int dz \left[\frac{1}{2}(c^2 - v^2)\,(u')^2 - \frac{1}{6}\alpha d^3\,(u')^3 - \cos u\right]. \tag{8.61}$$

Substituting a simple SG-type ansatz $u_{\mathrm{SG}}(z) = 4\tan^{-1}\exp(-z/d_{\mathrm{eff}})$ into Eq. (8.61), we obtain

$$E(d_{\mathrm{eff}}) = 4\frac{c^2 - v^2}{d_{\mathrm{eff}}} + \frac{2\pi}{3}\alpha\frac{c^3}{d_{\mathrm{eff}}^2} + 4d_{\mathrm{eff}}. \tag{8.62}$$

Although the variational approach does not describe rigorously the kink tails because of neglecting the discreteness effects, it allows us to find analytically the shape of the kink core and, therefore, to calculate approximately the kink velocity. Indeed, looking for extrema of the function $E(d_{\mathrm{eff}})$, we come to the equation $E'(d_{\mathrm{eff}}) = 0$, or

$$\left(\frac{d_{\mathrm{eff}}}{d}\right)^3 = \left[1 - \left(\frac{v}{c}\right)^2\right]\left(\frac{d_{\mathrm{eff}}}{d}\right) + \frac{\pi}{3}\alpha, \tag{8.63}$$

For the harmonic interaction, $\alpha = 0$, Eq. (8.63) has a solution for $|v| < c$ only, which describes relativistic narrowing of the SG kink, $d_{\mathrm{eff}} = d\left[1 - (v/c)^2\right]^{1/2}$. But for the anharmonic interaction, $\alpha > 0$, Eq. (8.63) has a solution for *any* kink velocity v, including supersonic velocities $|v| > c$. We emphasize that *supersonic excitations are possible for kinks (local compressions) only.*

Considering the kink as a rigid quasiparticle, the kink effective mass can be introduced as (see Sect. 3.1) $m_{\mathrm{k}} = a_s^{-1}\int dz\,[u'(z)]^2 = 4/\pi d_{\mathrm{eff}}$. Then, assuming that kink's parameters at nonzero F are the same as those for the $F = 0$ case, the steady-state kink velocity can be found approximately from the equation $v_{\mathrm{k}} = F/m_{\mathrm{k}}\eta = \pi c F d_{\mathrm{eff}}(v_{\mathrm{k}})/4\eta d$. Thus, Eq. (8.63) can be rewritten in the form

$$\left[1 + \left(\frac{\pi F}{4\eta}\right)^2\right]\left(\frac{d_{\mathrm{eff}}}{d}\right)^3 = \frac{d_{\mathrm{eff}}}{d} + \frac{\pi}{3}\alpha. \tag{8.64}$$

Numerical solution of Eq. (8.64) allows us to find the function $v_{\mathrm{k}}^{(\mathrm{var})}(F)$ that is shown by the dashed curve in Fig. 8.19 together with the dependence $v_{\mathrm{k}}(F)$ obtained by solution of the discrete equation of motion (8.44). One can see that in the anharmonic model we always have $v_{\mathrm{k}}^{(\mathrm{var})} > c$ at $F \to F_s = 1$, and although the simulation velocity is lower than $v_{\mathrm{k}}^{(\mathrm{var})}$ due to additional damping of the moving kink because of phonon radiation, the discrete kink still may reach a supersonic velocity. Thus, the variational approach predicts that the supersonic kinks may be expected in the *anharmonic* FK model only.

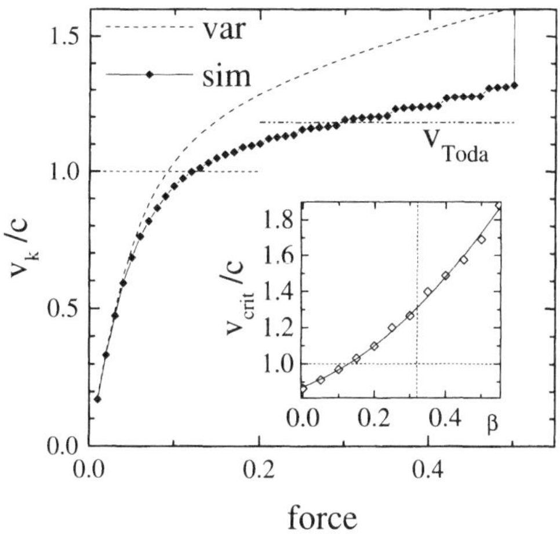

Fig. 8.19. Velocity v_k of the single kink versus the force F for $\beta = 1/\pi$, $g = 1$, and $\eta = 0.05$. The solid curve is for simulation results, and the dashed curve, for variational approximation. Inset: the critical kink velocity v_{crit} as a function of the anharmonicity parameter β at the fixed force $F = 0.5$.

Returning back to the *discrete* model, recall that in the classical FK model the driven kink cannot reach even the sound velocity, because it becomes unstable at $v_{crit} < c$ and the system goes to the "running" state (see above Sect. 8.5). However, in the anharmonic FK model, the critical kink velocity may exceed the sound speed. The dependence of v_{crit} on the anharmonicity parameter β is shown as inset in Fig. 8.19.

To study the double kink with the help of the variational approach, let us consider it as a sum of two single kinks separated by a distance r,

$$u_2(z) = u_{SG}(z - r/2) + u_{SG}(z + r/2). \tag{8.65}$$

Substituting the ansatz (8.65) into the functional (8.61), we obtain the effective energy $E(d_{eff}, r)$ which now depends on two parameters d_{eff} and r. Looking for a minimum of $E(d_{eff}, r)$ over d_{eff} at r fixed, one can check that for the classical FK model, $\alpha = 0$, the function $E(r) \equiv \min_{d_{eff}} E(d_{eff}, r)$ is a monotonically decreasing function of r, i.e., the kinks are repelled from one another. On the other hand, for the anharmonic FK model the function $E(r)$ has a minimum at some $r = r_{min} < \infty$, so two kinks attract one another and thus have to couple into a double kink. The "dissociation" energy of the double kink is very small at subsonic velocities, but becomes high enough at supersonic velocities (see inset in Fig. 8.20). The "size" r_{min} of the double kink decreases with $|v|$ increasing. Thus, although the variational approach with the SG-type ansatz is too crude, it nevertheless predicts that the multiple kinks could be stable for high (supersonic) kink velocities.

As is well known, the SG kinks of the same topological charge always repel from one another. The same is true for static ($v = 0$) kinks of the discrete FK model, including the anharmonic ($\alpha \neq 0$) model [319], contrary to the variational approximation which mistakenly predicts a weak attraction at $|v| < c$. Thus, it must exist a threshold kink velocity v_1 such that at small velocities $0 \leq v < v_1$ the steady-state solution of the model corresponds to

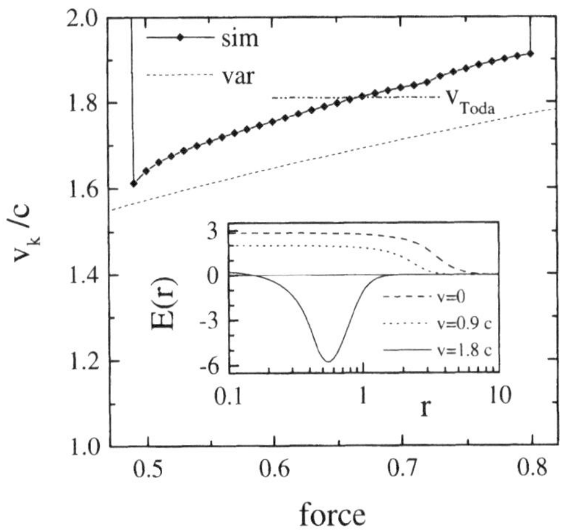

Fig. 8.20. The same as in Fig. 8.19 for the double kink. Inset: the effective energy $E(r) = \min_d E(d, r)$ of the double kink as function of subkink's separation r for fixed kink velocities $v_k = 0$, $0.9\,c$, and $1.8\,c$.

the 2π-kink, while at high velocities $v > v_1$, it corresponds to the double (4π-) kink (if v_1 is lower than the kink velocity at $F = F_s$, that is true at low enough values of η). Indeed, the simulation results presented in Fig. 8.20 demonstrate that the double kink is stable within the force interval $0.5 < F < 0.8$ but becomes unstable at higher as well as smaller forces, while the 2π-kink is stable for $F < 0.5$ only (a more careful calculations show that the regions of existence of the single and double kinks in the discrete model may overlap, see Braun *et al.* [719] for details). Similarly one could expect the existence of a second threshold velocity v_2 such that at $v > v_2$ the steady-state solution will correspond to the 6π-kink, *etc.*

Finally, note that both supersonic kinks and multiple kinks remain stable at nonzero temperatures as well. Notice also that at high forces the kink velocity is close to that of the Toda soliton [270]. Indeed, the shape of the Toda soliton is characterized by the "jump" $\Delta u = 2\mu a_s/\beta$, where μ is a parameter coupled with the soliton velocity by the relationship $v_{\rm Toda} = c\,\sinh(\mu a_s)/\mu a_s$. In the presence of the external substrate potential the jump Δu must be equal to pa_s for the p-kink; thus we obtain $\mu = p\beta/2$. In particular, for the anharmonicity parameter $\beta = 1/\pi$ we have $v_{\rm Toda}/c = (\sinh p)/p \approx 1.18$ for the 2π-kink and $v_{\rm Toda}/c \approx 1.81$ for the 4π-kink correspondingly. Thus, the supersonic and multiple kinks may be considered as Toda solitons "disturbed" by the external periodic potential.

8.7 Locked-to-Sliding Transition

8.7.1 Commensurate Ground States

If $T = 0$ and the ground state is trivial (i.e., $\theta = 1$ or $\theta = 1/q$ with an integer q), then the locked-to-running transition takes place at $F = F_s$ only. However, if $T > 0$, then the transition occurs at a lower threshold force, $F_f \approx F_{\rm crit} < F_s$. The detailed behavior at the locked-to-running transition is clear from Fig. 8.21, where the atomic trajectories just at the transition point are shown [190]. As one can see, the scenario starts with the creation of a kink-antikink pair. The kink and antikink move in opposite directions, quasielastically collide (because of the periodic boundary conditions), and quite soon a new kink-antikink pair is created in the tail of the primary kink and then another pair in the tail of the new kink. This process continues resulting in an avalanche-like growth of the kink-antikink pair concentration, finishing in the totally running state.

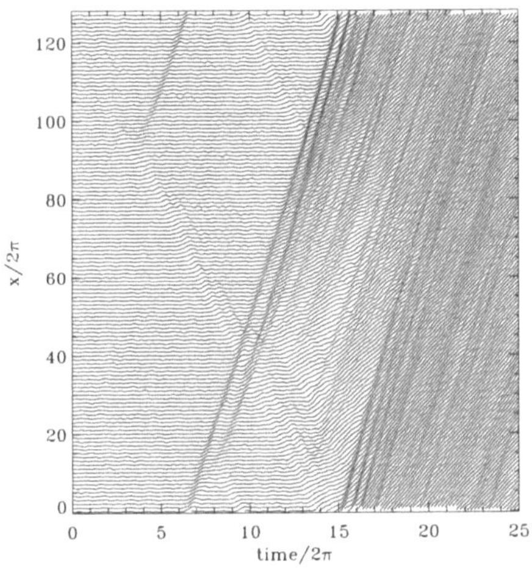

Fig. 8.21. Atomic trajectories for the commensurate FK model with exponential interaction at $F = 0.53$ and $T = 0.7$ ($\theta = 1$, $g_{\rm eff} = 1$, $\eta = 0.1$, $\beta = 1/2\pi$, $N = 128$).

Thus, in a finite system at small T, when the structure is *commensurate* and there are no kinks in the GS, then the first event is the thermally activated and force stimulated creation of a kink-antikink pair. On the other hand, in an infinite system there are always a nonzero concentration of kink-antikink pairs. The further scenario is the same as for a single fast kink, except that now the transition starts earlier, because we do not have to wait until the first kink-antikink pair is created in the tail of the fast kink.

8.7.2 Complex Ground States and Multistep Transition

The dimensionless atomic concentration $\theta = N/M$ (the ratio between the number of particles N and the number of available sites M) in the FK system plays a crucial role since it defines the number of geometrical kinks. These excitations can be defined for any background *commensurate* atomic structure $\theta_0 = p/q$, where p and q are relative prime integers. If the concentration θ slightly deviates from the background value θ_0, the ground state of the system corresponds to large domains with background commensurate coverage θ_0, separated by localized incommensurate zones of compression (expansion) called kinks (antikinks).

When the external force increases, the FK system with a non-trivial GS exhibits a hierarchy of first-order dynamical phase transitions from the completely immobile state to the totally running state, passing through several intermediate stages characterized by the running state of collective quasiparticle excitations, or kinks of the FK model. As an example, let us first consider the $\theta = 21/41$ case when the mass transport along the chain is carried out by trivial kinks constructed on the background of the $\theta_0 = 1/2$ structure. As the average distance between the kinks is large (equal to $41\, a_s$ in the ground state), the kink-kink interaction is small, and the atomic flux is restricted by the overcoming of kinks over the PN barriers [see Fig. 8.22(a)]. When the driving force F increases, the PN barriers are lowered [simultaneously with the original barriers of $V_{\rm tot}(x)$], resulting in the increase of the single kink mobility. Thus, at zero temperature the crossover from the locked $B = 0$ state to the kink-running state takes place at the force $F \approx F_{tk} = C\pi\varepsilon_{\rm PN}/a_s$, where the factor $C \approx 1$ depends on the shape of the PN potential. The mobility in the kink-running state is $B \approx \theta_k B_f$, where $\theta_k = 1/41$ is the dimensionless kink concentration.

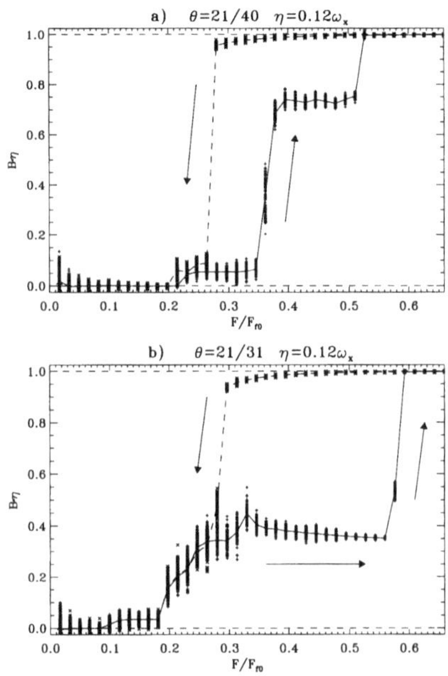

Fig. 8.22. The mobility $B = \langle v \rangle / F$ versus the force F for the underdamped ($\eta = 0.12$) FK model with exponential interaction ($g_{\rm eff} = 0.58$) (a) for $\theta = 21/40$ (trivial kinks on the background of trivial $\theta_0 = 1/2$ structure), and (b) for $\theta = 21/31$ (superkinks on the background of the complex $\theta_0 = 2/3$ structure).

The further scenario depends on the value of the damping coefficient η. At very low frictions, $\eta < 0.05\,\omega_0$, there is no intermediate stages, because the running kinks destroy themselves as soon as they start to move: they will immediately cause an avalanche driving the whole system to the totally running state of atoms. Thus, when the force increases, the system jumps from the locked state directly to the running state at the force F_{tk}.

At larger frictions, $\eta > 0.05\,\omega_0$, the above-mentioned intermediate stages with running kinks exist. A mechanism of the second abrupt increase of the mobility depends on η too. At low frictions the "forward" force F_f is friction-dependent, $F_f \approx m_k \eta v_{\rm crit}$. The transition to the running state is due to instability of fast kinks when they reach the critical velocity. On the other hand, at higher frictions the transition takes place when the force exceeds a certain threshold connected with the vanishing of the energy barrier for creation of additional kink-antikink pairs in the system similarly to the over-damped case described above and demonstrated in Fig. 8.7(a). The external force lowers the saddle-point energy for the creation of the kink-antikink pair determined by Eq. (8.36), thus increasing the rate of kink-antikink pairs generation. Therefore, the number of mass carriers in the system increases leading to the increase of the mobility. A crossover from the friction-dependent threshold to the friction-independent one was discussed by Paliy *et al.* [721] and Braun *et al.* [720].

Between the kink-running stage and the totally running state there may be a specific "traffic-jam" regime which will be discussed below in Sect. 8.9.

This qualitative picture holds also for a more complex atomic coverage like $\theta = 21/31$ [721] (see Fig. 8.22(b) for $N = 105$ and $M = 155$). In this case the state of running trivial kinks is preceded by the state of running superkinks. The ground state in this case corresponds to domains of the complex $\theta_0 = 2/3$ commensurate structure, separated by *superkinks* with an average spacing $30\,a_s$ between them. On the other hand, the $\theta = 2/3$ structure can be viewed as a dense lattice of *trivial* kinks defined on the background of the $\theta_0 = 1/2$ structure. This specificity clearly manifests itself in the $B(F)$ dependence. During the force-increasing process, now there are two sharp steps of increasing of the mobility B. The first one, at $F = F_{sk} \approx 0.08\,F_s$, corresponds to the situation where the superkinks start to slide, whereas the second step, occurring at $F = F'_{tk} \approx 0.18\,F_s$, corresponds to the transition of the trivial kinks to the running state. The critical force for superkinks F_{sk} is independent on friction, while the force F'_{tk} increases with the friction η and reaches the constant value $F'_{tk} \approx 0.24\,F_s$ at high friction, which almost exactly matches the value $F_{tk} \approx 0.23\,F_s$ for the case of $\theta = 21/40$ coverage. Indeed, because the running superkinks excite phonons, this may be interpreted as an effective increase of system temperature, and the second step on the $B(F)$ dependence has to occur either at $F \approx F_{PN}^{(tk)}$ or at $F \approx C'' \left[(\eta + \eta_{sk})/\omega_{PN}^{(tk)}\right] F_{PN}^{(tk)}$ whichever is lower, where we should take the trivial kink threshold force F_{tk} as that for the Peierls-Nabarro force $F_{PN}^{(tk)}$.

The mobility at the intermediate superkink-sliding steady states may be roughly estimated as $B \approx \theta_{sk} B_{sk}$, where θ_{sk} is the dimensionless concentration of superkinks, $B_{sk} \approx (m_{sk}\eta_{sk})^{-1}$ is the mobility of a single superkink, m_{sk} is its mass, and the friction $\eta_{sk} = \eta + \eta_{\rm int}$ consists of the external damping η and the intrinsic friction $\eta_{\rm int}$ emerged due to excitation of phonons as was discussed above in Sect. 7.2.2. However, the calculation of the superkink mobility B_{sk} is too complicated problem. An alternative approach proposed by Hentschel *et al.* [722], is to look for a solution of motion equations (8.44) in an approximate form,

$$x_l(t) = X(t) + \psi_l(t), \tag{8.66}$$

i.e. to split the dynamics into a center of mass (c.m.) contribution $X(t)$ and a spatiotemporal fluctuation $\psi_l(t)$, where $\langle \psi_l(t) \rangle_x = 0$ by construction (here $\langle \ldots \rangle_x$ stands for spacial average, and $\langle \ldots \rangle_t$, for temporal average). Neglecting higher harmonics of the washboard frequency $\omega_{\rm wash} = (2\pi/a_s)\, V$, where $V = \langle \dot{X}(t) \rangle_t$ is the average velocity of the steady-state motion, the c.m. coordinate may be approximately taken in the form

$$X(t) \approx Vt + B \sin(\omega_{\rm wash} t). \tag{8.67}$$

The fluctuating contribution is described approximately by the quasi-linear equation

$$\ddot{\psi}_q + \eta \dot{\psi}_q + \left(\omega_q^2 + C \cos X_q\right) \psi_q = 0, \tag{8.68}$$

where $\omega_q = 2\sqrt{g}\,\sin(qa_A/2)$ is the phonon spectrum of the elastic chain without the substrate potential, $C \approx \left(1 + 2\langle\langle \psi_l^2 \rangle_x \rangle_t\right)^{-1/2}$, and the spacial Fourier transform has been performed [$\psi_q = N^{-1} \sum_l^N \psi_l \exp(-iqla_A)$, $q = (j - N/2)\Delta q$, $\Delta q = 2\pi/Na_A$, and $j = 1, \ldots, N$ is integer, so that $|q| < \pi/a_A$]. Then, the oscillation of $X(t)$, Eq. (8.67), leads in the lowest approximation to excitation, due to parametric resonance in Eq. (8.68), of only one phonon mode with a frequency $\omega \sim \frac{1}{2}\omega_{\rm wash}$ and an amplitude b (see, however, Sect. 5.2.2). The amplitudes B and b have to be found by substitution Eq. (8.67) into Eq. (8.68) and then self-consistent solution of the resulting equation. Finally, equating the average energy input into the chain (per one time unit), $E_{\rm input} = FV$, to the average dissipation due to the c.m. motion, ηV^2, its temporal fluctuations, $\eta \left\langle \left(\dot{X}(t) - V\right)^2 \right\rangle_t$, and the spacial fluctuations, $\eta \langle\langle \dot{\psi}_l^2 \rangle_x \rangle_t$, one can find the velocity V and then the system mobility as $B \approx B_f / \left(1 + \frac{1}{2}B^2 + \frac{1}{8}b^2\right)$.

The kink-antikink nucleation regime for the classical FK model at $T = 0$ was studied in detail by Strunz and Elmer [712]. They called this regime the "fluid-sliding" state (FSS), while the totally running state was called the "solid-sliding" state (SSS). Calculating velocity correlation functions and Lyapunov spectra, Strunz and Elmer [712] have shown that the FSS is a spatiotemporally chaotic state. Although the velocity distribution function in the FSS may be well fitted by the Gaussian curve (the Maxwell distribution), the FSS is far away from a thermal equilibrium state, because the equipartition theorem is violated: the average energy of a phonon mode depends on its wavevector. The transition from the kink-dominated regime to the FSS (nonequilibrium "melting") is similar to the first-order phase transition. It is interesting that the dependence of the threshold force for this transition on the damping coefficient η is close to the $F_2(\eta)$ dependence (8.13) of the single-atom system of the $T \to 0$ limit. The backward FSS-to-pinned transition (nonequilibrium "freezing") is a nucleation process, and the final "frozen" state often contains domains of (two) different atomic densities.

If the system temperature is nonzero, then all transitions are smeared out, and also the first step on the $B(F)$ dependence from the creep (thermally activated) (super-) kink motion to the (super-) kink-sliding state should occur earlier, at $F \approx C'' \, (\eta/\omega_{PN}) \, F_{PN}$ instead of the PN force of the $T = 0$ case.

Finally, Fig. 8.23 (see also Fig. 5.9) demonstrates the $B(F)$ dependence for the concentration $\theta = 144/233 \approx 0.618$ which is close to the incommensurate concentration $\theta_{gm} = (\sqrt{5} - 1)/2$ ("the most irrational number", e.g., see Hardy and Wright [723]). Recall that for an incommensurate concentration, the $T = 0$ ground state of the system may correspond to two different states depending on the magnitude of the elastic constant g. For low g, $g < g_{\text{Aubry}}$ (where $g_{\text{Aubry}} \approx 1.0291926$ for $\theta = \theta_{gm}$) the GS is pinned, while for $g > g_{\text{Aubry}}$ the GS is sliding and has no activation barrier for motion. In the case of the pinned GS, $g < g_{\text{Aubry}}$, the mobility is zero at small forces. When F increases, the system reaches the Aubry point owing to the lowering the barriers of the inclined substrate potential, and at the force $F = F_{PN}$ the system exhibits nonzero mobility. When the external force F is further increased, the velocity shows a slow increase, which should consist of an infinite countable number of steps as for Devil's staircase but it is very difficult to resolve this behavior in detail. On the other hand, above the Aubry transition, $g > g_{\text{Aubry}}$, the mobility B is nonzero and close to the maximum value B_f at $F \to 0$. With increasing of the force in this case, the mobility decreases, and the c.m. velocity as a function of F again exhibits several steps. In both cases the steps correspond to the velocity locking effect and emerge due to excitation of phonons by parametric resonances as was briefly described above in Sect. 5.2.2. The $T = 0$ dynamics of the system at small forces may be described, at least with an accuracy of numerical simulation, by a uniquely defined dynamical hull function, while the transitions between the steps, by instabilities of the corresponding velocity-locked states [712]. Then, at higher forces, the Devil's staircase-like behavior is destroyed and, after the intermediate state corresponded to the kink-antikink nucleation regime (the spatiotemporal chaotic state), the system undergoes a sharp transition to the totally running state, so that the mobility abruptly increases to the maximum value $B \approx B_f$.

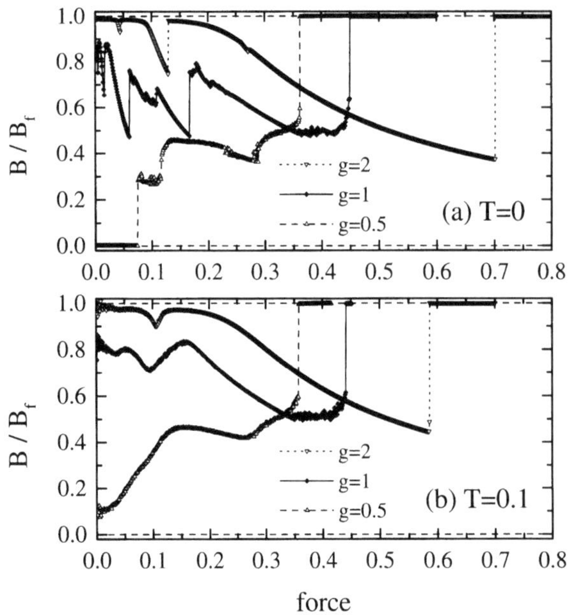

Fig. 8.23. Mobility B vs. the (adiabatically-slow increasing) driving force F for (a) zero and (b) nonzero temperature for the "golden–mean" concentration ($N = 144$, $M = 233$) and $g = 0.5$ (dashed), 1.0 (solid), and 2.0 (dotted).

For a nonzero temperature, the dependence $B(F)$ is smoothed [see Fig. 8.23(b)]. When the GS is pinned, $g < g_{\text{Aubry}}$, the system demonstrates hysteresis which is very similar to that of the $\theta = 2/3$ case of Fig. 8.22(b).

8.8 Hysteresis

Thus, the FK model with a commensurate GS or with an incommensurate GS but with $g < g_{\text{Aubry}}$, when the $F = 0$ ground state is pinned, demonstrates the hysteresis when the dc external force is varied. For a nontrivial GS, $\theta \neq 1/q$, the multiple steps during the force increasing process can be explained by the hierarchy of depinning: first, the geometrical kinks, then the "force excited" kinks and finally the atoms. During the force decreasing process, the system passes through several intermediate phases. Ariyasu and Bishop [724] have studied the *sliding-to-locked* transition for the underdamped SG model. The authors showed that the system proceeds first through a series of "cavity-mode" states, and then a series of kink-antikink wave train states (in an infinite system this sequence should be infinite, so the transition is continuous). A similar scenario is exhibited by the discrete FK chain [190]. For example, on the dependencies $B(F)$ shown in Fig. 8.24, one can clearly resolve several step; each step is characterized by an approximately constant velocity (note that the discrete system differs from the continuous SG one in that the former may have resonances with phonons, and this may result in locking of some states).

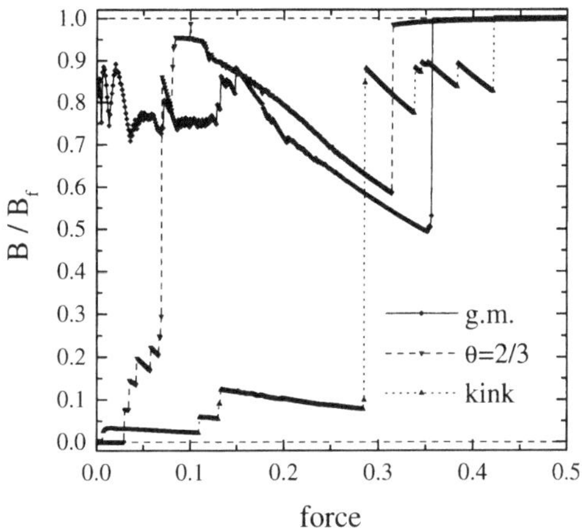

Fig. 8.24. Mobility B vs. F for the backward sliding-to-locked transition (when F decreases with an adiabatically slow rate) for the classical FK model with $g = 1$ at $T = 0$ for different concentrations: one antikink ($N = 144$, $M = 145$, dotted curve), $\theta = 2/3$ ($N = 144$, $M = 216$, gray), and the "golden-mean" concentration ($N = 144$, $M = 233$, solid).

Although at a nonzero temperature, $T > 0$, there should be no hysteresis in a one-dimensional system, simulations show that the hysteresis does exist. The width of the hysteretic loop decreases with temperature, the forward and backward threshold forces change with T as $F_f(0) - F_f(T) \propto \sqrt{T}$ and $F_b(T) - F_b(0) \propto \sqrt{T}$ [190], but the hysteresis does not disappear with thermal fluctuations. The width of the hysteretic loop, however, depends on a rate of force changing as usual in first-order phase transitions. Typical hysteretic dependencies $B(F)$ are presented in Fig. 8.25 for the harmonic and exponential interaction correspondingly. Four curves in Fig. 8.25 were calculated for four rates of F changing, which differ in ten times one from the next one. As seen, a width of the loop decreases with decreasing of R, but a well defined hysteresis still exists even for the smallest rate $R = 5.3\,10^{-7}$. Thus, the simulation shows that the hysteresis exists for any nonzero rate of force changing.

To explain the hysteresis analytically, Braun *et al.* [717] studied a simplified model based on the Fokker-Planck approach to the FK model. Using the MF approximation, the authors have shown that in the system of interacting atoms the hysteretic loop should be much larger and should survive at much higher temperatures than for a single Brownian atom. Indeed, due to concerned motion of atoms in the FK model, its atoms alone cannot exhibit bistability, the system must be transformed from one steady state to another as a whole. The Fokker-Planck approach shows that when the chain is in the low-mobility state, this state is more stable (comparing with the noninteracting system) because of effective decreasing of temperature, $T_{\rm eff} < T$; local fluctuations with high-velocity atoms are suppressed due to interatomic interaction. The transition to the running state begins only when the force F reaches the critical threshold F_f and the motion of topological excitations (kinks) becomes unstable. On the other hand, if the system is in the running state and the force is then decreased, the running state remains stable because

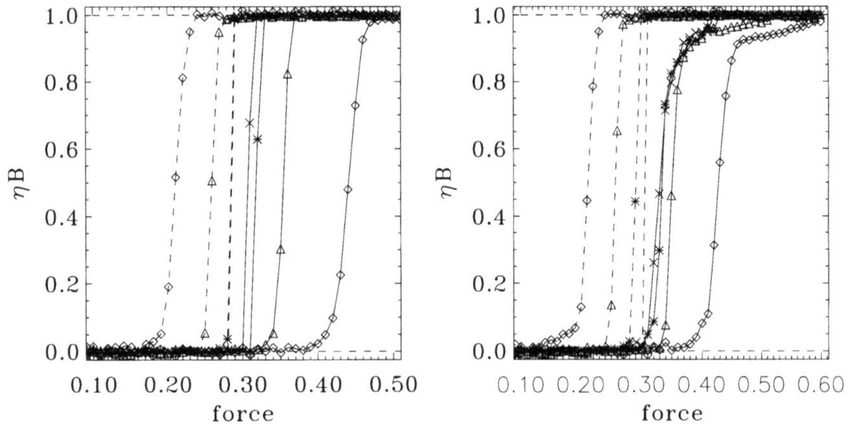

Fig. 8.25. Left: Hysteresis $B(F)/B_f$ for the standard FK model. Solid curves correspond to the force increasing, and dashed–to the force decreasing process. Four curves are for different rates of the force change $R = \Delta F/3 i_R \tau_0$, where $\Delta F = 0.01$, $\tau_0 = 2\pi$ is the characteristic period of the system, and i_R is marked by different symbols: diamonds: $i_R = 1$, triangles: $i_R = 10$, asterisks: $i_R = 10^2$, and crosses: $i_R = 10^3$. Parameters are: $\theta = 2/3$, $N = 256$, $g = 0.1$, $T = 0.1$, and $\eta = 0.1$. Right: The same for exponential interaction (8.55) with $\beta = 1/\pi$.

local fluctuations with low-velocity atoms (jams) are suppressed again due to the interaction. The effective driving force F_{eff} in the running state is larger than F, and the transition to the locked state may begin only when a jam of a critical size (which may be mesoscopically large for a strong interatomic interaction) appears for the first time.

8.9 Traffic Jams

Comparing the hysteretic curves of Fig. 8.25(left) for the standard FK model with those of Fig. 8.25(right) calculated for the anharmonic (exponential) interaction, one can see the following essential difference between them. For the harmonic interaction, the system goes directly from the low-mobility ("locked") state to the high-mobility ("running") state. Although the system may be found in steady states with intermediate values of B, e.g. $B/B_f \sim 0.5$, these states always correspond to a *homogeneous* state on a spacial scale larger than the lattice constant a_s. On the other hand, for anharmonic interaction between the atoms, the system passes through intermediate states which are *spatially inhomogeneous*. In this type of steady states, the system splits into two qualitatively different regions, which differ by atomic concentration and velocities. A typical picture of atomic trajectories is presented in Fig. 8.26. One can clearly distinguish "running" regions, where atoms move with almost maximum velocities, and "traffic-jam" regions, where atoms are almost immobile. In Fig. 8.26(right), which shows a small portion extracted from Fig. 8.26(left), one can see dynamics of a single jam. The jam grows from its left-hand side due to incoming atoms, which stop after collisions with the jam and then join to the jam. From the right-hand side, the jam shortens, emitting atoms to the right-hand-side running region. The velocity probability presents a two-bells shape, corresponding to static and moving atoms: this is a *coexistence regime*. Besides, in Fig. 8.26(right) one can see also a detailed scenario of jam's dynamics: when an incoming atom collides with the jam, it creates a kink (local compression) in the jam. This kink then runs to the right-hand side of the jam and stimulates there the emission of the atom into the right-hand-side running domain. A detailed study [717] shows that the traffic-jam state corresponds to the strange attractor of the system and therefore it is a stable steady state. Moreover, the traffic-jam state remains stable at nonzero temperatures, at least for small enough T.

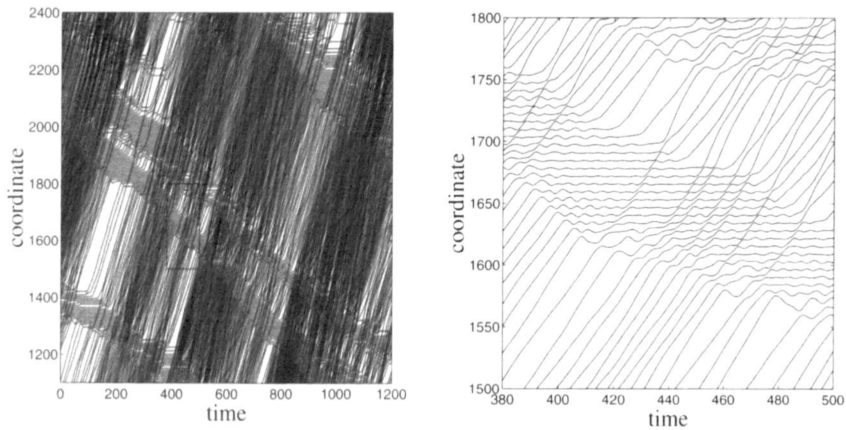

Fig. 8.26. Atomic trajectories for the exponential interaction with $\beta = 1/\pi$ at the fixed force $F = 0.33$. Other parameters are the same as in Fig. 8.25. The rectangle is shown enlarged on the right, and it shows a single jam.

However, because the TJ regime is not observed for the classical FK model with the harmonic interatomic interaction, one has to find *for what system parameters the system demonstrates the transition from the homogeneous state to the inhomogeneous traffic-jam regime.* To study this phenomenon, Braun *et al.* [717] took for the interaction of nearest neighboring atoms the Toda (exponential) potential $V(x) = V_0 e^{-\beta x}$, so that the characteristic radius of interaction is $r = \beta^{-1}$, and calculated the dependence $B(\eta)$ for a fixed value of the driving force F. Besides, they calculated the velocity correlation function,

$$K_l = \langle\langle(\dot{x}_{i+l} - \dot{x}_i)^2\rangle\rangle, \tag{8.69}$$

which helps to distinguish a homogeneous steady state from inhomogeneous ones.

It was found that when η decreases, the system passes from the low-mobility locked state (LS) to the high-mobility running state (RS). For the harmonic interaction, this transition occurs in one step, and the correlation functions $K_l(\eta)$ exhibit a peak just at the transition point. For the exponential interaction, the transition proceeds, on the contrary, by two steps, first the system passes to an intermediate state characterized by a plateau with $0 < B < B_f$, and only then with further decreasing of η, the running state with $B \approx B_f$ is reached finally. This intermediate state always corresponds to the steady state with jams. At the same time the correlation function $K_1(\eta)$ exhibits two peaks, one at the transition to the inhomogeneous traffic-jam state (JS), and the second at the transition to the running state. From the definition (8.69) one can see that the value of K_1 should be proportional to the number of jams in the system, because the velocities of nearest-neighboring atoms may differ essentially only at the boundaries separating the running and jam domains. Therefore, one can use the fact of existence of two peaks on the dependence $K_1(\eta)$ as an indication of the jam state, while the positions of these peaks show the parameter range for JS existence.

Recall that the dimensionless elastic constant, which is the main parameter of the classical FK model, is defined as $g = a_s^2 V''(a_A)/2\pi^2 \varepsilon_s$, so that for the potential (8.55) it is equal to $g = V_0 \beta^2 \exp(-\beta a_A)$. Making a series of runs for different values of the parameter β and keeping at the same time the value of the elastic constant g fixed, so that the limit $\beta \to 0$ corresponds to the harmonic interaction (the standard FK model), while the limit $\beta \to \infty$ describes a hard-core gas, allows us to plot the phase diagram on the (η, β)-plane by extracting the positions of maxima of $K_1(\eta)$ for every value of β. The phase diagram is presented in Fig. 8.27.

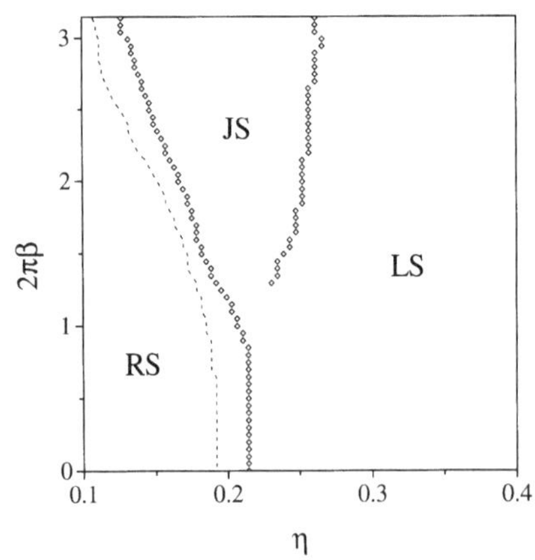

Fig. 8.27. Phase diagram for the (η, β)-plane at fixed value of g, where $N = 256$, $g = 0.1$, $F = 0.5$, and $T = 0.1$; LS is the locked state, JS is the steady state with jams, and RS is the running state. Diamonds correspond to maxima of $K_1(\eta)$, and the dashed curve corresponds to $B(\eta) = 0.9\, B_f$.

The simulation results show that the transition to the steady state with jams emerges for the exponential interaction (8.55) for $\beta > a_s^{-1}$ only. Thus, *the transition to the traffic-jam state emerges for short-range interatomic interaction only*, when the radius of interaction is smaller than the period of the external periodic potential, and only for small damping coefficients in an interval just preceding the transition to the running state. In this context Braun *et al.* [717, 725] proposed a generalized lattice-gas model with two states of atoms. This traffic-jam LG model is characterized by a nonlinear dependence of mobility on the jump probability, exhibits hysteresis, and describes the organization of immobile atoms into compact domains (jams). Besides, a MF theory for this LG model has been developed, which describes *kinetics* of traffic jams.

Finally, *traffic jams* emerge in the driven 1D system, if the following conditions are fulfilled: (i) It must be an external (substrate) potential; (ii) The motion must be underdamped, the particles should have two different states, the locked state and the running state. In the FK model the bistability exists due to inertia of atoms.

These two conditions are already sufficient for *existence* of the traffic-jam state. Indeed, if one prepares by hands the initial configuration with a jam and then abruptly applies the driving force, this state may survive in the classical FK model even at nonzero temperatures. Strunz and Elmer [712] have studied steady sliding states characterized by domains of different atomic densities (which may be considered as a generalization of the TJ state discussed above) in the classical FK model at $T = 0$. However, the *transition* from the homogeneous state to the traffic-jam state at slow variation of model parameters takes place only if two more conditions are satisfied:

(i) It must be some randomness in the system. Already the intrinsic chaos, which always exists due to nonintegrability of the discrete FK model, is sufficient for existence of the traffic-jam behavior (although in the $T = 0$ case we have to start from a random initial configuration). The simplest way to introduce a chaos into the system is to use Langevin motion equations with $T > 0$; in this case the traffic-jam state emerges for any initial state. Of course, the temperature should not be too large, the thermal energy must be lower than the energy of interatomic interaction (otherwise the behavior will be the same as for the system of noninteracting atoms) as well as it must be lower than ε_s (otherwise the behavior will be the same as for the system without external potential);

(ii) The interparticle interaction has to be anharmonic. Already the hard-core potential, when the atoms do not interact at all except they cannot occupy the same well of the substrate potential, is sufficient to produce the traffic-jam behavior. Thus, one might expect no transition to the traffic-jam state for the harmonic interatomic interaction. However, the situation is more subtle: there is no transition to the traffic-jam state for *atoms* in the standard FK model, but the *kinks* still may be organized in jams, because for any short-ranged interatomic interaction, the interaction between the kinks is always exponential.

8.10 Periodic Forces: Dissipative Dynamics

Finally, let us consider briefly the FK system driven by a (spatially uniform) periodic force $F(t) = F(t + \tau_{\rm ext})$, where $\tau_{\rm ext}$ is its period so that the ac frequency is $\Omega = 2\pi/\tau_{\rm ext}$. Due to complexity of the problem we first consider the overdamped case, following the survey by Floría and Mazo [487]. The equations of motion,

$$\dot{x}_l = g\,(x_{l+1} + x_{l-1} - 2x_l) - \sin x_l + F(t), \tag{8.70}$$

have the following symmetry: if $\{x_l(t)\}$ is the solution of equation (8.70) which corresponds to the configuration $\{x_l(t_0)\}$ of initial conditions, then the trajectory

$$\sigma_{i,j,m}\{x_l(t)\} \equiv \{x_{l+i}(t - m\tau_{\rm ext}) + ja_s\}\,, \tag{8.71}$$

where i, j and m are arbitrary integers, is the solution corresponding to the initial configuration $\sigma_{i,j,m}\{x_l(t_0)\}$. The trajectory $\{x_l(t)\}$ is *resonant* if there is a triplet (i, j, m) of integers such that the trajectory is invariant under the transformation (8.71),

$$\{x_l(t)\} = \sigma_{i,j,m}\{x_l(t)\}. \tag{8.72}$$

The average velocity of a resonant solution is

$$\bar{v} = (\Omega/m)(iw + j)a_s/2\pi, \tag{8.73}$$

where w is the window number. When w is irrational, then one can find a unique *minimal triplet* with the property that i, j and m have no common factors. For a rational window number, $w = r/s$ (r and s are co-prime integers), a minimal triplet is such that $ir + js$ and m have no common factors (the triplet is not unique in this case, but the integer m is defined uniquely). Then, the resonant velocity is called *harmonic* if $m = 1$, and *subharmonic* whenever $m > 1$. The main question of interest for ac driven systems is the one about existence and structural stability (i.e. robustness against a change of parameters) of the resonant solutions (8.72).

To get insight into the problem of synchronization of the FK chain to external periodic driving, let us consider first a simple case when the force $F(t)$ consists of constant force of duration $t_{\rm on}$, followed by time intervals $t_{\rm off}$ of zero force ($t_{\rm on} + t_{\rm off} = \tau_{\rm ext}$),

$$F(t) = \begin{cases} F \text{ if } \ 0 < t < t_{\rm on}, \\ 0 \text{ if } \ t_{\rm on} < t < t_{\rm on} + t_{\rm off}, \end{cases} \tag{8.74}$$

and assume that $t_{\rm off}$ is large enough, so that during this time interval the chain relaxes to some stable static configuration, before the next pulse is on. In this case we can use the result obtained above that the long-term average velocity $\bar{v}$ is independent of the initial configuration.

Now let us begin the analysis with the case of a commensurate structure with the window number $w = r/s$ and choose as initial conditions the zero-force GS configuration. If the amplitude of pulses is small, $F < F_{PN}$, or if $F > F_{PN}$ but the duration $t_{\rm on}$ is not large enough to bring the configuration out of the basin of attraction (in the s-dimensional configurational space corresponding to zero force, see Fig. 8.28) of the initial GS configuration, the configuration will relax, when the pulse is off, to the initial configuration. Thus, at small F the state with zero average velocity, $\bar{v} = 0$, is locked. However, if $F > F_{PN}$ and $t_{\rm on}$ is large enough, the system will relax to some stable configuration $\{x_l'\}$ during the $t_{\rm off}$ interval. Due to the preservation of the rotational order under the overdamped dynamics and the uniqueness (up to translation on the distance a_s and relabelling) of the GS configuration, $\{x_l'\}$ must be of the form $\{x_l'\} = \{x_{l+i} + ja_s\}$ with i and j integers. Consequently, the average velocity is necessarily given by Eq. (8.73) with $m = 1$ and corresponds to the harmonic resonance. *The average velocity* $\bar{v}$ *of a commensurate structure under periodic pulsed forcing* (8.74) *can only take the discrete values* (8.73) *corresponding to harmonic* ($m = 1$) *resonances.*

This "dynamical mode locking" can be explained in the framework of the s-dimensional configurational space as shown in Fig. 8.28. During the pulse on, the system moves along a certain trajectory in the configurational space. When the pulse is off, the system relaxes to the ground state associated with the basin of attraction to which the end point of that trajectory belongs. Consider now two forces with different amplitudes F_1 and F_2. If these am-

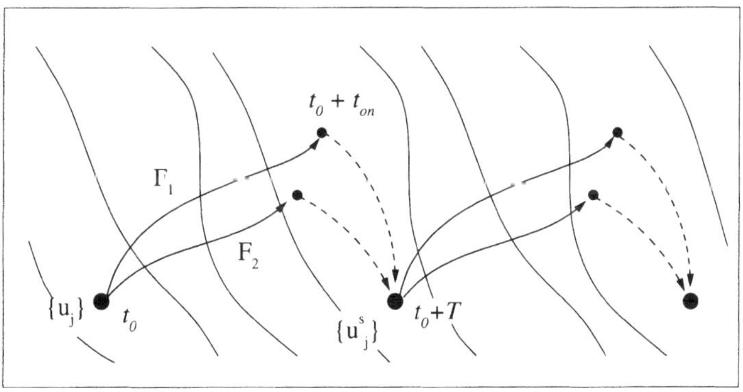

Fig. 8.28. Mode locking under a periodic pulsed force: schematic sketch of a s-dimensional configurational space portioned in basins of attraction associated with each ground state [487].

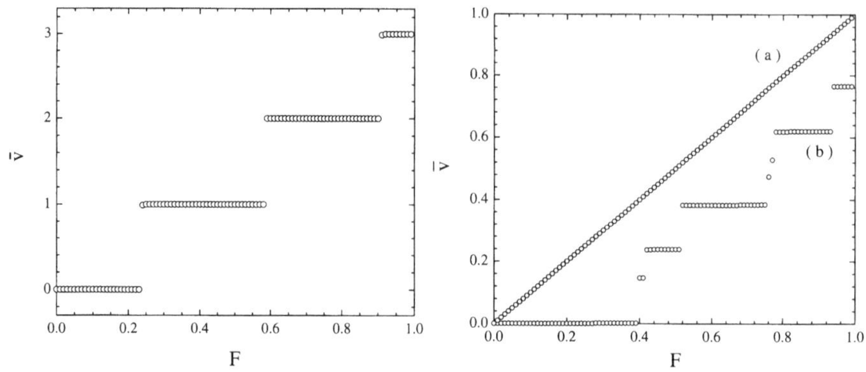

Fig. 8.29. Average velocity $\bar{v}$ as a function of the amplitude F of the pulsed periodic force (8.74) with $t_{\rm on} = 1$ and $t_{\rm off} = 20$. Left panel is for the commensurate structure with $w = 2/3$ and $g = 0.5$. Right panel is for the "incommensurate" structure with $w = 34/55$ and two values of g: (a) $g = 10 > g_{\rm Aubry}$ (above the Aubry transition) and (b) $g = 0.25 < g_{\rm Aubry}$ (below the Aubry transition) [487].

plitudes are so close to one another that the end points of both trajectories belong to the same basin of attraction, then the average velocity must be the same too, $\bar{v}(F_1) = \bar{v}(F_2)$. Thus, for a commensurate structure $\bar{v}(F)$ should be a piecewise constant non-decreasing function as shown in Fig. 8.29(left).

A similar behavior should exhibit an incommensurate structure below the Aubry transition ($g < g_{\rm Aubry}$), when the GS is pinned. This can be understood if one considers the IC GS as a limit of a sequence of commensurate ground states with rational approximants w_n to the desired irrational window number w. Now, however, the function $\bar{v}(F)$ corresponds to Devil's staircase as shown by curve (b) in Fig. 8.29(right).

The mode locking is only possible if the set of ground states is discrete. If this set is continuous [as it is for the IC GS above the Aubry transition, when the GS can be shifted on an arbitrary distance due to existence of the Goldstone (sliding) mode], no dynamical mode locking is possible, and $\bar{v}(F)$ should be a continuous strictly increasing function as shown by curve (a) in Fig. 8.29(right).

For a sinusoidal driving force,

$$F(t) = \bar{F} + F_{\rm ac} \cos(\Omega t), \tag{8.75}$$

the system behavior is more involved. Now the system has two competing frequency scales: the ac driving frequency Ω and the washboard frequency $\omega_{\rm wash} = (2\pi/a_s)\,\bar{v}$ due to the motion driven by the dc force $\bar{F}$. In the case of a commensurate structure with an *integer* window number w the system exhibits harmonic resonances only [$m = 1$ in Eq. (8.73)] analogously to the system driven by pulsed force as was described above [726, 727]. However, for rational *noninteger* values of $w = r/s$, when there are $s > 1$ atoms

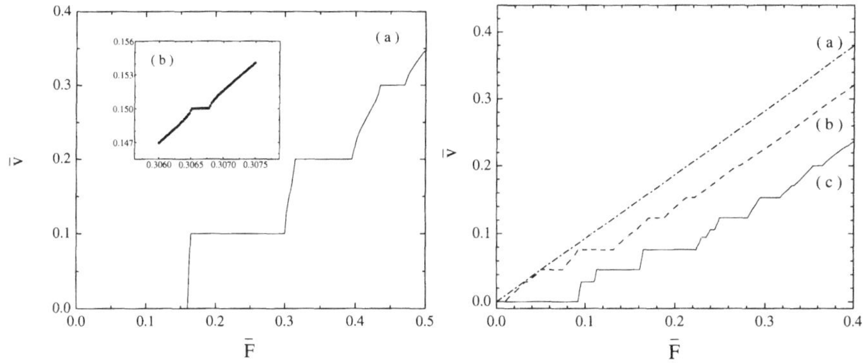

Fig. 8.30. Average velocity $\bar{v}$ versus the dc force $\bar{F}$ under a sinusoidal driving force (8.75) with $F_{\rm ac} = 0.2$ and $\tau_{\rm ext} = 2\pi/\Omega = 5$. Left panel is for the commensurate structure with $w = 1/2$ and $g = 0.25$. Only harmonic steps are resolved in (a), while the inset (b) shows a subharmonic $m = 2$ step. Right panel is for an incommensurate structure for three different values of $K \equiv g^{-1}$: (a) $K = 1.4$ (dot-dashed line), (b) $K = 2.8$ (dashed curve), and (c) $K = 4.0$ (solid curve) [487].

per one elementary cell of the GS structure, simulations demonstrate the existence of subharmonic resonances with $m > 1$ in Eq. (8.73) as shown in Fig. 8.30(left) [728, 729]. Moreover, in the latter case the system has unlocked (nonresonant) solutions. The analysis based on the Floquet theory suggests that when $\bar{F}$ changes, the locking-unlocking transition of a commensurate structure occurs due to a saddle-node bifurcation and, thus, it should be characterized by a critical exponent of $\frac{1}{2}$ for the average velocity (see details in Ref. [487]).

The incommensurate structure for the golden-mean window number $w = (\sqrt{5} - 1)/2$ driven by the sinusoidal force (8.75) was studied by Floría and Falo [730]. At low values of $K \equiv g^{-1} \ll 1$, the average velocity $\bar{v}$ is an increasing function of the average external force $\bar{F}$, $\bar{v} \sim \bar{F}$ [see curve (a) in Fig. 8.30(right)]. In this regime, where no locking to resonant velocities is observed, the computed trajectories, at any value of $\bar{v}$, can be expressed in terms of a two-variable *dynamical hull function* $h(x, y)$ as

$$x_l(t) = h(lwa_s + \bar{v}t + \alpha,\ \Omega t + \beta), \tag{8.76}$$

where α and β are arbitrary phases, and the function $\phi(x, y) = h(x, y) - x$ is periodic of period $a_s = 2\pi$ in the first variable and of period 2π in the second variable. The dynamical hull function $h(x, y)$ changes continuously as the parameters g, $\bar{F}$, $F_{\rm ac}$ and Ω vary, and approaches the analytical solution $h^{(0)}(x, y) = x + (F_{\rm ac}/\Omega)\sin y$ in the $K \to 0$ limit. Thus, the effect of a small periodic substrate potential on the ac-driven motion of incommensurate structures is just a smooth modulation of the solution at the integrable limit.

As K increases, however, $\bar{v}(\bar{F})$ starts to develop "plateau" (steps) at resonant velocities (8.73) [see curves (b) and (c) in Fig. 8.30(right)]. Simulations suggest that for each resonant value of $\bar{v}$ there is a critical value $K_c(\bar{v})$ above which there is an interval of $\bar{F}$ values for which the velocity locks at that resonant value. Moreover, at each particular value of $K_c(\bar{v})$, the corresponding hull function $h(x, y)$ becomes non-analytic. Therefore, the transition at $K_c(\bar{v})$ is a *dynamical Aubry transition.* Below this transition, $g < g_c(\bar{v}) \equiv 1/K_c(\bar{v})$, where the threshold $g_c(\bar{v})$ depends on the model parameters w, g and Ω, mode locking occurs, while above the transition, $g > g_c(\bar{v})$, there is no mode locking, and the steady state is described by an analytical two-variable hull function. The phase diagram for a particular choice of model parameters is presented in Fig. 8.31.

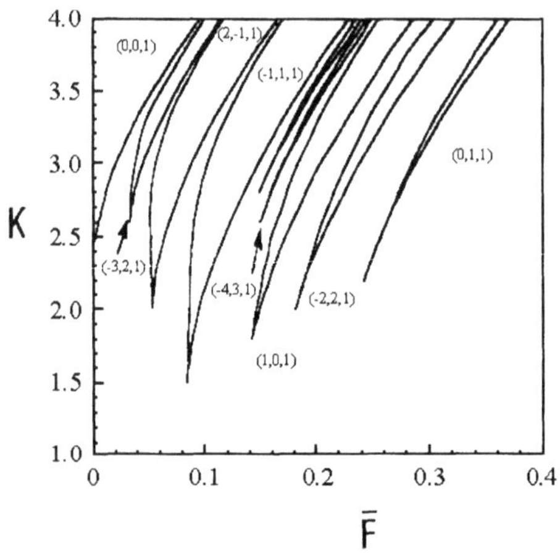

Fig. 8.31. Phase diagram of the different states in the $(K, \bar{F})$ plane for an incommensurate structure under a sinusoidal driving force (8.75) with $F_{\rm ac} = 0.2$ and $\tau_{\rm ext} = 2\pi/\Omega = 5$. The triplets (i, j, m) correspond to locking at the resonant velocities (only tongues for a few resonant velocities are shown) [487].

Thus, while the standard ("static") Aubry transition is the border point between pinned and sliding incommensurate structures, the dynamical Aubry transition separates, for a given velocity value $\bar{v}$, the locking regime, where the value of $\bar{v}$ is robust against small variation of the average external force, from the unlocked regime of motion of the incommensurate structure at the velocity $\bar{v}$. From this perspective, the motion of incommensurate structures under periodic pulses (8.74) described above, may be considered as a limiting case of sinusoidal forcing (8.75), in which the dynamical Aubry transition occurs at a unique value of g for all resonances, and coincides with the standard Aubry transition, $g_c(\bar{v}) = g_{\rm Aubry}$.

8.11 Periodic Driving of Underdamped Systems

Finally, let us describe in brief the results known for the ac driven underdamped FK model. In the case of the *additive* oscillating force with the amplitude A and the frequency Ω, the SG equation takes the form

$$\frac{\partial^2 u}{\partial t^2} - \frac{\partial^2 u}{\partial x^2} + \eta \frac{\partial u}{\partial t} + \sin u = A \sin(\Omega t + \phi). \tag{8.77}$$

We begin with a more simple situation when there are only a *single kink* in the system. Let us first consider a very simple example of motion of a single atom of mass m under the action of a force $F(t)$,

$$m\ddot{X}(t) + m\eta\dot{X}(t) = F(t), \tag{8.78}$$

where $X(t)$ is the atomic coordinate and η is the viscous damping. If the a.c. force is switched on at the time moment $t = 0$, i.e. $F(t < 0) = 0$ and $F(t \geq 0) = \mathrm{Re}\,(F_0 e^{i\Omega t})$ [here $F_0 = -imAe^{i\phi}$ is complex to incorporate the phase ϕ], then the undamped ($\eta = 0$) solution of Eq. (8.78) takes the trivial form,

$$X(t) = X_0 + V_0 t - \mathrm{Re}\left[\left(F_0/m\Omega^2\right) e^{i\Omega t}\right], \tag{8.79}$$

where the directed atomic velocity $V_0 = \dot{X}(0) + \mathrm{Im}\, F_0/m\Omega$ is determined by the initial velocity $\dot{X}(0)$ and the amplitude and phase of the oscillating force $F(t)$ at $t = 0$. Thus, in a general case we have $V_0 \neq 0$, and pure oscillatory atomic motion occur for certain "resonant" conditions only, when $\dot{X}(0) = -\mathrm{Im}\, F_0/m\Omega$. It is evident, however, that for any nonzero damping, $\eta > 0$, the directed motion will decay during a transient time $\tau \sim \eta^{-1}$, so that in the steady state the atom will only oscillate around a given site.

Coming back to the perturbed SG equation (8.77) and using the simple collective-coordinate approach described above in Sect. 8.4, we arrive at the following equation on the kink's coordinate $X(t)$ [731]

$$\ddot{X}(t) + \eta \left[1 - \dot{X}^2(t)/c^2\right] \dot{X}(t) = F(t)/m, \tag{8.80}$$

where m is the kink rest mass, c is the sound speed, and the a.c. driving force $F(t)$ is given by

$$F(t) = \sigma A \left[1 - \dot{X}^2(t)/c^2\right]^{3/2} \sin(\Omega t + \phi), \tag{8.81}$$

$\sigma = \pm 1$ being the kink topological charge. Thus, the kink motion is qualitatively the same as that for the simple single-atomic case described above, and exactly coincides with that in the non-relativistic limit $|\dot{X}(t)| \ll c$. This result was derived analytically and confirmed with simulation by Quintero and Sánchez [731] (the special cases of $\phi = 0$ and $\pi/2$ for the zero initial kink velocity, $\dot{X}(0) = 0$, was first considered by Olsen and Samuelsen [733]).

The kink behavior under the additive a.c. force becomes more interesting, if the kink has its internal (shape) mode with a frequency ω_B, which is a typical situation for a generalized FK model, e.g. for nonsinusoidal substrate potential or nonharmonic interatomic interaction. This question was studied by Quintero *et al.* [615] on the example of the ϕ^4 model, where exactly one shape mode exists. Indeed, they do observed a strong resonance, but at the driving frequency $\Omega = \omega_B/2$ and not at the normal resonant frequency $\Omega = \omega_B$ as one may expect. The explanation is that the a.c. force does not act directly on the shape mode (because of symmetry reasons), but rather it excites the shape mode indirectly via the excitation of the translational motion which then couples to the shape mode. The coupling of the internal and translational degrees of freedom leads finally to an erratic/chaotic kink motion at the resonance because of the interchange of energy between the kinetic energy of the kink center and the internal oscillations of the kink shape.

The discreteness of the FK model brings new effects in the kink motion under the a.c. force. Indeed, now the kink moves in the periodic Peierls-Nabarro potential. Thus, even if one neglects by radiation effects and the softness of the kink shape, the motion equation corresponds to the damped a.c. forced pendulum. Depending on the model parameters and initial conditions, such a system admits a rich variety of different behavior, including the so called running trajectories, which are attractors with a nonzero average velocity. The underdamped FK model under the a.c. driving was studied numerically by Martinez *et al.* [735]. For a particular case of relatively low damping ($\eta \sim 0.1\,\omega_0$) and driving frequency ($\Omega \sim 0.5\,\omega_{\rm PN}$) the authors observed the following scenario of system behavior when the amplitude A increases. At low A the kink is trapped in the PN valley and oscillates near its bottom with the driving frequency. At higher A the kink begins to jump to adjacent wells of the PN potential. Such type of motion is associated with the type-I intermittency (the kink stays for a long time in one well, then suddenly jumps to another well and stays there, again for a long time). The frequency of these jumps increases with A until a diffusive kink's motion is reached. The further increase of A finally leads to a mode-locking regime, in which the average kink velocity is exactly given by $v_{\rm k} = (p/q)\,a_s\Omega/2\pi$ with some co-prime integer numbers p and q. An important issue is that such a directed motion appears when the amplitude of the driving is about an order of magnitude *below* the depinning d.c. force for the same model parameters. Note also that if one averages over all initial conditions, the net kink velocity is zero due to symmetry of the model, the directed motion with positive velocity has the same probability as that with the negative one. However, if the system symmetry is violated in some way, the preferred direction of motion will emerge as will be described in the next Chapter 9.

The case when the underdamped SG system ($\eta \sim 0.05\,\omega_0$) is driven simultaneously by the d.c. and a.c. forces, was studied by Fistul *et al.* [736]. For a low driving frequency, $\Omega \sim 0.03\,\omega_0$, the authors found a strong decreasing of the kink velocity even for a relatively low amplitudes of the a.c. driving ($A < 1$). This effect may be interpreted as an effective additional "nonlinear damping" of the kink motion. The analytical estimates were found in a good agreement with both the direct numerical simulation and the experimental results for a long annular Josephson junction with one trapped fluxon.

In the non-driven case, if the damping η is nonzero, then the breather annihilates, and its energy is dissipated into radiation. But if the system is a.c. driven by the additive force, then the radiation losses are compensated by the external driving, so that the breather is stabilized, and its frequency becomes modulated by the driving [737, 738]. A more subtle scenario emerges when the SG breather is driven by the additive a.c. force in absence of damping. Using the same collective-coordinate approach as above, Quintero and Sánchez [731] have shown that in this case there exist a threshold amplitude A_c of the oscillation force (the value of A_c depends on the breather frequency), such that a small perturbation ($A < A_c$) allow breathers to exist (with the frequency modulated by the external driving), while at higher perturbation, $A > A_c$, the breather either dissociates transforming into a kink-antikink pair, or it annihilates emitting radiation if the breather frequency was close to the phonon band. Note that a similar effect of the breather stabilization exists for the parametric driving as well [739].

The latter result suggest a mechanism of kink-antikink pairs generation through an intermediate stage of breather excitation. Recall that if the GS corresponds to the trivial commensurate structure ($\theta = 1$), then the kinks in the system emerge due to thermally activated creation of kink-antikink pairs.

Finally, let us describe briefly the case when the system is driven by a plane wave. From plasma physics (or physics of electron beams) it is well

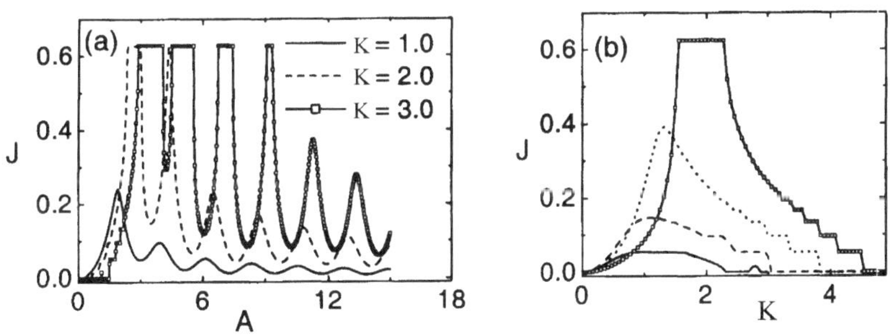

Fig. 8.32. Average current J of the model (8.82) versus (a) the wave amplitude A for different values of K (solid curve for $K = 1$, dash curve for $K = 2$, and solid curve with square symbols for $K = 3$) and (b) the barrier height K for different values of A (solid curve for $A = 1$, dash curve for $A = 1.5$, dotted curve for $A = 2$, and solid curve with square symbols for $A = 2.5$). The other parameters are: $g = 1$, $\Omega/2\pi = 0.1$, and $\delta = \phi = 0.09$ [740].

known that a charge (e.g., an electron) may be catched by an electromagnetic wave and travel together with it with the same velocity. One could expect a similar effect in the FK model. Indeed, such an effect was observed numerically by Zheng and Cross [740]. The authors studied the overdamped motion equation

$$\dot{x}_l - g\,(x_{l+1} - 2x_l + x_{l+1}) + K \sin x_l = A \cos(\Omega t - 2\pi l \phi) \tag{8.82}$$

with periodic boundary condition, when the number of geometrical kinks inserted into the chain is determined by the mismatch parameter $\delta = (a_s - a_A)/a_s$ ($a_s = 2\pi$ is the period of the substrate potential and a_A is the mean interatomic distance). Evidently, there are no directed motion in the cases of $\delta = n$ (no kinks to be driven) or $\phi = n$ (the case of standing wave) (here n is an integer, $n = 0, \pm 1, \ldots$). The steady state current $J = \langle \dot{x}_l(t) \rangle$ of the model (8.82) has the following symmetry: $J(\delta, \phi) = J(1-\delta, 1-\phi) = -J(\delta, 1-\phi)$ $= -J(1-\delta, \phi)$. Thus, we also must have zero current for $\delta = n + 1/2$ or $\phi = n + 1/2$.

The results of simulation are shown in Fig. 8.32. Notice that there is a threshold amplitude of the driving (an analog of the PN barrier) for the motion to start. The maximum current is achieved at $|\delta| \approx |\phi| \approx 0.12$. For high enough values of the height of the substrate potential the current may reach the maximal value $J = \Omega$ (see plateaus in Fig. 8.32(a). As a function of K, the current scales as $J \propto K^2$ at small K and exhibits mode-locking resonant steps at high K as shown in Fig. 8.32(b). Because any dynamical state of the overdamped FK model must be rotationally ordered, one can show that the current at the resonant steps is described by the relation $J = [(j\delta + k)/(j\phi_1 + m)]\Omega$ with integers j, k and m in accordance with the simulation results.

9 Ratchets

The original FK model is based on a symmetric on-site potential. However, if the system symmetry is broken in some way, then novel effects associated with so-called *ratchet* dynamics appear. The system driven by an external perturbation is known to exhibit ratchet dynamics if in the long-time limit the average current, defined as

$$\langle \dot{x} \rangle = \lim_{t \to \infty} [x(t) - x(0)]/t,$$

does not vanish when *all* macroscopic ("coarse grained") external perturbations, static forces and gradients (of temperature, concentration, chemical potential, etc.) vanish after averaging over space, time, and statistical ensembles. Ratchet systems have attracted a lot of attention because they are used to explain the physics of molecular motors and molecular pumps, along with the possibilities they open for various applications of nanotechnology.

9.1 Preliminary Remarks

To exhibit the ratchet dynamics, the system *must* satisfy the following two conditions:

(i) *The system must be out of thermal equilibrium*, simply because the equilibrium system cannot exhibit a systematic drift in a preferential direction due to the second law of thermodynamics. Therefore, the system must be driven by an external perturbation, for example, by an external force $F(t)$ with zero average, $\langle F(t) \rangle = 0$ (the forcing may be either periodic or stochastic, but in the latter case it must correspond to a color noise). We have mentioned already in Sec. 8.11 that under the periodic force, the kink may demonstrate the directed motion. However, if one averages over all initial conditions in that case, the average velocity will be zero, of course, due to symmetry of the system. Thus, we need one more condition.

(ii) *The space-time symmetry of the system must be violated*, either due to asymmetric substrate potential, or because of asymmetric driving, or both. The rigorous rules have been formulated by Flach *et al.* [741] and Reimann [742] and they will described below in Sect. 9.2.1.

If both these conditions are satisfied, then the system can exhibit the ratchet behavior. The next questions that emerge, are:

1. What is the direction of motion and how it can be manipulated? For some simple models one can predict or guess the drift direction from symmetry reasons, but often it depends on all the details of the substrate potential and the external force.
2. What is the "efficiency" of the such ratchet machine? Namely, how the current depends on the driving for a given ratchet model?

There are many various ways to organize the ratchet behavior. Indeed, a class of asymmetric systems is much larger than the subclass of symmetric ones. In fact, the question is not when the ratchet behavior emerges, but rather when the systematic drift is absent due to (sometimes hidden) symmetry of the system.

A typical substrate potential used in ratchet models, is the co-sinusoidal one with an addition of the second harmonic (see Fig. 9.1),

$$V(x) = -\cos x + \mu \cos(2x + \varphi). \tag{9.1}$$

As for the forcing $F(t)$, it may correspond, for example, to (i) a sinusoidal function, (ii) a symmetric dichotomous noise, (iii) an Ornstein-Uhlenbeck process (the exponentially correlated noise), (iv) a symmetric *shot* noise, *etc.*, or an arbitrary linear combination thereof.

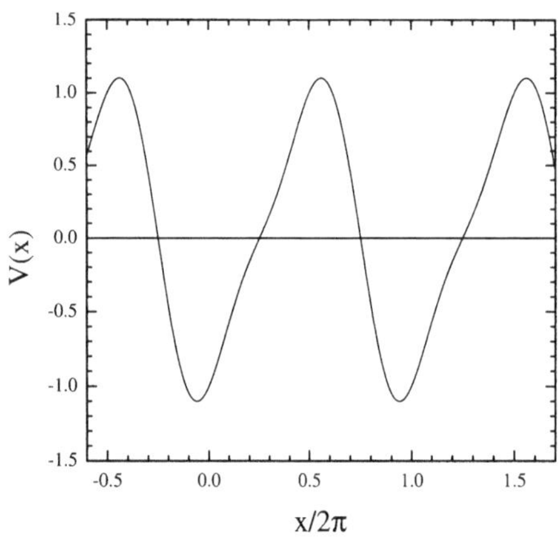

Fig. 9.1. The ratchet potential (9.1) for $\mu = -1/4$ and $\varphi = \pi/2$.

For a simpler case of a single atom in an external potential, we can divide different types of the ratchet systems into *diffusional ratchets* and *inertial ratchets*. When the model describes a system of interacting atoms, additional collective effects may appear due to interaction; the latter are considered below in Sect. 9.3 for the example of the FK model.

9.2 Different Types of Ratchets

9.2.1 Supersymmetry

First of all, following the work by Riemann [742] (see also Refs. [743, 744]), we formulate the general conditions when the directed motion *does not exist*. We start with the following simple definitions of the symmetry relations:

1. The periodic substrate potential is *symmetric*, if for all x
$$V(-x) = V(x). \tag{9.2}$$
2. The periodic potential is called *supersymmetric*, if for all x
$$V(x + a_s/2) = -V(x). \tag{9.3}$$
3. The external periodic force $F(t) = F(t+\tau)$ is called *symmetric*, if for all times t
$$F(t + \tau/2) = -F(t). \tag{9.4}$$
4. The periodic forcing is called *supersymmetric*, if it is antisymmetric for all t ,
$$F(-t) = -F(t). \tag{9.5}$$

These definitions assume appropriate shifts, e.g. Eq. (9.3) means that $V(x + a_s/2) = -V(x) + \Delta V$, but we may always put $\Delta V = 0$ because of irrelevancy of additive constants for the potential. Analogously, Eq. (9.5) means that there exists Δt such that $F(-t) = -F(t + \Delta t)$, and then Δt may be made zero by a shift of the time scale. The last two definitions remain valid for a stochastic forcing (unbiased and stationary) as well, if we rewrite, for example, Eq. (9.5) in the form

$$\langle F(-t_1)F(-t_2)\dots F(-t_n)\rangle = (-1)^n \langle F(t_1)F(t_2)\dots F(t_n)\rangle \tag{9.6}$$

and assume that it is valid for all integers $n \geq 1$ and all times $t_1, t_2, \dots, t_n$.

Now, the main statement is the following: for the motion equation with the additive driving,

$$\eta\, \dot{x}(t) + V'(x) = F(t) + \delta F(t), \tag{9.7}$$

where $\dot{x} \equiv dx/dt$, $\delta F(t)$ is the Gaussian white noise, *the directed motion is absent,* $\langle \dot{x} \rangle = 0$, *only if either* <u>*both*</u> *the substrate potential* $V(x)$ *and the driving* $F(t)$ *are symmetric, or if* <u>*both*</u> *of them are supersymmetric.* For all other combinations, $\langle \dot{x} \rangle \neq 0$ in a general case.

The case of both symmetric potential and driving force does not require a proof, the absence of the directed motion follows simply from symmetry reasons. As for the case when both the potential and the driving are supersymmetric, the proof is the following: introducing $z(t) = x(-t) + a_s/2$, we have $\dot{z}(t) = -\dot{x}(-t)$ so that $\langle \dot{z} \rangle = -\langle \dot{x} \rangle$. On the other hand, taking into account the supersymmetry conditions (9.3) and (9.5), one can verify that $z(t)$ satisfies the same equation (9.7) as $x(t)$, so that $\langle \dot{z} \rangle = \langle \dot{x} \rangle$. Thus, we arrive at $\langle \dot{x} \rangle = 0$. This proof is based on the fact that only *stationary* stochastic processes appear in Eq. (9.7).

Typical combinations of the potential and driving force are shown in Fig. 9.2: nonzero current appears in the cases (a) and (b), but it vanishes for the case (c) when both potential and driving force are supersymmetric.

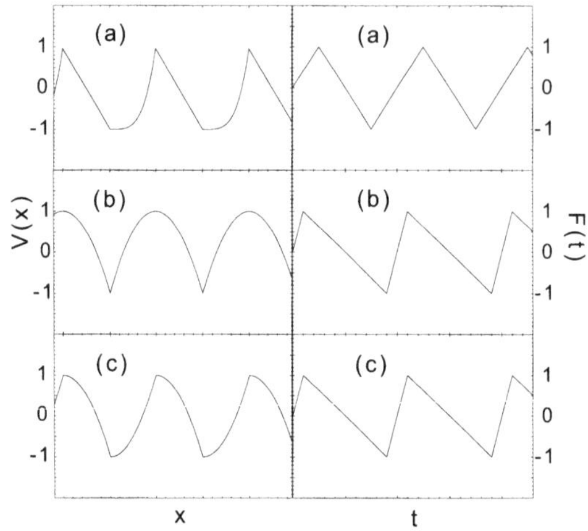

Fig. 9.2. Three combinations of the periodic substrate potential $V(x)$ and the external driving force $F(t)$: (a) $V(x)$ is asymmetric (but not supersymmetric) and $F(t)$ is symmetric (and at the same time supersymmetric), $\langle \dot{x} \rangle \neq 0$; (b) $V(x)$ is symmetric (but not supersymmetric) and $F(t)$ is supersymmetric (and asymmetric), $\langle \dot{x} \rangle \neq 0$; (c) both $V(x)$ and $F(t)$ are supersymmetric and $\langle \dot{x} \rangle = 0$ [742].

Next, we consider the case of parametric driving,

$$\eta \, \dot{x}(t) + V' \left[x(t), F(t) \right] = \delta F(t), \tag{9.8}$$

where $V'(x, F) = \partial V(x, F)/\partial x$. The *symmetry criterion* now takes the form

$$V \left[-x, F(t) \right] = V \left[x, F(t) \right], \tag{9.9}$$

while the *supersymmetry criterion* has to be rewritten as follows,

$$V \left[x + a_s/2, F(-t) \right] = -V \left[x, F(t) \right], \tag{9.10}$$

and again, *if either of the criteria (9.9) or (9.10) are met, a vanishing current in (9.8) is the consequence*, while in any other case, a nonzero current is generically expected.

9.2.2 Diffusional Ratchets

Now we consider different kinds of *diffusional ratchets*, when inertia effects can be neglected. Most of the models considered below, exhibit the existence of a threshold value of driving, below which no average directed current exists. The explanation of this threshold is the following [745]. In the linear response limit, which should work for a small amplitude of the driving, one may expand the external perturbation into a Fourier series. Then, due to linearity of this limit, the net current follows simply by summing up the contributions from all modes. But because each Fourier mode is symmetric, the resulting current should be zero.

Another general remark is that the current $\langle \dot{x} \rangle$ should approach zero in the $T \to \infty$ limit, since the effect of a finite potential or driving will be overruled by the noise. Therefore, for the models where the current also vanishes in the $T \to 0$ limit, a bell-shaped current-temperature curve is expected. Moreover, in the same way of reasoning one should expect the existence of optimal values of the driving frequency Ω and amplitude A, at which the current reaches its maximal values.

Finally, to estimate the efficiency of a particular ratchet scheme, one can add the constant force $F_{\rm dc}$ to the right-hand part of equation (9.7) or (9.8) and then to look for the so called *stopping force* F_s, at which the ratchet's directed current becomes zero. A larger is $|F_s|$, the more efficient is the ratchet, e.g., a larger "cargo" it can "bring" with itself.

Thermal Ratchets

A simple model of the *thermal ratchets* (also known as *Brownian motors*; *Brownian rectifiers* and *stochastic ratchets* are also in use) is described by the equation

$$\eta\, \dot{x}(t) + V'(x) = \delta F(t), \tag{9.11}$$

where the Gaussian white noise is subjected to periodic variations with period τ, $\langle F(t)F(0)\rangle = 2\eta k_B T(t)\delta(t)$ with $T(t+\tau) = T(t)$. For an asymmetric substrate potential $V(x)$ the steady state solution of Eq. (9.11) is characterized by nonzero current, $\langle \dot{x} \rangle \neq 0$. The mechanism of its operation is explained in Fig. 9.3: Brownian particles, initially located at the point x_0 in a minimum of $V(x)$, spread out when the temperature is switched to a high value. When the temperature jumps back to low values, most particles get captured again in the basin of attraction of x_0, but also substantially in that of $x_0 + a_s$ (hatched area in Fig. 9.3), and a net current of particles to the right ($\langle \dot{x} \rangle > 0$) results. An efficiency of the thermal ratchet may be estimated with the help of perturbation analysis [746] which yields at small τ

$$\langle \dot{x} \rangle \approx C\tau^2 \int_0^{a_s} dx\, V'(x)\, [V''(x)]^2, \tag{9.12}$$

where $C > 0$ is a constant which depends on the shape of functions $V(x)$ and $T(t)$. Evidently, the current should also tend to zero in the limit $\tau \to \infty$, when the system is in thermal equilibrium. Thus, the maximum current is achieved at an optimal value of the period $\tau \sim \eta^{-1}$.

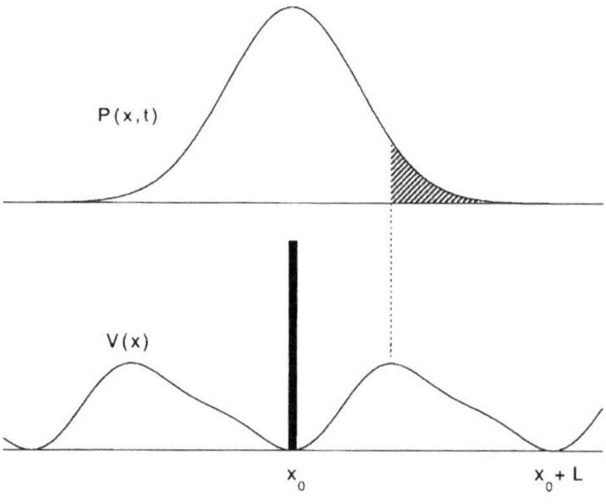

Fig. 9.3. The working mechanism of the thermal ratchet (9.11). Practically the same mechanism is at work when the temperature is kept fixed (but $T > 0$) and instead, the substrate potential is turned "on" and "off", as shown below in Fig. 9.4 [745].

This very simple model of the thermal ratchet (9.11) already exhibits the so called *current inversion* effect. Namely, for a specially chosen shape of the substrate potential (for particular examples of $V(x)$ with the current inversion effect, see Ref. [745]) the current may be negative at small τ and become positive at larger values of τ. Thus, if one varies the parameter τ, the current changes its sign. The same will be true for variation of any other model parameter, for example, the friction coefficient η. This effect opens the way of particle segregation: for example, Brownian particles with different sizes will have different friction coefficients and thus move in opposite directions being exposed to the same thermal environment and the same substrate potential. Moreover, if one takes into account inertia effects, such a particle separation mechanism could be inferred with respect to the particle masses, as well as for other dynamically relevant particle properties.

The arguments presented above can be generalized in the following way [746]: for a ratchet model of the type (9.8), and possibly also with an x- and/or t-dependent temperature, which depends on an arbitrary model parameter μ, there exists the shape of the substrate potential $V_{\text{ci}}\left[x, F(t)\right]$ such that the current $\langle \dot{x} \rangle$ as a function of μ exhibits a current inversion at some value μ_0, but *if and only if* there exist two potentials $V_1\left[x, F(t)\right]$ and $V_2\left[x, F(t)\right]$ with opposite currents $\langle \dot{x} \rangle$ for the reference value μ_0. To prove this, we have to construct the potential $V_\lambda\left[x, F(t)\right] = \lambda V_1\left[x, F(t)\right] + (1 - \lambda)\, V_1\left[x, F(t)\right]$, which parametrically depends on λ. Under the assumption that the current $\langle \dot{x} \rangle$ changes continuously with λ, it follows that it vanishes at a certain intermediate value $\lambda = \lambda_0$. Then the model with the potential $V_{\text{ci}}\left[x, F(t)\right] = V_{\lambda_0}\left[x, F(t)\right]$ must exhibit the current inversion with changing the parameter μ. The same is true for the ratchet model (9.7) with additive force driving, where in the same way one has to construct the driving $F_\lambda(t) = \lambda F_1(t) + (1 - \lambda)\, F_2(t)$.

If the temperature does not depend on time but changes periodically with space, $T(x + a_s) = T(x)$, then the model is called the *Seebeck ratchet*. In a general case, the current is nonzero for this model, if and only if

$\int_0^{a_s} dx\, V'(x)/T(x) \neq 0$ (see Ref. [745] and references therein). Note that in a mathematical sense, both the thermal ratchet and the Seebeck ratchet are equivalent to a fluctuating potential ratchet considered below.

Pulsating Ratchets

The parametrically driven ratchets (also called *pulsating ratchets*) are described by Eq. (9.8). A subclass of pulsating ratchets is the so-called *fluctuating potential ratchets*, when the potential has the form

$$V[x, F(t)] = [1 + F(t)]\; V(x), \tag{9.13}$$

i.e. the forcing $F(t)$ describes a "perturbation" to the "unperturbed" substrate potential $V(x)$ without affecting its spatial periodicity. A rather special but important example of pulsating ratchets is the so called "*on-off*" *ratchet*, when $F(t)$ in Eq. (9.13) can take only two possible values, e.g. $F(t) = -1$ (the potential is "off") and $F(t) = 0$ (the potential is "on"). Schematic picture of operating of the on-off pulsating ratchet at nonzero temperature $T > 0$ is shown in Fig. 9.4. Typically, the "natural" current direction in pulsating ratchets is given by the sign of the steep potential slope. Notice also that when the temperature is zero in the model (9.8), (9.13), then the motion equation $\eta\,\dot{x} + [1 + F(t)]\; V'(x) = 0$ reduces to the simple form $\eta\,\dot{y} = 1 + F(t)$ with the help of the substitution $y(x) = -\int^x dx'/V'(x')$. A similar simplification is also possible when both forces $F(t)$ and $\delta F(t)$ in the model (9.8), (9.13) correspond to the same kind of stochastic process, for example, they are described by two (mutually uncorrelated) Ornstein-Uhlenbeck processes with the same correlation time τ_c.

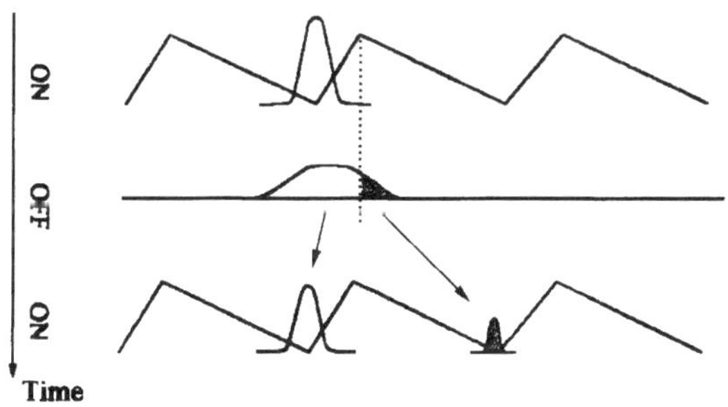

Fig. 9.4. Schematic picture of operating of the on-off pulsating ratchet at $T > 0$. The probability distribution of one particle broadens when the substrate potential is off; this produces directed motion in the hardest direction when the potential is switched on again [747].

Another subclass of pulsating ratchets, called *travelling potential ratchets*, has potential of the form

$$V[x, F(t)] = V[x - F(t)]\,, \tag{9.14}$$

where $F(t)$ is a periodic or stochastic function. [In the case when $F(t)$ is unbounded, e.g. $F(t) = c_{\text{wave}}t$, the substitution $y = x - F$ reduces the problem to the dc driven one, $\eta\dot{y} + V'(y) = -\eta c_{\text{wave}} + \delta F(t)$, so that the particle is captured by the travelling wave and moves with the same (*Brownian surfer*) or lower (*Brownian swimmer*) velocity in the same direction. However, if inertia effects are taken into account and the noise is colored, then the current inversion may occur [748].] This potential never satisfies the symmetry criterion (9.9), independently of whether the potential $V(x)$ itself is symmetric or not. Thus, one could expect the ratchet behavior even for the symmetric shape of the function $V(x)$. In fact, the substitution $y = x - F$ transforms the ratchet (9.14) into rocked ratchets considered below in Sect. 9.2.2.

An example of a generalization of this type of ratchets has been given by Porto *et al.* [749]. The substrate potential was composed of two waves that move relative each other,

$$V[x, F(t)] = V_{\alpha_1}(x) + V_{\alpha_2}[x - F(t)], \tag{9.15}$$

and the wave's shape $V_\alpha(x)$ was chosen in a simple form of a piecewise linear function defined on the interval $(0, 1]$,

$$V_\alpha(x) = \begin{cases} -1 + 2x/\alpha, & \text{if } 0 < x \le \alpha < 1, \\ 1 - 2(x-\alpha)/(1-\alpha), & \text{if } 0 < \alpha < x \le 1, \end{cases} \tag{9.16}$$

which then was periodically continued over the whole x axis. Except the symmetric case of $\alpha_1 = \alpha_2 = 1/2$, the system exhibits the ratchet behavior. Indeed, from the sequence of images in Fig. 9.5 one can see that for monotonically increasing F the atom stays at the same position in average (see

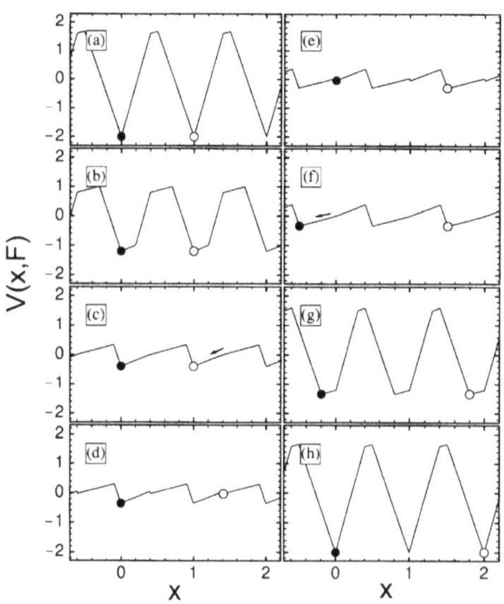

Fig. 9.5. Time evolution of the pulsating ratchet potential (9.15), (9.16) with $\alpha_1 = 2/5$ and $\alpha_2 = 1/2$. The snapshots are taken at (a) $F = 0$, (b) $F = 0.2$, (c) $F = 0.4$, (d) $F = 0.41$, (e) $F = 0.49$, (f) $F = 0.5$, (g) $F = 0.8$, and (h) $F = 1$. In parallel, the respective positions of the particle in the "quasistatic" limit are shown, both for $\dot{F}(t) > 0$ [e.g., $F(t) = t$, full circles, the time evolves from (a) to (h)] and $\dot{F}(t) < 0$ [e.g., $F(t) = -t$, open circles, the time evolves in the opposite direction from (h) to (a)]. The arrows indicate the direction of irreversible motion of the particle which occurs between snapshots (e) and (f) (full circle) and (d) and (c) (open circle) [749].

filled circles on the images), while for monotonically decreasing forcing, the atom moves to the left (open circles on the images). Thus, for any unbiased forcing $F(t)$, either periodic or stochastic, if its maximum amplitude exceeds 1 (the period of the substrate potential), the atom will move to the left only, and such a behavior does not depend essentially on details of the shape of the potential $V_\alpha(x)$. The irreversibility of the motion emerges due to existence of points of instability in the (x, F) plane, where a local minimum of the asymmetric potential $V(x, F)$ ceases to exist. When the minimum disappears, a particle located at such a point performs an *irreversible* motion jumping in the direction given by the potential's asymmetry (see details in Ref. [749]).

The further generalization which allows the current inversion, is based on two potentials moving in opposite directions, $V_\lambda\left[x, F(t)\right] = \lambda V_1(x + c_{\rm wave}t) + (1-\lambda)\, V_2(x - c_{\rm wave}t)$ with a proper choice of the functions $V_1(x)$ and $V_2(x)$ and the parameter λ [745].

Rocked Ratchets

Next, let us consider the so called *rocked ratchets*, also called the *tilting ratchets*, i.e. the ratchets driven by an additive periodic force. A simple consideration is possible in the adiabatic limit, when the variation of the force $F(t)$ in Eq. (9.7) is so slow that the system reaches the steady state for each value of the force. In this case the flow is zero at zero temperature (or exponentially small if $T > 0$) until $F(t)$ exceeds the maximum slope of the substrate potential in the given direction; after that the current takes a value close to the maximum $F(t)/\eta$. Thus, when the periodic force is symmetric, the net current (averaged over the period τ) is nonzero, if and only if the substrate potential is spatially asymmetric and the amplitude A of the forcing is larger than the threshold value A_0 determined by the slopes of $V(x)$. One can easy understand that the atom in this case will preferentially move in the direction where it has to climb a shallower side of the substrate potential, i.e. the current direction is just opposite to the "natural" direction of a fluctuating potential ratchet. If $T > 0$, then the stochastic force aids the escape of the Brownian particle over the potential barriers, therefore the directed motion begins at lower values of the amplitude of the oscillating force.

The limit of fast oscillating force, $\tau = 2\pi/\Omega \to 0$, is more complicated for analytical consideration. When $F(t)$ is symmetric and $V(x)$ is asymmetric, the approach based on the perturbation theory gives the result [745]

$$\langle \dot{x} \rangle = \frac{2\tau^4 Y \mathcal{N}}{\eta^5} \int_0^{a_s} \frac{dx}{a_s}\, V'(x) \left[V'''(x)\right]^2, \tag{9.17}$$

while when $F(t)$ is asymmetric and $V(x)$ is symmetric, the current is given by

$$\langle \dot{x} \rangle = \frac{\tau^4 \mathcal{N}}{4\,\eta^5} \left\{ Y_- \int_0^{a_s} \frac{dx}{a_s} \left[V'''(x)\right]^2 + \frac{1}{2}\beta Y_+ \int_0^{a_s} \frac{dx}{a_s} \left[V''(x)\right]^3 \right\}. \tag{9.18}$$

In Eqs. (9.17), (9.18) $\mathcal{N}$ is the normalization constant,

$$\mathcal{N} = \left[\left(\int_0^{a_s} \frac{dx}{a_s} e^{-\beta V(x)} \right) \left(\int_0^{a_s} \frac{dx}{a_s} e^{+\beta V(x)} \right) \right]^{-1}, \tag{9.19}$$

$\beta = (k_B T)^{-1}$, and the amplitudes Y, $Y_\pm$ are given by the integrals

$$Y = \int_0^1 dh \, [f_2(h)]^2 , \quad Y_\pm = \int_0^1 dh \, [f_0(h) \pm 2 f_2(h)] \, [f_2(h)]^2 \tag{9.20}$$

over the functions $f_i(h)$ defined through the recurrent relation $f_i(h) = \int_0^h ds \, f_{i-1}(s) + \int_0^1 ds \, s f_{i-1}(s)$ $(i = 1, 2, \ldots)$ with $f_0(h) = F(\tau h)$. According to these results, the rocked ratchets are very ineffective, all contributions up to the order τ^3 are zero.

The sign of the current in Eq. (9.17) is dictated by that of the steepest slope of $V(x)$ (for simple enough substrate potentials) independently of details of the symmetric driving $F(t)$. Thus, the "natural" current direction of the fast and slow rocked ratchets are opposite, and the current inversion with variation of the driving frequency Ω is typical for rocked ratchets.

As an example, let us consider a simple case of deterministic ratchet, when $\delta F(t) \equiv 0$ in Eq. (9.7), the substrate potential has the typical asymmetric ("ratchet") shape given by Eq. (9.1), and the driving force is pure sinusoidal,

$$F(t) = A \sin(\Omega t). \tag{9.21}$$

For this model, the solution of Eq. (9.7) is bound at small driving amplitudes, $A < A_0(\Omega)$, so that it approaches a function which is periodic in time for large times [750]. For $A > A_0(\Omega)$, the motion can become unrestricted, and the average velocity assumes an asymptotic value (independent of the initial conditions) of the form $\langle \dot{x} \rangle = (m/n)\, \Omega$ with integers m and n (recall that $a_s = 2\pi$ in our system of units). Such a behavior, when the current as a function of the driving amplitude, displays a structure of constant plateaux separated by discontinuous jumps, is typical for deterministic rocked ratchets in the overdamped limit (see Ref. [745] and references therein).

For the ratchet potential (9.1) with $\mu = -1/4$ and $\varphi = \pi/2$, the net flux is always positive $(m, n > 0)$, i.e. the particle moves towards the shallow side of the potential in the case of $T = 0$. However, if the temperature is nonzero, a current inversion takes place for sufficiently large frequency Ω in accordance with the perturbative result (9.17).

Fluctuating-Force Ratchets

Finally, if the force $F(t)$ in Eq. (9.7) corresponds to a stochastic process, then the model belongs to the *fluctuating force ratchets* type, which may be considered as another subclass of *tilting ratchets*. As we have mentioned already, the ratchet effect may occur in the case of color noise only, when the correlation time $\tau_c = \frac{1}{2} \int_{-\infty}^{+\infty} dt \, \langle F(t)F(0) \rangle / A$ (where A defines the noise amplitude, $A = \langle F^2(0) \rangle$), is nonzero.

The case of slow fluctuating force, $\tau_c \to \infty$, may be described by adiabatic approximation in the same way as above in Sect. 9.2.2 for the rocked ratchets. On the other hand, in the case of fast fluctuating force, $\tau_c \to 0$, the ratchet velocity can be found with the help of perturbation theory which leads to the expression (see Ref. [745] and references therein)

$$\langle \dot{x} \rangle = -\tau_c^3 \frac{\beta A \mathcal{N}}{\eta} \left\{ Y_1 \int_0^{a_s} \frac{dx}{a_s} \, V'(x) \left[V''(x) \right]^2 + A\beta^2 Y_2 \int_0^{a_s} \frac{dx}{a_s} \left[V'(x) \right]^3 \right\}, \tag{9.22}$$

where Y_1 and Y_2 are dimensionless and τ_c-independent coefficients determined by the specific noise under consideration, and the normalization constant $\mathcal{N}$ is given by Eq. (9.19). For example, for the stochastic force with the exponentially-correlated noise (the so called Ornstein-Uhlenbeck process), $\langle F(t)F(0) \rangle = (D/\tau_c) \exp(-|t|/\tau_c)$, one has $Y_1 = 1$ and $Y_2 = 0$. It is interesting that in this case the current directions in two limits, $\tau_c \to \infty$ and $\tau_c \to 0$, are just opposite one another, so that the current inversion takes place upon variation of the correlation time τ_c. As follows from Eq. (9.22), the fluctuating force is not too effective to produce the current, the first nonzero contribution appears in the third order of τ_c only.

9.2.3 Inertial Ratchets

Incorporating of the inertia term $m\ddot{x}$ into the motion equation makes the system dynamics much more complicated due to increasing of dimensionality of the phase space. For the one-atom ratchet driven by an additive force, the motion equation now is the following,

$$\ddot{x}(t) + \eta \, \dot{x}(t) + V' \left[x(t) \right] = F(t) + \delta F(t), \tag{9.23}$$

where $V(x)$ is periodic with the period a_s and $F(t)$ is periodic with the period τ. Now the current is zero in the case of symmetric potential and driving, but not so for the supersymmetric ones as it was in the overdamped case.

Several papers are devoted to the subclass of inertial ratchets, namely to the *deterministic* inertial ratchets, when the temperature is zero, i.e. $\delta F(t) \equiv 0$ in Eq. (9.23) (see Ref. [745] and references therein). In this case a stochastic behavior may appear in certain parameter regions of the model. Another difficulty is that now the long-time average current depends on the initial conditions in a general case. As a consequence, the current as a function of model parameters may show a quite complex behavior, including multiple inversions even for a simple shape of the substrate potential.

The adiabatic regime of the deterministic ratchet with the substrate potential (9.1) driven by the oscillating force (9.21) was studied by Borromeo

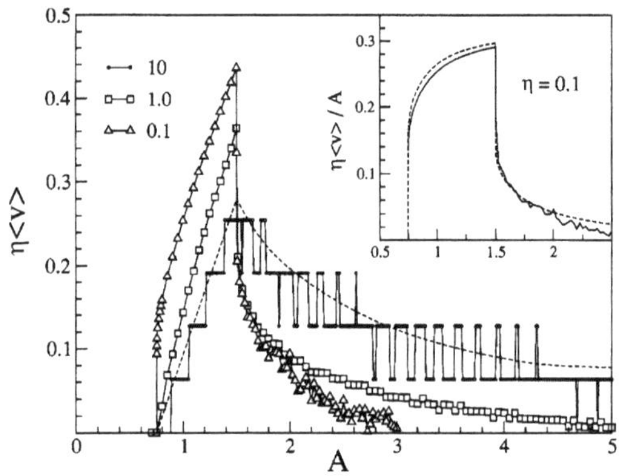

Fig. 9.6. The average velocity $\langle \dot{x} \rangle$ (times the damping η) as a function of the amplitude A of the oscillating force (9.21) of frequency $\Omega = 0.01$ for the ratchet potential (9.1) and different values of the damping coefficient η. The dashed curve represents the overdamped adiabatic limit ($\eta = 10$, $\Omega = 10^{-4}$). Inset: the numerical data for $\eta = 0.1$ (solid curve) are compared with analytical dependence (9.24) shown by dashed curve [751].

et al. [751]. The calculated current-force dependencies for different values of the damping coefficient η are shown in Fig. 9.6. For damping values in the range $0.07 < \eta < 0.5$ these dependencies demonstrate the universal behavior displayed in the inset of Fig. 9.6 and described by the function

$$\langle \dot{x} \rangle = \frac{A}{\eta} \cdot \begin{cases} \mu(A, A_-) & \text{if} \quad A_- < A < A_+, \\ \mu(A, A_-) - \mu(A, A_+) & \text{if} \quad A > A_+. \end{cases} \tag{9.24}$$

Here $\mu(A, A') \propto 1 + \sqrt{1 - (A'/A)^2}$, and the amplitudes A_- and A_+ correspond to the depinning thresholds of the (right/left) tilted potential $V(x) \mp xA$. Equation (9.24) was derived under the assumption that the particle gets depinned when $F(t) > A_-$, then runs with the velocity $v(t) \propto F(t)/\eta$, and gets pinned again as soon as $F(t) < (\eta/\omega_0)\, A_-$ according Risken's ideology of Sect. 8.2.2 (see details in Ref. [751]). Thus, these results show that the rectification efficiency of the ratchet grows as η^{-1} when the damping decreases, but only up to a certain critical damping $\eta^* \sim 0.07$. On further decreasing of η below η^*, the trajectories become extremely irregular and the ratchet current drops monotonically towards zero.

Some rigorous results are known in the Hamiltonian limit of zero damping and thermal noise, when the motion equation takes the form

$$\ddot{x}(t) + V'\left[x(t)\right] = F(t). \tag{9.25}$$

Following the paper by Flach *et al.* [741], let us call the system be S_a-symmetric, if both the substrate potential and the driving force satisfy the following two conditions,

$$\begin{cases} V(-x + \Delta x) = V(x) & \text{and} \\ F(t + \tau/2) = -F(t), \end{cases} \tag{9.26}$$

where Δx is some constant [notice that the second equation coincides with the symmetric condition of the overdamped case, see Eq. (9.4)]. In this case, if $x(t)$ is a solution of Eq. (9.25), then the trajectory $z(t) = -x(t+\tau/2) + \Delta x$ satisfies the same equation as well.

Next, let us call the system be S_b-symmetric, if the force $F(t)$ satisfies the condition $F(-t+\Delta t) = F(t)$ with some constant Δt. Now the trajectory $z(t) = x(-t+\Delta t)$ satisfies Eq. (9.25) provided $x(t)$ is the solution of this equation.

In both cases, the momentum changes its sign under the symmetry transformation, $\dot{z}(t) = -\dot{x}(t)$. Therefore, if the system has either the S_a symmetry or the S_b symmetry, then the net current averaged over *all* initial conditions, must be zero, $\langle \dot{x} \rangle = 0$. In all other cases one could expect generically a nonzero average current, provided there are no other hidden symmetries that change the sign of momentum.

If one considers a given particular trajectory in the system with the described above symmetry, than the result $\langle \dot{x} \rangle = 0$ for the given trajectory is expected as well, provided the initial condition $\{x(0), \dot{x}(0)\}$ is part of a stochastic layer which also contains an initial condition with $\dot{x} = 0$. The reason is based on the assumption of ergodicity of stochastic layers. On the other hand, if the initial condition $\{x(0), \dot{x}(0)\}$ for the given trajectory is not part of a stochastic layer which also contains an initial condition with $\dot{x} = 0$, then generically $\langle \dot{x} \rangle \neq 0$ even if the symmetry conditions described above are fulfilled.

Though it may be difficult in practice, in principle the entire phase space of the Hamiltonian dynamics (9.25) can be decomposed into its different regular and ergodic components, each of them characterized by its own particle current $\langle \dot{x} \rangle$. Schanz *et al.* [752] have shown that the particle current "fully averaged" according to the uniform (microcanonical) phase space density, can be written as

$$\int_0^\tau dt \int_0^{a_s} dx \int_{-\infty}^{\infty} dp\, \dot{x} = \lim_{p_0 \to \infty} \int_0^\tau dt \int_0^{a_s} dx \int_{-p_0}^{p_0} dp\, \frac{\partial H}{\partial p}. \qquad (9.27)$$

Since the dynamics is Hamiltonian, $H = \frac{1}{2}p^2 + V(x) - xF(t)$, it follows that the microcanonically weighted average velocity over all ergodic components is equal to zero. An immediate implication of this "sum rule" is that a necessary requirement for directed transport in a generic Hamiltonian system is a mixed phase space (with coexisting both regular and stochastic regions), because the microcanonical distribution is the unique invariant density in this case, and it is always approached in the long time limit. Thus, in this approach the directed transport has its origin in the "unbalanced" currents within the regular islands.

Another explanation of the Hamiltonian ratchets is based on ballistic flights in the vicinity of the boundary of stochastic layers and regular islands, as suggested by Denisov *et al.* [753, 754]. Both explanations are correct in principle, but practical applications restrict them to certain limits, the sum rule– to the case of resonances deeply embedded in the chaotic sea (very short flights, practically indistinguishable from chaotic diffusion, but the sum in the sum rule then converges fast), and the ballistic flight to the case of strong resonances and very efficient partial barrier presence due to cantori (very long flights, very well distinguishable from short time chaotic diffusion, easy to obtain corresponding probability distribution functions).

In the limit of fast oscillating driving, the average particle velocity for the equation (9.25) may be approximately calculated by separating the slow and fast contributions to the trajectory $x(t)$ [741]. In particular, for the symmetric substrate potential $V(x) = -\cos x$ but asymmetric driving

$$F(t) = A_1 \cos(\Omega t) + A_2 \cos(2\Omega t + \phi), \tag{9.28}$$

it was found that $\langle \dot{x} \rangle \propto (A_1^2 A_2/\Omega^3) \sin\phi$. The ac input (9.28) leads to the *mixing* of the two harmonics of frequencies Ω and 2Ω in the *nonlinear response* regime. Notice that the current vanishes if $A_1 = 0$ or $A_2 = 0$ or $\phi = 0, \pi$, when the mentioned above symmetries exist.

Coming back to Eq. (9.23) with $\eta > 0$, one can see that only the S_a-symmetry may survive for nonzero damping, and it will result in zero average velocity due to the same reasons as above. Finally, the symmetry reasons described above shows also the way how one can reverse the current of the ratchet (9.25). The first method is to use $F(-t)$ instead of $F(t)$ in this equation, while the second method, which also operates for systems with a nonzero damping η, is to change simultaneously the potential and the driving, $V(x) \to V(-x)$ and $F(t) \to -F(t)$.

We notice, however, that for small η the time reversal symmetry will be recovered again, i.e. Eq. (9.28) will hold approximately. Thus the zeroes of the current will be close to $\phi = 0, \pi$. The same is true for the supersymmetry in the overdamped case when η is large. Thus, even if symmetries are formally broken but one is close to a limit where they hold, variation of parameters like the relative phase ϕ will show the partial recovery of such a symmetry.

9.3 Solitonic Ratchets

All ratchet models described above should also operate for the case of systems of interacting atoms, e.g. the ratchet concepts should remain valid for the FK model, at least in the limit of weak interaction, $g \ll 1$. Moreover, the similar concepts can be introduced for the case of strongly interacting atoms when the ground state has only one kink; the latter can be treated as a quasi-particle, especially in the overdamped limit. However, some new effects appear due to interaction, e.g. if we take into account that kinks are 'soft' quasi-particles and the driving force may modify their parameters. We notice that the first application of ratchet ideas to the soliton-bearing systems was suggested by Marchesoni [755].

9.3.1 Symmetry Conditions

The symmetry reasons described above for the single-particle ratchets, should work for collective ratchets as well, because the broken symmetries cannot be restored by additional interactions between the atoms [741]. As an example, Flach *et al.* [756] considered the SG rocked ratchet

$$\frac{\partial^2 u}{\partial t^2} + \eta \frac{\partial u}{\partial t} - \frac{\partial^2 u}{\partial x^2} + V'_{\text{sub}}(u) = F(t) + \delta F(x,t), \tag{9.29}$$

where $V_{\text{sub}}(u)$ is periodic with the period $a_s = 2\pi$ and has minima at $u = l a_s$, $F(t)$ is periodic with the period $\tau = 2\pi/\Omega$, and $\delta F(x,t)$ is the Gaussian spatially uncorrelated white noise. Introducing the energy density $\rho(x,t) = \frac{1}{2}\left(u_t^2 + u_x^2\right) + V_{\text{sub}}(u)$, where the indices stand for the corresponding partial derivatives, and using the continuity equation $\rho_t + j_x = 0$, we obtain for the energy current $J(t) = \int_{-\infty}^{\infty} dx\, j(x,t)$ the following expression (here we assume that $\eta = 0$), $J(t) = -\int_{-\infty}^{\infty} dx(\partial u/\partial t)(\partial u/\partial x)$. One can see that Eq. (9.29) is invariant under the space inversion, $x \to -x$, while the current J changes its sign under this transformation. Thus, if the topological charge Q of a solution $u(x,t)$ (where Q is defined by $Q = \left[u(-\infty,t) - u(\infty,t)\right]/a_s$) is zero, $Q = 0$, then the time-averaged energy current for the such solution vanishes exactly, $\langle J(t) \rangle = 0$. On the other hand, in the case when the system has a nonzero number of geometrical kinks owing to the boundary conditions, $Q \neq 0$, the current will vanish only if *both* the substrate potential and the driving are symmetric, i.e. if $V_{\text{sub}}(u)$ satisfies the condition (9.2) and $F(t)$ satisfies the condition (9.4). Indeed, the system is characterized by the kink-antikink symmetry in this case (also due to harmonic interparticle interaction in Eq. (9.29)) and, therefore, it is invariant under the combined transformation $x \to -x$ and $u \to -u + Qa_s$, which does not change the topological charge of the system.

Both analytical and numerical results based on the collective coordinate theories suggest [757] that, especially in the underdamped limit the internal (or shape) mode of the kink has to be taken into account, thus the effect can not be reduced always to a single-particle problem.

9.3.2 Rocked Ratchets

Thus, if any of the symmetry conditions, either the reflection symmetry (9.2) of the substrate potential, or the shift symmetry (9.4) of the a.c. driving is violated, one should expect a nonzero average current (provided there are no other hidden symmetry). For example, for the co-sinusoidal substrate potential and the driving given by Eq. (9.28), the current should be nonzero if both A_1 and A_2 are nonzero and $\phi \neq 0$ or π. Numerical simulation by Flach *et al.* [756] (see Fig. 9.7) confirmed that these predictions are correct even for nonzero damping case (with decreasing of η the zeros of J will approach the

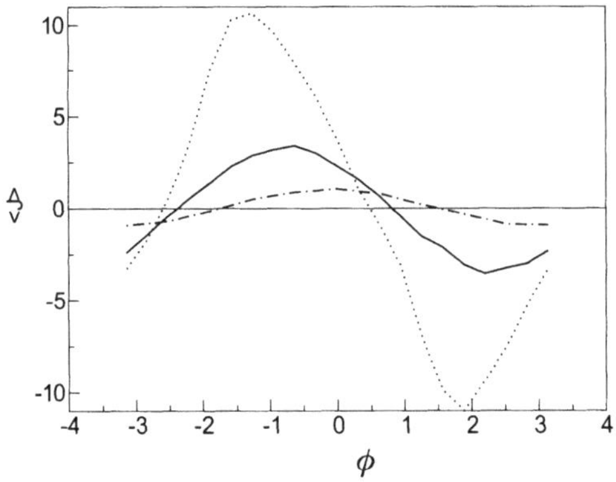

Fig. 9.7. Time-averaged energy current J vs. the phase shift ϕ for the ratchet (9.28), (9.29) with one kink ($Q = 1$) and the parameters $A_1 = A_2 = 0.2$, $\Omega = 0.1$ and $k_B T = 0.01$. Dotted curve is for $\eta = 0.01$, solid curve, for $\eta = 0.05$, and dash-dotted curve, for $\eta = 0.2$ (in the latter case the curve was scaled in five times to make it visible) [756].

$\phi = 0, \pi$ positions). Moreover, we see that the current changes its sign with variation of the phase shift ϕ, so that one can manipulate by the direction of the energy current by changing ϕ.

Another way to get a directed kink transport is to break the reflection symmetry (9.2) of the substrate potential. This case was considered by Costantini *et al.* [758]. The authors studied kink motion for the model (9.29) with the ratchet substrate potential (9.1) driven by the additive pure oscillating force (9.21) at zero temperature ($\delta F(x,t) \equiv 0$). It was found that in the overdamped limit the average kink velocity scales as $\langle v_{\rm k} \rangle \propto A^2$ [note that at zero temperature the directed kink motion should appear above a some threshold value of the driving force, $A > A_0(\Omega)$]. To explain such a dependence, Costantini *et al.* [758] noted that the shape of total potential $V_{\rm eff}(u) = V_{\rm sub}(u) - uF(t)$ experienced by kink, oscillates with the driving frequency. Thus, so should do the kink's effective mass as well, and the amplitude of kink's mass oscillation is $\delta m/m \propto A$ (notice that for the symmetric substrate potential we have $\delta m/m \propto A^2$). Then, due to asymmetry of the ratchet potential, a kink with the positive velocity, i.e. for $F(t) > 0$, should be more massive (and slower) than the kink with the negative velocity; hence, in average we have $\langle v_{\rm k} \rangle \sim \langle \delta\, [F/\eta m(t)] \rangle \propto (A/\eta m^2)\, \delta m \propto A^2$.

In the overdamped limit, e.g. for $\eta > 10$, the simulation has shown that the kink velocity scales with the damping coefficient η as $\langle v_{\rm k} \rangle \propto \eta^{-5}$. Such a dependence may be explained as appearing due to nonadiabatic effects because of a delay of kink's shape response to the driving frequency Ω, e.g., if $\delta m(\Omega)/\delta m(0) \sim \left[1 + (\eta\Omega)^2\right]^{-2}$. When the damping decreases further, the kink velocity reaches a maximum (at $\eta \gtrsim 0.1$) and then decreases again as shown in Fig. 9.8. The authors related this effect to the relativistic nature of the kink dynamics, which follows from the relation $\langle v_{\rm k} \rangle \approx \langle v(t)/\sqrt{1 + [v(t)/c]^2} \rangle$ with $v(t) = F(t)/\eta m(t)$.

In the underdamped limit, $\eta \ll 1$, the curves of Fig. 9.8 approach a plateau as η decreases below a characteristic value of the order of Ω. This effect was explained as appearing due to phonon radiation. Indeed, at extremely low η the kink will reach relativistic velocities for any small driving. Then, the relativistic decreasing of kink's width will result in strong phonon radiation due to discreteness effects, so that the effective damping coefficient $\eta_{\text{eff}} = \eta + \eta_{\text{ph}}$ experienced by kink, becomes almost independent on η. At low damping and high enough frequency of the driving ($\Omega > 0.2$), the solitonic ratchet exhibits the *current reverse* phenomenon (see Fig. 9.8 for the $\Omega = 0.5$ case). This interesting effect opens the way of manipulating kink's motion by ac forces.

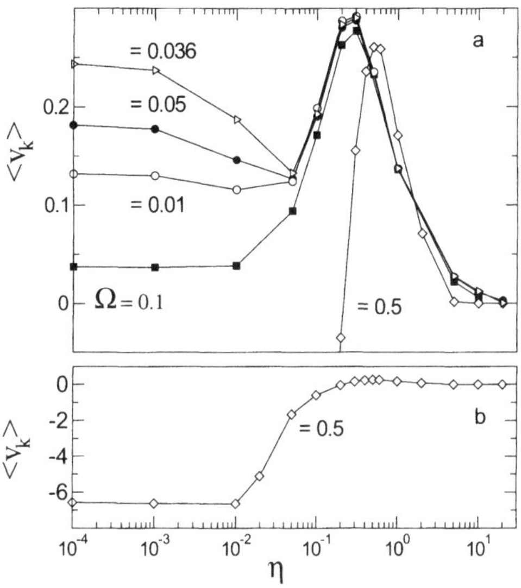

Fig. 9.8. Average kink velocity $\langle v_{\text{k}} \rangle$ versus the damping coefficient η on the ratchet potential (9.1) for the the ac force with the amplitude $A = 0.3$ and different values of the frequency Ω [758].

Finally, the dependence of the kink velocity on the driving frequency Ω demonstrates a strong resonance effect as shown in Fig. 9.9. To explain it appearance, note that the ratchet potential (9.1) with the parameters $\mu = 1/4$ and $\varphi = \pi/2$ admits the existence of kink's shape mode with the frequency $\omega_B \approx 0.8\,\omega_{\min} \approx 1.0562 < \omega_{\min}$. This frequency almost exactly matches with the resonance frequency observed in simulation, $\Omega \approx \omega_B$. It is interesting to recall that for the SG kink, the resonance was found to occur at $\Omega \approx \omega_B/2$ (see Sect. 8.11).

The same model has also been studied by Salerno and Quintero [759] with similar numerical results (although the interpretation was different: the authors claim that the directed transport emerges owing to nonzero damping η which mixes the *translational* and *internal* (or *shape*) modes of the kink; however, this contradicts the Costantini *et al.* results shown in Fig. 9.8). Salerno and Quintero [759] have shown also that the kink ratchet behavior is quite robust and survives at $T > 0$ as well. Finally, notice that the efficiency of

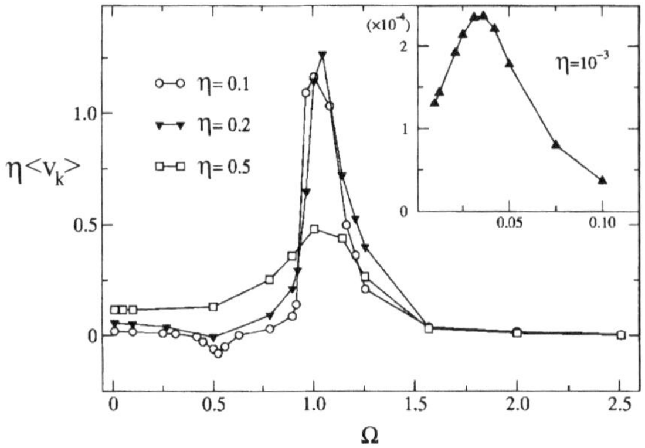

Fig. 9.9. Average kink velocity $\langle v_k \rangle$ (times η) versus Ω on the ratchet potential (9.1) for different values of the damping coefficient η for the ac force amplitude $A = 0.3$. Inset: the extremely underdamped case $\eta = 10^{-3}$ in the adiabatic regime of small Ω [758].

the such rachet machine is rather low, the oscillation-induced directed kink current is never even comparable with that achieved under the action of the dc force of the same amplitude.

When the reflection symmetry (9.2) of the substrate is broken, then the directed kink transport can be achieved by solely stochastic force $\delta F(x,t)$, provided it corresponds to a *color* noise, e.g., $\langle \delta F(x',t)\delta F(x,0)\rangle = 2\eta k_B T \delta(x' - x)\, g(t)$, with the correlator $g(t) \neq \delta(t)$ [755]. Indeed, let in the unperturbed system (i.e. in the absence of the external force) the kink shape is described by a function $u_k(x - X)$, where $X(t)$ is the coordinate of its center. Recall that the kink shape depends on the form of the substrate potential through Eq. (3.47) of Sect. 3.3. Then, using a simple collective-coordinate approach (based on the energy conservation arguments) in the nonrelativistic limit (e.g., for the overdamped case, $\eta \gg \omega_0$), Marchesoni [755] has shown that the kink's center coordinate is governed by the equation

$$m\ddot{X} + m\eta\dot{X} = F_g + \xi(t), \tag{9.30}$$

where m is the kink mass, $\xi(t)$ corresponds to a Gaussian noise, and the net force F_g is given by

$$F_g = -2(k_B T/m) \int_0^{\infty} d\tau'\, g(\tau')\, h(\tau') \tag{9.31}$$

with

$$h(\tau') \approx \int_{-\infty}^{\infty} dx\, u_{k,x}(x,0)\, u_{k,xx}(x,\tau'). \tag{9.32}$$

One can check that if the system symmetry is restored, either when the stochastic force corresponds to the white noise [for $g(t) = \delta(t)$ we have $F_g \propto h(0) \propto u_{k,x}^2(x)|_{-\infty}^{+\infty} = 0$], or when the substrate potential is symmetric [for $V_{\rm sub}(-u) = V_{\rm sub}(u)$ we have $h(\tau') = 0$ due to $u_{k,x}(x,t) = u_{k,x}(-x,t)$ and $u_{k,xx}(x,t) = -u_{k,xx}(-x,t)$], then the net force is zero, $F_g = 0$. In any other case, however, $F_g \neq 0$ generically, and the kink drifts with the average velocity $\langle \dot{X} \rangle = F_g/m\eta$ in the direction determined by a given shape of $V_{\rm sub}(x)$.

For the correlated noise with the correlator $g(t) = (1/2\tau_c)e^{-|t|/\tau_c}$, we can find that $F_g \propto \tau_c$ for $\tau_c \to 0$ (the white-noise limit) and $F_g \propto \tau_c^{-1}$ for $\tau_c \to \infty$. Thus, there exists an optimum value of τ_c (which depends on the particular shape of the substrate potential) when the current is maximum.

9.3.3 Pulsating Ratchets

For the system of interacting atoms with the parametric driving (9.13), the following new effect emerges: as was demonstrated by Floría *et al.* [747], the directed motion may exist solely due to the interaction, i.e. at $T = 0$. The mechanism of its operating is shown schematically in Fig. 9.10 for a simple case of the "on-off" ratchet, i.e. when $F(t)$ in Eq. (9.13) takes the values $F(t) = -1$ (the potential is "off") and $F(t) = 0$ (the potential is "on").

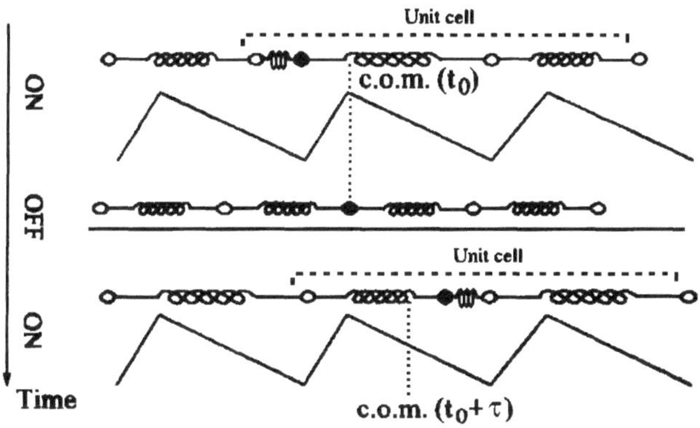

Fig. 9.10. Schematic of the zero-temperature solitonic "on-off" pulsating ratchet (compare with the $T > 0$ flashing ratchet in Fig. 9.4). The collective interaction between atoms produces a net flow when the asymmetric substrate potential is switched off and on. The atom mainly responsible for the flow in this cycle is highlighted. The dotted line marks the center of mass of the unit cell (three atoms per two periods of the potential) at t_0 and $t_0 + \tau$ to see the one-cycle advance [747].

As an example, Floría *et al.* [747] considered the overdamped deterministic motion of the chain in the potential

$$U(\{u_l\}) = \sum_l \{[1 + F(t)]\, V(u_l) + V_{\rm int}(u_{l+1} - u_l)\}, \qquad (9.33)$$

where $V(x)$ has an asymmetric shape and $V_{\rm int}(x)$ is a convex function (e.g., the harmonic interaction $V_{\rm int}(x) = \frac{1}{2}gx^2$). Under a rather general conditions this model allows a rigorous qualitative and even quantitative analysis.

Indeed, let $V(x)$ be the periodic function with the period a_s and the height K with one maximum and one minimum per period, the maximum being located at $x = 0$ and the minimum, at $x = a^* = \frac{1}{2}a_s + ra_s$, where r characterizes the asymmetry of the ratchet potential ($r = 0$ for the symmetric substrate potential). Then, let $F(t)$ be the periodic function with the period $\tau = \tau_{\text{on}} + \tau_{\text{off}}$ (the potential is on during the time τ_{on} and it is off for the time τ_{off}), and assume that the times τ_{on}, τ_{off} are long enough so that the chain can reach the equilibrium state. Under these assumptions, the chain's state is uniquely determined and rotationally ordered.

When the potential is off, the chain takes the equidistant configuration, $u_l = la_A + X_1$, where X_1 describes the center of mass (c.o.m.) position and a_A is determined by the atomic concentration $\theta = a_s/a_A$ [or the window number $w = a_A (\operatorname{mod} a_s)$] for the case of periodic boundary conditions. Let the time moment t_0 corresponds to the end of the potential-off semi-cycle. During the potential-on semi-cycle, $t_0 < t < t_0 + \tau_{\text{on}}$, the atoms will move to the nearest minima of the substrate potential. Then, during the next potential-off semi-cycle, the chain again takes the equidistant configuration, but now with the c.o.m. coordinate X_2. Thus, the chain advances the distance $\Delta X = X_2 - X_1$, and the current is given by $J = \Delta X/\tau$.

For *rational* values of the density, $\theta = s/q$ (s and q are irreducible integers), there is a well defined 1D map $X_{n+1} = \mathcal{M}(X_n)$ with the following properties: (*i*) $\mathcal{M}$ is non-decreasing (if $r > 0$), (*ii*) $\mathcal{M}$ is step-periodic, $\mathcal{M}(X + 1/s) = \mathcal{M}(X) + 1/s$, and (*iii*) $|\mathcal{M}(X) - X| \le C < 1$. The asymptotic flow J is then given by $J = \lim_{n\to\infty} [\mathcal{M}^n(X) - X]/n$, and it does not depend on the initial conditions. In the limit of very strong pinning, $K \to \infty$, the chain configuration at the end of potential-on semi-cycle is given by $u_l = a^* + a_s \operatorname{int}(lw + \alpha)$, where $\operatorname{int}(x)$ denotes the integer part of x, w is the window number, and α is an arbitrary phase. Thus, the map $\mathcal{M}(X)$ in this limit is explicitly given by $\mathcal{M}(X) = [r + 1/(2s)]\, a_s$ for $0 \le X < a_s/s$ [for other values of X one can use the step-periodicity (*ii*)], and the flow is $J = (a_s/s) \operatorname{int}(sr + 1/2)$. For *irrational* window numbers w, and the incommensurate limit we have $J = ra_s$ which is independently on w. Thus, the incommensurate structures display the ratchet collective effect in the limit of strong pinning for any asymmetric substrate potential, $r \neq 0$.

The results of numerical calculation are presented in Fig. 9.11. As expected, at a fixed w, the flow decreases with K, because intermediate pinned structures become less asymmetric. Analogously to the high pinning case, the incommensurate limit seemingly defines a continuous function $J(w)$ as shown on the left panel in Fig. 9.11. At commensurate values of w the flow shows point discontinuities (the right panel of Fig. 9.11), and the $J(w)$ dependence has a "crest" behavior.

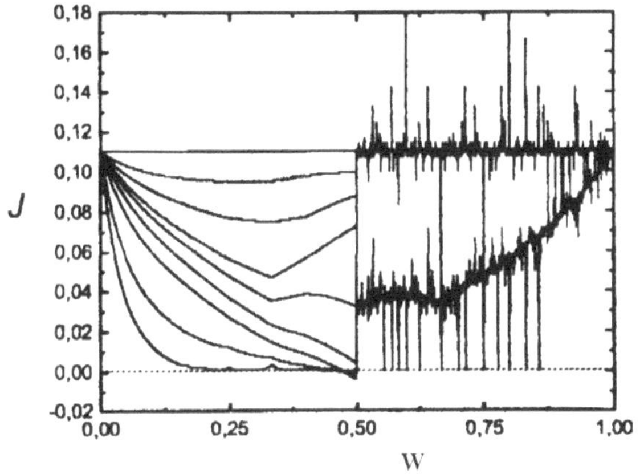

Fig. 9.11. Current J/a_s vs. the window number w for the asymmetric substrate potential with $r = 0.11$ and different K. Left: $J(w)$ for irrational w and, from top to bottom, $K = \infty, 6, 5, 4.5, 4, 3, 2, 1$ and 0.5. Right: incommensurate densities for $K = \infty, 4$ [747].

9.4 Experimental Realizations

One of the most accessible realization of a solitonic ratchet was demonstrated by using the Josephson junctions. Recall that in this case the JJ current corresponds to the driving force, and the measured voltage, to the kink velocity.

A realization of the one-kink ratchet scheme, based on a continuous 1D long Josephson junction of annular shape, has been proposed by Goldobin *et al.* [760]. An effective ratchet potential for the fluxon emerges by applying an external magnetic field and choosing a properly deformed shape of the junction, or by a deposition of a suitably shaped "control line" on the top of the junction in order to modulate the magnetic flux through it as shown in Fig. 9.12 [761, 762]. The experimentally measured voltage for this type of device is shown in Fig. 9.13. Depending of the amplitude of oscillating forcing $I_{\text{a.c.}}$, the fluxon may be pinned (region I in Fig. 9.13), exhibit a directed motion (region II), or follow the external driving force at large amplitudes

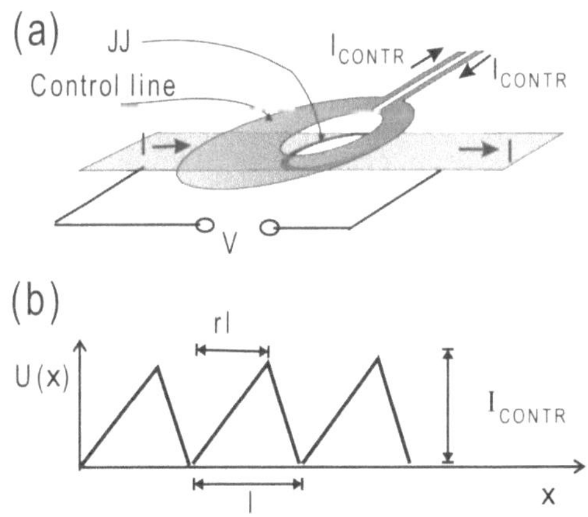

Fig. 9.12. (a) A long annular Josephson junction with a control line generating a sawtooth-like magnetic field. (b) The effective potential experienced by a fluxon trapped in the junction when the control current is turned on [762].

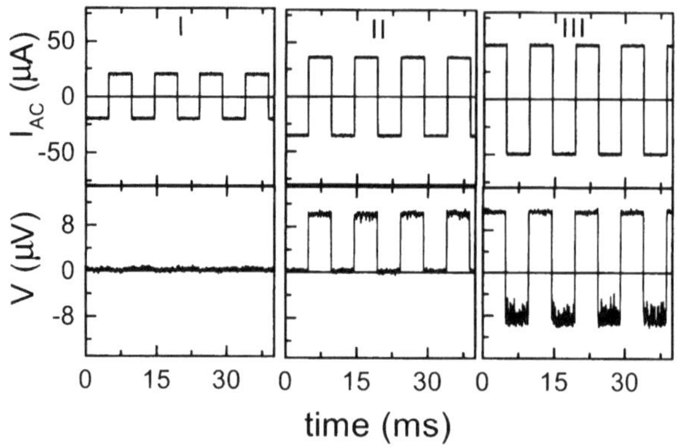

Fig. 9.13. Oscilloscope time traces of forcing current and measured voltage for the Josephson junction of Fig. 9.12(a) with the asymmetric ratchet potential of Fig. 9.12(b) (the asymmetry parameter $r = 0.62$) in the adiabatic regime ($\Omega = 100$ Hz) for the overdamped system ($T = 6.5$ K) for amplitude of forcing signal in three regions: $I_{\mathrm{a.c.}} < 25\ \mu$A (I), $I_{\mathrm{a.c.}} \sim 35\ \mu$A (II), and $I_{\mathrm{a.c.}} > 50\ \mu$A (III) [762].

of driving (region III). Such a diode effect was also observed for stochastic (colored) forcing as well. Moreover, when the temperature is decreased up to $T = 4.2$ K, the system comes into underdamped regime, and the "current reversal" effect is observed [763].

Then, Falo *et al.* [764] proposed to use a parallel array of JJ's with alternating critical currents and junctions' areas to model the *discrete* FK chain. In this system the effective substrate potential for fluxons can be chosen ratchet-shaped. The experimental realization of such type of device has been done by Trias *et al.* [765].

Finally, we would like to mention a realization of the rocked ratchet scheme proposed by Lee *et al.* [766]. Here a ratchet potential was proposed to construct by modulation of the thickness of the *superconducting film*. Then, the application of an alternating current to a superconductor can induce a systematic directed *vortex motion*. Thus, by an appropriate choice of the ratchet-shaped pinning potential, the rocking ratchet scheme can be exploited to remove unwanted trapped magnetic flux lines from the thin superconducting film. Quantitative estimates show that thermal fluctuations are practically negligible in this application of the rocking ratchet model.

10 Finite-Length Chain

When the interatomic potential is attractive a chain of a finite number of particles can exist in a stable state. Then we can study both the ground state and the dynamics of such a finite-length chain of a fixed number of particles. This kind of problem has applications in surface diffusion and the physics of atom clusters. This short chapter is devoted to some of these problems.

10.1 General Remarks

When the interatomic potential $V_{\rm int}(x)$ has an attractive branch, a *finite* FK chain, i.e. a chain consisting of N interacting atoms on a substrate potential, can exist in a stable state. For such finite-length chains the following two interesting questions emerge:

1. The ground state of the model. This question is important, e.g., in investigation of crystal growth;
2. The mobility of finite chains. In particular, this is interesting for study of surface diffusion of adsorbed clusters.

For the standard FK chain with $V_{\rm int}(x) = \frac{1}{2} g\,(x - a_{\rm min})^2$ and $V_{\rm sub}(x) = (1 - \cos x)$, a finite-length system is characterized by *three parameters*, N, g, and $a_{\rm min}$ (we recall that, in our system of units, $m_a = 1$, $a_s = 2\pi$ and $\varepsilon_s = 2$). The variation of the parameter $a_{\rm min}$ can be restricted by the interval $\pi \leq a \leq 2\pi$, because the potential energy (5.10) does not change provided the transformation of the equilibrium distance, $a_{\rm min} \to \pm a_{\rm min} + 2\pi$, and the positions of the atoms, $x_l \to \pm x_l + 2\pi l$, are done simultaneously.

Introducing the displacements u_l in the standard way, $x_l = l a_s + u_l$, we obtain the motion equations of a finite chain in the following form,

$$\ddot{u}_1 + \sin u_1 - g\,(u_2 - u_1 - a_s P) = 0, \tag{10.1}$$

$$\ddot{u}_l + \sin u_l - g\,(u_{l+1} - 2u_l + u_{l-1}) = 0, \quad 2 \leq l \leq N-1, \tag{10.2}$$

$$\ddot{u}_N + \sin u_N + g\,(u_N - u_{N-1} - a_s P) = 0, \tag{10.3}$$

where $\dot{u} \equiv du/dt$, and the misfit parameter $P \equiv (a_{\rm min}/a_s - 1)$ is considered within the finite interval $-1/2 \leq P \leq 0$. The equations can be easily extended to include the effects of anharmonic interaction [767].

A new feature which is inherent in a finite chain, is that it is always pinned, the activation energy E_a for the chain translation is always positive. For example, even for the chain of rigidly coupled atoms, $g = \infty$, the value E_a is equal to [176]

$$E_a = 2\frac{|\sin(Na_{\min}/2)|}{\sin(a_{\min}/2)}, \tag{10.4}$$

so that it vanishes only for the isolated values $a_{\min} = a_n^{\infty}$, where

$$a_n^{\infty} = 2\pi - n(2\pi/N), \quad n = 1, 2, \ldots, \mathrm{int}(N/2). \tag{10.5}$$

For odd N values the finite chain exhibits an analog of Aubry transition, when the minimum frequency of the vibration spectrum vanishes. However, the sliding (Goldstone) mode does not exist in the finite chain.

10.2 Ground State and Excitation Spectrum

10.2.1 Stationary States

The stationary-state configurations of the chain follow from solutions of the system of equations (10.1)–(10.3), where we have to put $\ddot{u}_l = 0$ for all l. In a general case, this system can be solved numerically only. The method of solution [768, 769] consists in reducing the set (10.1) to (10.3) to a single nonlinear equation for one variable, say u_1, which then is solved numerically within the interval $-\pi < u_1 \leq \pi$.

To classify different stationary solutions $\{x_l^{(0)}\}$, it is necessary to perform the normal-mode analysis, i.e. to find eigenvalues λ_n of the elastic matrix $\mathbf{A} \equiv \{A_{ll'}\}$, $A_{ll'} = \left(\partial^2 H/\partial x_l \partial x_l'\right)_{x_l = x_l^{(0)}}$,

$$\mathbf{A}\widetilde{\mathbf{u}}^{(n)} = \lambda_n \widetilde{\mathbf{u}}^{(n)}, \tag{10.6}$$

where $\widetilde{\mathbf{u}}$ is the N-dimensional vector, $\widetilde{\mathbf{u}} = \{\widetilde{u}_l\}$, and $\widetilde{u}_l = x_l - x_l^{(0)}$ is the atomic displacement away from the stationary state. The symmetric matrix $\mathbf{A}$ can be reduced to a diagonal form with the help of a unitary matrix $\mathbf{T}$,

$$\mathbf{A} = \mathbf{TBT}^{-1}, \;\; \mathbf{B} \equiv \{B_{mn}\}, \;\; B_{mn} = \lambda_n \delta_{mn}, \tag{10.7}$$

$$\widetilde{\mathbf{u}}^{(n)} = \mathbf{T}\mathbf{v}^{(n)}, \;\; \mathbf{v}^{(n)} \equiv \{v_m^{(n)}\}, \;\; v_m^{(n)} = v_n \delta_{mn}. \tag{10.8}$$

Then, close to a stationary state, the total potential energy U of the chain converts to the canonical form,

$$U \simeq E_{\mathrm{GS}}\left(\left\{x_l^{(0)}\right\}\right) + \frac{1}{2}\sum_{n=1}^{N} \lambda_n v_n^2. \tag{10.9}$$

From the physical viewpoint, only the following stationary states are of interest: the minima of the potential energy ($\lambda_n > 0$ for all n), and the saddle points with one eigenvalue being negative, and others, positive ($\lambda_1 < 0 < \lambda_2 < \ldots < \lambda_N$). The neighboring local minima can always be connected by a saddle trajectory, i.e. by a curve which passes through the saddle point and is determined by solution of the following system of differential equations,

$$\frac{\partial x_l}{\partial \tau} = -\frac{\partial U}{\partial x_l}, \quad l = 1, 2, \ldots, N, \tag{10.10}$$

where τ is a parameter along the trajectory. The saddle trajectory is the curve of steepest descent; at the stationary points the direction of the saddle trajectory is defined by the eigenvector $\pm\tilde{\mathbf{u}}^{(1)}$ corresponded to the minimum eigenvalue λ_1.

The stationary state of the system with the lowest potential energy is the ground state. Evidently, the ground state of the finite FK chain is infinitely degenerated since the substitution $x_l \to x_l + 2\pi$ transforms one ground state to the neighboring one. We will call the "adiabatic" trajectory the saddle trajectory which connects the neighboring ground states of the system. Every state of the system on the adiabatic trajectory can be associated with a unique parameter, the coordinate X, defined as $X = \sum_{l=1}^{N} x_l$ according to Bergmann and co-workers [146]. Thus, the potential energy of the system is described by the function

$$E(X) = U\left(\{x_l\}\right)_{x_l \in \text{adiab. tr.}}, \tag{10.11}$$

which is periodic with the period

$$b = 2\pi N. \tag{10.12}$$

If two ground states can be connected by several different saddle trajectories, the trajectory with the minimum value of activation energy E_a, defined as

$$E_a = \max[E(X)] - \min[E(X)], \tag{10.13}$$

should be taken as the adiabatic trajectory.

As the system moves along the adiabatic trajectory, the kinetic energy of the chain, $K = \frac{1}{2} m_a \sum_{l=1}^{N} (\partial x_l / \partial t)^2$, takes the form

$$K = \frac{m_a}{2} \sum_{l=1}^{N} \left(\frac{\partial x_l}{\partial X} \frac{\partial X}{\partial t} \right)^2 = \frac{m}{2} \left(\frac{\partial X}{\partial t} \right)^2, \tag{10.14}$$

where the effective mass of the chain is equal to

$$m(X) = m_a \sum_{l=1}^{N} \left(\frac{\partial x_l}{\partial X} \right)^2. \tag{10.15}$$

Using the definition (10.15), it is easy to show that

$$m \simeq \begin{cases} 1 & \text{if } g \to 0, \\ 1/N & \text{if } g \to \infty. \end{cases} \tag{10.16}$$

Finally, it can be shown [176] that near a stationary state (the minimum-energy state or a saddle state) with a coordinate X_s, the potential energy of the system can be represented as

$$E(X) \simeq E(X_s) \pm \frac{1}{2} m\omega_s^2 (X - X_s)^2, \tag{10.17}$$

where

$$\omega_s^2 = \pm\lambda_1 > 0. \tag{10.18}$$

In what follows we will use the following notations. Let us numerate the minima of the external potential in such a manner that the first atom of the chain occupies the potential well with the number 1, and the last (N-th) atom, the well with the number M. Then, we can designate different stationary states of the system by the integer $n = N - M$, and additionally by the index "s" for symmetric or "a" for asymmetric states (see Fig. 10.1). Moreover, to distinguish the states corresponding to local minima of the potential energy from the saddle ones, we will highlight the former by underscoring ($\langle \underline{n} \rangle$), and the latter, by overscoring ($\langle \bar{n} \rangle$). Evidently, at $a_{\min} = 2\pi$ the state $\langle \underline{0} \rangle_s$ always corresponds to the ground state. As the parameter $a_{\min}$ decreases, the GS configuration is changed in the following consequence,

$$\langle \underline{0} \rangle_s \to \langle \underline{1} \rangle \to \langle \underline{2} \rangle \to \ldots \to \langle \underline{n}^* \rangle, \tag{10.19}$$

where

$$n^* = \text{int}\left[\frac{(N-1)}{2}\right]. \tag{10.20}$$

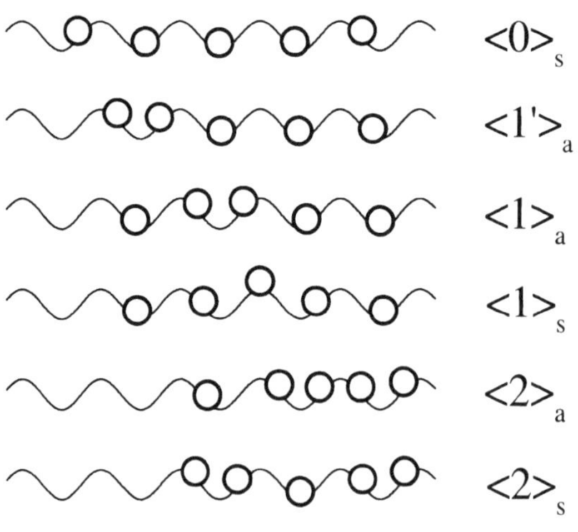

Fig. 10.1. Stationary-state configurations of the finite FK chain with $N = 5$ [176].

10.2.2 Continuum Approximation

If $g \gg 1$ and $N \gg 1$, we may use the continuum limit approximation, $la_s \to x$, $u_l \to u(x)$, so that Eqs. (10.1)–(10.3) reduce to the pendulum equation

$$\frac{d^2u}{d\widetilde{x}^2} = \sin u, \tag{10.21}$$

where $\widetilde{x} = x/d$ and $d = a_s\sqrt{g}$. A periodic solution of Eq. (10.21) has the form [see Eq. (1.21)]

$$u(\widetilde{x}) = \pi + 2\sin^{-1}\mathrm{sn}(-\sigma(\widetilde{x} - \beta)/k, k), \tag{10.22}$$

where $\sigma = -\mathrm{sign}P$, and k is the modulus, $0 < k \leq 1$. Recall that Eq. (10.22) describes a regular sequence of kinks (if $\sigma = +1$) or antikinks (for $\sigma = -1$) of width $d_{\mathrm{eff}} = kd$ separated by the distance R,

$$R = d\, 2k\mathbf{K}(k) \tag{10.23}$$

(in dimensionless units $\widetilde{d} = 1$ and $\widetilde{R} = 2k\mathbf{K}(k)$). In what follows we assume that the point $\widetilde{x} = 0$ is at the center of the chain.

For the finite chain of length $L = Na_s$ the solution (10.22) can be characterized by an integer n,

$$n = \mathrm{int}(L/R), \tag{10.24}$$

which is equal to the number of kinks in the finite chain. In our notations this state is denoted as $\langle n \rangle$. From symmetry reasons, for odd n, $n = 1, 3, \ldots$, the solution should describe the state with a kink situated at the center of the chain, so that we have to put $\beta = 0$ in Eq. (10.22). Otherwise, for even n, $n = 0, 2, \ldots$, we have to take $\beta = k\mathbf{K}(k)$, so that at the middle of the chain the density of excess atoms $\rho(x) = -(1/a_s)du/dx$ takes its minimum.

The modulus k in Eq. (10.22) should be determined by the free-end boundary conditions, $du/d\widetilde{x}|_{\widetilde{x}=\pm\frac{1}{2}\widetilde{L}} = Pd$ [see Eq. (5.106) in Sect. 5.3.1], which now leads to the relation

$$\mathrm{dn}\left(\frac{\widetilde{L}}{2k}, k\right) = \begin{cases} 2\sqrt{1-k^2}/k|P|d & \text{if } n = 0, 2, \ldots, \\ \frac{1}{2}k|P|d & \text{if } n = 1, 3, \ldots \end{cases} \tag{10.25}$$

From Eq. (10.25) it follows that k should lie within the interval

$$\left[1 + \left(\frac{1}{2}Pd\right)^2\right]^{-1/2} < k < \min\left[1, \left(\frac{1}{2}|P|d\right)^{-1}\right]. \tag{10.26}$$

Therefore, for given values L, g, and $a_{\min}$ Eq. (10.25) has solutions for a finite set of k values, $k = k_n$, where $n_{\min} < n < n_{\max}$ [770]. All the solutions correspond to metastable and unstable configurations except the one with

$n_{(GS)}$ and $k_n^{(GS)}$ which corresponds to the ground state and provides the absolute minimum of the potential energy of the finite chain.

The changing of the GS configuration with increasing of the misfit $|P|$ at fixed values of L and g is shown schematically in Fig. 10.2. Evidently, at $P = 0$ the state $\langle 0 \rangle$ with $k^{(GS)} = 1$ is the GS [see Fig. 10.2(a)]. As $|P|$ increases, the structure of the ground state changes passing consequently through the configurations $\langle 0 \rangle \to \langle 1 \rangle \to \ldots \to \langle n^* \rangle$. Such a behavior may be qualitatively explained in the following way. Let us interpret the free end of the chain as a "fraction" of a kink [771], and recall that kinks repel each other. When $|P|$ increases from zero up to a value P_1 ($P_1 > P_{FM}$, $P_1 \to P_{FM} \equiv 2/\pi^2\sqrt{g}$ as $N \to \infty$), the energy of the state $\langle 0 \rangle$ increases due to repulsion of its free ends, see Fig. 10.2(b). At $|P| = P_1$ the energy of creation of an additional kink, ε_k, becomes equal to zero, and at $P_1 < |P| < P_2$ (where $\varepsilon_k < 0$) the state with one kink is the GS since the creation of a second kink is energetically unfavorable due to kink-kink repulsion [see Figs. 10.2(c) and 10.2(d)]. With further increasing of $|P|$ within the interval $P_2 < |P| < P_3$ the state with two kinks becomes the GS [see Fig. 10.2(e)], and so on up to the state $\langle n^* \rangle$. The value P_n corresponded to the transition from the GS $\langle n - 1 \rangle$ to the GS $\langle n \rangle$, may be determined from the condition $nR = L$, which leads to the relation [176]

$$(P_n d)^2 \left(|P_n| - \frac{n}{N} \right)^{1/2} = \left(\frac{n}{2N} \right)^{1/2} \tag{10.27}$$

which, however, is valid provided $|P|d \gg 1$ only.

The kink's terminology used above, is useful as long as a kink can be "inserted" into the chain, i.e. whenever $g \ll g^*$, where the value of g^* is determined from the condition $2d = L$, so that

$$g^* = \frac{1}{4} N^2. \tag{10.28}$$

Otherwise, i.e. at $g \gg g^*$, the atomic chain can be approximately considered as a chain of rigidly coupled atoms.

10.2.3 Discrete Chains

The discrete FK chain can be investigated numerically only [176, 768, 772, 773] except the simplest cases of $N = 2$ and $N = 3$ [176]. Many qualitative features described above in the continuum limit approximation, are preserved for the discrete chain too, at least for large enough values of the parameter g. A new feature appearing in the discrete chain, is the existence of an intrinsic pinning caused by the PN potential. This results in that the finite FK chain is always pinned ($E_a > 0$) for any $g \neq \infty$, due to pinning of the chain's free ends [458].

When N is even as well as when N is odd but n is even, the GS configuration is symmetric (e.g., see the configurations $\langle 0 \rangle_s$ and $\langle 2 \rangle_s$ in Fig. 10.1). A

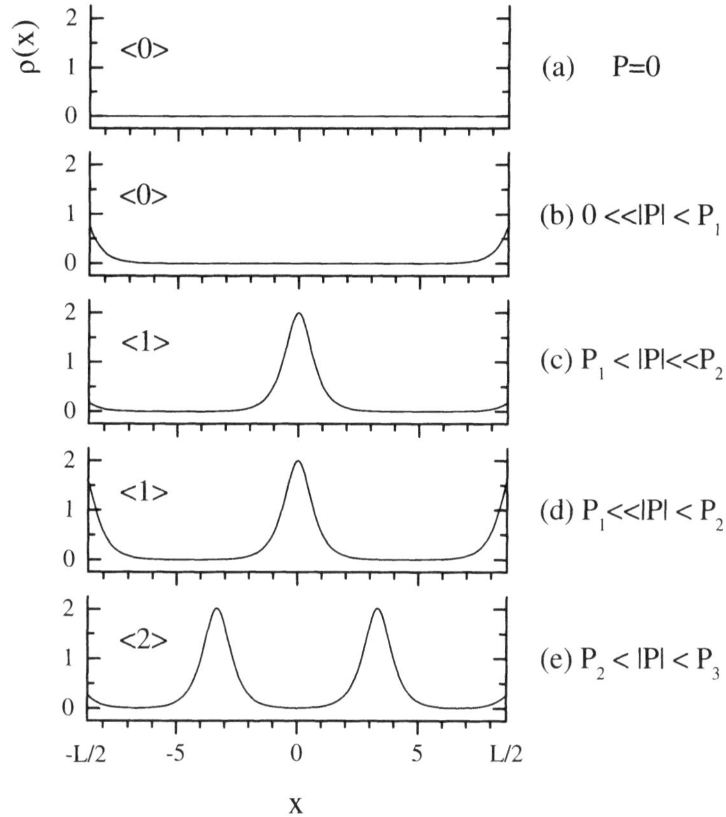

Fig. 10.2. Density of excess atoms $\rho(x) = -u_x(x)/a_s$ in the GS of the finite chain for fixed g and L and different values of the misfit P within the interval $-1/2 < P \leq 0$.

more complicated is the case when both N and n are odd. In this case the GS is symmetric for large enough g values, $g > g_A$, and the configuration $\langle n \rangle_s$ has a kink exactly at the middle of the chain; this kink is kept on the top of the PN potential by repulsive forces from the free ends (if $n = 1$) or other kinks (when $n = 3, 5, \ldots$). In this state the middle atom is in a top position (see the configuration $\langle 1 \rangle_s$ in Fig. 10.1). As the interatomic potential decreases below a critical value g_A, the kink-kink repulsive interaction becomes insufficient to keep the top configuration, and the energy of the chain is reduced by the central atom slipping away from the maximum of $V_{\text{sub}}(x)$. Thus, the state $\langle n \rangle_s$ becomes the unstable (saddle) state, while the new GS is asymmetric and double degenerated (e.g., the state $\langle 1 \rangle_a$ in Fig. 10.1). This breaking symmetry transition may be interpreted as the "finite-Aubry" transition by breaking of analyticity [490].

Besides the ground state, there exist metastable configurations, the number of which increases approximately linearly with the length of the chain, i.e., the density of configurational excitations is approximately constant [490]. The number of metastable states is greatly increased for $g < g_A$. Computer simulation shows that the energy difference per atom between the GS and the highest-energy metastable state is nearly constant. The average separation of particles a_A in the metastable configurations covers a range of values below and above the GS value, the interval of allowed a_A values decreases to zero as $g \to \infty$.

10.2.4 Vibrational Spectrum

Each stable configuration $\langle n \rangle$ has its own spectrum of linear excitations which consists of N frequencies determined by the eigenvalues λ_j of Eq. (10.6),

$$\omega_j^2 = \lambda_j, \quad j = 1, 2, \ldots, N. \tag{10.29}$$

This spectrum may be explained with the help of the quasiparticle-gas approach. First, it consists of $(N-n)$ phonon frequencies which form the phonon zone $(\omega_0, \omega_{\max})$. Note, however, that two of these modes split out from the bottom of phonon zone and are transformed into the localized "end" (or "surface") modes with frequencies which are separated from the phonon branch by a gap [771].

When $n \neq 0$, the n modes are taken away from the phonon zone and are transformed into n Goldstone (or PN for the discrete chain) kink modes. Due to interaction between kinks, these modes form their own spectrum. Recall that for an infinite chain in the IC state the kink collective modes form a spectrum with the acoustic dispersion law [see Eq. (5.74)],

$$\omega(\kappa) = c_s \kappa, \tag{10.30}$$

where $\kappa \leq \pi a_s / R$, R being the mean distance between kinks, and c_s is determined by Eq. (5.75). For the finite chain this spectrum consists of n discrete frequencies, and the wave vector κ in Eq. (10.30) is limited from below by the chain size, $\kappa \geq \pi a_s / L$. Therefore, the collective kink modes have a minimum value $\omega_G \simeq c_s \pi a_s / L$. A more accurate calculation gives the expression [770]

$$\omega_G \simeq \frac{\sqrt{1-k^2}}{k} \, \mathrm{sn} \left(\frac{\pi \sqrt{g}}{N} \frac{k \mathbf{K}(k)}{\mathbf{E}(k)}, k \right). \tag{10.31}$$

For a stable configuration the frequency of the lowest vibrational mode ω_G is always positive except the case when the configuration becomes unstable with changing model parameters. For example, for the "finite-Aubry" transition with decreasing of g at the point $g = g_A$ the GS configuration $\langle \underline{n} \rangle_s$ splits into three configurations: the saddle configuration $\langle \bar{n} \rangle_s$ and two new GS configurations $\langle \underline{n} \rangle_{a,\mathrm{left}}$ and $\langle \underline{n} \rangle_{a,\mathrm{right}}$ which are the mirror images of one another. Clearly that at the point $g = g_A$ the gap in the vibrational spectrum vanishes, and the chain has a soft mode ($\omega_G = 0$) corresponding to its instability [490]. As was shown by Braiman *et al.* [774, 775], near the transition the minimum frequency behaves as

$$\omega_G \propto |g - g_A|^{1/2}. \tag{10.32}$$

At low g the discrete FK chain has "chaotic" configurational excitations, the metastable states with randomly pinned kinks. As was shown by Martini *et al.* [771], vibrational spectrum of these states is approximately continuous.

Besides linear excitations, the infinite FK chain has essentially nonlinear excitations, or breathers. Recall that in the discrete FK chain these excitations have a finite lifetime owing to radiation damping (see Chapter 4). Similar excitations exist in the finite FK chain as well, but now they emerge only when the energy of excitation exceeds some threshold energy ε_c, where $\varepsilon_c \to 0$ when $N \to \infty$ [776]. An interesting nonlinear excitation which is specific for a finite chain, is the "end" nonlinear mode [777, 778] which may be represented as a half of the breather, the center of which is kept at the chain's end. This excitation describes the oscillation of a kink near the chain end, such that at each reflection from the free end, the kink is transformed into the antikink, then again into the kink, etc. In the $P = 0$ case the nonlinear end-mode is described in the continuum limit approximation by the equation [777]

$$u(\widetilde{x}, t) = 4 \tan^{-1} \left[\left(\frac{\kappa}{k'} \right)^{1/2} \mathrm{sn}(\beta\tau, \kappa)\, \mathrm{dn}(\alpha\widetilde{x}, k) \right], \tag{10.33}$$

where τ is the dimensionless time, $k' = \sqrt{1-k^2}$, $\kappa' = \sqrt{1-\kappa^2}$, $\alpha = [1 + (k')^2 + (k'/\kappa)(1+\kappa^2)]^{-1/2}$, $\beta = \alpha(k'/\kappa)^{1/2}$, the parameter κ is determined by the chain length, $\widetilde{L} = \mathbf{K}(k)/\alpha$, while the modulus k determines the frequency of the end mode,

$$\omega = \pi\alpha\sqrt{k'}\left[2\sqrt{\kappa}\mathbf{K}(\kappa)\right]^{-1}. \tag{10.34}$$

The mode (10.33) exists as far as the energy of excitation ε lies in the interval $\varepsilon_c < \varepsilon < \varepsilon_c'$, where

$$\varepsilon_c = \pi E_k / 2\widetilde{L}, \tag{10.35}$$

$$\varepsilon_c' = E_k \mathbf{E}(\lambda)\mathbf{K}(\lambda)/\widetilde{L}, \tag{10.36}$$

$E_k = 8\sqrt{g}$ being the kink energy, and the parameter λ is determined by the equation

$$\mathbf{K}(\lambda)\,(1 + \sqrt{1-\lambda^2}) = \widetilde{L}, \tag{10.37}$$

so that at $\widetilde{L} \gg 1$ it follows $\varepsilon_c' \simeq E_k(1 - 4e^{-\widetilde{L}}) \simeq E_k$. The condition $\varepsilon < \varepsilon_c'$ follows from the requirement that the amplitude of kink oscillations should not exceed a half of the chain length.

10.3 Dynamics of a Finite Chain

10.3.1 Caterpillar-Like Motion

When the energy of excitation ε exceeds the second threshold energy ε_c', two nonlinear end modes (10.33) overlap at the middle of the chain and are transformed into one united mode [777]

$$u(\widetilde{x},\tau) = 4\tan^{-1}\left[(\kappa'/k')^{1/2}\,\mathrm{dn}(\alpha\widetilde{x},k)\,\mathrm{sn}(\beta\tau,\kappa)/\mathrm{cn}(\beta\tau,\kappa)\right], \qquad (10.38)$$

where now $\alpha = [(1-k'/\kappa')(1-k'\kappa')]^{-1/2}$, $\beta = \alpha(k'/\kappa')^{\frac{1}{2}}$, and the parameters κ and k are determined by the relationships $\widetilde{L} = \mathbf{K}(k)/\alpha$ and $\varepsilon = E_k\mathbf{E}(k)\alpha$. The solution (10.38) describes oscillations of a kink between two free ends with the frequency

$$\omega = \pi\alpha(k'/\kappa')^{1/2}/\mathbf{K}(\kappa). \qquad (10.39)$$

At each reflection of the kink from a free end, the kink is transformed into the antikink, and the end atom is shifted to the right on the value $2a_s$. So, for one period $T = 2\pi/\omega$ of kink travelling along the chain, the kink runs to the right, reflects from the right free end transforming into the antikink, then the antikink runs to the left and reflects from the left free end transforming back to the kink, then it again runs to the right, etc. During this cycle the chain as a whole is pulled over the distance $2a_s$ to the right. Thus, the chain moves with the mean velocity

$$v = 2a_s/T = a_s\omega/\pi. \qquad (10.40)$$

This motion looks like a crawling caterpillar. A similar motion was observed by Stoyanov [779] in computer simulations, and for a weak-coupling limit of the standard FK model it was studied by Braiman *et al.* [780]. Kwaśniewski et al. [781] have shown also that the collision of the kink with the chain's free end is accompanied with a strong radiation, so that the kink reflection is observed for stiff enough chains, $g \geq 1$, while for the case of a weak interatomic interaction, $g < 1$, the kink disappears when it reaches the chain's end.

In the notations introduced above, the caterpillar motion proceeds along the trajectory

$$\ldots \to \langle 0\rangle \to \langle 1\rangle \to \langle 0\rangle \to \langle -1\rangle \to \langle 0\rangle \to \langle 1\rangle \to \ldots \qquad (10.41)$$

Note that this motion takes place in the special case of $g \gg 1$ and $P = 0$, when the PN relief and the damping of kink's motion can be neglected, and also the kink and antikink creation energies are equal one another. For a general case the caterpillar motion is carried out along the trajectory

$$\ldots \to \langle n\rangle \to \langle n+1\rangle \to \langle n\rangle \to \langle n-1\rangle \to \langle n\rangle \to \langle n+1\rangle \to \ldots \qquad (10.42)$$

and may be described in a similar way except that the kink (antikink) is to be replaced by a superkink (super-antikink). Note also that the caterpillar motion requires the threshold (activation) energy E_a which is equal approximately to the kink (superkink) creation energy E_k (or E_{sk}).

10.3.2 Adiabatic Trajectories

The adiabatic trajectory links one ground state $\langle \underline{n} \rangle$ with a coordinate X_i and the neighboring GS with $X_f = X_i + b$, and passes through saddle and metastable states $\langle n+1 \rangle$ or $\langle n-1 \rangle$ depending on which of the trajectories requires less activation energy. In the limiting cases $g = 0$ or $g = \infty$ the calculation of characteristics of this trajectory is trivial [176]. Here we describe the results of numerical study for an arbitrary N and $0 < g < \infty$ [772, 773, 176].

Recall that the ground state $\langle \underline{0} \rangle_s$ should be transformed into the states $\langle \underline{1} \rangle, \langle \underline{2} \rangle, \ldots, \langle \underline{n}^* \rangle$ when the parameter $a_{\min}$ is reduced. Therefore, we can define regions $O_0, O_1, \ldots, O_{n^*}$ on the parametric plane $(a_{\min}, g)$ such that in the region O_n the state $\langle \underline{n} \rangle$ is the ground one. For example, for $N = 4$ the region O_0 (with the ground state $\langle \underline{0} \rangle_s$) consists of the regions a–c in Fig. 10.3, and the region O_1 (with the ground state $\langle \underline{1} \rangle_s$) involves the regions d–i. The regions O_{n-1} and O_n are separated by the curve $[a_n]$ (see curve $[a_1]$ in Fig. 10.3), where the value a_n is equal to that value of $a_{\min}$ which satisfies the equation $E(\langle \underline{n-1} \rangle) = E(\langle \underline{n} \rangle)$, so that the region O_n is determined by the inequality $a_{n+1} < a_{\min} < a_n$. The curve $[a_1]$ is defined by the FvdM limit,

$$g_{[a_1]} \simeq \begin{cases} (a_{\min} - \pi)/2\pi & \text{if } g \ll 1, \\ (4/\pi)^2/(2\pi - a_{\min})^2 & \text{if } g \gg 1, \end{cases} \tag{10.43}$$

while the curves $[a_n]$ for $n \geq 2$ are determined by Eq. (10.27) for $g \gg 1$ and converge to the curve (10.43) for $g \ll 1$ because the energy of kink-kink repulsion is exponentially small for small g.

Every region O_n is in turn divided by the curve $[\tilde{a}_n]$ (see curve $[\tilde{a}_1]$ in Fig. 10.3) into two subregions, the "right" one, O_n^- $(\tilde{a}_n < a_{\min} < a_n)$, where the motion is carried out along the trajectory

$$\langle \underline{n} \rangle \to \langle n-1 \rangle \to \langle \underline{n} \rangle, \tag{10.44}$$

and the "left" subregion, O_n^+ $(a_{n+1} < a_{\min} < \tilde{a}_n)$, where the adiabatic trajectory passes through the state $\langle n+1 \rangle$,

$$\langle \underline{n} \rangle \to \langle n+1 \rangle \to \langle \underline{n} \rangle. \tag{10.45}$$

For $N = 4$ the region O_n^- involves subregion d–f in Fig. 10.3, and the region O_1^+ consists of subregions g–i. According to the definition of adiabatic trajectory, the curve $[\tilde{a}_n]$ is determined by the condition that the activation energies for the two trajectories, Eq. (10.44) and Eq. (10.45), are equal one another. Note that the straight line $a_{\min} = 2\pi$ corresponds to the curve $[\tilde{a}_0]$, and the straight line $a_{\min} = \pi$ coincides with the curve $[a_{n^*+1}]$ for even N, and with the curve $[\tilde{a}_n^*]$ for odd N.

In the region O_0 there is the curve $[1s]$ (see Fig. 10.3), to the right of which (i.e. at $a_{[1s]} < a_{\min} < 2\pi$) the state $\langle 1 \rangle_s$ is the saddle one, and to the left of which the state $\langle 1 \rangle_s$ corresponds to a local minimum of the system energy.

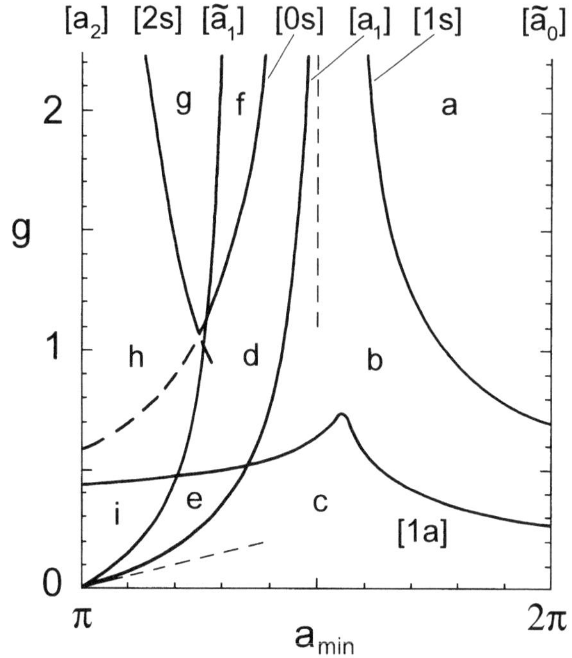

Fig. 10.3. Plane of system parameters $(a_{\min}, g)$ for the FK chain with $N = 4$. Definition of curves see in text. The letters a–i designate the regions with different adiabatic trajectories: (a) $\langle \underline{0} \rangle_s \to \langle \bar{1} \rangle_s \to \langle \underline{0} \rangle_s$, (b) $\langle \underline{0} \rangle_s \to \langle \bar{x} \rangle \to \langle \underline{1} \rangle_s \to \langle \bar{x} \rangle \to \langle \underline{0} \rangle_s$, (c) $\langle \underline{0} \rangle_s \to \langle \bar{x}' \rangle \to \langle \underline{1} \rangle_a \to \langle \bar{x}'' \rangle \to \langle \underline{1} \rangle_s \to \langle \bar{x}'' \rangle \to \ldots$, (d) $\langle \underline{1} \rangle_s \to \langle \bar{x} \rangle \to \langle \underline{0} \rangle_s \to \langle \bar{x} \rangle \to \langle \underline{1} \rangle_s$, (e) $\langle \underline{1} \rangle_s \to \langle \bar{x}' \rangle \to \langle \underline{1} \rangle_a \to \langle \bar{x}'' \rangle \to \langle \underline{0} \rangle_s \to \langle \bar{x}'' \rangle \to \ldots$, (f) $\langle \underline{1} \rangle_s \to \langle \bar{0} \rangle_s \to \langle \underline{1} \rangle_s$, (g) $\langle \underline{1} \rangle_s \to \langle \bar{2} \rangle_s \to \langle \underline{1} \rangle_s$, (h) $\langle \underline{1} \rangle_s \to \langle \bar{x} \rangle \to \langle \underline{2} \rangle_s \to \langle \bar{x} \rangle \to \langle \underline{1} \rangle_s$, (i) $\langle \underline{1} \rangle_s \to \langle \bar{x}' \rangle \to \langle \underline{1} \rangle_a \to \langle \bar{x}'' \rangle \to \langle \underline{2} \rangle_s \to \langle \bar{x}'' \rangle \to \ldots$

Similarly, in each region O_n^- (O_n^+) there exists the same curve $[(n-1)s]$ (or $[(n+1)s]$), to the left (right) of which the state $\langle \overline{n-1} \rangle_s$ (or $\langle \overline{n+1} \rangle_s$), and to the right (left) of which the state $\langle \underline{n-1} \rangle_s$ (or $\langle \underline{n-1} \rangle_s$) exist (see the curve $[0s]$ in the region O_1^-, and the curve $[2s]$ in the region O_1^+ in Fig. 10.3). In particular, the curve $[0s]$ is determined by the metastable FvdM limit P_{ms} [see Eq. (5.109) in Sect. 5.3.1],

$$g_{[o_s]} \simeq 4/(2\pi - a_{\min})^2 \quad \text{if} \quad g \gg 1. \tag{10.46}$$

Finally, in Fig. 10.3 the curve $[1a]$ is plotted, below which an additional stationary state $\langle \underline{1} \rangle_a$ appears; this state corresponds to the local minimum of the system energy (more precisely, the state $\langle x \rangle$ is splitted into three ones: $\langle \bar{x}' \rangle$, $\langle \underline{1}_a \rangle$, and $\langle \bar{x}'' \rangle$). Generally, there are curves $[na]$ ($n = 1, 2, \ldots, \operatorname{int}[(N-1)/2]$) below which the stationary states $\langle \underline{n} \rangle_a$ exist; in the case of odd N, these curves are merged with the curves $[ns]$ as

$a_{\min} \to a_n^\infty$ [176]. For odd N this splitting corresponds to the "finite-Aubry" transition described above.

The adiabatic motion of a chain can be described now in the following way [176, 770, 772, 773]. For the commensurate case of $a_{\min} = 2\pi$, on the onset of the motion a kink is introduced from the left end of the chain, then this kink moves along the chain to the right and is annihilated at the right end. According to the work of Kovalev [777], the creation of a kink at the free end of a semi-infinite chain (with the coordinate $y = 0$) and its subsequent motion to the right can be qualitatively viewed as the creation of a kink-antikink pair at the point $y = 0$ of an infinite chain, and their subsequent motion in opposite directions (kink to the right, $y_k = y$, and antikink to the left, $y_{\bar{k}} = -y$). Therefore, the energy of the system with one kink can be rewritten in the form

$$V(x) \simeq \varepsilon_k - \frac{1}{2} v_{\text{int}}(2y) - \frac{1}{2} v_{\text{int}}(4\pi N - 2y), \tag{10.47}$$

where the last two terms can be interpreted as the energy of attraction of the kink to the two free ends of the chain. Thus, the potential energy of the adiabatic motion of the chain, $E(X)$, can be approximately presented by the expression

$$E(X) = V(X) + V_{PN}(X), \tag{10.48}$$

where the second term describes the kink motion along the chain in the periodic Peierls-Nabarro potential, $V_{PN}(X) \simeq \frac{1}{2}\varepsilon_{PN}(1 - \cos X)$. In the case $g \to 0$ we have $\varepsilon_{PN} \to 2$ and $\varepsilon_k \to 0$, and the function $E(X)$ has $(N-1)$ local minima. As g increases, $\varepsilon_{PN} \to 0$, while the energy ε_k increases up to its saturation at $g \geq g^*$. At the same time the local minima of the function $E(X)$ disappear; at first (at $g \leq 1$, when $2d \simeq 2\pi$) the two minima which are the nearest neighbors to chain ends, then the minima more removed from the ends, and so on. Finally, when at $g \geq g^*$ the kink cannot be "inserted" into the chain, the function $E(X)$ has only one maximum for the state $\langle \bar{1} \rangle_s$, and the "solitonic" terminology becomes unsuitable.

In a general case of $a_{\min} \neq 2\pi$ with the ground state $\langle \underline{\mathrm{n}} \rangle$ $(n \neq 0)$, the adiabatic motion (10.45) can be viewed as the creation of an extra kink at the left end of the chain, and then, after a number of displacements, the annihilation of the extreme right kink at the right end of the chain (see Fig. 10.4a). For the trajectory (10.44) with an intermediate state $\langle n-1 \rangle$ the sequence of events is inverse: at first, the extreme right kink leaves the chain, and then a new kink is created at the left end of the chain (see Fig. 10.4b). Otherwise, the behavior of the system at $a_{\min} \neq 2\pi$ is qualitatively similar to that described above. Note that this motion may also be interpreted as "creation" → "motion along the chain" → "annihilation" of an extra superkink or super-antikink. Therefore, the activation energy of the chain motion can be approximately represented as

$$E_a \simeq \Delta\varepsilon_n(a_{\min}, g) + \varepsilon_{PN}(g), \tag{10.49}$$

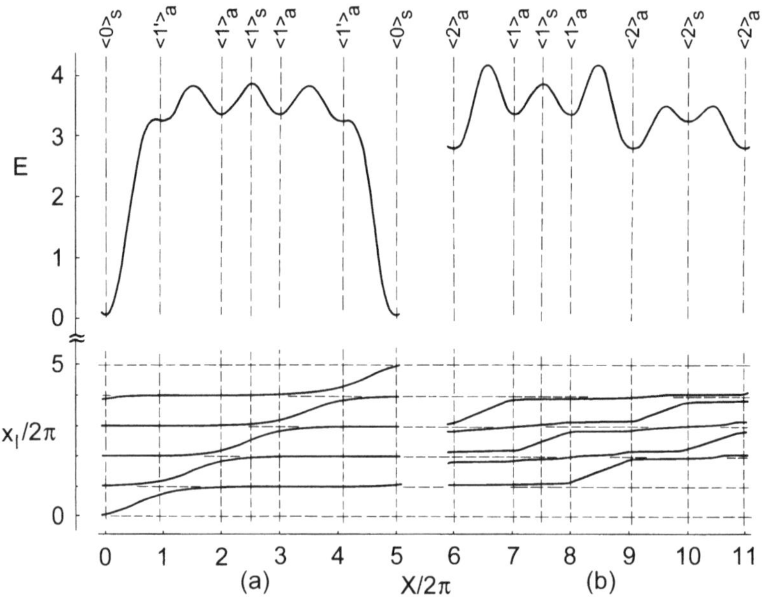

Fig. 10.4. Trajectories of adiabatic motion of the five-atomic FK chain for system parameters: $g = 0.3$ and (a) $a_{\min} = 1.90\pi$, and (b) $a_{\min} = 1.01\pi$ [176].

where

$$\Delta\varepsilon_n = \min\left\{\left|E\left(\langle n+1\rangle\right) - E\left(\langle n\rangle\right)\right|, \left|E\left(\langle n-1\rangle\right) - E\left(\langle n\rangle\right)\right|\right\} \qquad (10.50)$$

is equal approximately to the energy of creation of the superkink or super-antikink depending on which of the two energies is lower.

Thus, the function $E_a(a_{\min})$ reaches its maximum at the commensurate case ($a_{\min} = 2\pi$), has local maxima at $a_{\min} = \widetilde{a}_n$ and local minima at $a_{\min} = a_n$ (see Fig. 10.5). In vicinity of the curves $[a_n]$ there are regions

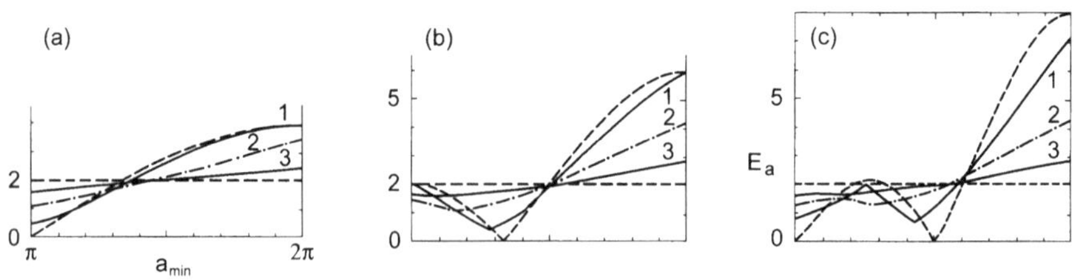

Fig. 10.5. Dependencies of activation energy E_a on parameter $a_{\min}$ for (a) $N = 2$, (b) $N = 3$, and (c) $N = 4$. Broken curves correspond to values $g = \infty$ and $g = 0$, while solid curves, to values $g = 1$ (curve 1) and $g = 0.1$ (curve 3), and chain curves, to values $g = 1/\pi$ for $N = 2$, and $g = 0.3$ for $N = 3$ and $N = 4$ [176].

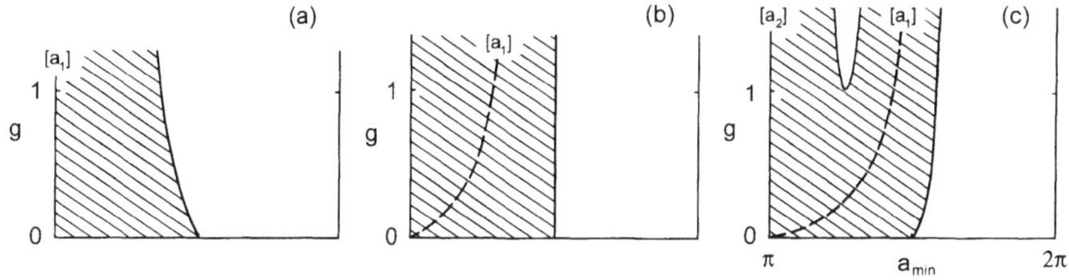

Fig. 10.6. Parametric plane $(a_{\min}, g)$ for (a) $N = 2$, (b) $N = 3$, and (c) $N = 4$. Hatched areas show regions where the activation energy for adiabatic motion of N-atomic chain is lower than that for a single atom [176].

on the parametric plane $(a_{\min}, g)$ (see hatched areas in Fig. 10.6) where the activation energy for the motion of the chain is lower then that for the motion of an isolated atom. Note that at a fixed value of g the functions $E_a(a_{\min})$ is only weakly dependent on the chain length N as long as the inequality $2d \ll L$ is satisfied [i.e. if $N \gg N^*$, where the value of N^* is determined by Eq. (10.28), $N^* = 2\sqrt{g}$]. In particular, at $g \ll 1$ the dependence $E_a(a_{\min})$ has the form

$$E_a \simeq 2 + \pi g(a_{\min} - 3\pi/2) \tag{10.51}$$

for any value of $N \geq 2$ [176].

The described "kink-flip" mechanism of chain motion along the adiabatic trajectory is to be realized in the overdamped case, when the energy exchange of the chain with the substrate is strong. In the opposite underdamped case, when the external friction is very small or absent at all, and the intrinsic friction is low enough (as it should be at $g \gg 1$ and $N \gg 1$), due to inertia of the chain the "caterpillar" trajectory (10.42) may be more preferable because it is more "straight" in the phase space. The activation energy of this trajectory is determined by Eq. (10.49) where, however, $\Delta\varepsilon_n$ is now equal to the maximum of ε_{sk} and $\varepsilon_{\overline{sk}}$, so the "caterpillar" motion is more probable for the $a_{\min} = \widetilde{a}_n$ case. Note also that if the chain moves with a finite velocity v, *the dynamic Peierls-Nabarro potential* will differ from the adiabatic one, because at velocities $v \geq b\omega_n$ the atoms of the chain will have no time to adjust to the substrate potential, and this is equivalent to an effective decreasing of the external potential amplitude.

10.3.3 Diffusion of Short Chains

In applications to realistic systems, the energy exchange between different degrees of freedom of the FK chain as well as between the chain and the substrate should be taken into account. Phenomenologically, this exchange leads to a friction with some coefficient η which describes the rate of energy exchange. Thus, the motion of the chain can be approximately described by the Langevin equation

$$m\ddot{X} + m\eta X + dE(X)/dX = \delta F(t), \tag{10.52}$$

where m is given by Eq. (10.15) and the stochastic force δF satisfies the fluctuation-dissipation theorem, $\langle \delta F(X,t)\, \delta F(X,t') \rangle = 2\eta(X) m(X) k_B T \delta(t - t')$. For times $t > \eta^{-1}$, the solution of the Langevin equation (10.52) corresponds to diffusional motion of the system with the diffusion coefficient

$$D = Rb^2, \quad R = R_0 \exp(-E_a/k_B T), \tag{10.53}$$

where R is the rate of escape of the system from the bottom of the potential well of $E(X)$. Neglecting a complicated form of the function $E(X)$ (namely, ignoring the existence of local minima), the energy $E(X)$ can be approximated by the Peyrard–Remoissenet potential [186]

$$E(X) \simeq \frac{1}{2} E_a V_{\mathrm{PR}}\left(\frac{2\pi X}{b}\right), \quad V_{\mathrm{PR}}(y) = \frac{(1-r)^2 - (1-\cos y)}{(1 + r^2 + 2r\cos y)}, \tag{10.54}$$

where r is a parameter ($|r| < 1$). Then the eigenfrequencies $\widetilde{\omega}_0$ [at the minimum of the potential $E(X)$] and $\widetilde{\omega}_s$ (at the saddle point), $\widetilde{\omega}_{0,s}^2 = \pm m_{0,s}^{-1} E''(X)_{X=0,b/2}$, are coupled by the following relation,

$$m_0 \widetilde{\omega}_0^2 m_s \widetilde{\omega}_s^2 = \left(E_a/2N^2\right)^2. \tag{10.55}$$

Excluding an exotic case of extremely low friction, the pre-exponential factor R_0 for the potential (10.54) is determined by the Kramers theory,

$$R_0 \approx \begin{cases} \widetilde{\omega}_0/2\pi & \text{if } \eta \le \widetilde{\omega}_s, \\ \widetilde{\omega}_0 \widetilde{\omega}_s/2\pi\eta & \text{if } \eta > \widetilde{\omega}_s. \end{cases} \tag{10.56}$$

Using Eqs. (10.53), (10.55) and (10.56), we obtain the expression for the diffusion coefficient,

$$D = D_0 \exp(-E_a/k_B T), \tag{10.57}$$

where

$$D_0 \simeq \begin{cases} 2\pi N^2 \widetilde{\omega}_0 & \text{if } \eta \le \widetilde{\omega}_s, \\ \pi E_a/\eta\sqrt{m_0 m_s} & \text{if } \eta > \widetilde{\omega}_s. \end{cases} \tag{10.58}$$

If the potential (10.54) has a shape of deep narrow wells separated by wide barriers (i.e. if $r < 0$), we can approximately take $E_a \simeq \frac{1}{2} m_0 \widetilde{\omega}_0^2 \pi^2$, so that the following estimation for the value of D_0 is obtained,

$$D_0 \simeq \begin{cases} 2\pi N^2 \widetilde{\omega}_0 & \text{if } N < N^{**}, \\ 2\pi \widetilde{\omega}_0^2/\eta & \text{if } N > N^{**}, \end{cases} \tag{10.59}$$

where

$$N^{**} = (\widetilde{\omega}_0/\eta)^{1/2}. \tag{10.60}$$

Thus, for the commensurate case ($a_{\rm min} = 2\pi$) at a fixed value of g, both the activation energy E_a (up to the saturation at $N \geq N^*$) and the pre-exponential factor D_0 (up to the saturation at $N \geq N^{**}$) increase with the increasing of the chain's length N. On the other hand, at fixed values of N and g the factors E_a, $\widetilde{\omega}_0$, and D_0 change "in-phase" if the equilibrium distance $a_{\rm min}$ is changed. So, in both cases the theory predicts the existence of a "compensation" effect, i.e. the values E_a and D_0 are changed simultaneously in the same direction. The diffusion coefficient D can either increase or decrease with the increase in the chain length N, the former situation takes place at $a_{\rm min} \simeq \widetilde{a}_n$, and the latter, at $a_{\rm min} \simeq a_n$. The phenomenological theory given above was used by Braun [176] to describe the diffusion of small clusters absorbed on crystal surfaces.

10.3.4 Stimulated Diffusion

A mobility of the finite FK chain can be drastically enhanced by applying the external time-periodic field with appropriately chosen frequency [770]. A promising candidate is the frequency ω_G, Eq. (10.31), the lowest frequency in the acoustic branch of kinks collective motion, because in this case all kinks vibrate in phase. The external driving will increase the chain energy, making the transition from the GS to metastable states more probable. But the metastable state has its own vibration spectrum which is different from that of the GS, and the resonance is destroyed. After spending some time in a metastable state the system returns to the GS where it is again in resonance with the external field. In this way the external field with "resonance" frequency stimulates the chain to "oscillate" between the GS and neighboring metastable states and makes it moving along the substrate. The molecular dynamics simulation [770] confirms the existence of "resonance-induced" mobility: the time-periodic external field of the frequency ω_G and a small amplitude (10^{-3} in our dimensionless units) significantly enhances the chain's random walks. It is important that varying the frequency of the exciting field one can vary the size of the cluster exhibiting stimulated mobility, because the frequency ω_G depends on N.

10.4 Nonconvex Potential

A realistic pair-wise interatomic potential such as the Morse potential (5.138), has a finite energy of dissociation, i.e. its attractive branch tends to zero with increasing the interatomic distance. In Sect. 5.4.3 it was shown that the FK model with the Morse potential has two peculiarities as compared with the standard FK model. First, the symmetry between the kink (local contraction of the chain) and the antikink (local extension) is violated. As compared against the kink, the antikink is characterized by a lower value of ε_k, and by higher values of ε_{PN} and m. As a result, the curves $[a_n]$ and $[\widetilde{a}_n]$ on the parametric plane $(a_{\min}, g)$ will be shifted. For example, at $a_{\min} = 2\pi$ the trajectory $\langle 0 \rangle \to \langle -1 \rangle \to \langle 0 \rangle$ will have a lower activation energy than the trajectory $\langle 0 \rangle \to \langle +1 \rangle \to \langle 0 \rangle$. Second, for the potential (5.138) the interatomic attractive force reaches a maximum at the distance $x = a_{\rm inf}$; beyond the inflection point $a_{\rm inf}$ the potential $V_{\rm int}(x)$ becomes concave, and the force reduces with increasing of the distance between the atoms. In a result, there exist an infinite number of metastable states, which correspond to the chain ruptured into two, three, ..., N parts (or single atoms). Recall that for negative misfits, $P \leq P_{\rm inf} < 0$ [i.e., for $a_{\rm inf} \leq a_s$, see Eq. (5.146)], the free end parts of the Morse–FK chain are always modulated (distorted) irrespective of the value $V_{\min}$ of the Morse potential (5.138). As a result, the metastability limit P_{ms}^{-} above which the kink-free chain (i.e., the state $\langle 0 \rangle$) cannot exist as a metastable configuration, disappears provided $V_{\min}$ is low enough (in Fig. 10.3 this effect will manifest itself as the break of the curve $[0s]$ at low enough g). Besides, at $P < P_{\rm inf}$ and large enough $V_{\min}$, the middle part of the chain is modulated too. Clearly that the finite-length FK chain should behave analogously.

Recall also that the fixed-density infinite chain cannot exist in the ground state $\langle 0 \rangle$ (i.e., without kinks) provided $P \leq P_{FM}^{-}$ (or $a_{\min} \leq a_{FM}^{-}$) because it will be ruptured into two semi-infinite parts. A similar situation may emerge for the finite FK chain in the metastable state $\langle 0 \rangle$ when its ends are fixed, for example, due to pinning at impurities. However, the rupture of the finite chain of N atoms has to take place provided $a_{\min} \leq a_{FM}^{-} - \delta a(N)$, where $\delta a(N)$ may be estimated as $\delta a(N) \simeq (a_{\rm inf} - a_{\min})/N$.

The described effect is quite important in growth of epitaxial layers with $a_{FM}^{-} - \delta a(N_0) < a_{\min} < a_{FM}^{-}$. At initial stages of growth, short expanded "islands" of sizes $N < N_0$ arises in states coherent with the substrate. But with increasing of the islands owing to incorporation some more atoms, the state $\langle 0 \rangle$ becomes metastable when $N > N_0$, so that it should be transformed either to the true GS with uniformly distributed dislocations (kinks) or to a metastable state by cracking to shorter chains depending on which of these processes needs lower activation energy [512, 513].

Another interesting effect predicted by Milchev and Markov [262] is that thermally migrating clusters may break into smaller fractions if they traverse configurations with abnormally stretched bonds. Thus, the substrate may act as a "filter", only the clusters smaller than some critical size will survive thermally induced migrations while the rest will decay into smaller fractions.

11 Two-Dimensional Models

One of the very important extensions of the classical FK model goes beyond its one-dimensional nature. In particular, the two-dimensional FK model describes a planar lattice of interacting atoms in an external periodic potential which can be either one- or two-dimensional. This chapter discusses some problems related to two-dimensional FK models.

11.1 Preliminary Remarks

In a general case of higher dimensions, the FK model describes a lattice of interacting atoms placed onto a multi-dimensional periodic potential. Even the case of two dimensions is extremely complicated. One of important new aspects of the two-dimensional (2D) model is that now not only the $T = 0$ ground state but also the $T \neq 0$ equilibrium state may be ordered exhibiting a number of different phases and phase transitions between them. Not going into specific details, below we discuss the existence of two possible equilibrium phases in 2D lattices.

When the substrate potential is absent, $V_{\rm sub}(x, y) \equiv 0$, the truly crystalline structure is impossible in a 2D system, and the system is disordered at any $T \neq 0$ [516], [782]–[784]. But at low enough temperatures, $T < T_{cr}$, the equilibrium state of the system is characterized by a quasi-long-range order (known as the "algebraic" order), and the corresponding phase is called *a floating crystal* [785]–[788]. In this phase the translational correlation function falls to zero according to a power law with increasing of the distance (instead of the exponential law in the liquid phase). The floating phase exhibits a shear stiffness. Besides, in this phase the orientational order (i.e., the orientations of bonds between nearest neighbors) is long-ranged as in usual crystals [786]. Clearly that at high temperatures, $T > T_{c\phi}$, the system must be totally disordered, so that both (translational and orientational) correlation functions should decay according to exponential laws. The floating phase melts by the well-known Kosterlitz-Thouless dislocation mechanism [789]–[791], and the transition from the "floating phase" to the "liquid phase" proceeds in two steps [792]. First, at $T \geq T_{cr}$ the translational quasi-long-range order is destroyed due to nucleation of dislocation pairs (i.e., the pairs consisting of two dislocations with opposite Burgers vectors) with their subsequent dissociation. Simultaneously the orientational long-range order is transformed into the orientational quasi-long-range order. Second, at $T \geq T_{c\phi}$ the orientational quasi-long-range order is destroyed due to creation of disclination pairs. Both these phase transitions are continuous. The intermediate phase at $T_{cr} < T < T_{c\phi}$ is called the hexatic liquid phase. It was observed, in particular, in molecular dynamics simulations for the system of atoms which repel according to Coulomb's law [793]. Note also that the random nonequilibrium ("frozen") impurities destroy the floating phase [794, 795].

In the opposite limiting case, when the substrate potential significantly exceeds the interatomic interactions, the system can be considered in the framework of a lattice-gas model. In this case the system exhibits a variety of structures including crystalline (commensurate), floating (incommensurate), and liquid (totally disordered) phases depending on the substrate symmetry, interaction law, concentration θ and temperature T, and associated phase transitions with θ or T variation may be continuous as well as discontinuous. In some particular cases the model admits an exact solution (for example, the Ising, Baxter, and Potts models), otherwise it can be studied by various methods including Monte Carlo simulation, low- and high-temperature series, finite-size transfer matrix method, etc. The description of properties of the lattice-gas model with short-range interactions may be found in review papers by Selke *et al.* [796, 797] (see also Ref. [48]; the case of long-range anisotropic (dipole-dipole) interaction was studied in Refs. [798, 799]).

In a general case, the 2D FK model stays just between these two limiting cases. Evidently we cannot give a complete description of the problem, our goal here is only to overview few papers devoted to two-dimensional FK models. The FK model may be generalized to two-dimensional versions in two different ways leading to scalar or vector FK models. In *scalar models*, the atoms are arranged in a 2D array, but atomic motion is still one-dimensional. The scalar model may be used, in particular, to describe a layer of atoms adsorbed on a furrowed surface such as the (112) plane of b.c.c. crystal. Because the energy barrier for motion across furrows is much higher than that for motion along the furrows, one may assume that adsorbed atoms move along the channels only. However, the interaction between the atoms in different channels is not negligible in a general case. A natural way to study the scalar model is to consider it as a system of coupled one-dimensional FK chains. Then, the interaction between the chains will lead to an interaction between topological excitations (kinks) in the chains, and if the later is attractive, the kinks will be arranged in a domain wall (DW), so that such configuration may be considered as a "secondary" FK chain oriented perpendicularly to the "primary" chains. This approach is described in Sect. 11.2.

In a more general model we have to allow for atoms to move in more than one dimension. The simplest model of this class is the FK model with a transverse degree of freedom [800]. In this model the atoms are still arranged in a one-dimensional chain, but the substrate potential is two-dimensional, it is periodic in one dimension and parabolic in the transverse direction, and the atoms can move in both dimensions. In physical systems the transverse degree of freedom may correspond, e.g., to motion of adsorbed atoms perpendicularly to the surface, or to their displacements orthogonal to furrows in the case of adsorption on furrowed crystal planes. New interesting aspects of the model behavior emerge for repulsive interatomic interaction. Namely, when the magnitude of repulsion increases (or the curvature of the transverse potential decreases) above a certain threshold, the straight-line GS atomic configuration becomes unstable, and the system is transformed into a new "zigzag" GS. In turn, this completely modifies the excitation spectrum of the model, results is appearing of kinks of different kinds, changes the Aubry transition scenario, etc. The *zigzag* FK model is described in Sect. 11.3.

Finally, Sects. 11.4 and 11.5 are devoted to truly 2D *vector* FK models, where the atoms can move in two dimensions, and the substrate potential is periodic in two dimensions as well. Vector models incorporate the features of both scalar and zigzag models and exhibit a rich spectrum of behavior.

11.2 Scalar Models

In the first, more simple extension of the FK model, we may leave a scalar nature of the field variable u connected with atomic displacements, but allow the atoms to form a 2D array. Thus, now displacements u_{l_x,l_y} are numerated by two integers l_x and l_y which describe the position of an atom in the two-dimensional array. The kinetic energy K and the substrate potential energy $U_{\rm sub}$ remain practically the same as in the 1D model,

$$K = \frac{1}{2}\sum_{l_x,l_y} \dot{u}^2_{l_x,l_y}, \quad U_{\rm sub} = \frac{\varepsilon_s}{2}\sum_{l_x,l_y}\left(1-\cos u_{l_x l_y}\right). \tag{11.1}$$

As for the interatomic interaction, a direct generalization of the standard FK model leads to the form

$$U_{\rm int} = \sum_{l_x,l_y}\left[\frac{g_x}{2}\left(u_{l_x+1,l_y} - u_{l_x,l_y} - a_{sx}P_x\right)^2 + \right.$$
$$\left. + \frac{g_y}{2}\left(u_{l_x,l_y+1} - u_{l_x,l_y} - a_{sy}P_y\right)^2\right], \tag{11.2}$$

where g_x and g_y are the elastic constants, a_{sx} and a_{sy} are the lattice constants, and P_x and P_y are the misfit parameters in the x and y directions correspondingly. If $g_x = g_y$ and $a_{sx} = a_{sy}$, the model is isotropic, otherwise we come to the anisotropic two-dimensional FK model. In the continuum limit the problem (11.1) to (11.2) reduces to two-dimensional SG equation $\ddot{u} - \Delta u + \sin u = 0$ which, unfortunately, is not exactly integrable.

However, the "ball and spring" approximation (11.2) for the interatomic interaction is rigorous only if mutual displacements of atoms in nearest-neighboring chains are small comparing with a_s. A more realistic model was proposed by Braun *et al.* [46]. Let us call an isolated FK chain as the "x-chain". Mass transport along the x-chain is carried by kinks (x-kinks) which are characterized by the width d_0, mass m_0, rest energy ε_{k0}, and the Peierls-Nabarro barrier ε_{PN0} (the index "0" corresponds to the isolated x-chain). The interaction of atoms belonging to the neighboring x-chains, leads to an interaction between the chains. Let $v(\Delta x)$ is the interaction energy of two atoms which belong to the nearest neighboring (NN) chains and shifted with respect to one another by a distance Δx along the chain, and let $u_1(x)$ and $u_2(x)$ are the displacements of the atoms in different chains from their equilibrium positions corresponded to a commensurate 2D structure. The simplest expression for the energy of interaction between the chains in the continuum-limit approximation is the following,

$$U'_{\text{int}} = \frac{\alpha}{2} \int \frac{dx}{a_{sx}} \, [u_1(x) - u_2(x)]^2, \tag{11.3}$$

where α is the corresponding elastic constant. Expression (11.3), however, does not take into account the following two important circumstances. First, the energy should remain unchanged if one of the chains which has no kinks, is shifted for one period of the commensurate structure. Therefore, Eq. (11.3) should be modified as ($a_{sx} = 2\pi$)

$$U'_{\text{int}} = \alpha \int \frac{dx}{a_{sx}} \, \{1 - \cos\left[u_1(x) - u_2(x)\right]\}. \tag{11.4}$$

Second, both Eqs. (11.3) and (11.4) do not describe an interaction between two chains in identical but inhomogeneous states, i.e. when $u_1(x) = u_2(x) \neq 0$. Recalling that the density of "excess" atoms in an inhomogeneous state is equal to $\rho(x) = -u'(x)/a_{sx}$, the energy of their interaction can be described by the expression

$$U''_{\text{int}} = \iint \frac{dx\, dx'}{a_{sx}^2} \, u'_1(x)\, v(x - x')\, u'_2(x'). \tag{11.5}$$

If the potential $v(x)$ is short-ranged on the scale d_0, we may approximate it as $v(x) \approx \gamma a_{sx} \delta(x)$. Combining Eqs. (11.4) and (11.5), we finally come to the expression

$$U_{\text{int}}[u_1, u_2] = a_{sx}^{-1} \int dx \, \{\alpha \left[1 - \cos(u_1(x) - u_2(x))\right] + \gamma u'_1(x) u'_2(x)\}, \tag{11.6}$$

where the parameters α and γ are coupled with the potential $v(x)$ by the relationships

$$\alpha = \left.\frac{\partial^2 v(x)}{\partial x^2}\right|_{x=0}, \quad \gamma = \int \frac{dx}{a_{sx}} \, v(x). \tag{11.7}$$

Now we can consider the system of parallel FK chains. It is described by the Hamiltonian

$$\mathcal{H} = \sum_j \left\{ H[u_j] + U_{\text{int}}[u_j, u_{j+1}] \right\}, \tag{11.8}$$

where the index j numerates the chains, H is the Hamiltonian of an isolated chain, and the pairwise interchain interaction is described by Eq. (11.6). Let us consider the fixed-density system and assume that each FK chain contains one x-kink. For a weak interchain interaction we may neglect by a change of the kink shape from the SG form and substitute into Eq. (11.8) the ansatz $u_j(x,t) = u_{\text{SG}}\left(x - X_j(t)\right)$. Then, if we take into account that x-kinks move in the periodic PN potential with the amplitude ε_{PN}, we obtain the following effective Hamiltonian:

$$\mathcal{H}_{\text{eff}} = \sum_j \left[\varepsilon + \frac{m}{2}\left(\frac{dX_j}{dt}\right)^2 + V(X_j - X_{j-1}) + \frac{\varepsilon_{PN}}{2}(1 - \cos X_j) \right]. \tag{11.9}$$

Here $\varepsilon = 4d/\pi$, $m = 4/\pi d$, and the potential energy of interkink interaction is described by the expression

$$V(X) = \varepsilon \left[\alpha W_1(X/d) - \beta W_2(X/d) \right], \tag{11.10}$$

where $\beta = \gamma/d^2$,

$$W_1(Y) = (1 + Y/\sinh Y)\tanh^2(Y/2) \tag{11.11}$$

and

$$W_2(Y) = 1 - Y/\sinh Y. \tag{11.12}$$

Thus, at small X ($|X| \ll d$) the interaction is harmonic, $V(X) \simeq \frac{1}{2}GX^2$ with the effective elastic constant $G = m(\alpha - \beta/3)$, while at $|X| \to \infty$ we obtain $V(X) \to E_{\text{dis}}$, where $E_{\text{dis}} = \varepsilon(\alpha - \beta)$. Depending on the system parameters, the $T = 0$ ground state of the anisotropic scalar 2D FK system may correspond to one of the following configurations:

1. In the case of interatomic repulsion the x-kinks repel each other too, and for the atomic concentration slightly above or below that corresponding to a commensurate lattice, the x-kinks should form the structures of the c(2×2) type with continuously varying period along the X axis;
2. For a certain system parameters, when $G < 0$ but $E_{\text{dis}} > 0$, "oblique" chains of x-kinks should emerge;
3. Finally, in the case of the attractive interaction between the atoms, the GS corresponds to the y-chains of x-kinks. Namely this case is mostly studied in literature. Note that the direct generalization of the FK model, Eq. (11.2), leads to this situation only.

All these configurations have been observed experimentally in adsorbed systems (see references in Ref. [47]).

Different atomic configurations such as point defects, jogs and dislocations have been studied numerically in the framework of the generalized two-dimensional [801] and three-dimensional [802] scalar Frenkel-Kontorova models. An anisotropic crystal is modelled as a system of parallel harmonic chains, and the atoms from the nearest neighboring chains interact via the Lennard-Jones or Morse potential, so that the "substrate potential" for a given chain is produced by the atoms from the adjacent chains. An interesting result obtained in these simulations, is that two kinks belonging to the NN chains, weakly repel one another on long distances, but become attracting at short distances, so that they may produce a bound state. Such a behavior follows from Eq. (11.10) for the parameters $\alpha \sim 2\beta$.

All generalizations of the FK model which take into account only a *finite* number of neighboring atoms, always lead to an exponential interaction between local defects such as kinks. On the other hand, the elastic interaction (which is always present in a solid) of two defects separated by a distance R in a 3D crystal must decay according to the power law $\propto R^{-3}$. To solve this problem and describe point defects in the 3D lattice, Kovalev *et al.* [65] proposed to surround the FK chain by a "tube" of radius $b \sim a_s$ and consider the solid outside the tube as an elastic continuum medium. The elastic displacements of atoms of the solid outside the tube are described by the displacement field $w(x,y,z)$, which satisfy the equation $\lambda w_{xx} + \mu(w_{yy} + w_{zz}) = 0$, where λ is the modulus of compression along the FK chain and μ is the modulus of shear in the YZ plane (for a simple scalar model of the theory of elasticity). If we denote by $w_0(x)$ the displacement field on the "tube" surface, then the static equation for atoms of the FK chain takes the form $d^2 u_{xx}(x) = \sin[u(x) - w_0(x)]$ in the continuum limit approximation. This equation should be supplemented now by the boundary condition $\mu a_s^2 b\,(\partial w/\partial \rho)\,|_{\rho=b} = \sin(w_0 - u)$, where $\rho = (y^2+z^2)^{1/2}$. As a result, a generalized integro-differential SG-type equation [similar to Eq. (3.147) of Sect. 3.5.4 where, however, the long-range interaction was introduced artificially "by hands"] was derived and analyzed. In particular, it was shown that the energy of interaction between two kinks separated by the distance R behaves as $v_{\rm int}(R) \approx \varepsilon_k (d/R)^3$.

Scalar model for $\theta < 1$. In the $\theta = 1$ case considered above, one extra kink (antikink) inserted into a chain, corresponds to one extra atom (or vacancy). Therefore, if the atoms repel each other, the kinks should be repelled as well. However, for the $\theta = p/q$ reference structure, when p and q are integers and $p < q$, the situation is not so trivial, because now one extra atom corresponds to q kinks.

As an example, let us consider the 2D FK lattice with the rectangular symmetry at $\theta = 1/2$. For the repulsive interatomic interaction the GS of the system corresponds to the c(2×2) structure. Therefore, the atoms in a given chain feel a potential from the nearest neighboring chains which should be added to the substrate potential, so that the total effective external potential for the given chain corresponds to the double sin-Gordon (DSG) potential, i.e. it has the period $2a_s$. Recall that in an isolated FK chain with the concentration $\theta = 1/2$ one inserted atom produces two kinks which repel from one another. Now, in the 2D system, both these kinks are the DSG kinks. Besides, the atomic structure between the kinks does not correspond to the GS configuration but to a metastable configuration, because one kink produces the shift a_s while the period of the GS structure is $2a_s$. In a result, two DSG kinks belonging to the NN chains, will additionally attract one another with the interaction potential which is directly proportional to the distance Δx between the kinks.

Then, if the 2D system has two pairs of kinks, one pair in one chain and another in the nearest neighboring chain, these pairs will effectively attract one another, because an overlapping of the interkink regions reduces the system energy. Finally, if we insert one extra atom into *each* x-chain, we will produce two y-chains of x-kinks, which now repel from one another, so that at $\theta = 1/2 \pm \delta$ ($\delta \to 0$) the GS corresponds to a stripped structure of y-chains. Emphasize that the x-kinks in NN channels are attracting even when the atoms itself repel each other. However, this interkink attraction is not too "strong", now the interkink attractive potential is $\propto \Delta x$ and not $\propto (\Delta x)^2$.

An infinite y-chain cannot be broken. However, two nearest neighboring y-chains may be broken *simultaneously*. Besides, one may create a pair of *finite* y-chains, one consisting of x-kinks and another of x-antikinks, but these y-chains must form a *closed loop*.

For the concentration $\theta = p/q$ with $q \geq 3$ a situation is similar but more complicated. Clearly, all the effects mentioned above, essentially determine the system phase diagram at $T \neq 0$ as well as the system dynamics.

11.2.1 Statistical Mechanics

The equilibrium state of the scalar 2D FK model at $T \neq 0$ was studied mainly in the framework of the harmonic interaction (11.2). Recall that this variant of the model at low temperatures exhibits the formation of y-chains of x-kinks called usually by domain walls (DW), or plane solitons, or discommensuration lines, or linear defects. An anisotropic model with $g_y = \gamma g_x$ (γ is a parameter, $0 \leq \gamma \leq 1$) in Eq. (11.2) for the fixed-density case (i.e., by employing periodic boundary conditions in both directions) was investigated by molecular dynamics technique by Yoshida *et al.* [803, 804] and Kato *et al.* [805]. In particular, the simulations have shown that at low T the residual ("geometrical") DW-line as well as the DW width fluctuates [see Fig. 11.1(a)]. When temperature increases, the thermally nucleated kink-antikink pairs are appearing in the x-chains, and the coupling of these x-kinks belonging to the neighboring x-chains results in creation of "loops" [Fig. 11.1(b)]. As usual, with further increasing of T, the x-kinks as well as the y-chains are thermally

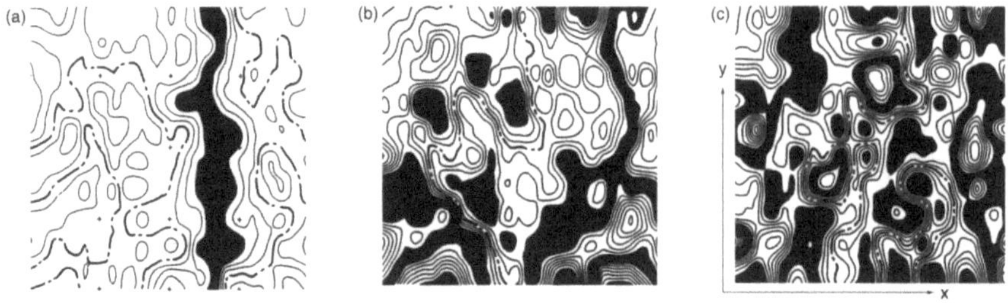

Fig. 11.1. Contour lines of atomic displacements for the isotropic scalar FK model at (a) $T = 0.05$, (b) $T = 0.2$, and (c) $T = 0.35$. The black tone (white tone) shows the regions around the top (bottom) of the substrate potential [805].

destroyed. Kink's contribution to the equilibrium properties of the system is similar to that obtained in the one-dimensional model. For example, the temperature dependence of the specific heat again exhibits a peak at $T \sim \varepsilon_k$, but a position of the peak depends now on the anisotropy parameter γ [803].

Thermodynamic properties of the system of parallel strips (domain walls) was widely studied analytically (see, e.g., monograph by Lyuksyutov *et al.* [48] and references therein). Recall briefly the main results of these studies. Let R_x is the period of the DW lattice in the x-direction, and $a_x = q_x a_{sx}$ is the period of the reference structure. At $T \neq 0$ the DW line is roughening due to thermal fluctuations. This increases the mean energy of repulsion of the neighboring walls because of anharmonicity of x-kinks' interaction. Besides, displacements of the domain walls are restricted by their "collisions", this leads to decreasing of the system entropy. Both factors give a contribution to the free energy (per unit length of the wall) which is equal to [806]–[810]

$$\Delta F \simeq (k_B T)^2 a_x^2 / R_x^2 \varepsilon_k a_{sy}. \tag{11.13}$$

Thus, the effective repulsion between walls follows the power law (11.13) even in the case of short-range interatomic interactions when the $T = 0$ ("mechanical") wall–wall interaction decays exponentially with R_x.

When R_x is commensurate with a_x, the $T = 0$ ground state is crystalline (the commensurate phase). But at $T \neq 0$ the DW line fluctuates, and at a temperature T_{depin} these fluctuations become equal to the PN period a_x. In a result, at $T > T_{\text{depin}}$ the domain walls are effectively unpinned, and the equilibrium state will correspond to the floating (incommensurate) phase. This phase transition occurs provided $R_x/a_x \geq 4$ [811, 812]. With further increasing of T the floating phase melts. If $q_x = 1$ or 2, the melting proceeds by the Kosterlitz-Thouless mechanism due to nucleation of dislocation pairs in the DW lattice (note that the transition occurs in only one step in this case). The theory of the process was developed by Lyuksyutov [813] and Coppersmith *et al.* [814].

On the other hand, if $T = 0$ but θ varies, the system passes through the infinite series of structures which form Devil's staircase. But for $T \neq 0$ most of high-order commensurate structures are floating or liquid (disordered). More details about the (T, θ) or (T, μ) phase diagrams may be found in monograph of Lyuksyutov *et al.* [48].

11.2.2 Dynamic Properties

Mass transport in the scalar FK model has been studied for a long time, because this is one of important problems in solid-state physics (e.g., we may mention the motion of dislocations, plasticity and ductility of metals, etc.). At high temperatures the dynamic properties can be studied by perturbation theory technique similarly to one-dimensional case (see Sect. 7.4). At low temperatures, however, the atomic flux is carried out by concerted motion of atoms driven, e.g., by an external force F or by a gradient of atomic concentration. Let us suppose that the driving dc force is applied along the x-direction, and discuss in brief how the formation of y-chains of x-kinks may modify the diffusion characteristics of the system. First of all recall that for noninteracting x-chains the activation energy for the motion along the x-direction is equal to the height of the PN relief, $\varepsilon_a \simeq \varepsilon_{PN0}$. Also it is known that a repulsion of x-kinks should result in a decrease of the activation barrier, $\varepsilon_a < \varepsilon_{PN0}$. On the other hand, in the case of attraction of x-kinks the barrier ε_a should increase usually. For example, a finite-length y-chain of x-kinks should move owing to creation of an y-kink at one end of the y-chain and its consequent motion along the Y axis to the another chain's end (see Sect. 10.3.2), so that $\varepsilon_a \simeq \frac{1}{2}E_{\text{pair}} + E_{PN}$. In the case of an infinite y-chain of x-kinks, the motion starts with creation of an y-kink–y-antikink pair and then proceeds by their motion in the opposite directions [see Fig. 11.2(a)], this scenario requires the energy $\varepsilon_a \simeq E_{\text{pair}} + E_{PN}$. Thus, at small driving forces F and $T \neq 0$ the atomic flux is a thermally activated process with activation energy ε_a determined by system parameters.

On the other hand, at $T = 0$ the atomic flux starts when the external force exceeds some critical (depinning) value F_c needed to overcome the activation barriers. This problem was studied for the model (11.1)–(11.2) in the case of small external damping by Pouget *et al.* [230]. Again, if it was one y-chain (domain wall) in the initial state, it will evolve according to the scenario described above [by nucleation of y-kink–y-antikink pairs, see Fig. 11.2(a)]. However, the following new interesting effect was observed in numerical simulation and explained by stability analysis technique: if the driven force is high enough ($F \geq 1.1$ in dimensionless units), nucleation of an additional "wall–antiwall" sequence behind the primary wall can occur, resulting in the formation of a "double layer" [see Fig. 11.2(b)]. It may be expected that this effect will be even more pronounced at $T \neq 0$ owing to thermally stimulated creation of loops. Kolomeisky *et al.* [815] have shown also that in the under-

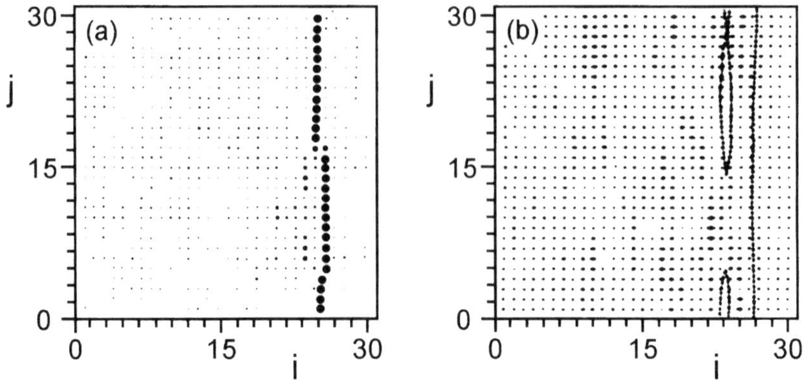

Fig. 11.2. Contour plot of u_{l_x,l_y} for the scalar FK model (11.1)–(11.2) driven by dc external force F in presence of external damping η. The size of the dots indicate the atomic velocity. Initial configuration corresponds to one DW (y-chain) in rest. Pictures are calculated by molecular dynamics technique for the underdamped case ($\eta = 0.085$ in dimensionless units) for two values of the driven force F: (a) $F = 1.1$ at $t = 400$ when a single y-kink–y-antikink pair is nucleated, and (b) $F = 1.2$ at $t = 300$ when loops are created in the wake of the propagating wall [230].

damped case a driven DW becomes unstable against spontaneous roughening for certain values of the driving force.

As is known from the dislocation theory, the motion of an extended object such as the y-chain (domain wall) essentially depends on the presence of impurities which may "pin" the y-chain. When a moving DW meets with a "stopper", for the further motion of the y-chain it should be either "tear off" the stopper (this needs to overcome the bonding energy of the chain's atom with the impurity), or the y-chain should be "broken" (leaving one of its atoms at the impurity site), the later requires the energy $E_{\rm dis}$. In connection with surface diffusion these questions were considered by Lyuksyutov and Pokrovsky [816].

11.3 Zigzag Model

The classical FK model as well as all its generalized versions discussed above in this book, had one serious restriction: the atoms were allowed to move in one dimension only. Thus, a natural further development of the model is to remove this restriction and to allow for the atoms to have two or more degrees of freedom. In the simplest case let us assume that every atom can move in two dimensions, say x and y, and place the atoms into a two-dimensional external potential which is periodic in one direction (x) and unbounded (for example, parabolic) in the transverse direction y so that the atoms are confined transversely. Thus, the substrate potential is $V_{\rm sub}(\mathbf{r}) = V_x(x) + V_y(y)$, where $V_x(x) = (\varepsilon_s/2)[1 - \cos(2\pi x/a_s)]$ and $V_y(y) = (m_a/2)\omega_{0y}^2 y^2$ Here ω_{0y} is the frequency of a single-atom vibration in the transverse direction. The limit $\omega_{0y} \to \infty$ corresponds to the standard FK model. As everywhere in the book, we use units such that $m_a = 1$, $\varepsilon_s = 2$, and $a_s = 2\pi$ so that the frequency of longitudinal vibrations $\omega_{0x} = (\epsilon_s/2m_a)^{1/2}(2\pi/a_s)$ is equal to 1.

The model with a transverse degree of freedom is not of academic interest only. First, this model may be considered as the simplest approximation for an *anisotropic* vector 2D FK model where we could expect that a small interaction between the nearest neighboring rows does not modify significantly the properties of the system. Second, the model may be considered as a first step in studying of the *isotropic* vector 2D FK model. Finally, the model is important itself, because various nonlinear phenomena in solid-state physics can be described by a model where a chain of interacting particles is placed into a "channel".

The model with the transverse degree of freedom still describes the 1D atomic chain. However, when we consider a *vector* model, we lose the main advantage of the 1D FK model in which the atoms are always strictly ordered and, therefore, they can be labelled in such a way that the atom l has always the atoms $l \pm 1$ as nearest neighbors. In a realistic 2D model the sequential order of the atoms can be changed by going through the second dimension. Also, because of losing of the sequential order, a rigorous vector model must take into account the interaction between *all* atoms in the system. Nevertheless, the ideology of topologically stable quasiparticles (kinks) of the classical FK model still remains very useful and instructive.

As long as the atoms use only an attractive branch of the interaction potential, static properties of the model are equivalent to those of the standard FK model. In a number of physical systems, however, the interatomic interaction is repulsive or has a repulsive branch at least. It can be due to Coulomb repulsion between ions in superionic conductors or between protons in hydrogen-bonded molecules, or due to Coulomb or dipole-dipole repulsion of atoms adsorbed on semiconductor or metal surfaces. Because qualitative results do not depend on the specific form of the interatomic potential provided that the repulsion is concave, below for concreteness we assume a Coulomb repulsion

$$V_{\rm int}(r) = V_0/r, \tag{11.14}$$

where V_0 characterizes the amplitude of repulsion. Clearly, for the repulsive interaction we have to study the fixed-density system, i.e. to impose periodic boundary conditions.

When the magnitude of the interatomic repulsion increases (or the curvature of the transverse potential decreases) above a certain threshold value, the trivial ground state [TGS, Fig. 11.3(a)] of the model undergoes a series of bifurcations. The first bifurcation always leads to a zigzag ground state [ZGS, Fig. 11.3(b)] and results in drastic change of system properties including a

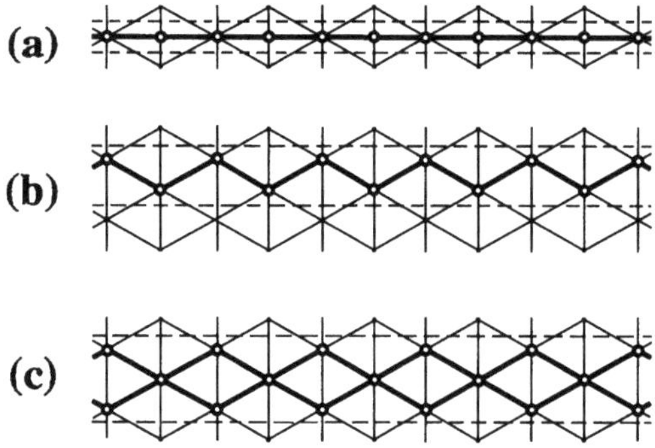

Fig. 11.3. Configurations of atoms in the 2D FK models with a transverse degree of freedom. Shown are separate stripes corresponding to: (a) "Trivial Ground State" (TGS); (b) "Zigzag Ground State" (ZGS), and (c) "Rhomboedric" configuration of atoms [817].

cusp in the effective elastic constant. For an incommensurate system, the bifurcations can interplay with the Aubry transition from a pinned to a sliding state. Moreover, the zigzag ground state can even cause multiple Aubry transitions (see Sect. 11.3.2). On the other hand, for a commensurate system the zigzag model may have more than one kind of kinks. In Sect. 11.3.3 we give a general scheme of classification of topological excitations for a complex ground state, and in Sect. 11.3.4 the kinks for the ZGS are discussed. Close to the bifurcation point, as well as above this point, dynamic properties of the zigzag FK model are deeply changed. All these issues are relevant for realistic systems as will be discussed briefly in Sect. 11.3.5.

11.3.1 Ground State

To find the GS configuration at $T = 0$, we have to look for the absolute minimum of the total potential energy of the system,

$$U = \sum_{i=1}^{N} \left[V_{\text{sub}}(\mathbf{r}_i) + \frac{1}{2} \sum_{i'=1}^{N^*} [V_{\text{int}}(|\mathbf{r}_i - \mathbf{r}_{i+i'}|) + V_{\text{int}}(|\mathbf{r}_i - \mathbf{r}_{i-i'}|)] \right], \quad (11.15)$$

where the index i labels the atoms, and the limit $N, N^* \to \infty$ is assumed. When the GS coordinates $\{\mathbf{r}_i^{(0)}\} \equiv \{x_i^{(0)}, y_i^{(0)}\}$ have been found, one can investigate a stability of the GS by calculating the phonon spectrum. The spectrum is determined by the eigenfrequencies of the elastic matrix $\mathbf{A}$ defined as

$$A_{ii}^{\alpha\alpha'} = \left(\frac{\partial^2 V_{\text{sub}}(\mathbf{r}_i)}{\partial u_i^\alpha \partial u_i^{\alpha'}} + \sum_{i'(i'\neq i)} \frac{\partial^2 V_{\text{int}}(|\mathbf{r}_i - \mathbf{r}_{i'}|)}{\partial u_{i'}^\alpha \partial u_{i'}^{\alpha'}} \right)_{\text{all } \mathbf{u}=0},$$

$$A_{ii'(i\neq i')}^{\alpha\alpha'} = \left(\frac{\partial^2 V_{\text{int}}(|\mathbf{r}_i - \mathbf{r}_{i'}|)}{\partial u_i^\alpha \partial u_{i'}^{\alpha'}} \right)_{\text{all } \mathbf{u}=0},$$

where the Greek indices correspond to Cartesian coordinates ($\alpha, \alpha' = x$ or y), and the vector $\mathbf{u}_i = \mathbf{r}_i - \mathbf{r}_i^{(0)}$ describes the displacement of the i-th atom from its equilibrium position. If the dimensionless concentration θ is rational so that the GS configuration is commensurate, the atomic structure is periodic, and it is convenient to make the Fourier transform

$$D_{mm'}^{\alpha\alpha'}(k) = \sum_{l=-\infty}^{\infty} A_{0,m;\,l,m'}^{\alpha\alpha'} e^{ikla}. \tag{11.16}$$

Here a is the period of the GS structure, and the atomic index i is split into two subindexes, $i = (l, m)$, where l labels the elementary cells while $m = 1, \ldots, s$ denotes the atoms within the cell. For each momentum k ($|k| \le \pi/a$), $\mathbf{D}(k)$ is a $2s \times 2s$ square matrix and the phonon spectrum consists of $2s$ branches labelled by an index j. The frequencies $\omega_j(k)$ are determined by the eigenvalue equation

$$\det\left[\omega^2(k)\mathbf{1} - \mathbf{D}(k)\right] = 0. \tag{11.17}$$

For a stable configuration all eigenfrequencies must be positive. When, for some model parameters, one of the frequencies vanishes, $\omega_j(k^*) = 0$, the corresponding configuration becomes unstable and evolves into a new configuration with the period $a^* = \pi/k^*$. The eigenvector associated to the vanishing frequency helps finding the new GS. This scenario corresponds to a continuous (second-order) phase transition. Besides, the model may also exhibit discontinuous (first-order) transitions when a model parameter (e.g., V_0) is changed. They occur when the energy of a metastable configuration becomes equal to the energy of the GS configuration at a transition point $V_0 = V_{\text{bif}}^{(m)}$, and beyond this point the metastable and GS configurations are exchanged.

In the standard FK model the dimensionless elastic constant

$$g = (a_s^2/2\pi^2\varepsilon_s)V_{\text{int}}''(a_A)$$

plays a central role. In the 2D model, the average elastic constant

$$g_{\text{eff}} = \frac{1}{N}\sum_{i=1}^{N} g_i \tag{11.18}$$

with

$$g_i = \frac{1}{2}\sum_{i'=1}^{N^*}\left[\frac{\partial^2}{\partial x_i^2}\left(V_{\text{int}}(|\mathbf{r}_i - \mathbf{r}_{i+i'}|) + V_{\text{int}}(|\mathbf{r}_i - \mathbf{r}_{i-i'}|)\right)\right]_{\text{all }\mathbf{u}=0}, \tag{11.19}$$

plays a similar role (for the standard FK model g_{eff} coincides with g).

Let us consider the simplest case of the trivial commensurate concentration $\theta = 1/q$ with an integer q, so that the average interatomic distance is $a_A = qa_s$. At small amplitude of interatomic repulsion the GS is trivial (TGS), and all atoms are situated at the bottoms of the corresponding wells, $x_i^{(0)} = ia_A$ and $y_i^{(0)} = 0$. The elementary cell of the system contains one atom only, and the phonon spectrum consists of two branches. The first branch corresponds to motion along the x direction,

$$\omega_1^2(k) = \omega_{0x}^2 + 2\sum_{l=1}^{\infty} V''_{\mathrm{int}}(la_A)[1 - \cos(kla_A)], \tag{11.20}$$

and the second branch, to motion along the y direction,

$$\omega_2^2(k) = \omega_{0y}^2 + 2\sum_{l=1}^{\infty}[V'_{\mathrm{int}}(la_A)/(la_A)][1 - \cos(kla_A)], \tag{11.21}$$

where $|k| \leq 1/2q$. Notice that the motions along the x and y directions are decoupled in the TGS.

When the magnitude of the interaction increases for repulsive interatomic interactions, $V'_{\mathrm{int}}(la_A) < 0$, the frequency $\omega_2(k)$ decreases and reaches zero at some critical value $V_0 = V_{\mathrm{bif}}$. If the interatomic repulsion decreases monotonically with increasing r [i.e. if $V'_{\mathrm{int}}(r)$ is always negative], the first instability emerges at the momentum $k = \pm 1/2q$. The corresponding bifurcation value V_{bif} can be determined from the equation

$$\omega_{0y}^2 + 4\sum_{p=0}^{\infty} \frac{V'_{\mathrm{int}}[(2p+1)a_A]}{(2p+1)a_A} = 0. \tag{11.22}$$

In particular, for the Coulomb repulsion (11.14) V_{bif} is equal to $V_{\mathrm{bif}} = \omega_{0y}^2 a_A^3/4C$, where $C \equiv \sum_{p=0}^{\infty}(2p+1)^{-3} = 1.05179\ldots$

Thus, for any monotonically decreasing interatomic repulsion the first bifurcation leads always to a *continuous* transition from the trivial ground state of Fig. 11.3(a) to the zigzag ground state of Fig. 11.3(b) with atomic coordinates $x_i^{(0)} = ia_A$, $y_i^{(0)} = (-1)^i b$. The amplitude b of the transverse atomic shifts is determined by the equation

$$\omega_{0y}^2 + 4\sum_{l=0}^{\infty} V'_{\mathrm{int}}(r_l)/r_l = 0, \quad r_l = [4b^2 + a_A^2(1+2l)^2]^{1/2}. \tag{11.23}$$

In the ZGS the value of the transverse splitting b increases with V_0 as is shown in Fig. 11.4(b). The elastic constant g_{eff} for the TGS increases linearly with V_0 up to the value $g_{\mathrm{bif}} \approx 9\,\omega_{0y}^2/16$. But after the bifurcation g_{eff} decreases, reaches the local minimum $g_{\min} \approx 4.7\,\omega_{0y}^2/16 \sim 0.5\,g_{\mathrm{bif}}$ and then rises again [see Fig. 11.4(c)]. The cusp of the elastic constant g_{eff} at the bifurcation point V_{bif} explains many remarkable properties of the model.

With further increase of V_0 (or decrease of ω_{0y}) the system undergoes next transitions to more complicated GS configurations. For a symmetric transverse potential such as the parabolic one, these transitions, contrary to the

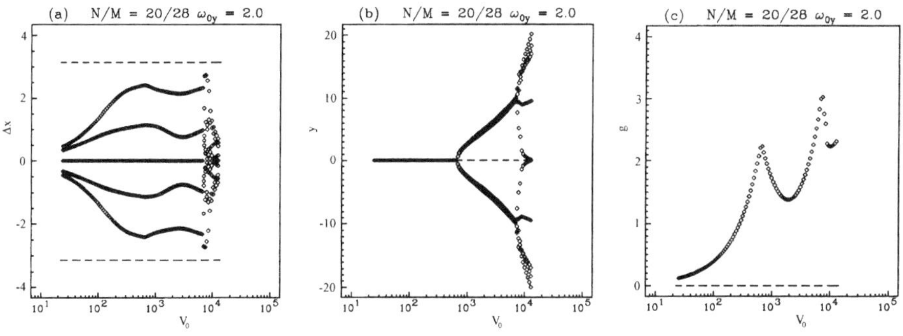

Fig. 11.4. (a) Longitudinal displacements Δx of atoms with respect to nearest minima of $V_x(x)$, (b) transverse displacements y, and (c) average elastic constant of the chain as functions of V_0 for $\theta = 5/7$ ($N = 20$, $M = 28$, $N^* = 60$) and $\omega_{0y} = 2$. The figure shows two bifurcations of the atomic structure [817].

first transition to the ZGS which was continuous, are discontinuous. Namely, close to a transition point both configurations, e.g. the zigzag and rhomboedric ones, are stable, but one of them corresponds to the GS while another to a metastable configuration with a higher energy. At the transition point, the energies of both configurations become equal one another. The sequence of the GS structures can be described qualitatively in the following way. When the longitudinal substrate potential is negligible (e.g., when $\omega_{0y} \gg \omega_{0x}$), the sequence of the ground state configurations may be viewed as broader and broader stripes cut out from the 2D hexagonal lattice as shown in Fig. 11.3, where the longitudinal lattice constant $a_{\rm cell}$ is determined by the atomic concentration θ [$a_{\rm cell} = (a_s/\theta)s'$, $s' = 2, 3, 4, 5, \ldots$ for the zigzag, rhomboedric, double zigzag, hexagonal, etc. configurations respectively], while the transverse atomic displacements evolve to adjust to the transverse potential $V_y(y)$. On the other hand, for an asymmetric transverse potential such as the Toda potential,

$$V_y(y) = \omega_{0y}^2 y_{\rm anh}^2 \left[\exp(-y/y_{\rm anh}) + (y/y_{\rm anh}) - 1\right], \tag{11.24}$$

where $\beta = y_{\rm anh}^{-1}$ is the anharmonic constant, not only the first bifurcation but also few higher-order ones, may be continuous [817].

11.3.2 Aubry Transitions

The cusp singularity of $g_{\rm eff}(V_0)$ in the model with transverse degree of freedom has also a strong effect on the Aubry transition in the incommensurate case. When the elastic constant g increases in the classical FK model, the incommensurate structure exhibits the Aubry transition which can be understood in terms of atoms' positions with respect to the maxima of the substrate potential. For small interactions (i.e. for $g < g_{\rm Aubry}$), the on-site potential dominates. The atoms tend to stay near the bottoms of the wells, and there is a forbidden region near the maxima. As a result, the atomic chain is pinned to the substrate because its translation requires moving the atoms up in the substrate potential, over the maxima. On the contrary, for $g > g_{\rm Aubry}$, the configuration is dominated by the interaction and all the values of the substrate potential are occupied by atoms, some of them being on the top of the barrier. When the atomic chain is translated over the substrate, some atoms go down in the potential while others go up and there exists no barrier to the translation.

The question which emerges is to what extent the introduction of a transverse degree of freedom can affect the Aubry transition. The answer can be derived from the behavior of the elastic constant $g_{\rm eff}$ versus V_0, as described above. Because the bifurcation point $V_{\rm bif}$ is determined by the curvature of the transverse substrate potential, $V_{\rm bif} \propto \omega_{0y}^2$, while the Aubry transition concerns essentially the longitudinal displacements, the relative positions of the two transitions can change depending on model parameters. The following three different scenario for the behavior of the system with increasing V_0 can be predicted:

(a) The case of ω_{0y} below a first threshold ω^*, $\omega_{0y} < \omega^*$, such that $V_{\rm bif} < V_{\rm Aubry}$. In this case the Aubry transition will not occur at all because the first bifurcation, to a zigzag state, which reduces the effective interatomic coupling, occurs before $g_{\rm eff}$ can reach the magnitude $g_{\rm Aubry}$ required for the Aubry transition. Taking $g_{\rm Aubry} \simeq 1$ for the golden mean $\theta_{\rm g.m.}$, the value ω^* can be estimated as $\omega^* \approx 4/3$;

(b) The case of large ω_{0y}, $\omega_{0y} > \omega^{**}$ (where ω^{**} is a second characteristic value), for which $V_{\rm bif} \gg V_{\rm Aubry}$ and $g_{\rm Aubry} < g_{\rm min} \equiv \min g_{\rm eff}(V_0)$ for all V_0 within the ZGS. In this case the Aubry transition is observed when V_0 reaches $V_{\rm Aubry}$ and the bifurcation which occurs later does not bring any qualitative change in the system behavior because the minimum of $g_{\rm eff}$ after the bifurcation is above $g_{\rm Aubry}$. Only a higher-order bifurcation (e.g., to the RGS) could bring a qualitative change. Taking $g_{\rm min} \simeq 0.5\, g_{\rm bif}$, the value ω^{**} can be estimated as $\omega^{**} \approx 4\sqrt{2}/3 \approx 1.9$ for $\theta = \theta_{\rm g.m.}$;

(c) For intermediate ω_{0y}, $\omega^* < \omega_{0y} < \omega^{**}$, when V_0 is increased, the system undergoes first the Aubry transition in which the pinned GS is transformed to a sliding GS. But then the bifurcation to the zigzag state can reduce the average elastic constant below $g_{\rm Aubry}$. The system undergoes a *reverse* Aubry transition and the lattice gets pinned to the substrate again. The further increase of $g_{\rm eff}$ which occurs after the minimum can cause again a direct Aubry transition restoring the sliding state, at least up to the second bifurcation.

These predictions were checked numerically by Braun and Peyrard [817] for the chain of $N = 34$ and $M = 47$, which is close to $\theta'_{\rm g.m.} = (3+\sqrt{5})/(5+\sqrt{5})$ equivalent to the golden mean for the Aubry transition. In the case of the transverse frequency $\omega_{0y} = 1$ for all values of V_0 the elastic constant did not reached the threshold value $g_{\rm Aubry} \simeq 1$, and the GS is pinned for all V_0. For the case of $\omega_{0y} = 2$ the Aubry transition took place in the TGS. Then, after the bifurcation the atomic displacements decreased slightly however, up to the second bifurcation, simulation showed atoms close to potential maxima, so the sliding phase persists up to the second bifurcation point. A more sensitive test was provided by the calculation of the linear response of the atomic chain to an external dc force F applied to all the atoms, along the direction of the x axis. Calculating the mean shift of the atoms, $\Delta x_{\text{shift}} = N^{-1}\sum_{i=1}^{N}(x_i - x_i^{(0)})$, with respect to their positions without the force, $x_i^{(0)}$, we can define a susceptibility as $\chi = \Delta x_{\text{shift}}/F$. One can expect $\chi \to \omega_{0x}^{-2}$ in the limit $V_0 \to 0$, and $\chi \to \infty$ in the sliding state. Because in a numerical simulation the value θ is always rational, the GS will stay slightly pinned even above the Aubry transition. The results of the simulation have shown a sharp drop of χ^{-1} when the Aubry transition was reached and then χ^{-1} stays practically equal to zero ($\chi^{-1} \approx 5\,10^{-4}$) for the whole range $V_{\text{Aubry}} < V_0 < V_{\text{bif}}^{(2)}$, so that the GS in this case is really very close to the sliding state in spite of the finiteness of the system.

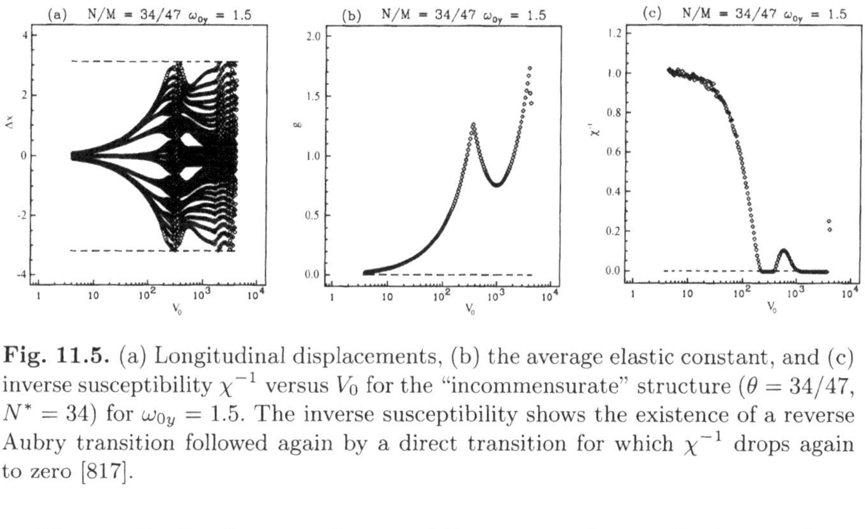

Fig. 11.5. (a) Longitudinal displacements, (b) the average elastic constant, and (c) inverse susceptibility χ^{-1} versus V_0 for the "incommensurate" structure ($\theta = 34/47$, $N^* = 34$) for $\omega_{0y} = 1.5$. The inverse susceptibility shows the existence of a reverse Aubry transition followed again by a direct transition for which χ^{-1} drops again to zero [817].

The results for the case of $\omega_{0y} = 1.5$ corresponding to the intermediate case (c), are shown in Fig. 11.5. As in the previous case, the Aubry transition takes place before the bifurcation, and then g_{eff} continues to rise to the value $g_{\text{bif}} = 1.189$ [see Fig. 11.5(b)]. But after the bifurcation which takes place at $V_{\text{bif}} = 381$, the elastic constant g_{eff} decreases and reaches the minimum $g_{\min} = 0.771$ which is lower than g_{Aubry} at $V_0 = 1079$. Fig. 11.5(b) shows that, in the vicinity of the minimum of the function $g_{\text{eff}}(V_0)$, the atomic displacements decrease strongly and there is not any more an atom on top of the maxima of $V_x(x)$. Thus, in the region near the minimum of $g_{\text{eff}}(V_0)$ the GS is pinned again. This is confirmed by the behavior of the susceptibility [Fig. 11.5(c)] which attests that one have observed the predicted reverse Aubry transition from the sliding to the pinned state at $V_0 = 578$ when $g_{\rm eff} = 0.898$. With further increase of V_0, after reaching the minimum the elastic constant increases again, and at $V_0 = 879$ when $g_{\rm eff} = 782$ the system undergoes a second direct Aubry transition before the second bifurcation. After the second bifurcation, the state becomes pinned forever.

11.3.3 Classification of Kinks

Next step of study of the zigzag model is to investigate its topological excitations, which are responsible for many aspects of the system behavior, in particular, for mass transport. Recall that the standard FK model has only two types of kinks, the kink which describes the minimally possible topologically-stable compression of the chain, and the antikink which describes the analogous expansion of the chain. The simplest case corresponds to the trivial ground state with $\theta = 1$, so that the lattice is Bravais and each minimum of the external potential is occupied by one atom in the GS. In this case the kink (antikink) configuration describes an extra atom (vacancy) inserted into the chain, when all other atoms are relaxed in order to adjust to the created local perturbation. But when $\theta = r/p$ (r and p being integers) with $r \neq p$, the elementary cell of the crystalline GS is non-Bravais (i.e., it contains more than one atom), and the situation becomes nontrivial even for the standard FK model. In particular, now the kink is characterized by a fractional atomic number p^{-1}, so that one additional atom inserted into the chain, produces p kinks. For the zigzag model the determination of a structure of topological excitations is a complicated problem. First, there may exist excitations of different kinds. Second, a question emerges how to find the *elementary* excitations which then may be used for construction of any other topological excitation. Below we describe, following the paper of Braun *et al.* [818], a general procedure which allows to find these elementary excitations for any complex GS.

Let us consider a complex GS of the FK model with $\theta = r:p$, so that the period of the GS structure is p (here we take the period of the external potential as the unit of length), and each elementary cell of the GS consists of r atoms (we use the notation $\theta = r:p$ instead of $\theta = r/p$ in order to emphasize that r and p for nontrivial GS may have a common divisor as, for example, in the ZGS, where $r = p = 2$). Then, the idea is to treat the complex $\theta = r:p$ GS of the chain as that consisting of r subsystems (subchains), each being characterized by the trivial structure. In a single subchain we may create kinks (subkinks) if we simply shift the right-hand side of the subchain for an integer number of periods of the substrate potential. In this way we may consider any topological excitation of the whole system as that constructed of subkinks. However, the subchains strongly interact with each other. Consequently, many combinations of subkinks are forbidden, because the right-hand side of the chain must correspond to a true GS configuration.

Thus, the problem reduces to looking for allowed combinations of subkinks, and then to distinguishing those combinations which correspond to elementary ones, so that any other combination can be constructed of the elementary ones.

The whole symmetry group $\mathcal{S}$ of the $\theta = r:p$ GS may be splitted into two subgroups, $\mathcal{S} = \mathcal{F} \bigotimes \mathcal{G}$. The first subgroup $\mathcal{F}$ is Abelian, $\mathcal{F} \equiv \{T^n\}$, where $n = 0, \pm 1, \ldots$ It is generated by the operator T which describes the translation of the chain as a whole for the distance p (i.e., for one period of the GS structure). The second subgroup $\mathcal{G}$ is the finite ("point") group which describes a local symmetry of the complex elementary cell. For the trivial GS, where all atoms are aligned to a line, $\mathcal{G}$ is the cyclic group consisting of p elements. $\mathcal{G}$ is generated by the operator G, $\mathcal{G} \equiv \{G^l\}$, $l = 0, 1, \ldots, p-1$ and $G^p = G^0 = 1$, where G corresponds to the translation of the chain as a whole for the unit distance (i.e., for one period of the substrate potential). On the other hand, when the GS is nontrivial, the point group $\mathcal{G}$ includes additionally the element J ($J^2 = 1$) which describes the "inversion" of the GS. For example, in the zigzag-FK model above the first bifurcation point, when the atoms in the GS are shifted from the line in the transverse direction, the action of J on the GS produces the "mirror image" of the state with respect to the chain's line. Thus, for the case when r and p are not relative prime and r is even, the GS is additionally doubly degenerated.

In a general case, in order to create a topological excitation in the chain, we have to choose an element of the whole symmetry group and to act by this operator on the GS thus obtaining a new GS configuration, and then to look for the kink configuration which links the old and new ground states, i.e. to find the minimum-energy configuration with the boundary conditions at infinities when the left-hand side of the chain is kept in the old GS, while the right-hand side, in the new GS.

Although a total number of topological excitations is infinite (but countable), all of them may be constructed of few kinds of "elementary kinks". To find the structure of elementary kinks, let us consider the whole system as that constructed of r subchains. The each subchain being considered independently from other subchains, has the trivial GS configuration, i.e. the elementary cell of the subchain contains one atom only. Analogously as it was done above, we can define the translation operators G_i and T_i, $i = 1, 2, \ldots, r$ acting on the i-th subchain only. Any element $S^{(\alpha)}$ of the whole symmetry group, $S^{(\alpha)} \in \mathcal{S}$, may now be presented as a product of elements of the subchain's subgroups $\mathcal{G}_i$ and $\mathcal{F}_i$,

$$S^{(\alpha)} = \prod_{i=1}^{r} G_i^{g_i^{(\alpha)}} \bigotimes T_i^{t_i^{(\alpha)}}, \tag{11.25}$$

where $g_i^{(\alpha)}$ and $t_i^{(\alpha)}$ are integers. But the contrary statement is not true, the set of all products $\prod_i G_i^g \bigotimes T_i^t$ exceeds the set $\mathcal{S}$. Indeed, because the subchains are strongly interacting, a relative arrangement of the subchains in the GS must not be violated, and this leads to a constrain on the admitted values of the integers $g_i^{(\alpha)}$ in Eq. (11.25). Namely, the following condition must be fulfilled for all subchains simultaneously:

$$g_i^{(\alpha)} \bmod p = g^{(\alpha)}, \quad i = 1, \ldots, r. \tag{11.26}$$

If we will name $g_i^{(\alpha)} \bmod p$ as the "color" of the i-th subchain, the condition (11.26) means that in the GS all subchains must have the same color $g^{(\alpha)}$.

In the way described above, we may construct any topological excitation of the system. Recalling that the operator G_i being applied to the right-hand side of the i-th subchain, creates the subkink in this subchain (and, analogously, the inverse operator G_i^{-1} creates the sub-antikink), we see that any topological excitation may be treated as that consisting of a corresponding set of subchain's subkinks. Because the GS of an isolated subchain is trivial and is characterized by the dimensionless concentration $\theta_i = 1/p$, a single subkink (sub-antikink) has the topological charge p^{-1} (or $-p^{-1}$). Therefore, a topological excitation of the whole system can be characterized by the topological charge $Q^{(\alpha)} = q_{\rm tot}^{(\alpha)}/p$, where

$$q_{\rm tot}^{(\alpha)} = \sum_{i=1}^{r} \left(g_i^{(\alpha)} + p t_i^{(\alpha)} \right). \tag{11.27}$$

Thus, the only question which remains still open, is how to classify the topological excitations, i.e. to select those excitations which may be considered as the simplest, or elementary ones. Taking into account that topological charges (11.27) are additive, so that the topological charge of a complex excitation is the sum of topological charges of the elementary excitations, it is not difficult to guess that the elementary excitations should correspond to those with minimum topological charges such as $q_{\rm tot} = 0$ and 1 or 2.

From Eqs. (11.26) and (11.27) it follows that $q_{\rm tot}$ may be presented as

$$q_{\rm tot} = rg + ph, \tag{11.28}$$

where r and p are the given integers determined by the concentration θ, the color g must be within the interval

$$1 \le g < p, \tag{11.29}$$

and h is an integer. So, the problem reduces to looking for such integers g [from the interval (11.29)] and h, which minimize the absolute value of $q_{\rm tot}$ defined by Eq. (11.27) for the given $\theta = r\!:\!p$. Let us proceed further in two steps.

Step 1. Suppose that the minimum $|q_{\rm tot}|$ is equal one. In this case Eq.(11.28) takes the form (we change here $h \to -h$ and put $q_{\rm tot} = -1$ for the sake of convenience)

$$ph = 1 + rg. \tag{11.30}$$

Suppose also that r and p are relative simple, so that the integer equation

$$ph = rg \tag{11.31}$$

has no solutions.

Let us put p points on a circumference and numerate them from 0 to $p-1$ as shown in Fig. 11.6. Then, let us begin from the point number 1, $a_0 = 1$, and make the anticlockwise revolution moving by "large" steps, each of r unit steps. After the first turn made in a few "large" steps, we come to a point a'_1 which is the most close one to the initial point a_0. Calculate the integer

$$s_1 = (p - a_0) \bmod r. \tag{11.32}$$

If $s_1 = 0$, then $a'_1 = 0$ (recall the points with numbers p and 0 coincide), and Eq. (11.30) has the solution with $h = 1$, while the color g is determined by the number of "large" steps during the first turn.

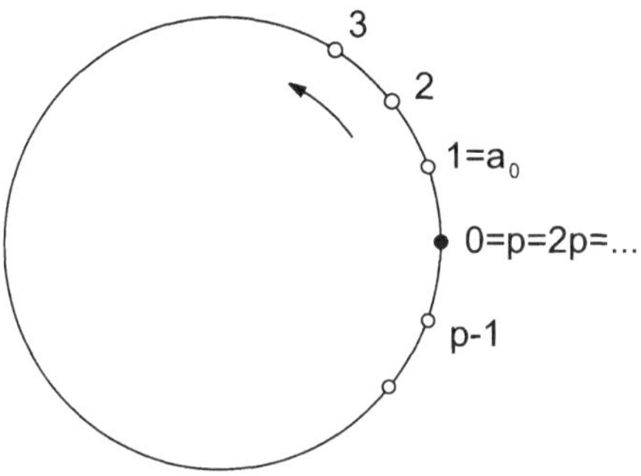

Fig. 11.6. Construction of elementary topological excitations.

Otherwise, if $s_1 \neq 0$, then $a'_1 = p - s_1$, and the next point in the anticlockwise direction is $a_1 = r - s_1 \neq a_0$ [to prove that $a_1 \neq a_0$, let us suppose that $a_1 = a_0$, and then we get $r - s_1 = 1$, $(p-1) \bmod r = r-1$, $p-1 = jr+r-1$ (j is an integer), and finally $p = (j+1)r$ which contradicts with the assumption that Eq. (11.31) has no solutions]. Then, starting from the point a_1, after the second revolution we come to a point a'_2. Calculate

$$s_2 = (p - a_1) \bmod r. \tag{11.33}$$

If $s_2 = 0$, we have $a'_2 = 0$, so that $h = 2$ in Eq. (11.30). Otherwise, the next point is $a_2 = r - s_2$. Again in the same way one can prove that $a_2 \neq a_1$ and also $a_2 \neq a_0$. Thus, after each turn we come to a new point in the circumference. But because the number of these points is finite (equal to p), after a finite number of steps we finally come to the point with the number 0, and the number of turns just gives the value of h. So, we have proved that Eq. (11.30) always has a solution provided Eq. (11.31) has no solutions, and have shown how to find it.

Step 2. Now let us suppose that Eq. (11.31) has a solution. It is easy to see that it is true, if and only if p and r have a common divisor $j_0 > 1$, i.e. if

$$p = p_0 j_0 \quad \text{and} \quad r = r_0 j_0. \tag{11.34}$$

Indeed, represent the product $\Pi = ph = rg$ as

$$\Pi = \underbrace{i_1 \ldots i_m}_{r} \underbrace{R \overbrace{i_{m+1} \ldots i_n}^{p}}_{g}, \tag{11.35}$$

where $i_1, \ldots, i_n$ are simple integers. From (11.35) we see that if r and p have no common divisors, it should be $g = Rp$ with $R > 1$, but this is forbidden due to the restriction (11.29).

Let j_0 corresponds to the greatest common divisor. If we now put $g = p_0$ and $h = -r_0$ in Eq. (11.28), we obtain $q_{\text{tot}} = 0$, i.e. we have found one kind of kinks with zero total topological charge. Putting $q_{\text{tot}} = 0$ in Eq. (11.27) and taking into account Eq. (11.26), one can see that the elementary excitations should correspond to the integers $g = p/2$ and $h = r/2$.

Besides, in the present case there exist also solutions with nonzero topological charge equal to $Q = j_0/p$. Their structure can be found from a solution of the equation

$$p_0 h = 1 + r_0 g \tag{11.36}$$

similarly as it was done above in Step 1, because now the integer equation $p_0 h = r_0 g$ has no solutions.

To summarize, when the point group $\mathcal{G}$ does not include the inversion operator J so that r and p are relative simple, the $\theta = r\!:\!p$ GS admits the existence of a single kind of elementary topological excitation, the SG-type kink with the topological charge $Q = 1/p$. The kink structure in this case may be found from a solution of the integer equation (11.30).

Otherwise, when r and p are not relative prime and have the greatest common divisor j_0, the $\theta = r\!:\!p$ GS supports the existence of two kinds of kinks. The first kind is the ϕ^4-type kink, it has the topological charge $Q = 0$, and its structure is characterized by the color $g = p/2$. The second kind of kinks is the SG-type kink with the topological charge $Q = j_0/p$, and its structure is determined by Eq. (11.36). We will call these two kinds of kinks as the "massive" kink (*MK*) and the "nonmassive" kink (*NMK*) correspondingly, because mass (charge) transport along the chain may be carried out only by kinks with nonzero topological charge, i.e. by the "massive" kinks. Note that in the zigzag model there is no constrains on a sequence of kinks of different kinds.

If we denote by k ($\bar{k}$) the subkink (sub-antikink) in a subchain, the elementary topological excitation K of the whole system may be represented as a set of r elements such as $K = \{g_1 k, g_2 k, \ldots, g_r k\}$. For example, Fig. 5.4 shows the structure of elementary kinks for the trivial $\theta = 3\!:\!5$ GS. Because

3 and 5 are relative simple, in this case we have only one kind of kinks, the SG-type kink with the topological charge $Q = 1/5$. Equation (11.30) in this case has the solution for $g = 2$ and $h = 1$, so that the kink structure is characterized by $g_1 = 2$, $g_2 = 2$ and $g_3 = -3$, or $K = \{2k, 2k, 3\bar{k}\}$. It is interesting that this structure essentially differs from that which might be expected from a naive approach. Indeed, if we, following the kink definition as the minimally possible compression of the chain, simply compress the chain by shifting its right-hand side for one period of the substrate potential to the left, we create the topological excitation with the topological charge $Q = 3/5$ and the structure $\{k, k, k\}$, which then has to be splitted into three elementary excitations,

To illustrate the conclusions made above, let us describe the kinks for the zigzag model. The ZGS is additionally doubly degenerated, $j_0 = 2$, and we have two kinds of kinks, the "massive" kink $MK = \{k, k\}$ with $Q = 2/p$ and the "nonmassive" kink $NMK = \{k, \bar{k}\}$ with $Q = 0$. In a more complicated case of the "rhomboidal" GS, e.g., for $\theta = 3{:}4$, which arises after the second bifurcation, the solution of Eq. (11.30) is $h = g = 1$, so that the kink structure is $K = \{\bar{k}, 3k, \bar{k}\}$.

Note that all topological excitations with the same total topological charge are identical from the topological point of view. For example, the "massive" kink for the "double-zigzag" GS has the structure $K = \{k, k, k, k\}$. But the configurations $K = \{k, 5k, 3\bar{k}, k\}$ and $K = \{3\bar{k}, 5k, 3\bar{k}, 5k\}$ describe the same topological excitation as well. Besides, any subchain may contain additionally any number of k-$\bar{k}$ pairs. All these configurations are different from the physical viewpoint, in particular, they may be characterized by different potential energies. One of them corresponds to minimum of the system potential energy, others may correspond to local minima or saddle configurations. Because the configurations with the same topological charge may be transformed to each other in a continuous way, the strategy developed above helps to look for possible trajectories of motion of a kink along the chain. Besides, owing to intrinsic structure of kinks for a complex GS, the kinks have to have intrinsic ("shape") modes which describe oscillations of the subkinks with respect to each other. Note also that the subkinks consisted the kink, are to be spatially bounded in a localized region, because a displacement of a single subkink from the region of the kink localization leads to increasing of the system energy linearly with this displacement. Thus, in this sense the subkinks remind quarks of the field theory, while the kink, an elementary particle constructed of the quarks.

11.3.4 Zigzag Kinks

Typical structures of the massive kink and antikink and the nonmassive kink for the $\theta = 1$ zigzag GS are shown in Figs. 11.7–11.8. To find the kink parameters analytically, first let us recall that for the one-dimensional FK model the motion equations in the continuum limit, $g \equiv V''_{\rm int}(a_s) \gg 1$, reduces to the sine-Gordon equation with a correction term due to anharmonicity of the interparticle interaction (for 1D kinks we use the sub-index "0"),

$$\frac{\partial^2 u}{\partial t^2} - D_0^2 \frac{\partial^2 u}{\partial x^2}\left(1 - \alpha_0 D_0 \frac{\partial u}{\partial x}\right) + \sin u = 0, \tag{11.37}$$

where

$$D_0 = 2\pi \left[V''_{\text{int}}(a_s) + 4V''_{\text{int}}(2a_s)\right]^{1/2}, \tag{11.38}$$

$$\alpha_0 = -\left(2\pi/D_0\right)^3 \left[V'''_{\text{int}}(a_s) + 8V'''_{\text{int}}(2a_s)\right], \tag{11.39}$$

and we had taken into account the interaction of nearest and next-nearest neighbors in order to make later a comparison with zigzag kinks. Anharmonicity of the interatomic potential breaks the symmetry between a kink and antikink,

$$m \simeq \frac{4}{\pi D_0}\left(1 - \frac{\pi}{6}\sigma\alpha_0\right), \tag{11.40}$$

$$\varepsilon_k \simeq \frac{1}{2\pi}\left\{8D_0 - 2\pi^2\sigma\left[2V'_{\text{int}}(a_s) + V'_{\text{int}}(2a_s)\right] - \frac{\pi}{3}\sigma\alpha_0 D_0\right\}. \tag{11.41}$$

It is clear that the results (11.38)–(11.41) remain valid for the zigzag model as long as the GS is one-dimensional.

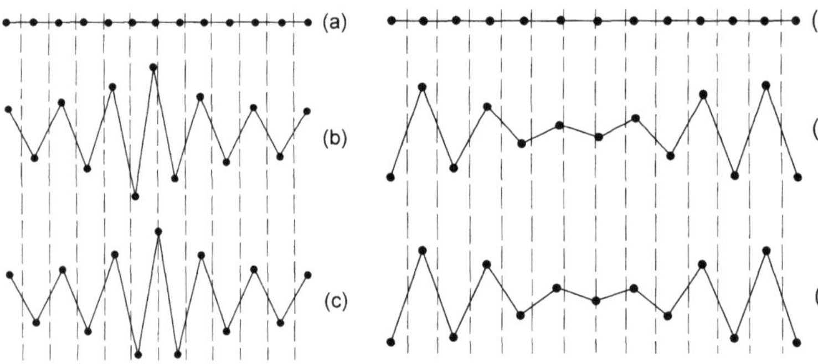

Fig. 11.7. Left: (a) One-dimensional kink and (b,c) massive kink on the background of ZGS [(b) describes the minimum-energy configuration and (c) corresponds to the saddle configuration]. The dashed lines show the positions of substrate maxima. Right: The same but for the case of a massive antikink [819].

To derive similar motion equations for kinks on the background of $\theta = 1$ ZGS, let us use, following the work of Braun *et al.* [819], the ansatz

$$x_l = la_s + u_l, \quad y_l = (-1)^l w_l, \tag{11.42}$$

and assume that the values of w_l slightly differ from b so that the parameter $z = (w_l + w_{l-1})^2 - 4b^2$ may be considered as a small one. Additionally, assume

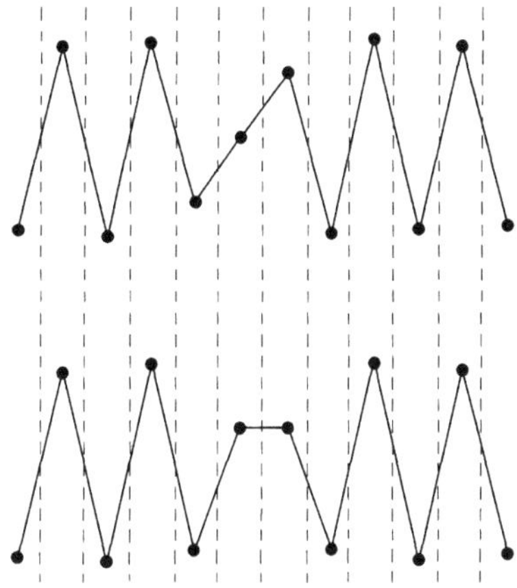

Fig. 11.8. Nonmassive kinks on the background of ZGS: (a) the minimum-energy configuration and (b) the saddle configuration [819].

that $|u_l - u_{l-1}| \sim z \ll a_s$, and expand the potential $V_{\rm int}(r)$ in a Taylor series around $r_0 = \sqrt{a_s^2 + 4b^2}$ up to the second order in small z. The reduced Hamiltonian may be written as

$$\begin{aligned} H = \textstyle\sum_l \big\{ &\tfrac{1}{2}\dot{x}_l^2 + \tfrac{1}{2}\dot{y}_l^2 + \tfrac{1}{2}\omega_{0y}^2 w_l^2 + (1 - \cos u_l) - \tfrac{1}{2}\pi\omega_{0y}^2(u_l - u_{l-1}) \\ &-\pi\omega_1^2(u_l - u_{l-2}) - \tfrac{1}{8}\omega_{0y}^2(w_l + w_{l-1})^2 + B_1(u_l - u_{l-1})^2 \\ &+B_2\left[(w_l + w_{l-1})^2 - 4b^2\right](u_l - u_{l-1}) + \tfrac{1}{8\pi}B_2\left[(w_l + w_{l-1})^2 - 4b^2\right]^2 \\ &-\tfrac{1}{8}\omega_1^2(w_l - w_{l-2})^2 + \tfrac{1}{2}g_1(u_l - u_{l-2})^2 \big\}, \end{aligned} \tag{11.43}$$

where $g = V''_{\rm int}(r_0)$, $g_1 = V''_{\rm int}(2a_s)$, $\omega_1^2 = -(2/a_s)V'_{\rm int}(2a_s)$, $B_1 = r_0^{-2}(4\pi^2 g - \omega_{0y}^2 b^2)$ and $B_2 = (\pi/4r_0^2)(\omega_{0y}^2 + 4g)$. In the continuum limit approximation, equations of motion corresponding to the Hamiltonian (11.43) may be written in the following form ($\widetilde{w} = w/b$):

$$-\delta_1^2 \frac{d^2 u}{dx^2} + \sin u = 16\pi B_2 b^2 \widetilde{w}\frac{d\widetilde{w}}{dx}, \tag{11.44}$$

$$-\delta^2 \frac{d^2\widetilde{w}}{dx^2} - \widetilde{w} + \widetilde{w}^3 = -\frac{2\pi^2}{b^2}\widetilde{w}\frac{du}{dx} - 3\pi^2\left[\widetilde{w}^2\frac{d^2\widetilde{w}}{dx^2} + \widetilde{w}\left(\frac{d\widetilde{w}}{dx}\right)^2\right] \tag{11.45}$$

(we dropped here the time derivatives since we are interesting in static properties only), where $\delta_1^2 = 8\pi^2(B_1 + 2g_1)$ and $\delta^2 = \pi^2\left[1 + (\pi/8b^2 B_2)(\omega_{0y}^2 - 4\omega_1^2)\right]$.

Therefore, for the *longitudinal* displacements the kink shape is described by a perturbed version of the SG equation while the *transverse* displacements are described by a perturbed ϕ^4-model. Indeed, let us suppose that in one direction the kink is absent. This may be reached if we demand a corresponding boundary condition in this direction and assume that one of the parameters, δ_1 or δ, is small (while the other one is large, which is necessary for the continuum limit approximation). For example, for the massive kink we should take $\delta \ll \delta_1$ and $\lim_{x\to\pm\infty} w(x) = 0$, then from Eq. (11.45) we obtain

$$w \simeq b - \frac{\pi^2}{b}\frac{du}{dx} \tag{11.46}$$

(a more rigorous approach was developed by Braun *et al.* [819]). The relation (11.46) remains valid until $b \simeq 1$. Substituting (11.46) into (11.44), we obtain the *perturbed* SG equation (11.37) but now with the parameters

$$D = 2\pi\left(4g_1 - \frac{1}{4}\omega_{0y}^2\right)^{1/2} \tag{11.47}$$

and

$$\alpha = -\frac{1}{D^3}\left[\left(\frac{\pi^2}{b}\right)^2(\omega_{0y}^2 + 4g) + (4\pi)^3 V_{\rm int}'''(2a_s)\right] \tag{11.48}$$

instead of D_0 and α_0, respectively. When α is small, the perturbation theory for the kink of SG equation can be applied, and the approximate kink solution of Eq. (11.37) can be written as

$$u(x) = 4\tan^{-1}\exp\left(-\frac{\sigma x}{D}\right) + \frac{4}{3}\alpha\frac{\tan^{-1}[\sinh(x/D)]}{\cosh(x/D)}. \tag{11.49}$$

When a kink moves along the adiabatic trajectory in the 2D model, its effective mass is defined, instead of Eq. (5.36) of the 1D model, by the expression

$$m(X) = m_a\sum_l\left[\left(\frac{\partial x_l}{\partial X}\right)^2 + \left(\frac{\partial y_l}{\partial X}\right)^2\right], \tag{11.50}$$

where X is the coordinate of the kink center. Using this definition, we obtain the effective mass of the massive kink,

$$m = \frac{4}{\pi D}\left(1 + \frac{\pi^4}{3D^2 b^3} - \frac{\pi}{6}\sigma\alpha\right). \tag{11.51}$$

Note that this expression differs from Eq. (11.40) by the second term which appears due to the *transverse degree of freedom*, it gives the main contribution to the kink mass after the leading term and *increases* the effective masses of both kink and antikink. Analogously one can calculate the rest energy of the massive kink,

$$\varepsilon_k = \frac{1}{2\pi}\left[8D + 2\pi^3\sigma(\omega_{0y}^2 + 4\omega_1^2) - \frac{\pi}{3}\sigma\alpha D\right]. \tag{11.52}$$

Unlike the case of the standard FK model, where the energies of kink and antikink were equal, in the zigzag model they are different. The most crucial difference in the energies of kink and antikink is due to the second term in Eq. (11.52), i.e. in the ZGS the kink-antikink symmetry is violated for the massive kinks even if we will use the harmonic interaction between atoms.

Moreover, contrary to the 1D model with anharmonic interactions, now the effective width of kink is *smaller* than that of antikink.

When only a nonmassive kink is present in the chain, we should take $\delta_1 \ll \delta$ and $\lim_{x\to\pm\infty} u(x) = 0$. Then from Eq. (11.44) we have $u \simeq 16\pi B_2 b^2 \widetilde{w}\widetilde{w}_x$ and, after the change of variables, $x \to \delta\widetilde{x}$, Eq. (11.45) is reduced to the perturbed ϕ^4 equation

$$-\frac{d^2\widetilde{w}}{d\widetilde{x}^2} - \widetilde{w} + \widetilde{w}^3 = -\gamma \left[\widetilde{w}^2 \frac{d^2\widetilde{w}}{d\widetilde{x}^2} + \widetilde{w} \left(\frac{d\widetilde{w}}{d\widetilde{x}} \right)^2 \right], \tag{11.53}$$

where the parameter $\gamma = (\pi^2/\delta^2)(32\pi B_2 + 3)$ is assumed to be small (for $\delta \gg 1$). Therefore, similarly to the previous case, we may apply the perturbation theory for the kink of ϕ^4 equation to obtain the approximate kink's form as follows,

$$\widetilde{w} = \sigma \left[\tanh\chi - \frac{\gamma}{2\cosh^2\chi} (\tanh\chi - \chi) \right], \tag{11.54}$$

where $\chi = x/\sqrt{2}\delta$. For the nonmassive kink, the transverse degree of freedom gives the same contribution to the kink mass as the correction term to its form, since $\gamma \sim 1/\delta^2$. Using (11.54) and the definition (11.50), we obtain the kink's mass as

$$m \simeq \frac{\sqrt{2}b^2}{3\pi\delta} \left[1 + \frac{\gamma}{10} + \frac{2}{7} \left(\frac{16\pi B_2 b}{\delta} \right)^2 \right]. \tag{11.55}$$

If we take into consideration that $\int_{-\infty}^{+\infty} dx\, u_x = 0$ due to boundary conditions, then in the continuum-limit approximation the system energy (11.43) may be rewritten in the form

$$E \simeq \frac{4b^4 B_2}{\pi^2} \int_{-\infty}^{\infty} dx \left[\frac{\delta^2}{2} \left(\frac{d\widetilde{w}}{dx} \right)^2 + \frac{1}{4}(1 - \widetilde{w}^2)^2 - \frac{\gamma\delta^2}{2} \widetilde{w}^2 \left(\frac{d\widetilde{w}}{dx} \right)^2 \right], \tag{11.56}$$

so that the energy of a single nonmassive kink is

$$\varepsilon_k \simeq \frac{16\sqrt{2}}{3} b^4 B_2 \delta \left(1 - \frac{3}{10}\gamma \right). \tag{11.57}$$

Figure 11.9 shows the numerical results of Braun *et al.* [819] for the kink energies when the frequency of transverse potential varies while the parameters of the interaction potential are kept fixed [in these calculations the exponential interaction, $V_{\text{int}}(r) = V_0 \exp(-\beta r)$, was used instead of the Coulomb one]. The continuum limit approximation (solid curves in Fig. 11.9) gives rather good estimations for MK (circles) when $\omega_{0y} < \omega_k^*$ and for $\overline{MK}$ (squares) when $\omega_{0y} < \omega^*$ (the value ω^* corresponds to the bifurcation of the TGS, and ω_k^*, to that for the chain with a kink), when the conditions $D \gg 1$

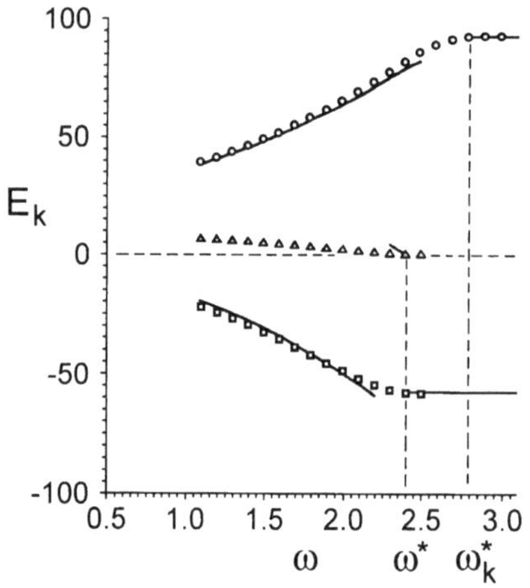

Fig. 11.9. Kink energies vs. transverse frequency ω_{0y} for massive kink (circles) and antikink (squares), and for nonmassive kinks (triangles). The interatomic repulsion is exponential [$V_{\rm int}(r) = V_0 \exp(-\beta r)$ with $V_0 = 200$ and $\beta = 0.3$]. Solid curves show the analytical results [819].

and $\alpha \ll 1$ are satisfied. Unfortunately, for *NMK* an accurate analytical estimation is available only in a very small region near ω^*, because the analytical approach assumes the smallness of b, but this latter condition is valid only in a small region (when $\gamma \ll 1$).

When the system is in contact with a thermal bath, the concentration of thermally created kinks is determined by their energies, $n_k \propto \exp(-\widetilde{\varepsilon}_k/k_B T)$, where $\widetilde{\varepsilon}_k = \frac{1}{2}[\varepsilon_k(\text{kink}) + \varepsilon_k(\text{antikink})]$. From Fig. 11.9 one can see that $\widetilde{\varepsilon}_k(NMK) < \widetilde{\varepsilon}_k(MK)$. Therefore, in the ZGS the concentration of nonmassive kinks is much higher than that of massive kinks, $n_k(NMK) \gg n_k(MK)$, and namely the nonmassive kinks will give the main contribution to *thermodynamic* properties of the quasi-2D system such as heat capacity, free energy, etc. On the contrary, *dynamic* properties of the chain (such as conductivity, diffusivity, etc.) are determined by massive kinks.

Unfortunately, close the bifurcation point, where $b \ll 1$, the analytical approach does not work. In the TGS but close to the transition point, the kink structure drastically changes: in the kink core region, where the chain is more compressed, the atoms begin to escape from the minima of $V_y(y)$ as shown in Fig. 11.10. The critical values of V_0 corresponding to the bifurcation to the ZGS are such that $V_{\rm bif}^{(\rm kink)} < V_{\rm bif}^{(\rm GS)} < V_{\rm bif}^{(\rm antikink)}$. For the kink case, the TGS–ZGS transition begins in the kink core region (i.e. in the region of local compression of the chain), and then the zigzag structure extends over the whole system. On the contrary, in the antikink case this transition ends in the antikink core region which is the region where the chain is stretched locally. The behavior of kink parameters can be understood in terms of the cusp of the average elastic constant $g_{\rm eff}(V_0)$. In particular, at $V_0 = V_{\rm bif}^{(\rm GS)}$ the kink core region is already transformed to the zigzag state so that $g_i^{(\rm kink)} < g_i^{(\rm GS)}$. On the other hand, the transition to the zigzag shape within the antikink core region occurs at larger V_0, so that $g_i^{(\rm antikink)} \sim g_i^{(\rm GS)}$. Because the value

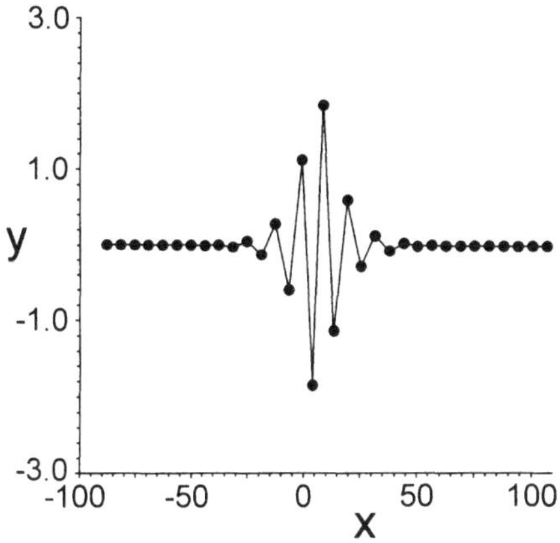

Fig. 11.10. Shape of the kink on the background of the trivial ground state close to the bifurcation point [819].

of the elastic constant g determines the parameters of topological excitations such as the effective kink mass m and the height of the Peierls-Nabarro potential ε_{PN}, it follows that, at and just above the bifurcation, we have $m^{(\mathrm{kink})} > m^{(\mathrm{antikink})}$ and $\varepsilon_{PN}^{(\mathrm{kink})} > \varepsilon_{PN}^{(\mathrm{antikink})}$, i.e. we came to the inequalities just opposite to those for the TGS.

The results of numerical calculation of the PN barriers (see Fig. 11.11) are in agreement with this qualitative consideration. Note that although in Fig. 11.11 the value of ε_{PN} seems go to zero when $\omega_{0y} > \omega^*$ for $\overline{MK}$ and $\omega_{0y} > \omega_k^*$ for MK, in fact these values for 1D kinks are only small but not zeroes: $\varepsilon_{PN} \simeq 2 \cdot 10^{-4}$ for $\overline{MK}$ and $\varepsilon_{PN} \simeq 5 \cdot 10^{-5}$ for MK so that $\varepsilon_{PN}(\overline{MK}) > \varepsilon_{PN}(MK)$. For the zigzag structure of kinks the PN barrier for MK is larger than that for $\overline{MK}$ in a wide range of the parameter ω_{0y}. Notice also an additional maximum of the function $\varepsilon_{PN}(\omega_{0y})$ associated with the appearance of the zigzag MK created on the background of TGS.

The PN barrier for NMK is shown in Fig. 11.11b by triangles. Recall that nonmassive kinks exist provided $\omega_{0y} < \omega^*$, i.e. for the ZGS only. It is interesting that the function $\varepsilon_{PN}(\omega_{0y})$ has a local minimum at $\omega_{0y} \simeq 1.31$. This effect is analogous to that known for a nonsinusoidal substrate potential (see Sect. 3.3.3). Namely, for $1.31 < \omega_{0y} < \omega^*$ the minimum of the potential energy is realized for the configuration shown in Fig. 11.8(a), while the configuration of Fig. 11.8(b) corresponds to the saddle one. However, at $1 < \omega_{0y} < 1.31$ the situation is just opposite: the configuration shown in Fig. 11.8(a) corresponds to the saddle state, while the configuration shown in Fig. 11.8(b), to the minimum-energy configuration. Moreover, close to the value $\omega_{0y} = 1.41$ both the configurations of Figs. 11.8(a,b) correspond to local maxima of energy, while the minimum-energy configuration is realized at an intermediate state.

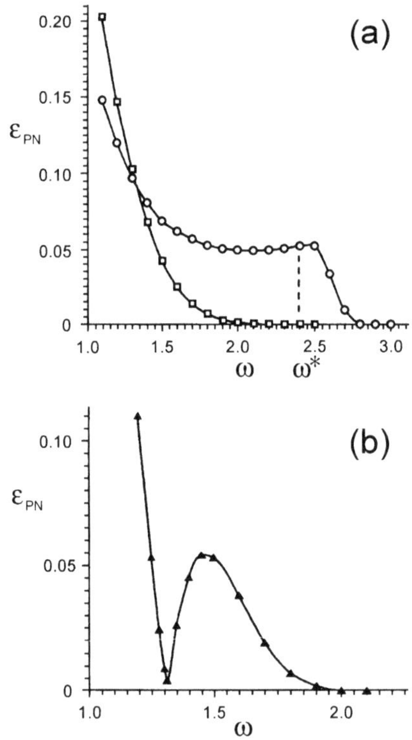

Fig. 11.11. Peierls-Nabarro barrier as function of the transverse frequency ω_{0y} for (a) massive kink (circles) and antikink (squares), and (b) nonmassive kink (triangles) for exponential interaction $[V_{\rm int}(r) = V_0 \exp(-\beta r)$ with $V_0 = 200$ and $\beta = 0.3]$ [819].

The existence of a transverse degree of freedom results in a qualitative change of the dependence of the diffusion coefficient on the atomic concentration. Recall that the chemical diffusion coefficient D_c coincides with the antikink diffusion coefficient if the concentration of atoms θ is slightly lower than the value $\theta = 1$, $D_c(\theta)\mid_{\theta=1-0} = D_c^- = D_{\bar{k}}$, because in this case the mass transport along the chain is carried out by antikinks. Otherwise, when $\theta = 1{+}0$, the mass transport is carried out by kinks, and $D_c(\theta)\mid_{\theta=1+0} = D_c^+ = D_k$. As long as chain's configuration is one-dimensional, the anharmonicity parameter α_0 is positive, and $\varepsilon_{PN}(\text{antikink}) > \varepsilon_{PN}(\text{kink})$. Because $D_k \propto \exp(-\varepsilon_{PN}/k_B T)$, this leads to the inequality $D_c^- < D_c^+$. Thus, the function $D_c(\theta)$ jump-like increases at $\theta = 1$. However, in the zigzag FK model the repulsion forces between the neighboring atoms in the region of a kink may exceed the forces which hold the atoms within the 1D line. In this case the atoms begin to escape out of the line, and the kink becomes less mobile than the antikink, $D_c^- > D_c^+$, so that the jump in the function $D_c(\theta)$ at $\theta = 1$ changes its sign. As seen from Fig. 11.11, a zigzag massive antikink (a vacancy) is more mobile compared with a zigzag massive kink (an additional atom in the chain) for a wide range of ω_{0y} values.

11.3.5 Applications

To estimate the value of the transverse frequency ω_{0y} in realistic physical systems, we may assume that, in a crystal, $V_y(y)$ results from a periodic potential $V_y(y) = \frac{1}{2}\varepsilon_{sy}[1 - \cos(2\pi y/a_{sy})]$, where ε_{sy} is the characteristic amplitude and a_{sy} is the characteristic distance for the transverse potential. For this shape of $V_y(y)$ we obtain $\omega_{0y}^2 = V_y''(0) = 2\pi^2\varepsilon_{sy}/a_{sy}^2$. For ions with a unit elementary charge e, $V_0 = e^2$ in Eq. (11.14). Then, putting $a_A = a_s/\theta$ into Eq. (11.22), we find that the TGS–ZGS bifurcation takes place when θ reaches a threshold value

$$\theta_{\text{bif}} = \left[\frac{\pi^2}{2}\left(\frac{a_s}{a_{sy}}\right)^2 \frac{\varepsilon_{sy}}{e^2/a_s}\right]^{1/3}. \tag{11.58}$$

Taking $a_s \sim a_{sy} \sim 3$Å and $\epsilon_{sy} \sim 0.1$ eV, we obtain $\theta_{\text{bif}} \sim 0.5$. This is a value which can easily be achieved in a real system such as atoms in the channel of a superionic conductor or protons in hydrogen-bonded chains. Thus, a usual picture of ions staying in line in a channel is too oversimplified, and the possibility of ionic motion in transverse directions could modify significantly some conclusions obtained within the framework of the standard FK model, for example those concerned with existing of sliding mode as was discussed in Sect. 11.3.2.

The accounting of transverse degrees of freedom is very important in modelling the adsorbed layers as well as in studies of the crowdion problem. In particular, the TGS–ZGS transition occurring with increasing of the atomic concentration, drastically decreases the mobility of topological excitations in these systems and, therefore, modifies the dependencies of mobility and diffusion coefficients on the concentration as was discussed in Sect. 11.3.4.

In applications to adsorbed systems, the zigzag model should be further generalized. *First,* if the transverse degree of freedom describes the atomic displacements perpendicular to the surface, this potential can no longer be treated as parabolic one. When an atom moves away from the surface, the transverse potential has to tend to a constant, the "vacuum level" for an atom taken away from the surface. As the potential must also exhibit a minimum near the surface to describe the sticking of the overlayer on the substrate, the transverse potential must be nonconvex. For example, the potential perpendicular to the surface may be taken in the Morse form,

$$V_y(y) = \varepsilon_d\left(e^{-\gamma y} - 1\right)^2, \tag{11.59}$$

which goes to the finite limit ε_d (known as the adsorption energy) when $y \to \infty$. The parameter γ determines the anharmonicity and it is related to the frequency ω_{0y} of a single-atom vibration in the normal direction by the relation $\omega_{0y}^2 = 2\gamma^2\varepsilon_d/m_a$. Contrary to the parabolic potential, the function (11.59) is nonconvex, i.e. it has an inflation point at $y = y_{\text{inf}} \equiv \gamma^{-1}\ln 2$, so that $\omega_{\rm eff}^2(y) \equiv V_y''(y) > 0$ for $y < y_{\rm inf}$, but $\omega_{\rm eff}^2(y) < 0$ at $y > y_{\rm inf}$. Besides, the total potential energy of a single atom near the substrate has now to be written as

$$V_{\rm sub}(x,y) = V_x(x)e^{-\gamma' y} + V_y(y). \tag{11.60}$$

The exponential factor in the first term of the right-hand side of Eq. (11.60) takes into account the decrease of influence of the surface corrugation as the atoms move away from the surface, so that $V_{\rm sub}(x,y) \to \varepsilon_d$ when $y \to \infty$.

The exponential factor $\exp(-\gamma' y)$ in Eq. (11.60) introduces a coupling between the x and y displacements. As the result, the atoms which are displaced from the minima of the substrate potential in the x direction, should additionally be forced in the transverse direction. It is easy to see that if an atom is shifted of Δx ($\Delta x < \frac{1}{2}a_s$) along the chain from the corresponding minimum of $V_{\rm sub}(x,y)$, at the same time it is submitted to a force in the y direction which shifts it in the normal direction of the quantity

$$\Delta y \approx \frac{1}{2}\gamma' \left(\frac{\omega_{0x}}{\omega_{0y}}\right)^2 (\Delta x)^2. \tag{11.61}$$

Therefore, a submonolayer structure of adsorbed atoms at $\theta \neq 1$, which may be considered as a kink or antikink superstructure with the period $a_\theta = a_s/|1-\theta|$, should be corrugated in the normal direction with the same period a_θ.

Second, the interaction potential should include both repulsive and attractive branches as, for example, the Lennard-Jones or Morse potentials, therefore it must be nonconvex too. The zigzag model generalized in this way, may be used for a qualitative explanation of adsorbed systems with more than one monolayer. Namely, with increasing of the atomic concentration, the atoms form first the 1D chain at the minimum of $V_y(y)$ which corresponds to a "first monolayer" of the adsorbed film. However, when the first layer becomes complete, further increasing of θ will lead to atomic configurations where new atoms have $y > y_{\rm inf}$ and they form the "second monolayer". As was shown by Braun and Peyrard [820], at some subtle but realistic choice of model parameters, an atom for the $\theta \sim 1$ case may have two stable positions, one corresponds to the atom inserted into the first layer, and another with this atom in the second layer. Thus, the model may exhibit the following "reconstructive growth scenario": at $\theta \leq 1$ the atoms fill the first layer, while at $\theta > 1$ new incoming atoms occupy the second layer. However, when the concentration θ increases above some threshold value $\theta = \theta'$, the incoming atoms continue to fill the second layer but, at the same time, they stimulate the escaping of atoms from the underlying first layer to the second layer, causing the reconstruction of the structure of the growing film. Then, for $\theta'' < \theta \leq 2$ (where θ'' is the second threshold value) the escaped atoms are pulled back into the first layer, and the film continues to grow in a usual way, new incoming atoms being placed over the completely filled first layer. Thus, within the interval $\theta' < \theta < \theta''$ the GS structure corresponds to a partially filled first layer, while at $1 \le \theta < \theta'$ and at $\theta'' < \theta < 2$, to a completely filled first layer. Such sequence of structures of adsorbed atoms was observed in the experiments for a lithium film growing on the (112) surface of tungsten or molybdenum [821, 822].

The same model was also used by Braun *et al.* [823] to describe the "solitonic-exchange" diffusion mechanism. When the first adsorbed layer is complete and new incoming atoms start to fill the second adlayer, the diffusion in the second layer follows usually the same laws as that in the first layer because the adatoms of the first monolayer play the role that substrate atoms played for the diffusion of the adatoms of the first layer, i.e. the first-layer adatoms create an external potential for the second-layer atoms. However, for the model described above the situation is more complicated owing to exchange of atoms between the first and second adlayers. As was described above, in some parameter range of the model the formation of a metastable defect in which an atom of the second layer penetrates into the first layer, is possible. In this metastable configuration the extra atom creates a localized solitonic excitation (kink), which usually has a very high mobility. Thus, even if the lifetime of the metastable configuration is small at a nonzero temperature, it may play the main role in the overall surface diffusion, generating an unusual temperature dependence of the diffusion coefficient. Such a diffusion mechanism is a *two-step exchange* diffusion mechanism. The main difference with one-step exchange diffusion is that the configuration with the adatom inserted into the first adlayer, now corresponds to a *metastable* state instead of the unstable (saddle) state as usually occur in the exchange mechanism. For this "exchange-solitonic" mechanism the diffusion follows the Arrhenius law at low temperatures, but the activation energy corresponds now to the difference in energies between the metastable state and the ground state (and not to the barrier for the transition from the second adlayer to the first one), while the preexponent factor is determined by the kink diffusion coefficient and essentially depends on temperature. A similar behavior was observed by Black and Tian [824] in a molecular dynamics experiment, where an isolated Cu adatom, which is diffusing on the Cu(100) surface, may enter the first substrate layer and create there a strain along a close-packed row. This localized excitation (crowdion) moves along the row for a distance of several lattice constants, and then the strain is relieved by an atom in the strained row popping out and returning to the surface. The simulation showed that this diffusion mechanism becomes important at high enough temperatures ($T \sim 900$ K for the Cu-Cu(100) adsystem).

11.4 Spring-and-Ball Vector 2D Models

In a general 2D FK model the atoms are allowed to move in both directions, so that a vector field variable, $\mathbf{u}_{l_x,l_y} \equiv \{u^x_{l_x,l_y}, u^y_{l_x,l_y}\}$, has to be associated with each atom. Vector models are closely connected with realistic physical objects such as atomic layers adsorbed on crystal surfaces. In a general case, now we should take into account the interaction between all atoms in the system. If, however, one wish to simplify the model taking into account the interaction of a small number of nearest neighbors only, we have to restrict artificially the consideration by the configurations with small atomic displacements within the corresponding elementary cells.

Thus, the model Hamiltonian for the truly 2D vector FK model takes the form $H = K + U_{\text{int}} + U_{\text{sub}}$, where the kinetic energy K is generalized trivially,

$$K = \frac{m_a}{2} \sum_{\mathbf{l}} |\dot{\mathbf{u}}_{\mathbf{l}}|^2. \tag{11.62}$$

For the sake of simplicity, one may assume that the interaction between the atoms is isotropic,

$$U_{\text{int}} = \frac{1}{2} \sum_{\mathbf{l} \neq \mathbf{l}'} V_{\text{int}}(|\mathbf{r}_{\mathbf{l}} - \mathbf{r}_{\mathbf{l}'}|), \tag{11.63}$$

where $\mathbf{r}_{\mathbf{l}}$ is the atomic coordinate and $V_{\text{int}}(r)$ is the pairwise potential which may be approximated by a harmonic function in the simplest case (*the spring-and-ball model*). The substrate potential energy is

$$U_{\text{sub}} = \sum_{\mathbf{l}} V_{\text{sub}}(\mathbf{r}_{\mathbf{l}}), \tag{11.64}$$

where $V_{\text{sub}}(\mathbf{r})$ describes the symmetry of the substrate (for example, the potential produced by the first surface layer of the substrate in the case of adsorbed films). In particular, for the rectangular substrate lattice with the $a_{sx} \times a_{sy}$ elementary cell the function $V_{\text{sub}}(x, y)$ may be modelled as

$$V_{\text{sub}}(x, y) = \frac{\varepsilon_x}{2} V_s(2\pi x/a_{sx}) + \frac{\varepsilon_y}{2} V_s(2\pi y/a_{sy})$$
$$+ \frac{1}{4}(\varepsilon_{\text{max}} - \varepsilon_x - \varepsilon_y)\, V_s(2\pi x/a_{sx})\, V_s(2\pi y/a_{sy}), \tag{11.65}$$

where ε_{max} is the maximum of the 2D potential, ε_x and ε_y are the activation (saddle) energies for motion in the x and y directions respectively, and $V_s(x)$ is a suitable periodic function normalized so that $V_s(0) = 0$ and $V_s(\pi) = 2$, for example, $V_s(x) = 1 - \cos x$. For the square lattice such as the (100) face of cubic crystals we have to put $a_{sx} = a_{sy} = a_s$ and $\varepsilon_x = \varepsilon_y = \varepsilon_s$, while for furrowed faces such as the (112) face of b.c.c. crystals we should take $a_{sx} < a_{sy}$ and $\varepsilon_x < \varepsilon_y$.

The triangular substrate lattices may be constructed in a similar way,

$$V_{\text{sub}}(x, y) = \frac{1}{2}\varepsilon \left\{ 1 - \cos(2\pi x/a_{sx}) \cos(\pi y/a_{sy}) + \frac{1}{2}\left[1 - \cos(2\pi y/a_{sy})\right] \right\}, \tag{11.66}$$

where the choice $a_{sy} = a_{sx}\sqrt{3}/2$ leads to triangular symmetry of the potential minima separated by isotropic energy barriers of height ε. The frequency of atomic vibration near the minima in this case is also isotropic, $\omega_{0x} = \omega_{0y} = (\varepsilon/2m)^{1/2}(2\pi/a_{sx})$. The potential (11.66) can be represented as $V_{\rm sub}(x,y) = \frac{1}{4}\varepsilon\,[3-\cos(\mathbf{k}_1\mathbf{r})-\cos(\mathbf{k}_2\mathbf{r})-\cos(\mathbf{k}_1\mathbf{r}+\mathbf{k}_2\mathbf{r})]$, where $\mathbf{r}=(x,y)$, and $\mathbf{k}_1$ and $\mathbf{k}_2$ are reciprocal vectors of the triangular lattice. A similar form of the potential is often used in studies of atomic layers adsorbed on isotropic triangular or hexagonal substrates [825, 826].

To classify different topological excitations of the vector 2D FK model, it is convenient to consider firstly a single atomic chain placed into the 2D substrate potential. When the whole chain lies within the same "channel" of the 2D potential, i.e. if $u_{\rm l}^y = 0$ at $l_x \to \pm\infty$ but $u_{\rm l}^x = 0$ at $l_x \to -\infty$ and $u_{\rm l}^x = \pm a_{sx}$ at $l_x \to \infty$, such topological excitation is called the *edge* kink (ek). On the other hand, when the chain's ends lie in different neighboring valleys of the potential, i.e. $u_{\rm l}^x = 0$ at $l_x \to \pm\infty$ but $u_{\rm l}^y = 0$ at $l_x \to -\infty$ and $u_{\rm l}^y = \pm a_{sy}$ at $l_x \to \infty$, such topologically stable excitation is called the *screw* kink (sk). The screw kink and antikink have the same parameters because they both correspond to a local extension of interatomic bonds. Usually, $\varepsilon_k(\mathrm{ek}) > \varepsilon_k(\mathrm{sk})$ but $\varepsilon_{PN}(\mathrm{ek}) < \varepsilon_{PN}(\mathrm{sk})$.

Now let us consider the two-dimensional array of atoms as a system of parallel atomic chains. If only one of the chains contains an additional atom, such a configuration is called the *crowdion*. But when the every chain contains one kink and the kinks belonging to nearest neighboring chains attract each other, they will form a dislocation line (domain wall, linear soliton, etc.). Depending on the type of kinks, the dislocation may be either edge or screw types. Note that the edge dislocation is similar to the domain wall of the scalar FK model considered above in Sect. 11.2.

In a general case a dislocation can be characterized by the Burgers vector (see, e.g., Ref. [827]). If $\mathbf{u} \equiv (u^x, u^y)$, the Burgers vector is defined as

$$\mathbf{b} \equiv \oint_C \frac{\partial \mathbf{u}}{\partial \mathbf{l}}\, d\mathbf{l}, \tag{11.67}$$

where the integral is taken over a closed contour C. If the contour contains no dislocations, $\mathbf{b}=0$. If the contour contains a dislocation, $\mathbf{b}$ is the closure failure in circumscribing the dislocation. For the screw dislocation $\mathbf{b}$ is parallel to the dislocation line, while for the edge dislocation $\mathbf{b}$ is perpendicular to the line. In a general case, a dislocation has both edge and screw components.

11.4.1 The Ground State

In the case of harmonic interatomic interaction, the kinks in the nearest neighboring chains attract one another, thus forming the chains of kinks, or domain walls. Let us firstly consider the $T=0$ ground state for the fixed-density case. When $\theta = 1$, i.e. the number of atoms is equal to the number of minima of $V_{\rm sub}(\mathbf{r})$, the GS is trivial, all atoms lie at the minima of $V_{\rm sub}(\mathbf{r})$. If $\theta \neq 1$ but a stress is applied in one direction only (namely, the periodic boundary conditions look as $\mathbf{u}_{l_x+N_x,l_y} = \mathbf{u}_{l_x,l_y} + N_w\mathbf{b}$ and $\mathbf{u}_{l_x,l_y+N_y} = \mathbf{u}_{l_x,l_y}$), the GS contains N_w parallel domain walls. This situation is similar to that considered above for the scalar 2D FK model. However, if $\theta \neq 1$ and the stress is applied in both directions, the GS will correspond to a cross grid of dislocations.

Snyman and Snyman [828] have investigated numerically the GS for the hexagonal substrate potential, modelling epitaxial monolayer structures on the (111) face of a f.c.c. metal. The symmetries of the substrate potential and the spring-and-ball monolayer are shown in Figs. 11.12(a,b). The computer monolayer structure for the 4% natural misfit is as depicted in Fig. 11.13, where four unit cells are shown. Two kinds of "coherent" adatom islands are formed separated from each other by strips of dislocated atoms. These computer results are perfectly consistent with that predicted by the elastic theory of dislocations.

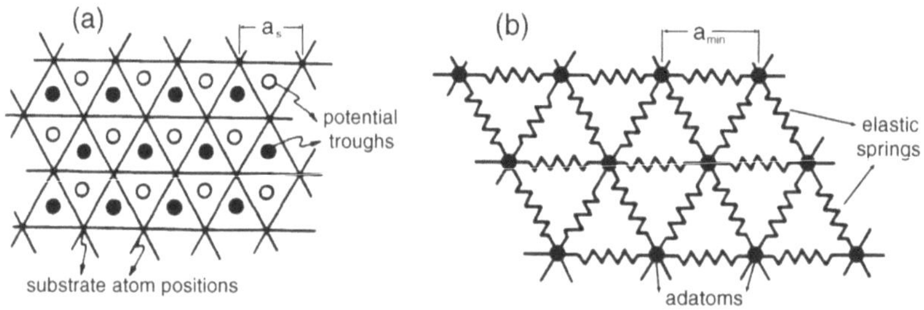

Fig. 11.12. (a) The substrate potential symmetry, and (b) the spring-and-ball overgrowth film for the hexagonal interfacial symmetry [828].

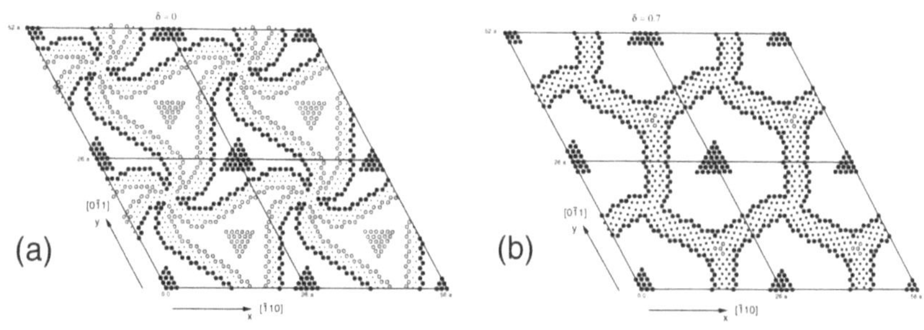

Fig. 11.13. Computer monolayer structures for $(a_{\min} - a_s)/a_s = 0.04$ and different values of δ: (a) $\delta = 0$, and (b) $\delta = 0.7$. The full circles indicate adatoms in proper positions and the open circles denote atoms occupying improper positions. The dots designate dislocated atoms in transition locations between proper and improper positions [828].

The investigation of the overgrowth layers with rectangular interfacial symmetry is more complicated because the simple spring-and-ball model is unstable with respect to shear deformations. Therefore the model should be improved by taking into account the interaction of next-nearest (diagonal) neighboring atoms. The corresponding elastic constants have to be selected in a way to satisfy the requirements of the elasticity theory to define correctly the shear modulus and Poisson ratio. Computer simulations [768, 829] in agreement with the elastic theory predictions show that the misfit is accommodated by a cross rectangular grid of edge dislocations. Similarly, molecular dynamics simulation for the hexagonal substrate symmetry predicts the formation of honeycomb network of domain walls [830].

The theoretical investigation of the GS configurations at low misfit parameters shows that the important parameter of the system is the energy $\varepsilon_{\text{cross}}$ of domain wall crossing. If wall intersections are energetically favorable ($\varepsilon_{\text{cross}} < 0$), a domain wall network should be created by a discontinuous phase transition as the misfit increases [831]. Otherwise, if $\varepsilon_{\text{cross}} > 0$, the theory predicts that a striped phase where walls are oriented only in one direction, should arise first [807, 832], and only later the DW grid arises with further increasing of the misfit, both phase transitions being continuous.

Now let us discuss briefly the $T \neq 0$ ground state. The striped DW lattice melts similarly to that of the scalar FK model. The melting of the honeycomb network was investigated by Abraham *et al.* [830] with the help of molecular dynamics. The simulation showed that with temperature increasing, distortion from the perfect honeycomb structure becomes more prevalent, characterized by significant fluctuations from the symmetry directions, wall thickening, and wall roughening (see Fig. 11.14). The attempts of an-

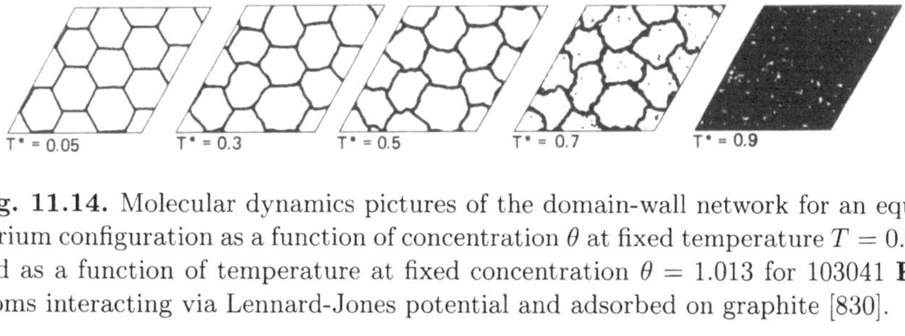

Fig. 11.14. Molecular dynamics pictures of the domain-wall network for an equilibrium configuration as a function of concentration θ at fixed temperature $T = 0.05$ and as a function of temperature at fixed concentration $\theta = 1.013$ for 103041 **Kr** atoms interacting via Lennard-Jones potential and adsorbed on graphite [830].

alytical solution of the problem were made by Bak *et al.* [831] (see also Refs. [48, 503, 833] and references therein).

Finally, at large misfit between the substrate and overlayer, the homogeneous incommensurate structures emerge. As the coverage θ increases, the angle between the substrate and the adatom structure may change abruptly by the discontinuous phase transition [834, 835]. However, when temperature varies but θ is kept fixed, the phase transitions may be continuous [836].

11.4.2 Excitation Spectrum

Analogously to the one-dimensional case, the two-dimensional FK model has an infinite number of metastable states (configurational excitations) provided the interatomic coupling is weak enough. These configurations can be again considered in the framework of the standard map, i.e. as trajectories of an artificial dynamical system with, however, two discrete "times" l_x and l_y (note that the evolution in one "time" variable depends on the other). Most of these configurational excitations are spatially chaotic. Approximating the substrate potential by a piece-wise harmonic function, the energy of the configuration can be calculated with the help of the Green function technique. Then, the density of metastable configurations and their contribution to the specific heat can be calculated. This program was elaborated by Uhler and Schilling [837] for the spring-and-ball rectangular FK model. The results explain the glassy-like behavior of the system without inherent disorder.

The phonon spectrum of the two-dimensional FK model has been investigated by the molecular dynamics technique by Black and Mills [838].

11.4.3 Dynamics

Dynamic properties of the spring-and-ball vector FK model have been studied by Lomdahl and Srolovitz [839, 840] with the help of the molecular dynamics technique. The authors solved the Langevin motion equations for the rectangular 30×30 FK model with $\theta = 1$ as the reference structure. For the initial configuration, they took the lattice with one edge or one screw dislocation line, and then the system evolution under the action of an external dc driving force (or the shear stress because the force is applied to the "adatoms" only) was investigated. The parameters of the model were chosen as $\varepsilon_k(\text{ek})/\varepsilon_k(\text{sk}) \approx 7$ and $\varepsilon_{\text{PN}}(\text{ek})/\varepsilon_{\text{PN}}(\text{sk}) \approx 10^{-5}$ which are close to normal conditions for crystals. The driving force was allowed to have both components, $\mathbf{F} = (F_x, F_y)$. The results obtained can be summarized as follows:

(i) At $T \neq 0$ the dislocation lines are thermally roughened similarly as in the scalar model. The roughness increases with temperature rising but decreases with increasing dislocation velocity. The screw dislocation has relatively smaller roughness (due to larger PN barriers) comparing with the edge dislocation;

(ii) Due to inequality $\varepsilon_{\rm PN}({\rm sk}) \gg \varepsilon_{\rm PN}({\rm ek})$, the screw dislocation moves much slower than the edge one. The motion of the screw is thermally activated, its velocity is controlled by the rate of thermal nucleation of kink-antikink pairs. Kinks on screws are edges and hence after nucleation they propagate along the screw with large velocity because of small PN barriers;

(iii) The moving dislocation is damped by an intrinsic friction due to radiation of phonons. The moving edge dislocation emits longitudinal phonons, i.e. the phonon field has the same component as the Burgers vector. However, a moving screw dislocation emits phonons of both fields, the high frequency transverse phonons (in the field of the Burgers vector) and the lower frequency, smaller amplitude and much more weakly damped longitudinal phonons in the field normal to **b**;

(iv) The examination of the temporal evolution of an initially circular dislocation loop shows that the loop expands in an asymmetric form: the edge components move much faster than do the screw components as it has to be expected due to inequality $\varepsilon_{\rm PN}({\rm sk}) \gg \varepsilon_{\rm PN}({\rm ek})$;

(v) For the chosen model parameters the thermal creation of dislocations (dislocation loops) is negligible. However, behind the moving edge dislocation, small dislocation loops with the same Burgers vector as the initial straight dislocation are nucleated [Fig. 11.15(b)]. These loops continue to grow until impingement [Fig. 11.15(c)]. At impingement the loops loose their identity and form into nominally straight dislocations [Fig. 11.15(d)]. The straight dislocation formed in this manner and closest to the initial dislocation has the same Burgers vector as the initial one. These two dislocations propagate through the crystal much in the manner of a single double Burgers vector

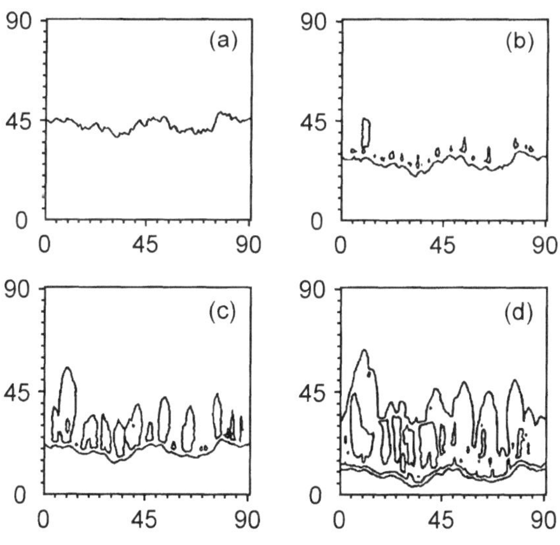

Fig. 11.15. Dislocation generation behind an edge dislocation moving under the action of an applied force $F_x = 0$, $F_y = 0.175$ and $k_B T = 0.15$ at $t = 0, 20, 30$, and 40, respectively [840].

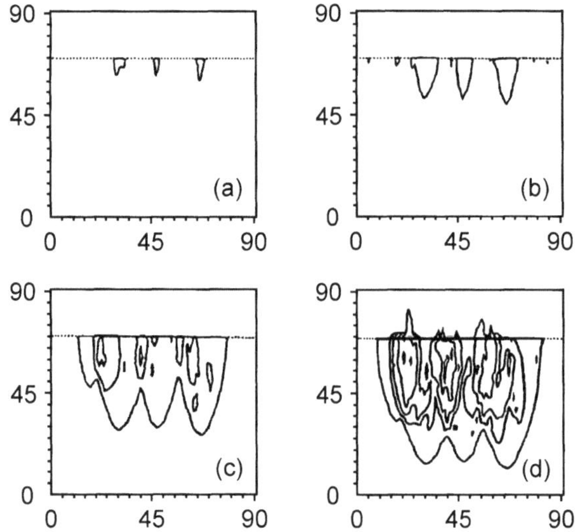

Fig. 11.16. Dislocation generation on a static screw dislocation (dashed line). The applied force is $F_x = 0$, $F_y = 0.175$, $k_BT = 0.15$, and times are 17.5 (a), 30 (b), 55 (c), and 70 (d). The solid curves correspond to dislocations with $\mathbf{b} = (0, a_s)$ [840].

dislocation. The loops nucleation is due to a cooperative effect between the phonons radiated from a moving dislocation and the thermal field.

Besides, the nucleation of dislocation loops on the immobile screw dislocation (pinned by the PN potential) was observed in the presence of the applied force (Fig. 11.16). While these loops are growing and unpinning, additional loops are nucleated. Nucleation of dislocation loops occurs more easily on the screw dislocation than in the bulk because for particular screw/edge orientations the energy of the adjacent screw/edge dislocation is reduced over that of the separated dislocations. Thus, while the homogeneous nucleation of dislocation loops is usually improbable, the described heterogeneous nucleation may occur at high enough stresses. At low stresses, however, dislocations may be produced by well known mechanisms such as Frank-Read sources, cross-slip, etc.

Finally, we have to mention the paper by Gornostyrev *et al.* [841], where the Langevin dynamics of the spring-and-ball model was simulated for the case of triangular substrate potential.

11.5 Vector 2D FK Model

The behavior of the *anisotropic* trully-vector 2D FK model is almost identical to that of the 1D FK model [190], so it behaves similarly to the scalar 2D FK model discussed above in Sect. 11.2. The only difference is in the transition to the final sliding state. The exact critical force depends slightly on the external conditions, which means that the transitions do not occur simultaneously in all the channels. When a channel has jumped to the sliding state, it enhances the probability for the neighboring channels to jump due to interatomic interactions. This first channel corresponds therefore to a nucleation event. On the contrary, the transition of a channel from the sliding state to the locked state is almost independent of neighboring channels [190].

The ground state and statistical mechanics of the *isotropic* 2D FK model can be described by the spring-and-ball model (see Sect. 11.4) with a good accuracy. The nonlinear dynamics of the isotropic model, however, has to be considered in the framework of the trully-vector 2D model, because mutual atomic displacements may be too large. Unfortunately, investigations of the isotropic 2D underdamped vector FK system are very limited, and there are still a number of open questions, in particular: (i) How does the transition begin for the commensurate $\theta = 1$ system; does it start with the creation of a DW–anti-DW pair (dislocation loop) or with the emergence of a running island? (ii) How does the transition proceed after the nucleation event? (iii) How does the mechanism of the transition depend on the atomic concentration and on the layer structure? (iv) If a DW already preexists in the system, is its motion stable, and up to what velocity? (v) Does a spatially nonuniform "traffic jam" steady state appear during the transition? In the next subsection we present some answer on these questions.

11.5.1 Locked-to-Sliding Transition

Braun *et al.* [842] considered a two-dimensional layer of particles with position vectors $\mathbf{u} = (u_x, u_y)$, subject to a periodic substrate potential (11.66) with the triangular symmetry. The periodic boundary conditions in both spatial directions x and y were used. The authors considered the cases of a pairwise exponential repulsion between the atoms,

$$V_{\text{int}}(r) = V_0 \exp(-\gamma r), \tag{11.68}$$

with the choice $\gamma = a_s^{-1}$, and the Lennard-Jones (LJ) interaction,

$$V_{\text{int}}(r) = V_0 \left[\left(\frac{a_s}{r}\right)^{12} - 2\left(\frac{a_s}{r}\right)^{6} \right], \tag{11.69}$$

where r is the interatomic separation. The exponential case is purely repulsive with no minimum in the potential. This situation corresponds, in particular, to atoms chemically adsorbed on a metal surface, when, due to breaking of the translational symmetry in the direction normal to the surface, the atoms have a nonzero dipole moment which leads to their mutual repulsion [36]. For the LJ interaction there is both attraction and repulsion, with a minimum in the potential at the atomic separation $r = a_s$. When the atoms are closely packed and experience only the repulsive branch of the interaction as, for example, in the case of a lubricant film confined between two substrates, both potentials are qualitatively the same. But when the layer has expanded regions, where the interatomic separation exceeds the equilibrium distance $r = a_s$, the LJ potential may lead to first-order phase transitions.

The equations of motion for the displacement vectors of the N atoms $\mathbf{u}_i$ $(1 \leq i \leq N)$ are given by

$$\ddot{v}_i + \eta \dot{v}_i + \frac{d}{dv_i}\left[\sum_{j(j\neq i)} V_{\mathrm{int}}(|\mathbf{u}_i - \mathbf{u}_j|) + U_{\mathrm{sub}}(\mathbf{u}_i)\right] = F^v + F^v_{\mathrm{rand}}, \qquad (11.70)$$

where $v = u_x$ or u_y, $F^v = F_x$ or F_y is the externally applied force (in the simulation the driving force was chosen to act only in the x direction, i.e. $F_x = F$, $F_y = 0$) and F^v_{rand} is the random force required to equilibrate the system to a given temperature, T. The dimensionless system of units was used ($m_a = 1$, $a_s = 2\pi$ and $\varepsilon = 2$, so that $\omega_0 = 1$ and the oscillation period of a particle in the substrate potential is $\tau_s = 2\pi$).

To characterize the system it is helpful to define the effective elastic constant , $g_{\mathrm{eff}} = a_s^2 V''(r_0)/2\pi^2\varepsilon$, where r_0 is the mean interatomic distance. This single number gives an indication of the strength of the elastic constant of the layer of atoms relative to the strength of the substrate potential. A value of g_{eff} much smaller than 1 indicates a relatively weakly coupled layer, whereas a value much larger than 1 describes a stiff atomic layer compared with the substrate depth. Both large and small g_{eff} correspond to physically relevant systems. For a monolayer adsorbed on a surface, one is typically in the limit of low g_{eff} [36]. A lubricant layer between two blocks of material corresponds to the limit of large g_{eff}. The fact that one has a block of material and not simply a free monolayer means that the layer immediately above the interface has a much larger effective elastic constant. Although the disturbances are limited to the layer of the block at the sliding interface, the other layers above help maintain the separation between the first layer and the lower block, reducing the effective substrate potential depth. Also, the other layers tend to lock the structure of the first layer, making it more rigid than a single monolayer.

For the case of a closely packed atomic layer , $\theta = 1$, the ground state of the atomic layer has a triangular structure commensurate with the substrate. The spectrum of the ideal triangular lattice is described by the functions $\omega_x^2(k) = \omega_g^2\left[3 - 2\cos(ak_x) - \cos(ak_x/2)\cos(a_y k_y)\right]$ and $\omega_y^2(k) = 3\omega_g^2\left[1 - \cos(ak_x/2)\cos(a_y k_y)\right]$, where $\omega_g = \sqrt{g_{\mathrm{eff}}}$ and $a_y = a\sqrt{3}/2$, so that for the transverse wave propagating in the x direction (i.e. $k_y = 0$) the group velocity is $v_T = \lim_{k_x \to 0} d\omega_y(k)/dk_x = \sqrt{3}a\omega_g/2\sqrt{2}$, while for the longitudinal wave the group velocity is $v_L = \lim_{k_x \to 0} d\omega_x(k)/dk_x = \sqrt{3}\, v_T$.

Because in the GS all atoms lie in substrate minima, the layer is strongly pinned. To initiate the motion, a topological defect (either a localized defect like crowdion or a dislocation loop) has to emerge first. The size of this defect must be of order $d \sim a_s\sqrt{g_{\mathrm{eff}}}$, so one could expect different scenarios for small and large values of g_{eff}. Indeed, around some threshold value of $g_{\mathrm{eff}} = g_1$ there is a crossover from one characteristic scenario to another in the early stages of the transition.

Stiff layer. In the regime $g_{\text{eff}} \gg g_1$ the transition from the locked to running state is mediated by the formation of an island of moving atoms in a sea of essentially stationary particles. The size of the moving island grows quickly in the direction of the driving force, and somewhat more slowly in the perpendicular direction. *Inside the island the atoms largely maintain their triangular structure* due to the stiffness of the atomic layer. Hence one sees areas of essentially perfect triangular lattice surrounded by a loop of partial dislocations. Due to the periodic boundary conditions, the island eventually joins up on itself. There is then a strip, oriented parallel to the driving force and bounded in the direction perpendicular to the driving force, in which particles move along the periodically-continued system. Outside this strip the particles are immobile. We shall denote such a state as a *river*. The river then broadens perpendicular to the driving direction until all atoms are moving.

The evolution of this scenario is shown in a series of snapshots in Fig. 11.17. The same scenario was observed for both interactions. However, the value of g_1 above which the described scenario is observed, is different for the LJ and exponential interactions. The LJ layer behaves more rigidly than the exponential case for the same value of g_{eff}. It was found that $0.1 < g_1 < 1$ for the LJ case and $g_1 \approx 1$ for the exponential case.

The size of the island nucleated was found to increase with increasing g_{eff}. As one increases g_{eff} further the characteristic nucleation island size increases towards the system size. For a very stiff layer of particles, $g_{\text{eff}} > g_2$, the transition occurs via the whole system of atoms beginning to move. The

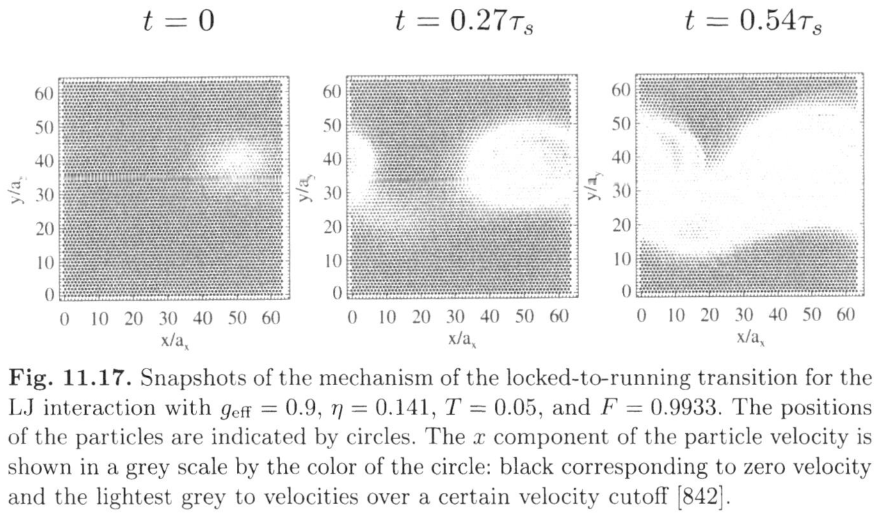

Fig. 11.17. Snapshots of the mechanism of the locked-to-running transition for the LJ interaction with $g_{\text{eff}} = 0.9$, $\eta = 0.141$, $T = 0.05$, and $F = 0.9933$. The positions of the particles are indicated by circles. The x component of the particle velocity is shown in a grey scale by the color of the circle: black corresponding to zero velocity and the lightest grey to velocities over a certain velocity cutoff [842].

value of g_2 naturally depends on the system size as well as on the details of $V_{\text{sub}}(x, y)$ and $V_{\text{int}}(r)$. In particular, g_2 increases with N, so that one can mediate the transition with larger and larger islands by increasing the system size. For the same system size and temperature, the LJ interaction again proved stiffer than the exponential interaction, having a smaller value of g_2.

Weakly bound layer, exponential interaction. For a weak interatomic interaction, $g_{\text{eff}} \ll g_1$, it was found that the locked-to-running transition was mediated by the formation of an avalanche of moving particles leaving behind a depleted, low-density region. *The atoms in the disturbed region do not show a regular structure*, so we may speculate that the beginning of the locked-to-running transition is caused by the nucleation of a "melted" island. Snapshots of the system for the exponential interatomic interaction at three times during the transition are shown in Fig. 11.18 for the case of $g_{\text{eff}} = 0.1$. A thermal fluctuation starts a small number of particles moving in the direction of the driving force. This fluctuation persists long enough to locally destroy the triangular ordering of atoms and allows the atoms to escape the substrate minima. The moving atoms are accelerated by the driving force, leaving a depleted region behind them. The size of the disturbed region quickly grows in the direction of the driving force as the moving atoms impinge on stationary particles in front of them. Thermal fluctuations and the unbalanced repulsive interaction left by the atoms that have moved off their lattice sites cause stationary atoms adjacent to the depleted region to be easily set into motion. A dense region of atoms is created at the leading edge of the disturbance as the mobile particles are able to move faster than the leading edge can propagate. Mobile particles are therefore slowed down as they approach the leading edge, producing a dense region of slower particles, in much the same way as traffic slows down on approaching a hindrance on a road. We shall call this kind of situation a *traffic-jam* effect. The bulge at

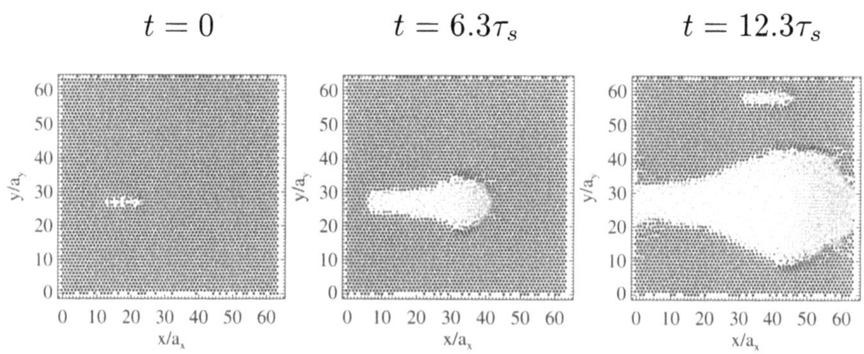

Fig. 11.18. Snapshots of the locked-to-running transition for the exponential interaction with $g_{\text{eff}} = 0.1$, $\eta = 0.1$, $T = 0.01$, and $F = 0.94$ [842].

the leading edge of the disturbance grows both parallel and perpendicular to the driving force until the domain of moving atoms joins up on itself due to the periodic boundary conditions. The formation of the bulge is clearly visible in the two later snapshots of Fig. 11.18 (notice that in this particular simulation a second nucleation event also takes place).

Weakly bound layer, LJ interaction. The scenario observed for the LJ interaction is slightly different. Snapshots for the LJ interaction with $g_{\text{eff}} = 0.023$ are shown in Fig. 11.19. Again, a thermal fluctuation allows a small number of particles to start to move in the x-direction. For the case shown in Fig. 11.19 this was $4-5$ neighboring particles in the same $y =$ Const. plane. The disturbed region quickly expands in the direction of the driving as in the case of the exponential interaction and again a depleted region is formed. In the LJ case the atoms in the wake of the disturbed region are set into motion by the attractive interaction between the particles at separations in excess of $r = a_s$. The effect of the moving particles creating a traffic jam-like, denser region is different for the LJ interaction than for the exponential interaction. In the latter case a very dense region formed and produced a bulging out of the disturbed region. As can be seen in Fig. 11.19, for the LJ interaction the increased density causes the disturbed region to grow not only in the direction of the driving force, but also preferentially along the other two slip planes of the triangular lattice. The density at the leading edge of the disturbed region does not become as large as for the exponential interaction, and instead of bulging out the disturbed region develops "barbs" at angles $\pm\pi/3$ relative to the driving direction. At later times the disturbed region joins up with itself due to the periodic boundaries.

One extra atom. For the commensurate case described above, the first event in the forward transition is the creation of the disturbed region due to thermal fluctuations. The rate of this process is controlled by the Boltzmann factor $\exp(-\varepsilon_{\text{act}}/T)$, where the activation energy ε_{act} is determined by the

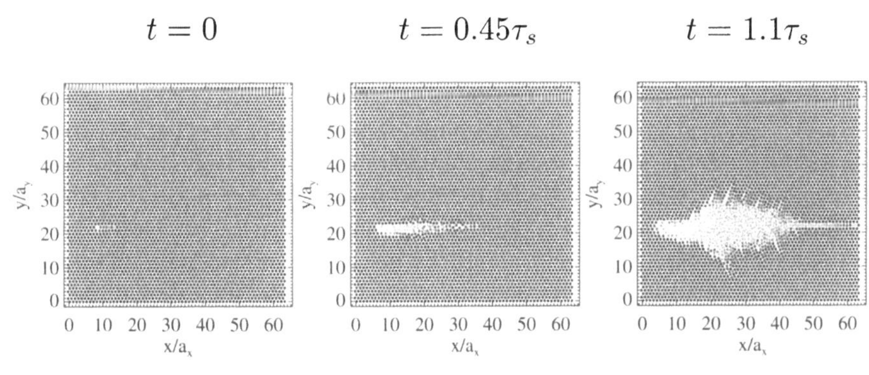

Fig. 11.19. Snapshots of the mechanism of the locked-to-running transition for the LJ interaction with $g_{\text{eff}} = 0.023$, $\eta = 0.2$, $T = 0.0025$, and $F = 0.9901$ [842].

driving force, for example $\varepsilon_{\text{act}} \propto (1-F)^{1/2}$ in the $g_{\text{eff}} \ll 1$ case. In the simulation, Braun *et al.* [842] (2001) used low temperatures to reduce "noise" in the snapshots and, thus, they had to use forces very close to the maximum value, $F = 1$. In order to check whether the scenario is different for lower forces, the authors also made simulations with one extra atom inserted into the commensurate structure. In the 1D FK model such a situation corresponds to one "geometrical" kink which easily moves through the lattice. In the isotropic 2D model, the configuration with one inserted atom surrounded by the "distortion cloud" (due to shifting of neighboring atoms around the inserted one) produces a localized configuration called crowdion. Similarly to the 1D model, the crowdion moves much more easily than atoms of the background commensurate structure. At small F the crowdion performs directed random walks in the x direction, but sometimes it may also jump to the nearest neighboring x-channel and push the atoms from their ideal positions. With increasing force, at some moment an additional crowdion–anti-crowdion pair is created, and subsequently the avalanche starts to grow. The subsequent scenario was found to be the same as described above for the fully commensurate case. The critical force F_c when the avalanche begins, however, is now much lower than in the commensurate case.

Later times. In all cases, the disturbed region eventually joins up on itself, due to the periodic boundary conditions. For relatively small square systems (e.g., for $N \leq 64 \times 64$) the moving domain did not have much opportunity to grow perpendicular to the driving force before joining with itself and so a *river* of moving particles was formed. In the middle of river the atoms move with almost the maximum velocity F_c/η and keep the ordered triangular structure. However, the boundary between the running and locked regions of the layer, which takes about ten atomic channels, is totally disordered. The width of the river was found to grow approximately linearly with time, at a rate that decreased with increasing damping or temperature, until the whole system was in the running state. This scenario is quite general, because even in an infinite system, neighboring disturbed regions arising from thermal fluctuations, will overlap with each other when they reach some critical size, as usual in percolation problems.

For larger square systems Braun *et al.* [842] observed that the moving domain in fact grows considerably both perpendicular and parallel to the driving force. At $T = 0$ the leading, approximately semicircular, edge of the moving domain was found to both grow in diameter at a rate v, and to propagate at speed c. Both v and c were found to be approximately equal to the longitudinal group speed v_L of the layer. As the moving domain then gains area by translational motion and growth, the mobility should increase as t^2.

11.5.2 "Fuse-Safety Device" on an Atomic Scale

Nonlinear dynamics of all models described above has one common feature: the average drift velocity $\langle v \rangle$ of the system is a *monotonically increasing* function of the applied force F. Below we describe the *nonmonotonic* scenario, when the system has a nonzero conductivity ($\langle v \rangle \neq 0$) at low forces and becomes locked ($\langle v \rangle = 0$) at higher forces. Such *fuse* scenario will be described for two simple mechanical models of interacting atoms having *more than one spatial dimension*:

(a) The zigzag FK mode, i.e. the chain of atoms of unit mass subjected to the 2D external potential sinusoidal with the amplitude $\varepsilon_s = 2$ and the period $a_s = 2\pi$ in the x direction (along the chain), and parabolic in the transverse direction y, $V_{\text{sub}}(x, y) = (1 - \cos x) + (1/2)\omega_y^2 y^2$;

(b) The 2D isotropic FK model, i.e. the array of atoms adsorbed on the periodic 2D substrate of the square symmetry, $V_{\text{sub}}(x, y) = 2 - (\cos x + \cos y)$.

For the model (a) an isolated atom near the potential minima has the frequency of x-vibration $\omega_x = 1$, while the frequency of transverse y-vibration is given by ω_y. For the model (b) the vibrations near the minima are isotropic, $\omega_x = \omega_y = 1$. If the force is applied along x, an isolated atom start to slide at the critical force $F_{\text{crit}} \equiv \pi\varepsilon_s/a_s = 1$. In both models we take the exponential interaction between the atoms, $V_{\text{int}}(r) = V_0 e^{-\beta r}$ (in simulation the value $\beta = 1/\pi$ was used), where the amplitude V_0 can be expressed for the sake of convenience as $V_0 = g(\pi e)^2$. The atomic motion is governed by the Langevin equations with the viscous friction η and the external dc force F applied in the x direction.

Let us consider the situation when each potential well is filled with one atom (the trivial commensurate structure), plus some excess atoms are added, which in the ground state form localized compressions, or *kinks*. For example, in the model (a), we place N atoms on the M wells, thus having $N_k = N - M$ kinks in the chain with dimensionless concentration $\theta = N/M$. In the model (b), this is repeated for every row $y = 2\pi n$, n being an integer.

The kinks are the effective mass carriers, since they start to slide at a force much smaller than $F_{\text{crit}} = 1$. The specific features of both 2D models introduced above, by contrast with the one-dimensional FK model, is that the extra spatial dimension allows for the transverse "y-orientation" of the kinks. The increasing force, applied along x, can turn the sliding kinks transversely so that they stop, and the system ends up locked. The main result, the kink velocity [defined as $v_k = \langle v \rangle N/N_k$] versus the external force F at different frictions η and concentrations θ, is shown in Fig. 11.20(a) for the model (a) as an example. The "fused" behavior can be seen clearly: an initially linear growth of v_k with F saturates as soon as the kinks reach certain critical velocity, followed by the drop of v_k to zero. Figure 11.20(b) reveals that this process is accompanied by the growing transverse y-displacements of the atoms.

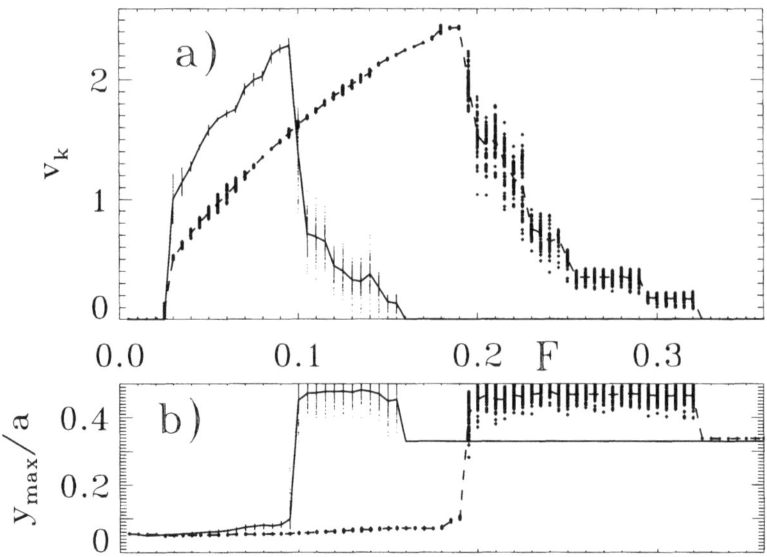

Fig. 11.20. The fuse scenario for the 2D isotropic FK model. (a) Kink velocity v_k vs. force F. Curves show the average values, while symbols are for the raw data from the simulation. Small dots and solid curves are for $\eta = 0.1$, bold dots and dashed curves are for $\eta = 0.2$. (b) The same as in (a) for the maximum transverse displacement of the atoms $y_{\max}$ defined as the maximum deviation of the y coordinate of an atom from the nearest line $y = 2\pi n$, n being an integer [843].

A detailed investigation of the fuse behavior may be found in the paper by Braun *et al.* [843]. A mechanism of the trapping lies in random transverse oscillations of the atoms in the cores of kinks. When the force grows, the amplitude of these oscillations sharply increases after the kinks reach certain critical velocity, and finally the mobile state is transformed into immobile one with the kinks oriented *transversely* to the applied force.

Generally, to exhibit such a peculiar scenario which resembles a "fuse safety device" on an atomic scale, the system should have the following ingredients: (i) a *mobile* ground state, (ii) a close metastable state, which is *immobile* at the same forces, and (iii) with increasing of the applied force F the system should be moved from the mobile state to the immobile one, thus locking itself.

12 Conclusion

A number of books have already been published in the field of nonlinear dynamics and solitons, and many volumes include an analysis of nonlinear models of low-dimensional solid-state physics developed for different applications. A majority of those books includes discussions of seemingly different physical phenomena, where nonlinear excitations become important, as well as some technical methods for solving particular nonlinear equations. The title of this book sounds a bit narrow, and, indeed, it is used more often in the narrow context of a classical model for dislocation dynamics. The spontaneous, legitimate reaction of a careless reader to this title can be summarized by the general question: *"Why study a particular model?"*

Our main effort in selecting the topics covered by this book was to demonstrate the unique generality of many of the basic physical concepts and methods. Unlike many other authors, we did not restrict ourselves to specific applications or a specific field of physics but, instead, tried to present a panoramic view on the fundamental physical concepts of low-dimensional nonlinear physics. Indeed, *The Frenkel-Kontorova Model* project provides a unique framework for realizing our goals, linking seemingly different fields of physics where nonlinear models become important for describing many physical properties. Working on different parts of this book over a period of more than ten years, we have been trying to follow the rapidly growing literature on this and related fields. To our great surprise, we find more and more examples of novel physical systems where the fundamental concepts and results based on those introduced for the FK model, and the analysis of its nonlinear excitations such as kinks and breathers, can be effectively applied.

It would be relevant to reflect some of the most recent developments, in order to expand the panoramic presentation of the physically important models given in Chap. 2. These applications range from the physics of vortex matter in high-T_c superconductors to microscopic engines on the atomic scale, and the number of relevant examples seems indeed endless. Below, we mention only a few of the most recent developments which bring new ideas and new model generalizations not mentioned previously in this book.

Vortex matter. In recent years, great interest has grown in the study of the static and dynamic properties of ensembles of vortices (so-called *vortex matter*) on the spatial profile of the pinning potential originating from defects in the underlying material. The behavior of the vortex lattice was investigated experimentally as well as theoretically in the presence of a great variety of these pinning landscapes, ranging from purely random to highly periodic. Several of these studies revealed the existence of static channels of easy vortex flow at currents just above the critical value. In addition, this critical current was shown to be proportional to the interaction strength between vortices inside the channel and those at the channel edges, as expressed by the shear modulus of the vortex lattice. A precise theoretical description of this phenomenon and of its dependence on the channel width is still lacking. Besseling *et al.* [844] studied the properties of static easy flow channels applying the FK model, and they demonstrated that a mismatch between channel width and lattice constants induces (point) defects in the channel leading to an almost vanishing shear strength. This can have important implications for the properties of vortex matter in a pinning potential with large spatial variations in strength.

Driven vortex lattices interacting with either random or periodic disorder have attracted growing interest due to the rich variety of nonequilibrium dynamic phases that are observed in these systems. These phases include the elastic and plastic flow of vortices which can be related to vortex-lattice order and transport properties. Periodic pinning arrays interacting with a vortex lattice are now attracting increasing attention as recent experiments with patterns of holes and magnetic dots have produced interesting commensurability effects and enhanced pinning. These systems are an excellent realization of an elastic lattice interacting with a periodic substrate that is found in a wide variety of condensed matter systems including charge-density waves, Josephson-junction arrays, and FK-type models of friction. An interesting aspect of periodic pinning arrays is how the symmetry properties of the array affect the transport properties when the vortex lattice is driven at different angles [845].

In many problems associated with the pinning of the vortex lattices, the effective substrate potential is aperiodic. Throughout the book, we mainly considered the case of the sinusoidal substrate potential and, although we often discussed a possible change of the results for a nonsinusoidal shape of the potential, the substrate was always assumed to be strictly periodic. An interesting generalization of the model emerges if the potential is *quasi-periodic*. This problem was studied, in particular, by van Erp *et al.* [846] and Vanossi *et al.* [847]. A model of this type may describe the surface of a quasi-crystal, for example, and also gives insight into the more complex problem of a random substrate potential. For a fixed-density chain, this model also demonstrates the Aubry transition from the locked state to the sliding state with an increase of the elastic constant g. In such a model, however, the sliding state appears for the so-called spiral-mean atomic concentration, while for golden-mean one the PN force is nonzero for any strength of the interaction, contrary to the classical FK model.

Novel topological defects. The most powerful application of the concepts associated with the FK model is to model complicated topological structures or defects in essentially two (and even three) dimensional systems. Magnetic systems are the most commonly used example. For example, Kovalev [75] proposed generalization of the FK model to describe antiferromagnetically (AFM) ordered elastic crystals, and he demonstrated that the FK model can describe complex topological defects in two-dimensional AFM crystals in the form of a bound state of a dislocation with a magnetic declination. An equivalence between some simple magnetic models and the FK model was formulated by Trallori [848], including the models of Fe/Cr(211) magnetic superlattices.

Recently, Korshunov [849] presented an argument for a kink-antikink unbinding transition in the fully frustrated XY model (on square or triangular lattices). This forces the phase transition associated with unbinding of vortex pairs to take place at a lower temperature than the other phase transition, associated with proliferation of the Ising-type domain walls. These results can be applicable to a description of superconducting junction arrays and wire networks in a perpendicular magnetic field, as well as for planar antiferromagnets with a triangular lattice.

Tchernyshyov and Pryadko [850] described the regime of strong coupling between charge carriers and the transverse dynamics of an isolated conducting stripe, such as those found in cuprate superconductors. A stripe is modelled as a partially doped domain wall in an AFM, and doped holes can lose their spin and create a new topological object, *a holon*, which can move along the stripe without frustrating the AFM environment. One aspect in which the holons on the domain wall differ from those in an ordinary one-dimensional electron gas is their transverse degree of freedom: a mobile holon always resides on a transverse kink (or antikink) of the domain wall. This gives rise to two holon flavors and to strong coupling between doped charges and transverse fluctuations of a stripe.

Inspired by the experimental observation that cuprate stripes tilt away from the crystal axis, Bosch *et al.* [851] proposed a new type of stripe phase in the heavy doping regime. The topological excitations associated with this doping are *fractionally charged kinks* that make the stripes fluctuate and tilt. The experimental doping dependence of the tilt angle can be used to determine the fundamental charge quantum of the stripes. They argued that the charged kinks themselves should order. It should therefore be possible to observe the super-lattice reflections of this kink lattice in diffraction experiments. On the one hand, the kinks carry charge and under the assumption that the screening length is of the order of, or larger than, the average kink separation, the kinks would tend to maintain a maximum separation, thereby forming a Wigner crystal of kinks.

Atomic scale engines. The FK model was frequently employed to explain the basic principles of operation of atomic scale engines such as biological motors, ratchet systems, molecular rotors, and molecular machinery in general. The main challenge for this kind of device is to explain how to transform energy to directed motion on such a small scale. In particular, Porto *et al.* [852] suggested a new approach for building microscopic engines on the atomic scale that move translationally or rotationally and can perform useful functions such as pulling a cargo. Characteristic of these engines is the possibility to determine dynamically the directionality of the motion. The approach is based on the transformation of the input energy to directed motion through a dynamical competition between the intrinsic lengths of the moving object and the supporting carrier. The main advantages of this innovative approach are (a) the same concept applies for both translational and rotational motions, (b) the directionality of motion is determined dynamically and does not require spatial asymmetry of the moving object or of the supporting carrier, (c) the velocity obtained can be varied over a wide range, independent of the direction, and (d) the engine is powerful enough to allow for transportation of a cargo.

Unfortunately, some other topics such as the thermal conductivity of low-dimensional systems or the so-called quantum FK model, were not described in the book. This demonstrates one more, how broad are the applications of the FK model!

What does one learn, in conclusion, from a study of this particular model? As long as the outcome of the investigation is a development of universal tools for studying nonlinear systems, we believe the reader can acquire the basic foundations in the interdisciplinary concepts and methods of both solid-state physics and nonlinear science. This book covers many important topics such as the nonlinear dynamics of discrete systems, the dynamics of solitons and their interaction, commensurate and incommensurate systems, and we believe the list of possible applications of the model will grow steadily in the near future. The reason for this is simple: realistic physical systems are often very complicated, and it is always important to develop the fundamental concepts of physics having in mind simpler models. *After all, this is what the field of physics is all about!*

13 Historical Remarks

In the summer of 1948 I asked one of my academic teachers, Ulrich Dehlinger, professor of theoretical physics at the Technische Hochschule Stuttgart (now Universität Stuttgart), to suggest to me a line of research for a *Diplom–Arbeit* in physics (an approximate equivalent of a master's thesis). It had to be a theoretical topic, since at that time, only three years after the end of World War II, laboratory space and experimental equipment were extremely short in the almost completely destroyed university. Even finding a promising topic constituted a major problem, since during the war Germany had been cut off from the international scientific literature almost completely, and new contacts had not yet been established.

Professor Dehlinger suggested to me to have a look at the paper by Frenkel and Kontorova [6], on which he and Kochendörfer had written a comment a few years earlier [853]. Both papers were based on a model that Dehlinger had developed as part of his *Habilitationsschrift*, submitted to the Technische Hochschule Stuttgart on May 14th, 1928, see Fig. 13.1. Dehlinger realized that the model admitted non-trivial mechanically stable solutions, for which he derived an approximate analytical expression. In modern parlance these '*Verhakungen*' would be called pairs of abrupt kinks of opposite sign or simply *kink pairs.* If the Prandtl–Dehlinger model is understood as a reduction from a two-dimensional model, as was intended by Dehlinger, his '*Verhakungen*' correspond to dislocation dipoles, the dissolution of which was held responsible for the recovery and/or recrystallization of plastically deformed metals. For a list of 'translation' from one model to the other see Ref. [854].

Ludwig Prandtl had developed essentially the same model already in 1912/13 but did not publish it until 1928 [1]. The approaches of Prandtl and Dehlinger to the solution of the basic equations were completely different, however. Dehlinger acknowledged the priority of Prandtl in conceiving the model but emphasized that his work was done independently.

My attention was immediately captured by Frenkel and Kontorova's by now well known analytical solution of a uniformly moving single kink, since it was obvious that the full model could not have any uniformly moving solutions. However, the physical meaning of the assumptions that Frenkel and Kontorova had to make in order to arrive at their 'solution' were not immediately clear. Particularly enigmatic was the introduction of a negative atomic mass. After some fruitless efforts to clarify the situation by searching for *exact* solutions, I realized that the model allowed for the crystal structure in two distinct ways, and that these may be handled separately, depending on the questions to be answered.

(i) The existence of atoms is taken into account by the *discreteness* of the model. A direct consequence of this is that the basic equation is a set *difference*-differential equations.

(ii) The force that two adjacent rows of atoms or, more generally, two neighbouring lattice planes, exert on each other is a *periodic function* of their relative displacement. In the model it is represented by a sine function whose period a is equal to an interatomic distance of the crystal structure.

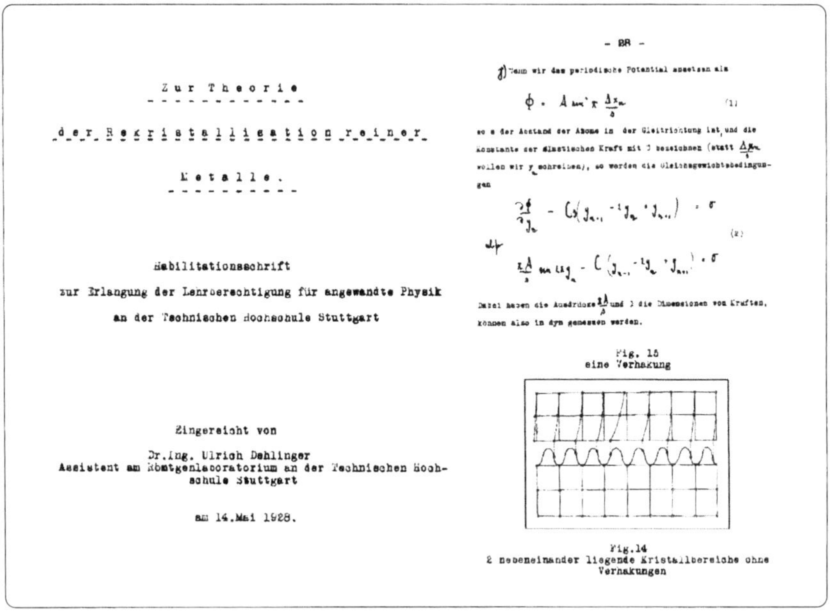
Zur Theorie

der Rekristallisation reiner

Metalle.

Habilitationsschrift

zur Erlangung der Lehrberechtigung für angewandte Physik

an der Technischen Hochschule Stuttgart

Eingereicht von

Dr.Ing. Ulrich Dehlinger

Assistent am Röntgenlaboratorium an der Technischen Hochschule Stuttgart

am 14.Mai 1928.

– 88 –

γ) Wenn wir das periodische Potential ansetzen als

$$\Phi = A \sin^2 \pi \frac{\Delta x_n}{a} \qquad (1)$$

wo a der Abstand der Atome in der Gleitrichtung ist, und die Konstante der elastischen Kraft mit C bezeichnen (statt $\frac{\Delta x_n}{a}$ wollen wir y_n schreiben), so werden die Gleichgewichtsbedingungen

$$\frac{\partial \Phi}{\partial y_n} - C(y_{n-1} - 2y_n + y_{n+1}) = 0 \qquad (2)$$

oder

$$\frac{\pi A}{a} \sin 2\pi y_n - C(y_{n-1} - 2y_n + y_{n+1}) = 0$$

Dabei haben die Ausdrücke $\frac{\pi A}{a}$ und C die Dimensionen von Kräften, können also in dyn gemessen werden.

Fig. 15
eine Verhakung

Fig.14
2 nebeneinander liegende Kristallbereiche ohne Verhakungen

Fig. 13.1. Front page of the *Habilitationsschrift* of U. Dehlinger with excerpts. *'Verhakungen'* are pairs of abrupt kinks of opposite sign or simply *kink pairs.*

In my *Diplomarbeit* I showed how to arrive in a systematic way at a *continuum model* that retained feature (ii) and thus, through the lattice parameter a, one of the important characteristics of the crystal structure. I derived the non-linear partial differential equation

$$\frac{\partial^2 u}{\partial z^2} - \frac{\partial^2 u}{\partial t^2} = \sin u, \qquad (13.1)$$

where z is a normalized spatial coordinate, t a normalized time, and $u = u(z,t)$ a normalized relative displacement of the two adjacent rows of atoms.

Equation (13.1) is Lorentz-invariant and, therefore, I was able to obtain uniformly moving solutions from the static solutions in terms of Jacobian elliptic functions; the Frenkel-Kontorova solution was a special case.

For my doctoral thesis, I chose two main topics. Equation (13.1) was much simpler than the original set of difference-differential equations. This led me to attempt an extension of the two-parameter family of solutions I had obtained in my *Diplomarbeit*. As a first step, I developed a perturbation-theory approach and used it to treat, among other topics, the influence of external forces and to solve the initial-value problem for the solutions I had derived earlier. The approach may be considered as a forerunner of the exact solution that was obtained after the solitonic properties of Eq. (13.1) had been discovered. Another outcome that has remained important till today was the modification of the vibrational spectrum of a crystal by the presence of kinks in dislocations. It plays a key role in the statistical thermodynamics of kinks, as discussed in Chaps. 6 and 7 of the present book.

The second main topic arouse from the controversy between Frenkel–Kontorova and Dehlinger–Kochendörfer on the question whether the Frenkel–Kontorova solution was related to the dislocation concept as developed by Taylor [855], Polanyi [856], and Orowan [857]. Dehlinger and Kochendörfer [853] strongly emphasized this relationship, which had been explicitly denied by Frenkel and Kontorova [6]. In the mean time, at the suggestion of Orowan, Peierls [142] had quantitatively treated a model which combined item (ii) with Taylor's continuum approach, which was based on the linearized theory of elasticity of isotropic media [855]. By trial and error, he had found the solution of the non-linear integro-differential equation that describes a single dislocation. Since, from the physics point of view, the Orowan–Peierls model described the same situation as Eq. (13.1) in the time-independent case, and since no systematic procedure of solving Peierls' integro-differential equation was (and still is) known, I developed a heuristic approach for deriving the set of solutions of Peierls' equation that correspond to the *Diplomarbeit* solutions [858].

The simplicity and the elegance of the results obtained led me to suspect that, in spite of the non-linearity, analytical solutions of Eq. (13.1) might exist beyond those I had found so far. In the physics teaching of that time, non-linear problems played almost no role, and if they turned up, they were usually treated by perturbation theory and linearization. I remembered that in the differential geometry of surfaces this was different, and that there *all* problems were non-linear except for trivial ones that had to do with planes. Sometime in the middle of 1950, when I browsed through the only book on differential geometry on my bookshelf [859], I found the casual remark that if on surfaces of Gaussian curvature $K = -1$ (the so-called pseudo-spherical surfaces) the asymptotic lines $p =$const., $q =$const. are used as parameter lines, the angle ω between the two sets of parameter lines obeys the equation

$$\frac{\partial^2 \omega}{\partial p \, \partial q} = \sin \omega. \tag{13.2}$$

Of course, I immediately realized that Eq. (13.2) is nothing but Eq. (13.1) written in light-cone coordinates. In the language of differential geometry, the 'physical coordinates'

$$u = p + q, \quad v = p - q \tag{13.3}$$

are the lines of curvature of the surface. The asymptotic lines p =const., q =const. are the *characteristic lines* of Eq. (13.1).

Blaschke's book said nothing about explicit solutions of Eq. (13.2). But since surfaces of constant curvature are among the simplest non-trivial objects of differential geometry, it was likely that detailed results had been obtained. I found access to these results in Bianchi's [860, 861] and Eisenhart's [862] textbooks on differential geometry. It turned out that the decisive step had been taken by Bianchi at the age of 23 in his *'tesi di abilitazione'* in 1879 [863]. From one solution of Eq. (13.2), Bianchi's *'trasformatione complementare'* generated another one. It is remarkable that Bianchi achieved this breakthrough entirely by geometrical reasoning. The analytical formulation of the *'trasformatione complementare'* was given only much later and turned out to be a special case of the Bäcklund transformation of Eq. (13.2). Even more important was the *'teorema di permutabilitá'* of Bianchi [864], since it allows us to obtain, without quadratures, a third solution of Eq. (13.2) when two solutions are given.

My attempt to trace the first appearance of Eq. (13.2) in the literature led me to the belief that Enneper [8] was the first to treat Eq. (13.2). I therefore introduced the name Enneper equation for Eqs. (13.1) and (13.2) [13]. In a recent very thorough investigation, Heyerhoff [865, 866] has shown that Bour [867] in his studies on surfaces derived Eq. (13.1) already in 1862, though not in the context of pseudospherical surfaces. Thus, a historically correct name for Eqs (13.1) and (13.2) would be *Bour–Enneper equation.*

By the end of the summer of 1950 it was clear that I had hit a gold lode but that, under the precarious economical and personal circumstances of those days, mining it before submitting my thesis was out of the question. Luckily, I found in Hans Donth a very capable student who took up the topic in his *Diplomarbeit.* We realized that in physical terms, Bianchi's *theorema di permitabilitá* meant that for Eqs. (13.1) and (13.2) the addition theorem of $\tan(u/4)$ played the same role as the superposition principle for linear equations. It is this superposition property that characterizes *solitonic partial differential equations.*

The first 'new' solution we discovered was the breather solution resulting from the superposition of two kinks of opposite sign moving in opposite directions with equal speed. The oscillatory and localized character of the breather raised the question of an appropriate nomenclature. Dehlinger and

Kochendörfer [853] had coined the expression *Eigenbewegungen* (characteristic motions) for the single-kink solutions of Eq. (13.1). They had recognized a certain analogy to the *Eigenschwingungen* (characteristic vibrations) of crystal lattices (fixed energy, corresponding to the eigenvalues of oscillators), and they had suspected the existence of modes that were intermediate between the kink solutions and the familiar small-amplitude vibrations. After we had discovered these 'missing links' and, moreover, that all these modes are superposable in the sense discussed above, I felt that the name *Eigenbewegungen*, indicating greater generality than *Eigenschwingungen*, was highly appropriate. In order to distinguish between the kink–type solutions and the breather–type solutions, I called the first class *translatorische Eigenbewegungen* and the second one *oszillatorische Eigenbewegungen*. Half a century later I still think that these names (with their English counterparts) would be better than the all-comprising expression 'soliton', which is indeed used for such a wide range of objects that it has to be redefined each time.

The great disappointment came when we presented our results to the scientific public and when I tried to arouse the interest of mathematicians and of physicists working in dislocation theory. A typical reaction was that we were considering a very special situation of no general interest, which was not worth being pursued further. The only encouragement from outside came from Elliot Montroll. Being a typical 'problem solver' with very wide interests, he would surely have enjoyed the impact of 'soliton theory' on many branches of science, had only he lived long enough. As a consequence of the disappointing reaction of the scientific community we stopped publishing our mathematical results (some of them have not appeared in the literature even now) and turned towards applications. Fortunately, this turned out to be a very fruitful line of research, too.

In 1952, I was invited to write two contributions to the newly conceived many-volume *Encyclopedia of Physics* of Springer, viz. one on lattice defects [868] and one on crystal plasticity [869]. Writing the first of these contributions in 1953/54 offered the opportunity not only to summarize our results and to relate them to the work of other authors but also to suggest possible relationships to experimental observations. The most important suggestions concerned, on the one hand, the strong increase of the critical shear stress at low temperatures of body-centered cubic metals such as molybdenum and of valence crystals such as the semiconductors with diamond crystal structure, and, on the other hand, the prediction of a mechanical relaxation effect associated with the thermally activated and stress-assisted formation of kink pairs. The first suggestion was fully endorsed by the experimental work of the 1960s and 1970s. With regard to the second suggestion, shortly before the manuscript went to press, the detailed publication of Bordoni's discovery of a mechanical low-temperature relaxation in various plastically deformed face-centered cubic metals [870] was brought to my attention. The (admittedly limited) observations appeared to fit the theoretical predictions.

So I included the interpretation of the Bordoni relaxation in terms of the kink-pair formation in my 1955 Encyclopedia contribution [868].

The kink interpretation of the Bordoni relaxation (and of similar relaxation phenomena that were discovered subsequently) became the starting point of a very important development. We realized that, in contrast to straight dislocations, kinks in dislocations are very suitable objects of statistical mechanics, in particular in the treatment of rate processes involving dislocations. With important contributions by H. Donth, F. Pfaff, P. Schiller, and G. Stenzel, the basic theory was worked out within the next decade (see Ref. [871], which contains also a summary of the attempts to carry Dehlinger's treatment of the discrete model further).

Equation (13.1) is not always the best starting point if taking into account the influence of an applied stress on the dislocation motion is essential. An example is provided by the treatment of the Gibbs free line energy of a dislocation in the saddle-point configuration of kink-pair generation. Following Eshelby [872], for this type of problem we may get physically correct answers if in Eq. (13.1) we replace $\sin u$ by $(1/2)du^2(1-u)^2/du = u(1-u)(1-2u)$ and add a constant term representing the external stress. A complete analytical treatment of the stress- and temperature-assisted generation rate of kink pairs in dislocations has been developed in this way under the assumption that the continuum description referred to above is applicable [873, 874].

The interaction between experiments and theory developed in a very fruitful way and is still going on. At present, it constitutes the main source of information on the Peierls barriers of dislocations [875, 876].

Stuttgart, November 2002 *Alfred Seeger*

References

1. L. Prandtl: Z. angew. Math. Mech. **8**, 85 (1928)
2. U. Dehlinger: Ann. Phys. (Leipzig) **2**, 749 (1929)
3. Ya. Frenkel, T. Kontorova: Phys. Z. Sowietunion **13**, 1 (1938)
4. T.A. Kontorova, Ya.I. Frenkel: Zh. Eksp. Teor. Fiz. **8**, 89 (1938)
5. T.A. Kontorova, Ya.I. Frenkel: Zh. Eksp. Teor. Fiz. **8**, 1340 (1938)
6. Ya. Frenkel, T. Kontorova: J. Phys. Acad. Sci. USSR **1**, 137 (1939)
7. Ph. Rosenau: Phys. Lett. A **118**, 222 (1986)
8. A. Enneper: Nachr. Köngl. Gesellsch. d. Wiss. Göttingen (1870), pp. 493-511
9. A.V. Bäcklund: *Om ytor med konstant negative krökning*, Lunds Universitets Ars-skrift XIX, **IV**, 1 (1882)
10. A. Kochendörfer, A. Seeger: Z. Physik **127**, 533 (1950)
11. A. Seeger, A. Kochendörfer: Z. Physik **130**, 321 (1951)
12. A. Seeger, H. Donth, and A. Kochendörfer: Z. Physik **134**, 173 (1953)
13. A. Seeger, In: *Continuum Models of Discrete Systems*, ed by E. Kröner, K.H. Anthony (University of Waterloo, Waterloo, Ontario 1980) p. 253
14. R. Döttling, J. Esslinger, W. Lay, A. Seeger, In: *Nonlinear Coherent Structures*, Lecture Notes in Physics, vol. 353, ed by M. Barthes, J. Léon (Springer-Verlag Berlin 1990) p. 193
15. N.J. Zabusky, M.D. Kruskal: Phys. Rev. Lett. **15**, 240 (1965)
16. C.S. Gardner, J.M. Greene, M.D. Kruskal, R.M. Miura: Phys. Rev. Lett. **19**, 1095 (1967)
17. J.K. Perring, T.H.R. Skyrme: Nucl. Phys. **31**, 550 (1962)
18. B.D. Josephson: Adv. Phys. **14**, 419 (1965)
19. A.C. Scott: Amer. J. Phys. **37**, 52 (1969)
20. M.J. Ablowitz, D. Kaup, A.C. Newell, H. Segur: Phys. Rev. Lett. **30**, 1262 (1973)
21. V.E. Zakharov, S.V. Manakov, S.P. Novikov, L.P. Pitaevsky: *Theory of Solitons* (Moscow Nauka 1980) [English Translation by Consultants Bureau, New York 1984]
22. Y. Hsu: Phys. Rev. D **22**, 1394 (1980)
23. J. Zadrodziński, M. Jaworski: Z. Phys. B **49**, 75 (1982)
24. M. Jaworski: Phys. Lett. A **125**, 115 (1987)
25. J. Rubinstein: J. Math. Phys. **11**, 258 (1970)
26. A.L. Martinez: Phys. Lett. A **114**, 285 (1986)
27. Yu.S. Kivshar, B.A. Malomed, Rev. Mod. Phys. **61**, 763 (1989)
28. S. Nakajima, T. Yanashita, Y. Onodera: Phys. Rev. B **45**, 3141 (1974)
29. M. Remoissenet: *Waves Called Solitons* (Springer Berlin Heidelberg New York 1999), Chap. 6

30. S. Dusuel, P. Michaux, M. Remoissenet: Phys. Rev E **57**, 2320 (1998)
31. F.C. Frank, J.H. van der Merwe: Proc. Roy. Soc. (London) A **198**, 205 (1949)
32. F.C. Frank, J.H. van der Merwe: Proc. Roy. Soc. (London) A **200**, 125 (1949)
33. W. Atkinson, N. Cabrera: Phys. Rev. **138**, 763 (1965)
34. A.M. Kosevich, A.S. Kovalev, In: *Radiation and Other Defects in Solids* (Inst. of Physics Publishers Tbilisi 1974) (in Russian)
35. A.S. Kovalev, A.D. Kondratyuk, A.M. Kosevich, A.I. Landau: phys. stat. sol. (b) **177**, 117 (1993)
36. O.M. Braun, V.K. Medvedev: Usp. Fiz. Nauk **157**, 631 (1989) [Sov. Phys.-Usp. **32**, 328 (1989)]
37. F. Liu, M. Lagally: Phys. Rev. Lett. **76**, 3156 (1996)
38. S.C. Erwin, A.A. Baski, L.J. Whitman, R.E. Rudd: Phys. Rev. Lett. **83**, 1818 (1999)
39. T.M. Jung, S.M. Prokes, R. Kaplan, J. Vac. Sci. Technol. A **12**, 1838 (1994)
40. A.A. Baski, S.C. Erwin, and L.J. Whitman: Surf. Sci. **423**, L265 (1999)
41. U. Harten, A.M. Lahee, J.P. Toennies, and Ch. Wöll: Phys. Rev. Lett. **54**, 2619 (1985)
42. M. Mansfield, R.J. Needs, J. Phys.: Cond. Matter **2**, 2361 (1990)
43. Y.N. Yang, B.M. Trafas, R.L. Siefert, and J.H. Weaver: Phys. Rev. B **44**, 3218 (1991)
44. P. Franzosi, G. Salviati, M. Seaffardi, F. Genova, S. Pellegrino, A. Stano: J. Cryst. Growth **88**, 135 (1988)
45. S.A. Kukushkin, A.V. Osipov: Surf. Sci. **329**, 135 (1995)
46. O.M. Braun, Yu.S. Kivshar, A.M. Kosevich, J. Phys. C **21**, 3881(1988)
47. O.M. Braun, Yu.S. Kivshar: J Phys. Cond. Mat. **2**, 5961 (1990)
48. I.F. Lyuksyutov, A.G. Naumovets, V.L. Pokrovsky, *Two-Dimensional Crystals* (Naukova Dumka Kiev 1988) [English Translation by Academic Press Boston 1992]
49. O.M. Braun, Yu.S. Kivshar, Phys. Rev. B **43**, 1060 (1991)
50. J.C. Hamilton, R. Stumpf, K. Bromann, M. Giovannini, K. Kern, H. Brune, Phys. Rev. Lett. **82**, 4488 (1999)
51. J.V. Barth, H. Brune, G. Ertl, R.J. Behm: Phys. Rev. B **42**, 9307 (1990)
52. C. Günther, J. Vrijmoeth, R.Q. Hwang, R.J. Behm: Phys. Rev. Lett. **74**, 754 (1995)
53. R. Blinc, A.P. Levanyuk: *Incommensurate Phases in Dielectrics* (North Holland Amsterdam 1986)
54. S.V. Dmitriev, K. Abe, T. Shigenari: J. Phys. Soc. Jpn. **65**, 3938 (1996)
55. S.V. Dmitriev, T. Shigenari, A.A. Vasiliev, K. Abe: Phys. Rev. B **55**, 8155 (1997)
56. F. Axel, S. Aubry, J. Phys. C **14**, 5433 (1981)
57. J.J.M. Slot, T. Janssen: Physica D **32**, 27 (1988)
58. Y. Ishibashi: J. Phys. Soc. Jpn. **60**, 212 (1991)
59. J. Hlinka, H. Orihara, and Y. Ishibashi: J. Phys. Soc. Jpn. **67**, 3488 (1988)
60. T. Janssen, *Incommensurate Phases in Dielectrics*, vol. 14.1, ed by R. Blinc, A.P. Levanyuk (North Holland Amsterdam 1986)
61. H.R. Paneth: Phys. Rev. **80**, 708 (1950)
62. Ya.I. Frenkel: *Introduction into the Theory of Metals* (Nauka Leningrad 1972) (in Russian)
63. W. Xiao, P.A. Greaney, D.C. Chrzan: Phys. Rev. Lett. **90**, 156102 (2003)

64. A.I. Landau, A.S. Kovalev, and A.D. Kondratyuk: phys. stat. sol. (b) **179**, 373 (1993)
65. A.S. Kovalev, A.D. Kondratyuk, A. M. Kosevich, A.I. Landau: Phys. Rev. B **48**, 4122 (1993)
66. M. Suezawa, K. Sumino: phys. stat. sol. (a) **36**, 263 (1976)
67. A. Sugiyama: J. Phys. Soc. Jpn. **47**, 1238 (1979)
68. W. Döring: Z. Naturf. **3**, 373 (1948)
69. R. Becker: J. Phys. Radium **12**, 332 (1951)
70. R.A. Cowley, J.D. Axe, M. Iizumi: Phys. Rev. Lett. **36**, 806 (1976)
71. D.A. Bruce: J. Phys. C **14**, 5195 (1981)
72. U. Enz: Helv. Phys. Acta **37**, 245 (1964)
73. H.J. Mikeska: J. Phys. C **11**, L29 (1978)
74. A.R. Bishop, W.F. Lewis: J. Phys. C **12**, 3811 (1979)
75. A.S. Kovalev: Low Temp. Phys. **20**, 815 (1994)
76. H.J. Mikeska, M. Steiner: Adv. Phys. **40**, 191 (1991)
77. J.K. Kjems, M. Steiner: Phys. Rev. Lett. **41**, 1137 (1978)
78. L.P. Regnault, J.P. Stirling: J. Phys. C **15**, 1261 (1982)
79. M. Steiner, K. Kakurai, J.K. Kjems: Z. Phys. B **53**, 117 (1983)
80. J.P. Boucher, J.P. Renard: Phys. Rev. Lett. **45**, 486 (1980)
81. J.P. Boucher, H. Benner, F. Devreux, L.P. Regnault, J. Rossat-Mignod, C. Dupas, J.P. Renard, J. Bouillot, and W.G. Stirling: Phys. Rev. Lett. **48**, 431 (1982)
82. H. Benner, H. Seitz, J. Weise, J.P. Boucher: J. Magn. Mat. **45**, 354 (1984)
83. H. Seitz, H. Benner: Z. Phys. B **66**, 485 (1987)
84. J. Cibert, Y.M. d'Aubigné: Phys. Rev. Lett. **46**, 1428 (1981)
85. H. Benner, J. Weise, R. Geick, H. Sauer: Europhys. Lett. **3**, 1135 (1987)
86. D.W. McLaughlin, A.C. Scott: Phys. Rev. A **18**, 1652 (1978)
87. N.F. Pedersen, In: *Josephson Effect—Achievments and Trends*, ed by A. Barone (World Scientific Singapore 1986)
88. R.D. Parmentier, In: *The New Superconducting Electronics*, ed by H. Weinstock and R.W. Ralston (Kluwer, Dordrecht 1993) p. 221.
89. N.F. Pedersen, A.V. Ustinov: Supercond. Sci. Technol. **8**, 389 (1995)
90. A.V. Ustinov, R.D. Parmentier, In: *Nonlinear Physics: Theory and Experiment*, ed by E. Alfinito, M. Boiti, L. Martina, and F. Pempinelli (World Scientific Singapore 1996) p. 582
91. A.V. Ustinov: Physica D **123**, 315 (1998)
92. M.B. Mineev, G.S. Mkrtchjan, V.V. Schmidt: J. Low Temp. Phys. **45**, 497 (1981)
93. N. Grønbech-Jensen, M.R. Samuelsen, P.S. Lomdahl, J.A. Blacburn: Phys. Rev. B **42**, 3976 (1990)
94. N. Grønbech-Jensen, O.H. Olsen, M.R. Samuelsen: Phys. Lett. A **179**, 27 (1993)
95. S. Sakai, P. Bodin, N.F. Pedersen: J. Appl. Phys. **73**, 2411 (1993)
96. R. Kleiner, F. Steinmeyer, G. Kunkel, P. Müller: Phys. Rev. Lett. **68**, 2394 (1992)
97. R. Kleiner, P. Müller: Phys. Rev. B **49**, 1327(1994)
98. J.U. Lee, J.E. Nordman, G. Hohenwarter: Appl. Phys. Lett. **67**, 1471 (1995)
99. P. Binder, A.V. Ustinov: Phys. Rev. E **66**, 016603 (2002)
100. A.V. Ustinov, T. Doderer, R.P. Huebener, N.F. Pedersen, B. Mayer, V.A. Oboznov: Phys. Rev. Lett. **69**, 1815 (1992)

101. H.S.T. van der Zant, T.P. Orlando, S. Watanabe, S.H. Strogatz: Phys. Rev. Lett **74**, 174 (1995)
102. S. Watanabe, H.S.J. van der Zant, S.H. Strogatz, T.P. Orlando: Physica D **97**, 429 (1996)
103. S. Yomosa: Phys. Rev. A **27**, 2120 (1983)
104. S. Yomosa: Phys. Rev. A **30**, 474 (1984)
105. S. Homma, S. Takeno: Progr. Theor. Phys. **72**, 679 (1984)
106. L.V. Yakushevich: Phys. Lett. A **136**, 413 (1989)
107. M. Peyrard, A.R. Bishop: Phys. Rev. Lett. **62**, 2755 (1989)
108. T. Dauxois, M. Peyrard, A.R. Bishop: Phys. Rev. E **47**, 684 (1993)
109. G.Z. Zhou, C.T. Zhang: Physica Scripta **43**, 347 (1991)
110. G. Gaeta, C. Reiss, M. Peyrard, T. Dauxois: Rivista Nuovo Cim. **17**, 1 (1994)
111. L.V. Yakushevich: *Nonlinear Physics of DNA* (Wiley Chichester 1998)
112. S.W. Englander, N. Kallenback, A. Heeger, J.A. Krumhansl, S. Litwin: Proc. Nat. Acad. Sci. **77**, 7222 (1980)
113. A.R. Bishop: J. Phys. C **11**, L329 (1978)
114. J.B. Boyce, B.A. Huberman: Phys. Rep. **51**, 189 (1979)
115. W. Dieterich, P. Fulde, I. Peschel: Adv. Phys. **29**, 527 (1980)
116. V.Ya Antonchenko, A.S. Davydov, A.V. Zolotaryuk: phys. stat. solidi (b) **115**, 631 (1983)
117. A.V. Zolotaryuk, K.H. Spatschek, E.W. Ladke: Phys. Lett. A **101**, 517 (1984)
118. A.V. Zolotaryuk: Teor. Mat. Fiz. **68**, 415 (1986) [Theor. Math. Phys. **68**, 916 (1986)]
119. M. Peyrard, St. Pnevmatikos, N. Flytzanis: Phys. Rev. A **36**, 903 (1987)
120. D. Hochstrasser, H. Bütner, H. Desfontaines, M. Peyrard: Phys. Rev. A **36**, 5332 (1988)
121. St. Pnevmatikos: Phys. Rev. Lett. **60**, 1534 (1988)
122. A.V. Zolotaryuk, St. Pnevmatikos: Phys. Lett. A **143**, 233 (1990)
123. St. Pnevmatikos, Yu.S. Kivshar, A.V. Savin, A.V. Zolotaryuk, M.J. Velgakis: Phys. Rev. A **43**, 5518 (1991)
124. A.V. Savin, A.V. Zolotaryuk: Phys. Rev. A **44**, 8167 (1991)
125. L.N. Christophorov, Yu.B. Gaididei: Phys. Lett. A **167**, 367 (1992)
126. P.L. Christiansen, A.V. Savin, A.V. Zolotaryuk: J. Comput. Phys. **134**, 108 (1997)
127. G. Kalosakas, A.V. Zolotaryuk, G.P. Tsironis, E.N. Economou: Phys. Rev. E **56**, 1088 (1997)
128. Yu.S. Kivshar: Phys. Rev. A **43**, 3117 (1991)
129. J. Pouget, G.A. Maugin: Phys. Rev. B **30**, 5306 (1984)
130. J. Pouget, G.A. Maugin: Phys. Rev. B **31**, 4633 (1995)
131. G.A. Maugin, A. Miled: Phys. Rev. B **33**, 4830 (1986)
132. B. Bhushan, J.N. Israelachvili, U. Landman: Nature **374**, 607 (1995)
133. J.B. Sokoloff: Surf. Sci. **144**, 267 (1984)
134. J.B. Sokoloff: Phys. Rev. B **42**, 760 (1990)
135. M. Weiss and F.-J. Elmer: Phys. Rev B **53**, 7539(1996)
136. M. Weiss and F.-J. Elmer: Z. Phys. B **104**, 55 (1997)
137. D. Cule, T. Hwa: Phys. Rev. Lett. **77**, 278 (1996)
138. D. Cule, T. Hwa: Phys. Rev. B **57**, 8235 (1998)
139. O.M. Braun, T. Dauxois, M.V. Paliy, M. Peyrard: Phys. Rev. Lett. **78**, 1295 (1997)

140. O.M. Braun, A.R. Bishop, J. Röder: Phys. Rev. Lett. **79**, 3692 (1997)
141. J. Röder, J.E. Hammerberg, B.L. Holian, A.R. Bishop: Phys. Rev. B **57**, 2759 (1998)
142. R. Peierls: Proc. Phys. Soc. (London) **52**, 34 (1940)
143. F.R.N. Nabarro: Proc. Phys. Soc. (London) **59**, 256 (1947)
144. V.L. Indenbom: Sov. Phys.- Cryst. **3**, 197 (1958)
145. V.L. Indenbom, A.N. Orlov: Sov. Phys.- Usp. **5**, 272 (1962)
146. D.J. Bergman, E. Ben-Jacob, Y. Imry, and K. Maki: Phys. Rev. A **27**, 3345 (1983)
147. R. Hobart: J. Appl. Phys. **36**, 1948 (1965)
148. R. Hobart: J. Appl. Phys. **37**, 3573 (1966)
149. V.L. Pokrovsky: J. Phys. (Paris) **42**, 761 (1981)
150. Y. Ishimori, T. Munakata: J. Phys. Soc. Jpn. **51**, 3367 (1982)
151. Y. Ishibashi, I. Suzuki: J. Phys. Soc. Jpn. **53**, 4250 (1984)
152. C.R. Willis, M. El-Batanouny, P. Stancioff: Phys. Rev. B **33**, 1904 (1986)
153. V.F. Lazutkin, I.G. Schachmannski, M.B. Tabanov: Physica D **40**, 235(1989)
154. S. Flach, K. Kladko: Phys. Rev. E **54**, 2912 (1996)
155. B. Joos: Solid State Commun. **42**, 709 (1982)
156. K. Furuya, A.M. Ozorio de Almeida: J. Phys. A **20**, 6211 (1987)
157. R. Hobart, V. Celli: J. Appl. Phys. **33**, 60 (1962)
158. R. Hobart: J. Appl. Phys. **36**, 1944 (1965)
159. J.F. Currie, S.E. Trullinger, A.R. Bishop, J.A. Krumhansl: Phys. Rev. B **15**, 5567 (1977)
160. P. Stancioff, C.R. Willis, M. El-Batanouny, S. Burdick: Phys. Rev. B **33**, 1912 (1986)
161. S. Flach, C.R. Willis: Phys. Rev. E **47**, 4447 (1993)
162. S. De Lillo: Nuovo Cimento B **100**, 105 (1987)
163. K. Kladko, I. Mitkov, and A.R. Bishop: Phys. Rev. Lett. **84**, 4505 (2000)
164. P.G. Kevrekidis, I.G. Kevrekidis, and A.R. Bishop: Phys. Lett. A **279**, 361 (2001)
165. O.M. Braun, Yu.S. Kivshar, I.I. Zelenskaya, Phys. Rev. B **41**, 7118 (1990)
166. N. Theodorakopoulos, W. Wünderlich, R. Klein: Solid State Commun. **33**, 213 (1980)
167. V.I. Al'shits: Fiz. Tverd. Tela **11**, 2405 (1969) [Sov. Phys.-Solid State **11**, 1947 (1970)]
168. V. Celli, N. Flytzanis: J. Appl. Phys. **41**, 4443 (1970)
169. V.I. Al'shits, V.L. Indenbom, A.A. Shtol'berg. Zh. Eksp. Teor. Fiz. **60**, 2308 (1971) [Sov. Phys. JETP **33**, 1240 (1971)]
170. S. Ishioka: J. Phys. Soc. Jpn. **34**, 462 (1973)
171. M. Peyrard, M.D. Kruskal: Physica D **14**, 88 (1984)
172. R. Boesch, C.R. Willis, M. El-Batanouny: Phys. Rev. B **40**, 2284 (1989)
173. A.M. Kosevich, *The Crystal Lattice* (Wiley Berlin 1999).
174. J.A. Combs, S. Yip: Phys. Rev. B **28**, 6873 (1983)
175. R. Boesch, C.R. Willis: Phys. Rev. B **39**, 361 (1989)
176. O.M. Braun: Surface Sci. **230**, 262 (1990)
177. R. Boesch, R.P. Stancioff, C.R. Willis: Phys. Rev. B **38**, 6713 (1988)
178. A. Igarashi, T. Munakata: J. Phys. Soc. Jpn. **58**, 4025 (1989)
179. A.V. Savin, Y. Zolotaryuk, J.C. Eilbeck: Physica D **138**, 267 (2000)
180. J.C. Eilbeck, R. Flesch: Phys. Lett. A **149**, 200 (1990)

181. M.M. Bogdan, A.M. Kosevich, In: *Nonlinear Coherent Structures in Physics and Biology*, ed by K.H. Spatschek, F.G. Mertens (Plenum, New York 1994) p. 373
182. A. Champneys, Yu.S. Kivshar: Phys. Rev E **61**, 2551 (2000)
183. G.L. Alfimov, V.G. Korolev: Phys. Lett. A **246**, 429 (1998)
184. O.M. Braun: Surface Sci. **213**, 336 (1989)
185. C.A. Condat, R.A. Guyer, M.D. Miller: Phys. Rev. B **27**, 474 (1983)
186. M. Peyrard, M. Remoissenet: Phys. Rev. B **26**, 2886 (1982)
187. M. Remoissenet, M. Peyrard: Phys. Rev. B **29**, 3153 (1984)
188. H. Segur: J. Math. Phys. **24**, 1439 (1983)
189. D.K. Campbell, J.F. Schonfeld, C.A. Wingate: Physica D **9**, 1(1983)
190. O.M. Braun, Yu.S. Kivshar, M. Peyrard: Phys. Rev. E **56**, 6050(1997)
191. Yu.S. Kivshar, D.E. Pelinovsky, T. Cretegny, M. Peyrard: Phys. Rev. Lett. **80**, 5032 (1998)
192. P.G. Kevrekidis, C.K.R.T. Jones: Phys. Rev. E **61**, 3114 (2000)
193. P.G. Kevrekidis, M.I. Weinstein: Physica D **142**, 113 (2000)
194. N.J. Balmforth, R.V. Craster, P.G. Kevrekidis: Physica D **135**, 212(2000)
195. D.K. Campbell, M. Peyrard, P. Sodano: Physica D **19**, 165 (1986)
196. J. Kurosawa: J. Appl. Phys. Suppl. **33**, 320 (1962)
197. W.T. Sanders: Phys. Rev. **128**, 1540 (1962)
198. J. Kratochvil, V.L. Indenbom: Czech. J. Phys. B **13**, 814 (1963)
199. S. Ishioka: J. Phys. Soc. Jpn. **36**, 187 (1974)
200. P. Anninos, S. Oliveira, R.A. Matzner: Phys. Rev. D **44**, 1147 (1991)
201. M. Peyrard, D.K. Campbell: Physica D **9**, 33(1983)
202. J.M. Speight, R.S. Ward: Nonlinearity **7**, 475 (1994)
203. S. Flach, Y. Zolotaryuk, K. Kladko: Phys. Rev. E **59**, 6105 (1999)
204. P. Bak: Rep. Progr. Phys. **45**, 587 (1982)
205. K. Maki, P. Kumar: Phys. Rev. B **14**, 118, 3920(1976)
206. J. Schiefman, P. Kumar: Phys. Scripta **20**, 435 (1979)
207. R. Giachetti, P. Sodano, E. Sorace, V. Tognetti: Phys. Rev. B **30**, 4014 (1984)
208. C.R. Willis, M. El-Batanouny, S. Burdick, R. Boesch: Phys. Rev. B **35**, 3496 (1987)
209. P. Sodano, M. El-Batanouny, C.R. Willis: Phys. Rev. B **34**, 4936 (1986)
210. S. Burdick, C.R. Willis, M. El-Batanouny: Phys. Rev. B **36**, 6920 (1987)
211. I.M. Lifshitz: Phys. (USSR) **7**, 215; 249 (1943)
212. I.M. Lifshitz: Phys. (USSR), **8**, 89 (1944)
213. M.A. Krivoglaz: Zh. Eksp. Teor. Fiz. **40**, 567 (1961) [Sov. Phys.-JETP **13**, 397 (1961)].
214. A.A. Maradudin: *Theoretical and Experimental Aspects of the Effects of Point Defects and Disorder on the Vibrations of Crystals* (Academic Press New York 1966)
215. S.A. Gredeskul, Yu.S. Kivshar: Phys. Rep. **216**, 1(1992)
216. K. Forinash, M. Peyrard, B.A. Malomed: Phys. Rev. E **49**, 3400 (1994)
217. D. Baeriswyl, A.R. Bishop: J. Phys. C **13**, 1403 (1980)
218. H. Reisinger, F. Schwabl: Z. Phys. B **52**, 151(1983)
219. Yu. Galpern, A. Filippov: Zh. Eksp. Teor. Fiz. **86**, 1527 (1984) [Sov. Phys. JETP **59**, 894 (1984)]
220. I. Markov, A. Trayanov: J. Phys. C **21**, 2475(1987)
221. A. Milchev: Phys. Rev. B **33**, 2062 (1986)

222. A. Milchev: Physica D **41**, 262 (1990)
223. B.A. Malomed, A. Milchev: Phys. Rev. B **41**, 4240 (1990)
224. M.B. Fogel, S.E. Trullinger, A.R. Bishop, J.A. Krumhansl: Phys. Rev. Lett. **36**, 1411 (1976)
225. M.B. Fogel, S.E. Trullinger, A.R. Bishop, J.A. Krumhansl: Phys. Rev. B **15**, 1578 (1977)
226. M.J. Rice: Phys. Rev. B **28**, 3587 (1984)
227. J.C. Fernandez, M.J. Goupil, O. Legrand, G. Reinish: Phys. Rev. B **34**, 6207 (1986)
228. P. Woafo, T.C. Kofané: Sol. State Comm. **89**, 261 (1994)
229. T.I. Belova, A.E. Kudryavtsev: Sov. Phys. - Usphekhi **40**, 359 (1997)
230. J. Pouget, S. Aubry, A.R. Bishop, P.S. Lomdahl: Phys. Rev. B **39**, 9500 (1989)
231. M. Salerno, Yu.S. Kivshar: Phys. Lett. A **193**, 263(1994)
232. C.T. Zhang: Phys. Rev. A **35**, 886 (1987)
233. M. Salerno: Phys. Rev. A **44**, 5292 (1991)
234. M. Salerno: Phys. Lett. A **167**, 43 (1992)
235. M. Ricchetty, W. Metzger, H. Heuman: Proc. Nat. Acad. Sci. **84**, 4610 (1988)
236. B.A. Malomed, A.A. Nepomnyashchy: Phys. Rev. B **45**, 12435 (1992)
237. A. Kenfack, T.C. Kofané: Solid State Comm. **89**, 513 (1994)
238. F. Zhang: Phys. Rev E **58**, 2558 (1998)
239. Yu.S. Kivshar, A. Sánchez, O.A. Chubykalo, A. M. Kosevich, L. Vázquez: J. Phys. A: Math. Gen. **25**, 5711 (1992)
240. S.A. Gredeskul, Yu.S. Kivshar, L.M. Maslov, A. Sánchez, L. Vázquez: Phys. Rev. A **45**, 8867 (1992)
241. P.J. Pascual, L. Vázquez: Phys. Rev. B **32**, 8305(1985)
242. P. Biller, F. Petruccione: Phys. Rev. B **41**, 2139 (1990)
243. P. Biller, F. Petruccione: Phys. Rev. B **41**, 2145 (1990)
244. F.G. Bass, Yu.S. Kivshar, V.V. Konotop, Yu.A. Sinitsyn: Phys. Rep. **157**, 63 (1988)
245. A. Sánchez, L. Vázquez: Int. J. Mod. Phys. B **5**, 2825 (1992)
246. J. Lothe, J.P. Hirth: Phys. Rev. **115**, 543(1959)
247. P.A. Kazantsev, V.L. Pokrovsky: Zh. Eksp. Teor. Fiz. **58**, 677 (1968) [Sov. Phys.-JETP **31**, 362 (1970)]
248. J.R. Patel, A.R. Chandhuri: Phys. Rev. **143**, 601 (1966)
249. V.N. Erofeev, V.I. Nikitenko: Fiz. Tverd. Tela **13**, 146 (1971) [Sov. Phys.- Solid State **13**, 116 (1971)].
250. B.V. Petukhov: Fiz. Tverd. Tela **13**, 1445 (1971) [Sov. Phys.- Solid State **13**, 1207 (1971)].
251. B.V. Petukhov: Fiz. Tverd. Tela **25**, 1822 (1983) [Sov. Phys.- Solid State **25**, 1048 (1983)].
252. V.M. Vinokur: J. Physique (Paris) **47**, 1425 (1986)
253. I.R. Sagdeev, V.M. Vinokur: J. Physique (Paris) **48**, 1395(1987)
254. J.A. Gonzaléz, J.A. Holyst: Phys. Rev. B **45**, 10338 (1992)
255. J.A. Gonzaléz, B. de A. Mello: Phys. Lett. A **219**, 226 (1996)
256. J.A. Holyst: Phys. Rev. E **57**, 4786 (1998)
257. J.A. Gonzaléz, A. Bellorín, L.E. Guerrero: Phys. Rev. E **60**, R37 (1999)
258. T.L. Einstein: CRC Crit. Rev. Solid State and Mater. Sci. **7**, 261 (1978)
259. O.M. Braun: Fiz. Tverd. Tela **23**, 2779 (1981) [Sov. Phys.- Solid State **23**, 1626 (1981)]

260. M.A. Vorotyntsev, A.A. Kornyshev, A.I. Rubinshtein: Dokl. Akad. Nauk SSSR **248**, 1321 (1979) [Sov. Phys. Dokl. **24**, 848 (1979)]
261. L.A. Bol'shov, A.P. Napartovich, A.G. Naumovets, A.G. Fedorus: Usp. Fiz. Nauk **122**, 125 (1977) [Sov. Phys.- Usp. **20**, 412 (1977)]
262. A. Milchev, I. Markov: Surface Sci. **136**, 503 (1984)
263. I. Markov, A. Milchev: Thin Solid Films **126**, 83 (1985)
264. O.M. Braun, F. Zhang, Yu.S. Kivshar, L. Vázquez: Phys. Lett. A **157**, 241 (1991)
265. F. Zhang: Phys. Rev. E **54**, 4325 (1996)
266. A. Milchev, G.M. Mazzucchelli: Phys. Rev. B **38**, 2808 (1988)
267. A. Milchev, Th. Fraggis, St. Pnevmatikos: Phys. Rev. B **45**, 10348 (1992)
268. M. Toda: J. Phys. Soc. Jpn. **22**, 431 (1967)
269. M. Toda: J. Phys. Soc. Jpn. **23**, 501 (1967)
270. M. Toda: *Theory of Nonlinear Lattices* (Springer-Verlag Berlin 1981)
271. A.V. Savin: Zh. Eksp. Teor. Fiz. **108**, 1105 (1995) [JETP **81**, 608 (1995)]
272. Y. Zolotaryuk, J.C. Eilbeck, A.V. Savin: Physica D **108**, 81 (1997)
273. A.M. Kosevich, A.S. Kovalev: Solid State Commun. **12**, 763(1973)
274. K. Konno, W. Kameyama, H. Sanuki: J. Phys. Soc. Jpn. **37**, 171 (1974)
275. C. Haas: Solid State Commun. **26**, 709 (1978)
276. R.B. Griffiths, W. Chou: Phys. Rev. Lett. **56**, 1929 (1986)
277. R.B. Griffiths, In: *Fundamental Problems in Statistical Mechanics–VII*, ed by H. van Beijeren (Elsevier Science Publishers Amsterdam 1990) p. 69
278. S. Marianer, L.M. Floria: Phys. Rev. B **38**, 12054 (1988)
279. J.E. Byrne, M.D. Miller: Phys. Rev. B **39**, 374 (1989)
280. M. Marchand, K. Hood, A. Caillé: Phys. Rev. Lett. **58**, 1660 (1987)
281. S. Takeno, S. Homma: J. Phys. Soc. Jpn. **55**, 65 (1986)
282. C.S.O. Yokoi, L.-H. Tang, W. Chou: Phys. Rev. B **37**, 2173 (1988)
283. S. Marianer, A.R. Bishop: Phys. Rev. B **37**, 9893 (1988)
284. G.R. Barsch, J.A. Krumhansl: Phys. Rev. Lett. **53**, 1069 (1984)
285. G.R. Barsch, B. Horovitz, and J.A. Krumhansl: Phys. Rev. Lett. **59**, 1251 (1987)
286. S. Marianer, A.R. Bishop, and J. Pouget, In: *Competing Interactions and Microstructures: Statics and Dynamics*, ed by R. Lesar, A.R. Bishop, R. Heffner, Springer Proceedings in Physics, Vol. 27 (Springer-Verlag Berlin 1988)
287. J.J.M. Slot, T. Janssen: J. Phys. A: Math. Gen. **21**, 3559 (1988)
288. S.K. Sarker, J.A. Krumhansl: Phys. Rev. B **23**, 2374 (1981)
289. M. Remoissenet, N. Flytzanis: J. Phys. C **18**, 1573 (1985)
290. M. Croitoru: J. Phys. A: Math. Gen. **22**, 845(1989)
291. P. Woafo, J.P. Kenne, T.C. Kofané: J. Phys.: Cond. Matter **5**, L123 (1993)
292. G.A. Baker: Phys. Rev. **122**, 1477 (1961)
293. A.M. Kac, E. Helfand: J. Math. Phys. **4**, 1078 (1973)
294. S.F. Mingaleev, Yu.B. Gaididei, E. Majerníkova, S. Shpyrko: Phys. Rev. E **61**, 4454 (2000)
295. V.L. Pokrovsky, A. Virosztek: J. Phys. C **16**, 4513 (1983)
296. J.C. Wang, D.F. Pickett: J. Chem. Phys. **65**, 5378 (1976)
297. M.B. Gordon, J. Villain: J. Phys. C **12**, L151 (1979)
298. I.F. Lyuksyutov: Zh. Eksp. Teor. Fiz. **82**, 1267(1982) [Sov. Phys.-JETP **55**, 737 (1982)].
299. A.L. Talapov: Zh. Eksp. Teor. Phys. **83**, 442 (1982)

300. F.D.M. Haldane, J. Villain: J. Physique (Paris) **42**, 1673 (1981)
301. Yu.M. Aliev, K.N. Ovchinnikov, V.P. Silin, S.A. Uryupin: Zh. Eksp. Teor. Fiz. **107**, 972 (1995) [JETP **80**, 551 (1995)]
302. G.L. Alfimov, V.M. Eleonsky, N.E. Kulagin, N.V. Mitzkevich: Chaos **3**, 405 (1993)
303. P. Rosenau, J.M. Hyman: Phys. Rev. Lett. **70**, 564(1993)
304. V.V. Konotop, S. Takeno: Phys. Rev. E **60**, 1001 (1999)
305. Yu.S. Kivshar: Phys. Rev. E **48**, 43 (1993)
306. P.T. Dinda, T.C. Kofane, M. Remoissenet: Phys. Rev. E **60**, 7525 (1999)
307. Yu.S. Kivshar, B.A. Malomed: Europhys. Lett. **4**, 1215 (1987)
308. K.Ø. Rasmussen, A.R. Bishop, N. Grønbech-Jensen: Phys. Rev. E **58**, R40 (1998)
309. F. Geniet, J. Leon: Phys. Rev. Lett. **89**, 134102 (2002)
310. D.A. Semagin, A.M. Kosevich, T. Shigenari: Physica B **316-317**, 170 (2002)
311. R. Boesch, M. Peyrard: Phys. Rev. B **43**, 8491 (1991)
312. S.V. Dmitriev, T. Shigenari, K. Abe, A.A. Vasiliev, A.E. Miroshnichenko: Comput. Mat. Sci. **18**, 303 (2000)
313. A.M. Kosevich and A.S. Kovalev: Sov. Phys.-JETP **67**, 1793 (1974)
314. M. Remoissenet: Phys. Rev. B **33**, 2386 (1986)
315. O. Bang, M. Peyrard: Physica D **81**, 9 (1995)
316. J. Denzler: Commun. Math. Phys. **158**, 397 (1993)
317. B. Birnir: Comm. Pure Appl. Math. **157**, 103 (1994)
318. J. Geike: Phys. Rev. E **49**, 3539 (1994)
319. O.M. Braun, Yu.S. Kivshar: Phys. Rep. **306**, 1 (1998)
320. P.G. Kevrekidis, A. Saxena, A.R. Bishop: Phys. Rev. E **64**, 026613 (2001)
321. S.V. Dmitriev, T. Shigenari, A.A. Vasiliev, A.E. Miroshnichenko: Phys. Lett. A **246**, 129 (1998)
322. S.V. Dmitriev, T. Miyauchi, K. Abe, T. Shigenari: Phys. Rev. E **61**, 5880 (2000)
323. A.E. Miroshnichenko, S.V. Dmitriev, A.A. Vasiliev, T. Shigenari: Nonlinearity **13**, 837 (2000)
324. S.V. Dmitriev, Yu.S. Kivshar, T. Shigenari: Phys. Rev. E **64**, 056613 (2001)
325. S.V. Dmitriev, Yu.S. Kivshar, T. Shigenari: Physica B **316-317**, 139 (2002)
326. Yu.S. Kivshar, F. Zhang, L. Vázquez: Phys. Rev. Lett. **67**, 1177 (1991)
327. S.V. Dmitriev, Yu.S. Kivshar: unpublished (2003)
328. A.M. Kosevich, A.S. Kovalev: Fiz. Nizk. Temp. **1**, 1544 (1975) [Sov. J. Low Temp. Phys. **1**, 742 (1975)]
329. A.D. Boardman, V. Bortolani, R.F. Willis, K. Xie, H.M. Mehta: Phys. Rev. B **52**, 12736 (1995)
330. Yu.S. Kivshar, B.A. Malomed: J. Phys. A: Math. Gen. **21**, 1553 (1988)
331. A.S. Kovalev, F. Zhang, Yu.S. Kivshar: Phys. Rev. B **51**, 3218 (1995)
332. J.A.D. Wattis, S.A. Harris, C.R. Grindon, C.A. Laughton: Phys. Rev. E **63**, 061903 (2001)
333. M.M. Bogdan, A.S. Kovalev, I.V. Gerasimchuk: Fiz. Nizk. Temp. **23**, 197 (1997) [Low Temp. Phys. **23**, 145 (1997)]
334. N.G. Vakhitov, A.A. Kolokolov: Izv. VUZov Radiofiz. **16**, 1020 (1973) [Radiophys. Quant. Electron. **16**, 783 (1973)]
335. A.A. Sukhorukov, Yu.S. Kivshar, O. Bang, J.J. Rasmussen, P.L. Christiansen: Phys. Rev. E **63**, 036601 (2001)

336. Yu.S. Kivshar, F. Zhang, and A.S. Kovalev: Phys. Rev. B **55**, 14265 (1997)
337. T. Fraggis, St. Pnevmatikos, E.N. Economou: Phys. Lett. A **142**, 361 (1989)
338. F. Zhang, Yu.S. Kivshar, B.A. Malomed, L Vázquez: Phys. Lett. A **159**, 318 (1991)
339. F. Zhang, Yu.S. Kivshar, L. Vázquez: Phys. Rev. A **45**, 6019 (1992)
340. F. Zhang, F., Yu.S. Kivshar, L. Vázquez: Phys. Rev. A **46**, 5214 (1992)
341. F. Zhang, Yu.S. Kivshar, L. Vázquez: J. Phys. Soc. Jpn. **63**, 466 (1994)
342. B.A. Malomed, D.K. Campbell, N. Knowles, and R.J. Flesh: Phys. Lett. A **178**, 271 (1993)
343. T.I. Belova, A.E. Kudryavtsev: Zh. Eksp. Teor. Fiz. **108**, 1489 (1995) [JETP **81**, 817 (1995)]
344. A. Tsurui: J. Phys. Soc. Jpn. **34**, 1462 (1973)
345. B.A. Malomed: J. Phys. A: Math. Gen. **25**, 755 (1992)
346. W. Chen, Y. Zhu, L. Lu: Phys. Rev. B **67**, 184301 (2003)
347. A.A. Maradudin, E.W. Montroll, G.H. Weiss: *Lattice Dynamics in the Harmonic Approximation* (Academic New York 1963)
348. P.W. Anderson: Phys. Rev. **109**, 1492 (1958)
349. A.S. Dolgov: Sov. Phys. Solid State **28**, 907 (1986)
350. A.J. Sievers, S. Takeno: Phys. Rev. Lett. **61**, 970 (1988)
351. S. Takeno, K. Kisoda, A.J. Sievers: Progr. Theor. Phys. Suppl. **No. 94**, 242 (1988)
352. J.B. Page: Phys. Rev. B **41**, 7835 (1990)
353. V.M. Burlakov, S.A. Kiselev, V.N. Pyrkov: Sol. State Commun. **74**, 327 (1990)
354. V.M. Burlakov, S.A. Kiselev, V.N. Pyrkov: Phys. Rev. B **42**, 4921 (1990)
355. S. Takeno, K. Hori: J. Phys. Soc. Jpn. **59**, 3037 (1990)
356. S.R. Bickham, A.J. Sievers: Phys. Rev. B **43**, 2339 (1991)
357. Yu.S. Kivshar: Phys. Lett. A **161**, 80 (1991)
358. R. Scharf, A.R. Bishop: Phys. Rev. A **43**, 6535 (1991)
359. S.R. Bickham, A.J. Sievers, S. Takeno: Phys. Rev. B **45**, 10344 (1992)
360. K.W. Sandusky, J.B. Page, and K.E. Schmidt: Phys. Rev. B **46**, 6161 (1992)
361. K. Hori, S. Takeno: J. Phys. Soc. Jpn. **61**, 4263 (1992)
362. T. Dauxois, M. Peyrard: Phys. Rev. Lett. **70**, 3935 (1993)
363. T. Dauxois, M. Peyrard, C.R. Willis: Phys. Rev. E **48**, 4768 (1993)
364. S.A. Kiselev, S.R. Bickham, A.J. Sievers: Phys. Rev. B **48**, 13508 (1993)
365. Yu.S. Kivshar: Phys. Lett. A **173**, 172(1993)
366. Yu.S. Kivshar: Phys. Rev. E **48**, 4132 (1993)
367. S. Flach, K. Kladko, C.R. Willis: Phys. Rev. E **50**, 2293 (1994)
368. O. Bang, M. Peyrard: Phys. Rev. E **53**, 4143 (1996)
369. S. Takeno, M. Peyrard: Physica D **92**, 140 (1996)
370. K. Forinash, T. Cretegny, M. Peyrard: Phys. Rev. E **55**, 4740 (1997)
371. A.J. Sievers, J.B. Page, In: *Dynamical Properties of Solids VII. Phonon Physics: The Cutting Edge*, ed by G.K. Horton, A.A. Maradudin (Elsevier Amsterdam 1995) p. 137
372. S.A. Kiselev, S.R. Bickham, A.J. Sievers: Comm. Cond. Mat. Phys. **17**, 135 (1995)
373. S. Flach, C.R. Willis: Phys. Rep. **295** 181 (1998)
374. R.S. MacKay, S. Aubry: Nonlinearity **7**, 1623(1994)
375. S. Aubry: Physica D **71**, 196 (1994)
376. R.S. MacKay: Physica A **288**, 174 (2000)

377. R. Bourbonnais, R. Maynard: Phys. Rev. Lett. **64**, 1397 (1990)
378. R. Bourbonnais, R. Maynard: Int. J. Mod. Phys. **1**, 233 (1990)
379. Ch. Claude, Yu.S. Kivshar, O. Kluth, K.H. Spatschek: Phys. Rev. B **47** 14228 (1993)
380. S. Flach: Phys. Rev. E **51**, 3579 (1995)
381. Yu.S. Kivshar, D.K. Campbell: Phys. Rev. E **48**, 3077 (1993)
382. D. Cai, A.R. Bishop, N. Grønbech-Jensen: Phys. Rev. Lett. **72**, 591 (1994)
383. R.S. MacKay, J.-A. Sepulchre: Physica D **82**, 243 (1995)
384. S. Aubry: Physica D **103**, 201 (1997)
385. J.-A. Sepulchre, R.S. MacKay: Nonlinearity **10**, 679 (1997)
386. S. Aubry: Ann. Inst. H. Poincaré, Phys. Théor. **68**, 381 (1998)
387. J.C. Eilbeck, P.S. Lomdahl, A.C. Scott: Physica D **16**, 318(1985)
388. J. Carr, J.C. Eilbeck: Phys. Lett. A **109**, 201 (1985)
389. A. Scott: Philos. Trans. R. Soc. (London) A **315**, 423 (1985)
390. H. Willaime, O. Cardoso, P. Tabeling: Phys. Rev. Lett. **67**, 970 (1991)
391. A.C. Scott, P.L. Christiansen: Phys. Scripta **42**, 257(1990)
392. O. Bang, J.J. Rasmussen, P.L. Christiansen: Physica D **68**, 169(1993)
393. O. Bang, J.J. Rasmussen, P.L. Christiansen: Nonlinearity **7**, 205(1994)
394. B.A. Malomed, M.I. Weinstein: Phys. Lett. A **220**, 91(1996)
395. B.C. Gupta, K. Kundu: Phys. Lett. A **235**, 176 (1997)
396. V.I. Bespalov, V.I. Talanov: Pis'ma Zh. Eksp. Teor. Fiz. **3**, 471 (1966) [JETP Lett. **3**, 307 (1966)]
397. T.B. Benjamin, J.E. Feir: J. Fluid. Mech. **27**, 417 (1967)
398. L.A. Ostrovskii: Zh. Eksp. Teor. Fiz. **51**, 1189 (1967) [Sov. Phys. JETP **24**, 797 (1967)]
399. A. Hasegawa: Phys. Rev. Lett. **24**, 1165 (1970)
400. Yu.S. Kivshar, M. Peyrard: Phys. Rev. A **46**, 3198 (1992)
401. I. Daumont, T. Dauxois, M. Peyrard: Nonlinearity **10**, 617(1997)
402. Yu.S. Kivshar, M. Salerno: Phys. Rev. E **49**, 3543 (1994)
403. V.I. Karpman: Pis'ma Zh. Eksp. Teor. Fiz. **6**, 829 (1967) [JETP Lett. **6**, 277 (1967)]
404. S. Flach, C.R. Willis: Phys. Rev. Lett. **72**, 1777 (1994)
405. W. Krolikowski, Yu.S. Kivshar: J. Opt. Soc. Am. B **13**, 876 (1996)
406. M.J. Ablowitz, J.F. Ladik: J. Math. Phys. **17**, 1011 (1976)
407. M. Salerno: Phys. Rev. A **46**, 6856 (1992)
408. A.A. Vakhnenko, Yu.B. Gaididei: Theor. Mat. Fiz. **68**, 350 (1986) [Theor. Math. Phys. **68**, 873 (1987)]
409. Yu.S. Kivshar: Phys. Rev. B **46**, 8652 (1992)
410. S.J. Putterman, P.H. Roberts: Proc. R. Soc. (London) **440**, 135 (1993)
411. G. Huang, J. Shen, H. Quan: Phys. Rev. B **48**, 16795 (1993)
412. Yu.S. Kivshar, W. Krolikowski, O.A. Chubykalo: Phys. Rev. E **50**, 5020 (1994)
413. B. Denardo, B. Galvin, A. Greenfield, A. Larraza, S. Putterman, W. Wright: Phys. Rev. Lett. **68**, 1730 (1992)
414. S.-Y. Lou, G. Huang: Mod. Phys. Lett. B **9**, 1231 (1995)
415. V. Bortolani, A. Franchini, R.F. Willis: Phys. Rev. B **56**, 8047 (1997)
416. M. Johansson, Yu.S. Kivshar: Phys. Rev. Lett. **82**, 85 (1999)
417. S. Takeno, M. Peyrard: Phys. Rev. E **55**, 1922 (1997)
418. F. Fischer: Ann. Physik **2**, 296 (1993)
419. S. Takeno: J. Phys. Soc. Jpn. **61**, 2821 (1992)

420. S. Flach, K. Kladko, S. Takeno: Phys. Rev. Lett. **79**, 4838 (1997)
421. S. Flach, C.R. Willis, E. Olbrich: Phys. Rev. E **49**, 836 (1994)
422. J.M. Tamga, M. Renmoissenet, J. Pouget: Phys. Rev. Lett. **75**, 357 (1995)
423. J. Pouget, M. Remoissenet, J.M. Tamga, 1993, Phys. Rev. E **47**, 14866 (1993)
424. S. Flach, K. Kladko, R.S. MacKay: Phys. Rev. Lett. **78**, 1207 (1997)
425. P.G. Kevrekidis, K.Ø. Rasmussen, A.R. Bishop: Phys. Rev. E **61**, 2006 (2000)
426. D. Bonart, A.P. Mayer, U. Schröder: Phys. Rev. Lett. **75**, 870 (1995)
427. J.-H. Choy, S.-J. Kwon, G.S. Park: Science **280**, 1589 (1998)
428. J.-P. Locquet, J. Perret, J. Fompeyrine, E. Mächler, J.W. Seo, G. Van Tendeloo, Nature **394**, 453 (1998)
429. J.L. Marín, J.C. Eilbeck, F.M. Russell: Phys. Lett. A **281**, 21 (2001)
430. J.L. Marín, J.C. Eilbeck, F.M. Russell: Phys. Lett. A **248**, 225 (1998)
431. S. Takeno, K. Kawasaki: J. Phys. Soc. Jpn. **60**, 1881 (1991)
432. S. Takeno, K. Kawasaki: Phys. Rev. B **45**, 5083 (1992)
433. R.F. Wallis, D.L. Mills, A.D. Boardman: Phys. Rev. B **52**, R3828 (1995)
434. R. Lai, A.J. Sievers: Phys. Rep. **314**, 147 (1999)
435. U.T. Schwarz, L.Q. English, A.J. Sievers: Phys. Rev. Lett. **83**, 223 (1999)
436. P. Binder, D. Abraimov, A.V. Ustinov, S. Flach, Y. Zolotaryuk: Phys. Rev. Lett. **84**, 745 (2000)
437. E. Trias, J.J. Mazo, T.P. Orlando: Phys. Rev. Lett. **84**, 741 (2000)
438. H.S. Eisenberg, Y. Silberberg, R. Morandotti, A.R. Boyd, J.S. Aitchison: Phys. Rev. Lett. **81**, 3383 (1998)
439. A.A. Sukhorukov, Yu.S. Kivshar, H.S. Eisenberg, Y. Silberberg: IEEE J. Quantum Electron. **39**, 31 (2003)
440. Yu.S. Kivshar and G.P. Agrawal: *Optical Solitons: From Fibers to Photonic Crystals* (Academic San Diego 2003), 560 pp.
441. F.M. Russell, Y. Zolotaryuk, J.C. Eilbeck, T. Dauxois: Phys. Rev. B **55**, 6304 (1997)
442. C. Baesens, R.S. MacKay: Physica D **69**, 59(1993)
443. S. Aubry, P.Y. Le Daeron: Physica D **8**, 381 (1983)
444. B. Mandelbrot: *Fractals* (Freeman San Francisco 1977)
445. W. Chou, R.B. Griffiths: Phys. Rev. B **34**, 6219 (1986)
446. C.S. Ying: Phys. Rev. B **3**, 4160 (1971)
447. J.B. Sokoloff: Phys. Rev. B **16**, 3367 (1977)
448. S. Aubry, In: *Solitons and Condensed Matter Physics*, ed by A.R. Bishop, T. Schneider, Solid State Sciences Vol. 8 (Springer Berlin 1978), p. 264
449. S. Aubry, In: *Some Non-Linear Physics in Crystallographic Structure*, Lecture Notes in Physics Vol. 93 (Springer Berlin 1979) p. 201
450. S. Aubry: Ferroelectrics **24**, 53 (1980)
451. S. Aubry: In: *Physics of Defects, Les Houches, Session XXXV*, ed by R. Balian, M. Kleman, J.P. Poirier (North-Holland, Amsterdam 1980), p. 431
452. S. Aubry, 1981, In: *Symmetries and Broken Symmetries*, ed by N. Boccara (Idset Paris 1981), p. 313
453. S. Aubry, In: *Numerical Methods in the Study of Critical Phenomena*, ed by J. Della Dora, J. Demongeot, B. Lacolle (Springer Berlin 1981), p. 78
454. S. Aubry, In: *The Devil's Staircase Transformation in Incommensurate Lattices*, Vol. 925 of Lecture Notes in Mathematics (Springer Berlin 1982), p. 221.
455. L.N. Bulaevsky, D.I. Chomsky: Zh. Eksp. Teor. Fiz. **74**, 1863 (1978)

456. G. Theodorou, T.M. Rice: Phys. Rev. B **18**, 2840 (1978)
457. V.L. Pokrovsky, A.L. Talapov: Zh. Exp. Theor. Fiz. **75**, 1151 (1978)
458. J.E. Sacco, J.B. Sokoloff: Phys. Rev. B **18**, 6549 (1978)
459. P. Bak: Phys. Rev. Lett. **46**, 791 (1981)
460. S. Aubry: Physica D **7**, 240 (1983)
461. W. Selke, In: *Phase Transitions and Critical Phenomena*, ed by C. Domb, J.L. Lebowitz (Academic Press New York 1992), p. 1
462. M. Hupalo, J. Schmalian, M.C. Tringides: Phys. Rev. Lett. **90**, 216106 (2003)
463. M. Peyrard, S. Aubry: J. Phys. C **16**, 1593 (1983)
464. S.L. Shumway, J.P. Sethna: Phys. Rev. Lett. **67**, 995(1991)
465. H.J. Schellnhuber, H. Urbschat, A. Block: Phys. Rev. A **33**, 2856 (1986)
466. H.J. Schellnhuber, H. Urbschat, J. Wilbrink: Z. Phys. B **80**, 305 (1990)
467. B. Lin, B. Hu: J. Stat. Phys. **69**, 1047 (1992)
468. J.D. Meiss: Rev. Mod. Phys. **64**, 795 (1992)
469. B.V. Chirikov: Atomnaya Energiya **6**, 630 (1959)
470. B.V. Chirikov: Phys. Rep. **52**, 263 (1979)
471. J.M. Greene: J. Math. Phys. **20**, 1183 (1979)
472. J.M. Greene, R.S. MacKay, F. Vivaldi, M.J. Feigenbaum: Physica D **3**, 468 (1981)
473. A.J. Lichtenberg, M.A. Lieberman: *Regular and Stochastic Motion* (Springer Berlin 1983).
474. H.G. Schuster: *Deterministic Chaos. An Introduction* (Physik-Verlag Weinheim 1984)
475. S.J. Shenker, L.P. Kadanoff: J. Stat. Phys. **27**, 631(1982)
476. R.S. MacKay: *Renormalization in Area-Preserving Maps* (World Scientific, Singapore 1993)
477. L.-H. Tang, R.B. Griffiths: J. Stat. Phys. **53**, 853(1988)
478. K. Sasaki, L.M. Floría: J. Phys. C **1**, 2179 (1989)
479. B. Joos, B. Bergersen, R.J. Gooding, M. Plischke: Phys. Rev. B **27**, 467 (1983)
480. E. Burkov, B.E.C. Koltenbach, L.W. Bruch: Phys. Rev. B **53**, 14179 (1996)
481. J.A. Ketoja, I.I. Satija: Physica D **104**, 239 (1997)
482. P. Tong, B. Li, B. Hu: Phys. Rev. B **59**, 8639 (1999)
483. B. Hu, B. Li, H. Zhao: Phys. Rev. E **61**, 3829 (2000)
484. B. Sutherland: Phys. Rev. A **8**, 2514 (1973)
485. W.L. McMillan: Phys. Rev. B **16**, 4655 (1977)
486. A.D. Novaco: Phys. Rev. B **22**, 1645 (1980)
487. L.M. Floría, J.J. Mazo: Adv. Phys. **45**, 505(1996)
488. R.C. Black, I.I. Satija: Phys. Rev. B **44**, 4089 (1991)
489. S.N. Coppersmith, D.S. Fisher: Phys. Rev. B **28**, 2566 (1983)
490. S.R. Sharma, B. Bergersen, B. Joos: Phys. Rev. B **29**, 6335 (1984)
491. L. de Seze, S. Aubry: J. Phys. C **17**, 389 (1984)
492. S. Aubry, P. Quemerais, In: *Low Dimensional Electronic Properties of Molybdenum Bronzes and Oxides*, ed by C. Schlenker (Kluwer Dordrecht 1984)
493. R.S. MacKay: Physica D **50**, 71 (1991)
494. O. Biham, D. Mukamel: Phys. Rev. A **39**, 5326 (1989)
495. L. Consoli, H.J.F. Knops, A. Fasolino: Phys. Rev. Lett. **85**, 302 (2000)
496. R.A. Guyer, M.D. Miller: Phys. Rev. A **17**, 1774 (1978)
497. R.A. Guyer, M.D. Miller: Phys. Rev. Lett. **42**, 718 (1979)
498. L.A. Bol'shov: Fiz. Tverd. Tela **13**, 1679 (1971)

499. P. Bak, R. Bruinsma: Phys. Rev. Lett. **49**, 249 (1982)
500. Ya.G. Sinai, S.E. Burkov: Uspechi Mat. Nauk **38**, 205(1983)
501. I. Dzyaloshynsky: Zh. Eksp. Teor. Fiz. **47**, 992 (1964)
502. W.L. McMillan: Phys. Rev. B **14**, 1496 (1976)
503. J. Villain, In: *Ordering in Strongly Fluctuating Condensed Matter Systems*, ed by T. Riste (New York Plenum 1980) p. 222
504. J.P. McTaque, M. Nielsen, L. Passel, In: *Ordering in Strongly Fluctuating Condensed Matter Systems*, ed by T. Riste (New York Plenum 1980) p. 195.
505. S. Iwabuchi: Prog. Theor. Phys. **70**, 941 (1983)
506. R.C. Black, I.I. Satija: Phys. Rev. Lett. **65**, 1 (1990)
507. I. Markov, A. Milchev: Surface Sci. **136**, 519 (1984)
508. I. Markov, A. Milchev: Surface Sci. **145**, 313(1984)
509. A. Milchev, I. Markov: Surface Sci. **156**, 392(1985)
510. S. Aubry, K. Fesser, A.R. Bishop: J. Phys. A **18**, 3157 (1985)
511. M.D. Miller, J.S. Walker: Phys. Rev. B **44**, 2792 (1991)
512. I. Markov, A. Trayanov: J. Phys. C **21**, 2475(1988)
513. I. Markov, A. Trayanov: J. Phys.: Cond. Mat. **2**, 6965(1990)
514. K. Ploog, W. Stolz, L. Tapfer, In: *Thin Film Growth Techniques for Low-Dimensional Structures* (New York, Plenum 1987) p. 5.
515. C.G. Goedde, A.J. Lichtenberg, M.A. Liberman: Physica D **59**, 200 (1992)
516. R.E. Peierls: Helv. Phys. Acta **7**, 81 (1934)
517. J.A. Krumhansl, J.R. Schrieffer: Phys. Rev. B **11**, 3535 (1975)
518. T. Schneider, In: *Solitons*, ed by S.E. Trullinger, V.E. Zakharov, V.L. Pokrovsky (Springer-Verlag Berlin 1986) Chap. 7
519. T. Tsuzuki, K. Sasaki: Progr. Theor. Phys. Suppl. **94**, 73 (1988)
520. T. Schneider, E. Stoll: Phys. Rev. B **22**, 5317 (1980)
521. R.A. Guyer, M.D, Miller: Phys. Rev. B **20**, 4748 (1979)
522. P.W. Anderson, B.I. Halperin, C.M. Varma: Philos. Mag. **25**, 1 (1972)
523. W.A. Phillips: J. Low Temp. Phys. **7**, 351 (1972)
524. S.N. Coppersmith: Phys. Rev. A **36**, 3375 (1987)
525. H.U. Beyeler, L. Pietronero, S. Strässler: Phys. Rev. B **22**, 2988(1980)
526. L. Pietronero, S. Strässler: Phys. Rev. Lett. **42**, 188 (1979)
527. P. Reichert, R. Schilling: Phys. Rev. B **32**, 5731 (1985)
528. F. Vallet, R. Schilling, S. Aubry: Europhys. Lett. **2**, 815(1986)
529. F. Vallet, R. Schilling, S. Aubry: J. Phys. C **21**, 67 (1988)
530. O.V. Zhirov, G. Casati, D.L. Shepelyansky: Phys. Rev. E **65**, 26220(2002)
531. L. Pietronero, W.R. Schneider, S. Strässler: Phys. Rev. B **24**, 2187(1981)
532. H.V. Löhneysen, H.J. Schink, W. Arnold, H.U. Beyeler, L. Pietronero, S. Strässler: Phys. Rev. Lett. **46**, 1213 (1981)
533. J.F. Currie, J.A. Krumhansl, A.R. Bishop, S.E. Trullinger: Phys. Rev. B **22**, 477 (1980)
534. S.E. Trullinger: Phys. Rev. B **22**, 418 (1980)
535. K. Sasaki: Phys. Rev. B **33**, 2214 (1986)
536. H. Takayama, K. Maki: Phys. Rev. B **20**, 5002 (1979)
537. H. Takayama, K. Maki: Phys. Rev. B **21**, 4558 (1980)
538. H. Takayama, K. Wada: J. Phys. Soc. Jpn. **50**, 3549 (1981)
539. N. Theodorakopoulos: Z. Phys. B **46**, 367(1982)
540. N. Theodorakopoulos: Phys. Rev. B **30**, 4071(1984)
541. N. Theodorakopoulos, E.W. Weller: Phys. Rev. B **37**, 6200 (1988)

542. E. Tomboulis: Phys. Rev. D **12**, 1678 (1975)
543. T. Miyashita, K. Maki: Phys. Rev. B **28**, 6733 (1983)
544. T. Miyashita, K. Maki: Phys. Rev. B **31**, 1836 (1985)
545. M. Fukuma, S. Takada: J. Phys. Soc. Jpn. **55**, 2701 (1986)
546. K. Sasaki: Prog. Theor. Phys. **70**, 593 (1983)
547. K. Sasaki: Prog. Theor. Phys. **71**, 1169 (1984)
548. S.E. Trullinger, K. Sasaki: Physica D **28**, 181 (1987)
549. C.R. Willis, R. Boesch: Phys. Rev. B **41**, 4570 (1990)
550. D.J. Scalapino, M. Sears, R.A. Ferrell: Phys. Rev. B **6**, 3409 (1972)
551. N. Gupta, B. Sutherland: Phys. Rev. A **14**, 1790 (1976)
552. M.J. Gillan, R.W. Holloway: J. Phys. C **18**, 4903(1985)
553. M.J. Gillan, R.W. Holloway: J. Phys. C **18**, 5705(1985)
554. J.E. Sacco, A. Widom, J.B. Sokoloff: J. Stat. Phys. **21**, 497 (1979)
555. W. Kohn: Phys. Rev. **115**, 809 (1959)
556. M. Abramowitz, I.A. Stegun: *Handbook of Mathematical Functions* (US Dept. of Commerce Washington 1972)
557. M. Croitoru, D. Grecu, A. Visinescu, V. Cionga: Rev. Roum. Phys. **29**, 853(1984)
558. S. Goldstein: Proc. R. Soc. Edinburgh T **49**, 210 (1929)
559. R.M. De Leonardis, S.E. Trullinger: Phys. Rev. A **20**, 2603 (1979)
560. Y. Okwamoto, H. Takayama, H. Shiba: J. Phys. Soc. Jpn. **46**, 1420 (1979)
561. K. Sasaki: Prog. Theor. Phys. **67**, 464 (1982)
562. K. Sasaki, T. Tsuzuki: Solid State Commun. **41**, 521 (1982)
563. K. Sasaki: Prog. Theor. Phys. **68**, 411 (1984)
564. K. Sasaki: Phys. Rev. B **33**, 7743 (1986)
565. M.J. Gillan: J. Phys. C **18**, 4885 (1985)
566. R.J. Baxter: *Exactly Solved Models in Statistical Mechanics* (Academic London 1982)
567. L. Tonks: Phys. Rev. **50**, 955 (1936)
568. R.M. De Leonardis, S.E. Trullinger: Phys. Rev. B **22**, 4558 (1980)
569. J.A. Holyst, A. Sukiennicki, 1984, Phys. Rev. B **30**, 5356 (1984)
570. C.R. Willis, M. El-Batanouny, R. Boesch, P. Sodano: Phys. Rev. B **40**, 686(1989)
571. R.M. De Leonardis, S.E. Trullinger: Phys. Rev. B **27**, 1867(1983)
572. M. Croitoru: J. Phys. A: Math. Gen. **20**, 1695 (1987)
573. J.R. Kenné, P. Woafo, T.C. Kofané: J. Phys.: Cond. Matter **6**, 4277 (1994)
574. C.W. Gardiner: *Handbook of Stochastic Methods for Physics, Chemistry and the Natural Sciences* (Springer-Verlag Berlin 1985)
575. H.C. Brinkman: Physica **22**, 29 (1956)
576. E.G. d'Agliano, W.L. Schaich, P. Kumar, H. Suhl: In: *Nobel Symposium 24. Collective Properties of Physical Systems*, ed by B. Lundquist (Academic New York 1973)
577. E.G. d'Agliano, P. Kumar, W. Schaich, H. Suhl: Phys. Rev. B **11**, 2122(1975)
578. R. Zwanzig, In: *Lectures in Theoretical Physics*, Vol. 3, ed by W.E. Brittin, B.W. Downs, J.Downs (Interscience, New York 1961)
579. R. Zwanzig: Phys. Rev. **124**, 983 (1961)
580. H. Mori: Prog. Theor. Phys. **33**, 423 (1965)
581. H. Mori, H. Fugisaka: Prog. Theor. Phys. **49**, 764(1973)
582. D. Forster: *Hydrodynamic Fluctuations, Broken Symmetry and Correlation Functions* (W.A. Benjamin London 1975)

583. R. Kubo: J. Phys. Soc. Jpn. **12**, 570 (1957)
584. R. Kubo: Rep. Prog. Phys. **29**, 255 (1966)
585. R.L. Stratonovich: *Topics in the Theory of Random Noise* (Gordon and Breach, New York 1967)
586. Yu.M. Ivanchenko, L.A. Zil'berman: Pis'ma JETF **8**, 189 (1968) [JETP Lett. **8**, 113 (1968)].
587. Yu.M. Ivanchenko, L.A. Zil'berman: Sov. Phys. JETP **28**, 1272 (1969)
588. V. Ambegaokar, B.I. Halperin: Phys. Rev. Lett. **22**, 1364 (1969)
589. V. Ambegaokar, B.I. Halperin: Phys. Rev. Lett. **23**, 274 (1969)
590. H. Risken, H.D. Vollmer: Phys. Lett. A **69**, 387 (1979)
591. H. Risken, H.D. Vollmer: Z. Phys. B **35**, 177(1979)
592. G. Wilemski: J. Stat. Phys. **14**, 153 (1976)
593. U.M. Titulaer: Physica A **91**, 321 (1978)
594. H. Risken, H.D. Vollmer, M. Mörsch: Z. Phys. B **40**, 343 (1981)
595. H. Risken: *The Fokker–Planck Equation* (Springer Berlin 1984)
596. V.O. Mel'nikov: Phys. Rep. **209**, 1 (1991)
597. P. Fulde, L. Pietronero, W.R. Schneider, S. Strässler: Phys. Rev. Lett. **35**, 1776 (1975)
598. W. Dieterich, I. Peschel, W.R. Schneider: Z. Phys. B **27**, 177 (1977)
599. W. Dieterich, T. Geisel, I. Peschel: Z. Phys. B **29**, 5 (1978)
600. H. Risken, H.D. Vollmer: Z. Phys. B **31**, 209(1978)
601. H. Risken, H.D. Vollmer: Z. Phys. B **33**, 297 (1979)
602. H.D. Vollmer, H. Risken: Z. Phys. B **34**, 313 (1979)
603. H.A. Kramers: Physica **7**, 284 (1940)
604. P. Hänggi, P. Talkner, M. Borkovec: Rev. Mod. Phys. **62**, 251(1990)
605. V.P. Zhdanov: Surface Sci. **214**, 289 (1989)
606. H. Mori: Prog. Theor. Phys. **34**, 399 (1965)
607. W.R. Schneider: Z. Phys. B **24**, 135 (1976)
608. T. Geisel: Phys. Rev. B **20**, 4294 (1979)
609. T. Munakata, A. Tsurui: Z. Phys. B **34**, 203(1979)
610. H. Haken: *Synergetics* (Springer-Verlag Berlin 1980)
611. E. Joergensen, V.P. Koshelets, R. Monaco, J. Mygind, M.R. Samuelsen: Phys. Rev. Lett. **49**, 1093 (1982)
612. F. Marchesoni: Phys. Lett. A **115**, 29 (1986)
613. M. Remoissenet: Solid State Commun. **27**, 681(1978)
614. M. Büttiker, R. Landauer: J. Phys. C **13**, L325(1980)
615. N.R. Quintero, A. Sánchez, F.G. Mertens: Phys. Rev. Lett. **84**, 871 (2000)
616. N.R. Quintero, A. Sánchez, F.G. Mertens: Phys. Rev. E **60**, 222 (1999)
617. F. Marchesoni: Phys. Rev. Lett. **74**, 2973(1995)
618. F. Petruccione, P. Biller: Phys. Rev. B **41**, 2145(1990)
619. M. Ogata, Y. Wada: J. Phys. Soc. Jpn. **54**, 3425(1985)
620. M. Ogata, Y. Wada: J. Phys. Soc. Jpn. **55**, 1252(1986)
621. C. Kunz: Phys. Rev. A **34**, 510 (1986)
622. C. Kunz: Phys. Rev. B **34**, 8144 (1986)
623. F. Marchesoni, C.R. Willis: Phys. Rev. A **36**, 4559(1987)
624. N. Theodorakopoulos, E.W. Weller: Phys. Rev. B **38**, 2749(1988)
625. V.G. Bar'yakhtar, B.A. Ivanov, A.L. Sukstanskii, E.V. Tartakovskaya: Teor. Mat. Fiz. **74**, 46 (1988)
626. S.M. Alamoudi, D. Boyanovsky, F.I. Takakura: Phys. Rev. E **57**, 919(1998)

627. Y. Wada, J.R. Schrieffer: Phys. Rev. B **18**, 3897(1978)
628. N. Theodorakopoulos, R. Klein: phys. stat. sol. (a) **61**, 107(1980)
629. N. Theodorakopoulos, R. Klein, In: *Physics in One Dimension*, ed by J. Bernasconi, T. Schneider (Springer-Verlag Berlin 1981)
630. K. Fesser: Z. Phys. B **39**, 47 (1980)
631. Y. Wada, H. Ishiuchi: J. Phys. Soc. Jpn. **51**, 1372(1982)
632. K. Sasaki, K. Maki: Phys. Rev. B **35**, 257(1987)
633. K. Sasaki, K. Maki: Phys. Rev. B **35**, 263(1987)
634. N. Theodorakopoulos: Z. Phys. **33**, 385 (1979)
635. B.A. Ivanov, A.K. Kolezhuk: Pis'ma Zh. Eksp. Teor. Fiz. **49**, 489(1989) [JETP Lett. **49**, 489 (1989)]
636. B.A. Ivanov, A.K. Kolezhuk: Phys. Lett. A **146**, 190(1990)
637. F. Marchesoni, C.R. Willis: Europhys. Lett. **12**, 491(1990)
638. L. Pietronero, S. Strässler: Solid State Commun. **27**, 1041 (1978)
639. C. Kunz, J.A. Combs: Phys. Rev. B **31**, 527(1985)
640. C. Cattuto, G. Costantini, T. Guidi, F. Marchesoni: Phys. Rev. B **63**, 94308(2001)
641. J.A. Combs, S. Yip: Phys. Rev. B **29**, 438(1984)
642. R.W. Holloway, M.J. Gillan: Solid State Commun. **52**, 705(1984)
643. T. Schneider, E. Stoll: Phys. Rev. B **23**, 4631(1981)
644. W.C. Kerr, D. Baeriswyl, A.R. Bishop: Phys. Rev. B **24**, 6566(1981)
645. S. Aubry: J. Chem. Phys. **64**, 3392 (1976)
646. W. Hasenfratz, R. Klein, N. Theodorakopoulos: Solid State Commun. **18**, 893(1976)
647. K. Kawasaki: Prog. Theor. Phys. **55**, 2029 (1976)
648. G.F. Mazenko, P.S. Sahni: Phys. Rev. B **18**, 6139(1978)
649. P.S. Sahni, G.F. Mazenko: Phys. Rev. B **20**, 4674(1979)
650. A.R. Bishop: Solid State Commun. **30**, 37 (1979)
651. A.R.Bishop: J. Phys. A **14**, 1417 (1981)
652. K.M. Leung, D.L. Huber: Solid State Commun. **32**, 127(1979)
653. E. Allroth, H.J. Mikeska: J. Phys. C **13**, L725(1980)
654. E. Allroth, H.J. Mikeska: Z. Phys. B **43**, 209 (1981)
655. J.P. Boucher, F. Mezei, L.P. Regnault, J.P. Renard: Phys. Rev. Lett. **55**, 1778(1985)
656. J.P. Boucher, L.P. Regnault, R. Pynn, J. Bouillot, J.P. Renard: Europhys. Lett. **1**, 415(1986)
657. K. Maki: J. Low Temp. Phys. **41**, 327 (1980)
658. H.J. Mikeska: J. Phys. C **13**, 2913 (1980)
659. W. Dieterich, I. Peschel: Physica A **95**, 208(1979)
660. E.H. Lieb, D.C. Mattis: *Mathematical Physics in One Dimension* (Academic New York 1966)
661. B.J. Ackerson: J. Chem. Phys. **69**, 684 (1978)
662. A.Z. Akcasu, J.J. Duderstadt: Phys. Rev. **188**, 479(1969)
663. D. Forster, P.C. Martin: Phys. Rev. A **2**, 1575 (1970)
664. G.F. Mazenko: Phys. Rev. A **3**, 2121 (1971)
665. C.D. Boley: Phys. Rev. A **5**, 986 (1972)
666. A.Z. Akcasu: Phys. Rev. A **7**, 182 (1973)
667. J. Bosse, W. Götze, M. Lücke: Phys. Rev. A **17**, 434(1978)
668. Y. Imry, B. Gavish: J. Chem. Phys. **61**, 1554(1974)

669. S. Alexander, P. Pincus: Phys. Rev. B **18**, 2011(1978)
670. A.R. Bishop, W. Dieterich, I. Peschel: Z. Phys. B **33**, 187(1979)
671. S.E. Trullinger, M.D. Miller, R.A. Guyer, A.R. Bishop, F. Palmer, J.A. Krumhansl: Phys. Rev. Lett. **40**, 206 (1978)
672. S.E. Trullinger, M.D. Miller, R.A. Guyer, A.R. Bishop, F. Palmer, J.A. Krumhansl: Phys. Rev. Lett. **40**, 1603 (1978)
673. T. Schneider, E.P. Stoll, R. Morf: Phys. Rev. B **18**, 1417(1978)
674. K.C. Lee, S.E. Trullinger: Phys. Rev. B **21**, 589 (1980)
675. H.D. Vollmer: Z. Phys. B **33**, 103 (1979)
676. M. Büttiker, R. Landauer: Phys. Rev. A **23**, 1397(1981)
677. M. Büttiker, R. Landauer: Phys. Rev. Lett. **43**, 1453(1979)
678. M. Büttiker, T. Christen: Phys. Rev. Lett. **75**, 1895(1995)
679. G.H. Vineyard: J. Phys. Chem. Solids **3**, 121(1957)
680. R. Landauer, J.A. Swanson: Phys. Rev. **121**, 1668(1961)
681. J.S. Langer: Ann. Phys. **41**, 108 (1967)
682. J.S. Langer: Phys. Rev. Lett. **21**, 973 (1968)
683. J.S. Langer: Ann. Phys. **54**, 258 (1969)
684. L. Gunther, Y. Imry: Phys. Rev. Lett. **44**, 1225 (1980)
685. O.M. Braun, Yu.S. Kivshar: Phys. Rev. B **50**, 13388 (1994)
686. P. Woafo, T.C. Kofane, A.S. Bokosah: Phys. Scripta **56**, 655(1997)
687. A.R. Bishop, S.E. Trullinger: Phys. Rev. B **17**, 2175 (1978)
688. P. Reimann, C. van der Broeck, H. Linke, P. Hänggi, J.M. Rubi, A. Pérez-Madrid: Phys. Rev. Lett. **87**, 10602 (2001)
689. H.D. Vollmer, H. Risken: Z. Phys. B **52**, 259 (1983)
690. D.E. McCumber: J. Appl. Phys. **39**, 3113 (1968)
691. E. Ben Jacob, D.J. Bergman, B.J. Matkowsky, Z. Schuss: Phys. Rev. A **26**, 2805 (1982)
692. M. Büttiker, E.P. Harris, R. Landauer: Phys. Rev. B **28**, 1268 (1983)
693. H.D. Vollmer, H. Risken: Z. Phys. B **37**, 343 (1980)
694. K. Voigtlaender, H. Risken: J. Stat. Phys. **40**, 397 (1985)
695. H. Risken, K. Voigtlaender: J. Stat. Phys. **41**, 825 (1985)
696. G. Costantini, F. Marchesoni: Europhys. Lett. **48**, 491 (1999)
697. A.A. Middleton: PhD. Thesis (Princeton University 1990)
698. A.A. Middleton, D.S. Fisher: Phys. Rev. Lett. **66**, 92 (1991)
699. A.A. Middleton, Phys. Rev. Lett. **68**, 670 (1992)
700. A.A. Middleton, D.S. Fisher: Phys. Rev. Lett. **47**, 3530 (1993)
701. L. Sneddon: Phys. Rev. Lett. **52**, 65 (1984)
702. S.N. Coppersmith, D.S. Fisher: Phys. Rev. A **38**, 6338 (1988)
703. A. Carpio, L.L. Bonilla: Phys. Rev. Lett. **86**, 6034 (2001)
704. Z. Zheng, G. Hu, B. Hu, Phys. Rev. Lett. **86**, 2273 (2001)
705. T. Munakata: Phys. Rev. A **45**, 1230 (1992)
706. F. Marchesoni: Phys. Rev. Lett. **73**, 2394 (1994)
707. A.V. Savin, G.P. Tsironis, A.V. Zolotaryuk: Phys. Lett. A **229**, 279 (1997)
708. A.V. Savin, G.P. Tsironis, A.V. Zolotaryuk: Phys. Rev. E **56**, 2457 (1997)
709. G. Costantini, F. Marchesoni: Phys. Rev. Lett. **87**, 114102 (2001)
710. A.V. Ustinov, M. Cirillo, B.A. Malomed: Phys. Rev. B **47**, 8357 (1993)
711. O.M. Braun, B. Hu, A. Zeltser: Phys. Rev. E **62**, 4235 (2000)
712. T. Strunz, F.-J. Elmer: Phys. Rev. E **58**, 1601, 1612.
713. J.-A. Sepulchre, R.S. MacKay: Physica D **113**, 342 (1998)

714. F. Marchesoni, M. Borromeo: Phys. Rev. B **65**, 184101 (2002)
715. A.R. Bishop, D.K. Campbell, P.S. Lomdahl, B. Horovitz, S.R. Phillpot: Synthetic Metals **9**, 223 (1984)
716. A.V. Ustinov, B.A. Malomed, S. Sakai: Phys. Rev. B **57**, 11691 (1998)
717. O.M. Braun, B. Hu, A. Filippov, A. Zeltser: Phys. Rev. E **58**, 1311 (1998)
718. O.M. Braun: Phys. Rev. E **62**, 7315 (2000)
719. O.M. Braun, Hong Zhang, Bambi Hu, and J. Tekic: Phys. Rev. E **67**, 066602 (2003)
720. O.M. Braun, T. Dauxois, M.V. Paliy, M. Peyrard, B. Hu: Physica D **123**, 357 (1998)
721. M. Paliy, M., O. Braun, T. Dauxois, B. Hu: Phys. Rev. E **56**, 4025 (1997)
722. H.G.E. Hentschel, F. Family, Y. Braiman: Phys. Rev. Lett. **83**, 104 (1999)
723. G.H. Hardy, E.M. Wright: *Introduction to the Theory of Numbers* (Oxford University Press Oxford 1979)
724. J.C. Ariyasu, A.R. Bishop: Phys. Rev. B **35**, 3207 (1987)
725. O.M. Braun, B. Hu: J. Stat. Phys. **92**, 629 (1998)
726. M.J. Renné, D. Polder: Rev. Phys. Appl. **9**, 25 (1974)
727. J.R. Waldram, P.H. Wu: J. Low Temp. Phys. **47**, 363 (1982)
728. M. Inui, S. Doniach: Phys. Rev. B **35**, 6244(1987)
729. F. Falo, L.M. Floría, P.J. Martínez, J.J. Mazo: Phys. Rev. B **48**, 7434(1993)
730. L.M. Floría, F. Falo: Phys. Rev. Lett. **68**, 2713 (1992)
731. N.R. Quintero, A. Sánchez: Phys. Lett. A **247**, 161(1998)
732. N.R. Quintero, A. Sánchez: Eur. Phys. J. B **6**, 133 (1998)
733. O.H. Olsen, M.R. Samuelsen: Phys. Rev. B **28**, 210 (1983)
734. N.R. Quintero, A. Sánchez, F.G. Mertens: Phys. Rev. E **62**, 5695 (2000)
735. P.J. Martinez, P. J., F. Falo, J.J. Mazo, L.M. Floria, A. Sánchez: Phys. Rev. B **56**, 87 (1997)
736. M.V. Fistul, E. Goldobin, A.V. Ustinov: Phys. Rev. B **64**, 92501 (2001)
737. P.S. Lomdahl, M.R. Samuelsen: Phys. Rev. A **34**, 664 (1986)
738. P.S. Lomdahl, M.R. Samuelsen: Phys. Lett. A **128**, 427 (1988)
739. N. Grønbech-Jensen, Yu.S. Kivshar, M.R. Samuelsen: Phys. Rev. B **43**, 5698 (1991)
740. Z. Zheng, M.C. Cross, G. Hu, Phys. Rev. Lett. **89**, 154102 (2002)
741. S. Flash, O. Yevtushenko, Y. Zolotaryuk: Phys. Rev. Lett. **84**, 2358 (2000)
742. P. Reimann: Phys. Rev. Lett. **86**, 4992 (2001)
743. O. Yevtushenko, S. Flach, Y. Zolotaryuk, A.A. Ovchinnikov: Europhys. Lett. **54**, 141 (2001)
744. S. Denisov, S. Flach, A.A. Ovchinnikov, O. Yevtushenko, Y. Zolotaryuk: Phys. Rev. E **66**, 041104 (2002)
745. P. Reimann: Phys. Rep. **361**, 57 (2002)
746. P. Reimann, R. Bartussek, R. Häussler, P. Hänggi: Phys. Lett. A **215**, 26 (1996)
747. L.M. Floría, F. Falo, P.J. Martínez, J.J. Mazo: Europhys. Lett. **60**, 174 (2002)
748. T. Bena, M. Copelli, C. Van den Broeck: J. Stat. Phys. **101**, 415 (2000)
749. M. Porto, M. Urbakh, J. Klafter, Phys. Rev. Lett. **85**, 491 (2000)
750. P. Jung, J.G. Kissner, P. Hänggi: Phys. Rev. Lett. **76**, 3436 (1996)
751. M. Borromeo, G. Costantini, F. Marchesoni: Phys. Rev. E **65**, 041110 (2002)
752. H. Schanz, M.-F. Otto, R. Ketzmerick, T. Dittrich: Phys. Rev. Lett. **87**, 70601 (2001)

753. S. Denisov, S. Flach: Phys. Rev. E **64**, 056236 (2001)
754. S. Denisov, J. Klafter, M. Urbakh, S. Flach: Physica D **170**, 131 (2002)
755. F. Marchesoni: Phys. Rev. Lett. **77**, 2364 (1996)
756. S. Flash, Y. Zolotaryuk, A.E. Miroshnichenko, M.V. Fistul: Phys. Rev. Lett. **88**, 184101 (2002)
757. M. Salerno, Y. Zolotaryuk: Phys. Rev. E **65**, 056603 (2002)
758. G. Costantini, F. Marchesoni, M. Borromeo: Phys. Rev. E **65**, 051103 (2002)
759. M. Salerno, N.R. Quintero: Phys. Rev. E **65**, 025602 (2002)
760. E. Goldobin, A. Sterck, D. Koelle: Phys. Rev. E **63**, 031111 (2001)
761. G. Carapella: Phys. Rev. B **63**, 054515 (2001)
762. G. Carapella, G. Costabile: Phys. Rev. Lett. **87**, 77002 (2001)
763. G. Carapella, G. Costabile, R. Latempa, N. Martucciello, M. Cirillo, A. Polcari, F. Filatrella, cond-mat/0112467 (2001)
764. F. Falo, P.J. Martinez, J.J. Mazo, S. Cilla: Europhys. Lett. **45**, 700 (1999)
765. E. Trias, J.J. Mazo, F. Falo, T.P. Orlando: Phys. Rev. E **61**, 2257 (2000)
766. C.-S. Lee, B. Janko, I. Derenyi, A.-L. Barabasi: Nature **400**, 337 (1999)
767. A. Vanossi, A. Franchini, V. Bortolani: Surf. Sci. **502-503**, 437 (2002)
768. J.A. Snyman, J.H. van der Merve: Surface Sci. **42**, 190(1974)
769. J.A. Snyman, J.H. van der Merwe: Surface Sci. **45**, 619(1974)
770. S. Stoyanov, H. Müller-Krumbhaar: Surface Sci. **159**, 49(1985)
771. K.M. Martini, S. Burdick, M. El-Batanouny, G. Kirczenow: Phys. Rev. B **30**, 492(1984)
772. I. Markov, V. Karaivanov: Thin Solid Films **61**, 115(1979)
773. I. Markov, V. Karaivanov: Thin Solid Films **65**, 361(1980)
774. Y. Braiman, J. Baumgarten, J. Jortner, J. Klafter: Phys. Rev. Lett. **65**, 2398(1990)
775. Y. Braiman, J. Baumgarten, J. Klafter: Phys. Rev. B **47**, 11159 (1993)
776. A.M. Kosevich, A.S. Kovalev: *Introduction to Nonlinear Physical Mechanics* (Naukova Dumka Kiev 1989)
777. A.S. Kovalev: Fiz. Tverd. Tela **21**, 1729 (1979)
778. A. Paweek, M. Jaworski, J. Zagrodzinski: J. Phys. A: Math. Gen. **21**, 2727 (1988)
779. S. Stoyanov, in: *Growth and Properties of Metal Clusters*, ed by J. Bourdon (Elsevier Amsterdam 1980)
780. Y. Braiman, F. Family, H.G.E. Hentschel: Phys. Rev. E **53**, R3005(1996)
781. A. Kwaśniewski, P. Machnikowski, P. Magnuszewski: Phys. Rev. E **59**, 2347(1999)
782. L.D. Landau: Zh. Eksp. Teor. Fiz. **7**, 627 (1937) [Phys. Z. Sowjetunion **11**, 26 (1937)]
783. N. Mermin, H. Wagner: Phys. Rev. Lett. **17**, 1133(1966); Erratum: Phys. Rev. Lett. **17**, 1307 (1966)
784. P.C. Hohenberg: Phys. Rev. **158**, 383 (1967)
785. B. Jancovici: Phys. Rev. Lett. **19**, 20 (1967)
786. N.D. Mermin: Phys. Rev. **176**, 250 (1968)
787. V.L. Berezinsky: Zh. Eksp. Teor. Fiz. **59**, 907 (1970)
788. V.L. Berezinsky: Zh. Eksp. Teor. Fiz. **61**, 1144 (1971)
789. J.M. Kosterlitz, D.J. Thouless: J. Phys. C **6**, 1181 (1973)
790. J.M. Kosterlitz: J. Phys. C **7**, 1046 (1974)
791. A.P. Young: Phys. Rev. B **19**, 1855 (1979)

792. B.I. Halperin, D.R. Nelson: Phys. Rev. B **19**, 2457(1979)
793. V.M. Bedanov, G.V. Gadiyak, Yu.E. Lozovik: Zh. Eksp. Teor. Fiz. **88**, 1622 (1985)
794. Y. Imry, S.K. Ma: Phys. Rev. Lett. **35**, 1399 (1975)
795. I.F. Lyuksyutov, M.V. Feigelman: Zh. Eksp. Teor. Fiz. **86**, 774 (1984)
796. W. Selke, K. Binder, W. Kinzel: Surface Sci. **125**, 74 (1983)
797. W. Selke: Phys. Rep. **170**, 213 (1988)
798. Yu.M. Malozovsky, V.M. Rozenbaum: Physica A **175**, 127 (1991)
799. V.M. Rozenbaum, V. M., V.M. Ogenko, A.A. Chuiko: Usp. Fiz. Nauk **161**, 79 (1991)
800. O.M. Braun, Yu.S. Kivshar: Phys. Rev. B **44**, 7694 (1991)
801. P.L. Christiansen, A.V. Savin, A.V. Zolotaryuk: Phys. Rev. E **57**, 13564 (1998)
802. A.V. Savin, J.M. Khalack, P.L. Christiansen, A.V. Zolotaryuk: Phys. Rev. B **65**, 054106 (2002)
803. F. Yoshida, Y. Okwamoto, T. Nakayama: J. Phys. Soc. Jpn. **50**, 1039 (1981)
804. F. Yoshida, Y. Okwamoto, T. Nakayama: J. Phys. Soc. Jpn. **51**, 1329 (1982)
805. H. Kato, H., Y. Okwamoto, T. Nakayama: J. Phys. Soc. Jpn. **52**, 3334 (1983)
806. A. Luther, V.L. Pokrovsky, J. Timonen, In: *Phase Transition in Surface Films* (Plenum New York 1980) p. 506.
807. V.L. Pokrovsky, A.L. Talapov: Zh. Eksp. Teor. Fiz. **78**, 269 (1980)
808. Y. Okwamoto: J. Phys. Soc. Jpn. **49**, 8 (1980)
809. T. Natterman: J. Phys.(France) **41**, 1251 (1980)
810. H.J. Schulz: Phys. Rev. B **22**, 5274 (1980)
811. J.V. Jose, L.P. Kadanoff, S. Kirkpatrick: Phys. Rev. B **16**, 1217 (1977)
812. P.B. Wiegman: J. Phys. C **11**, 1583 (1978)
813. I.F. Lyuksyutov: Pis'ma Zh. Eksp. Teor. Fiz. **32**, 593 (1980)
814. S.N. Coppersmith, D.S. Fisher, B.I. Halperin, P.A. Lee, W.F. Brinkman: Phys. Rev. Lett. **46**, 549 (1981)
815. E.B. Kolomeisky, T. Curcic, J.P. Straley: Phys. Rev. Lett. **75**, 1775 (1995)
816. I.F. Lyuksyutov, V.L. Pokrovsky: Pis'ma Zh. Eksp. Teor. Fiz. **33**, 343 (1981)
817. O.M. Braun, M. Peyrard: Phys. Rev. E **51**, 4999 (1995)
818. O.M. Braun, O.A. Chubykalo, T.P. Valkering: Phys. Rev. B **53**, 13877 (1996)
819. O.M. Braun, O.A. Chubykalo, Yu.S. Kivshar, L.Vázquez: Phys. Rev. B **48**, 3734 (1993)
820 O.M. Braun, M. Peyrard: Phys. Rev. B **51**, 17158 (1995)
821. V.K. Medvedev, A.G. Naumovets, T.P. Smereka: Surface Sci. **34**, 368(1973)
822. M.S. Gupalo, V.K. Medvedev, B.M. Palyukh, T.P. Smereka: Sov. Phys. Solid State **21**, 568 (1979)
823. O.M. Braun, T. Dauxois, M. Peyrard: Phys. Rev. B **54**, 313 (1996)
824. J.E. Black, Z.-J. Tian: Phys. Rev. Lett. **71**, 2445 (1993)
825. R.G. Caflisch, A.N. Berker, M. Kardar: Phys. Rev. B **31**, 4527 (1985)
826. K. Kern, G. Comsa, In: *Kinetics of Ordering and Growth at Surfaces*, ed by M.G. Lagally (Plenum Press New York 1990)
827. J.P. Hirth, J. Lothe: *Theory of Dislocations* (John Wiley New York 1982)
828. J.A. Snyman, H.C. Snyman: Surface Sci. **105**, 357(1981)
829. J.H. van der Merwe: J. Appl. Phys. **41**, 4725 (1970)
830. F.F. Abraham, W.E. Rudge, D. Auerbach, S.W. Koch: Phys. Rev. Lett. **52**, 445 (1984)
831. P. Bak, D. Mukamel, J. Villain: Phys. Rev. B **19**, 1610 (1979)

832. V.L. Pokrovsky, A.L. Talapov: Phys. Rev. Lett. **42**, 65 (1979)
833. F.S. Rys: Phys. Rev. Lett. **51**, 849 (1983)
834. A. Novaco, J.P. McTague: Phys. Rev. Lett. **38**, 1286 (1976)
835. G.V. Uimin, L.N. Tschur: Pis'ma JETF **28**, 20 (1978)
836. S.E. Burkov: Pis'ma JETF **29**, 457 (1979)
837. W. Uhler, R. Schilling: Phys. Rev. B **37**, 5787 (1988)
838. J.E. Black, D.L. Mills: Phys. Rev. B **42**, 5610(1990)
839. P.S. Lomdahl, D.J. Srolovitz: Phys. Rev. Lett. **57**, 2702 (1986)
840. D.J. Srolovitz, P.S. Lomdahl: Physica D **23**, 402(1986)
841. Yu.N. Gornostyrev, M.I. Katsnelson, A.V. Kravtsov, A.V. Trefilov: Phys. Rev. B **60**, 1013 (1999)
842. O.M. Braun, M.V. Paliy, J. Röder, A.R. Bishop: Phys. Rev. E **63**, 036129 (2001)
843. O.M. Braun, M. Paliy, B. Hu: Phys. Rev. Lett. **83**, 5206 (1999)
844. R. Besseling, R. Niggebrugge, P.H. Kes: Phys. Rev. Lett. **82**, 3144 (1999)
845. C. Reichhardt, F. Nori: Phys. Rev. Lett. **82**, 414 (1999)
846. T.S. van Erp, A. Fasolino, O. Radulescu, T. Janssen: Phys. Rev. B **60**, 6522 (1999)
847. A. Vanossi, J. Röder, A.R. Bishop, V. Bortolani: Phys. Rev. Lett. **63**, 017203 (2000)
848. L. Trallori: Phys. Rev. B **57**, 5923 (1998)
849. S.E. Korshunov: Phys. Rev. Lett. **88**, 167007 (2002)
850. O. Tchernyshyov, L.P. Pryadko: Phys. Rev. B **61**, 12503 (2000)
851. M. Bosch, W. van Saarloos, J. Zaanen: Phys. Rev. B **63**, 092501(2001)
852. M. Porto, M. Urbakh, J. Klafter: Phys. Rev. Lett. **85**, 6058 (2000)
853. U. Dehlinger, A. Kochendörfer: Z. Physik **116**, 576 (1940)
854. A. Seeger, H. Engelke, Theory of kink mobilities at low temperatures, in: *Dislocation Dynamics* eds A.R. Rosenfield, G.T. Hahn, S.L. Bement, Jr., R.I. Jaffe (McGraw-Hill New York 1969) pp. 623-650
855. G.I. Taylor: Proc. Roy. Soc. London A **145**, 362 (1934)
856. M. Polanyi: Z. Physik **89**, 660 (1934)
857. E. Orowan: Z. Physik **89**, 634 (1934)
858. A. Seeger: Z. Naturforschg **8a**, 246 (1953)
859. W. Blaschke, *Vorlesungen über Differential-Geometrie* Vol. I, Elementare Differentialgeometrie, 4-th ed. (Springer Berlin 1945)
860. L. Bianchi: *Lezioni di Geometri Differenziale*, 3rd ed (Spoerri Pisa 1920)
861. L. Bianchi: *Vorlesungen über Differentialgeometrie*, 2nd ed. (Teubner. Leipzig 1912)
862. L.P. Eisenhart: *A Treatise on the Differential Geometry of Curves and Surfaces* (Ginn Boston 1909)
863. L. Bianchi: Annali Scuola Norm. Sup. Pisa (1), **2**, 285 (1879)
864. L. Bianchi: Rend. Acc. Naz. Lincei (5) **8**, 484 (1899)
865. M. Heyerhoff: *Die frühe Geschichte der Solitonentheorie*, Dr. rer. nat. thesis (Universität Greifswald Greifswald 1997)
866. M. Heyerhoff, in: *Mathematikgeschichte und Unterricht I. Mathematik in Wandel*, ed by M. Tiepell (Franzbecker Hildesheim-Berlin 1998), pp. 295-305
867. E. Bour: Journal de l'École Imperiale Polytechnique **19**, cahier 39 (1862) pp. 1-148
868. A. Seeger: Theorie der Gitterfehlstellen, in: *Encyclopedia of Physics*, Vol. VII/1. ed by S. Flügge (Springer Berlin 1955)

869. A. Seeger: Kristallplastizität, in: *Encyclopedia of Physics*, Vol. VII/2. ed by S. Flügge (Springer Berlin 1958)
870. P.G. Bordoni: J. Acoust. Soc. Amer. **26**, 495 (1954)
871. A. Seeger, P. Schiller, Kinks and dislocation lines and their effects on the internal griction in crystals, in: *Physical Acoustics* Vol. III A. ed by W.P. Mason (Academic Press New York and London 1966), pp. 361-495
872. J.D. Eshelby: Proc. Roy. Soc. London A **266**, 222 (1962)
873. A. Seeger, Structure and diffusion of kinks in monoatomic crystals, in: *Dislocations 1984*, eds P. Veyssiére, L. Kubin, J. Castaing (CNRS Paris 1984) pp. 141-177
874. A. Seeger, Solitons and sattistical mechanics, in: *Trends in Applications of Pure Mathematics to Mechanics* (Lecture Notes in Physics 249) eds E. Kröner, K. Kichgässner (Springer Berlin 1986) pp. 114-155
875. A. Seeger: Z. Matellkde **93**, 760 (1992)
876. A. Seeger: Mater. Sci. Eng. A (in press)

Index

Texts and Monographs in Physics

Series Editors: R. Balian W. Beiglböck H. Grosse E. H. Lieb N. Reshetikhin H. Spohn W. Thirring

Essential Relativity Special, General, and Cosmological Revised 2nd edition By W. Rindler

The Elements of Mechanics By G. Gallavotti

Generalized Coherent States and Their Applications By A. Perelomov

Quantum Mechanics II By A. Galindo and P. Pascual

Geometry of the Standard Model of Elementary Particles By. A. Derdzinski

From Electrostatics to Optics A Concise Electrodynamics Course By G. Scharf

Finite Quantum Electrodynamics The Causal Approach 2nd edition By G. Scharf

Path Integral Approach to Quantum Physics An Introduction 2nd printing By G. Roepstorff

Supersymmetric Methods in Quantum and Statistical Physics By G. Junker

Relativistic Quantum Mechanics and Introduction to Field Theory By F. J. Ynduráin

Local Quantum Physics Fields, Particles, Algebras 2nd revised and enlarged edition By R. Haag

The Mechanics and Thermodynamics of Continuous Media By M. Šilhavý

Quantum Relativity A Synthesis of the Ideas of Einstein and Heisenberg By D. R. Finkelstein

Scattering Theory of Classical and Quantum N-Particle Systems By. J. Derezinski and C. Gérard

Effective Lagrangians for the Standard Model By A. Dobado, A. Gómez-Nicola, A. L. Maroto and J. R. Peláez

Quantum The Quantum Theory of Particles, Fields, and Cosmology By E. Elbaz

Quantum Groups and Their Representations By A. Klimyk and K. Schmüdgen

Multi-Hamiltonian Theory of Dynamical Systems By M. B laszak

Renormalization An Introduction By M. Salmhofer

Fields, Symmetries, and Quarks 2nd, revised and enlarged edition By U. Mosel

Statistical Mechanics of Lattice Systems Volume 1: Closed-Form and Exact Solutions 2nd, revised and enlarged edition By D. A. Lavis and G. M. Bell

Statistical Mechanics of Lattice Systems Volume 2: Exact, Series and Renormalization Group Methods By D. A. Lavis and G. M. Bell

Conformal Invariance and Critical Phenomena By M. Henkel

The Theory of Quark and Gluon Interactions 3rd revised and enlarged edition By F. J. Ynduráin

Quantum Field Theory in Condensed Matter Physics By N. Nagaosa

Quantum Field Theory in Strongly Correlated Electronic Systems By N. Nagaosa

Information Theory and Quantum Physics Physical Foundations for Understanding the Conscious Process By H.S. Green

Magnetism and Superconductivity By L.-P. Lévy

The Nuclear Many-Body Problem By P. Ring and P. Schuck

Perturbative Quantum Electrodynamics and Axiomatic Field Theory By O. Steinmann

Quantum Non-linear Sigma Models From Quantum Field Theory to Supersymmetry, Conformal Field Theory, Black Holes and Strings By S. V. Ketov

Texts and Monographs in Physics

Series Editors: R. Balian W. Beiglböck H. Grosse E. H. Lieb
N. Reshetikhin H. Spohn W. Thirring

The Statistical Mechanics of Financial Markets 2nd Edition By J. Voit

Statistical Mechanics A Short Treatise
By G. Gallavotti

Statistical Physics of Fluids
Basic Concepts and Applications
By V. I. Kalikmanov

Many-Body Problems and Quantum Field Theory An Introduction
By Ph. A. Martin and F. Rothen

Foundations of Fluid Dynamics
By G. Gallavotti

High-Energy Particle Diffraction
By E. Barone and V. Predazzi

Physics of Neutrinos
and Applications to Astrophysics
By M. Fukugita and T. Yanagida

Relativistic Quantum Mechanics
By H. M. Pilkuhn

The Geometric Phase in Quantum Systems
Foundations, Mathematical Concepts,
and Applications in Molecular
and Condensed Matter Physics
By A. Bohm, A. Mostafazadeh, H. Koizumi,
Q. Niu and J. Zwanziger

The Atomic Nucleus as a Relativistic System
By L.N. Savushkin and H. Toki

The Frenkel–Kontorova Model
Concepts, Methods, and Applications
By O. M. Braun and Y. S. Kivshar

Aspects of Ergodic, Qualitative and Statistical Theory of Motions
By G. Gallavotti, F. Bonetto and G. Gentile

MIX
Papier aus verantwortungsvollen Quellen
Paper from responsible sources
FSC® C105338

If you have any concerns about our products,
you can contact us on
ProductSafety@springernature.com

In case Publisher is established outside the EU,
the EU authorized representative is:
Springer Nature Customer Service Center GmbH
Europaplatz 3, 69115 Heidelberg, Germany

Printed by Libri Plureos GmbH
in Hamburg, Germany